A Guide to Lie Systems with Compatible Geometric Structures

A Guide to Lie Systems with Compatible Geometric Structures

Javier de Lucas
University of Warsaw, Poland

Cristina Sardón Muñoz
Instituto de Ciencias Matemáticas, Spain

NEW JERSEY • LONDON • SINGAPORE • BEIJING • SHANGHAI • HONG KONG • TAIPEI • CHENNAI • TOKYO

Published by

World Scientific Publishing Europe Ltd.
57 Shelton Street, Covent Garden, London WC2H 9HE
Head office: 5 Toh Tuck Link, Singapore 596224
USA office: 27 Warren Street, Suite 401-402, Hackensack, NJ 07601

Library of Congress Cataloging-in-Publication Data
Names: de Lucas, J. (Javier), 1981–   author. | Sardón Muñoz, Cristina, author.
Title: A guide to Lie systems with compatible geometric structures /
    Javier de Lucas (University of Warsaw, Poland) &
    Cristina Sardón Muñoz (Instituto de Ciencias Matemáticas, Spain).
Description: New Jersey : World Scientific, 2019. | Includes bibliographical references.
Identifiers: LCCN 2019015556 | ISBN 9781786346971 (hc)
Subjects: LCSH: Differential equations. | Lie algebras. | Geometry, Differential.
Classification: LCC QC20.7.D5 L8292 2019 | DDC 512/.482--dc23
LC record available at https://lccn.loc.gov/2019015556

British Library Cataloguing-in-Publication Data
A catalogue record for this book is available from the British Library.

Copyright © 2020 by World Scientific Publishing Europe Ltd.

All rights reserved. This book, or parts thereof, may not be reproduced in any form or by any means, electronic or mechanical, including photocopying, recording or any information storage and retrieval system now known or to be invented, without written permission from the Publisher.

For photocopying of material in this volume, please pay a copying fee through the Copyright Clearance Center, Inc., 222 Rosewood Drive, Danvers, MA 01923, USA. In this case permission to photocopy is not required from the publisher.

For any available supplementary material, please visit
https://www.worldscientific.com/worldscibooks/10.1142/Q0208#t=suppl

Desk Editors: Aanand Jayaraman/Jennifer Brough/Shi Ying Koe

Typeset by Stallion Press
Email: enquiries@stallionpress.com

Vitanda est improba siren desidia
–To Iza and Nadia

Preface

A Lie system is a non-autonomous system of first-order ordinary differential equations whose general solution can be written as an autonomous function, the superposition rule, of a generic family of particular solutions and a set of constants. Relevant types of Lie systems are non-autonomous linear systems of first-order ordinary differential equations and most types of Riccati equations are, e.g., matrix and conformal Riccati equations. Although being a Lie system is the exception rather than the rule in differential equations, their relevant applications in physical and mathematical problems justify their study.

The study of Lie systems can be traced back to the end of the 19th century when their analysis was initiated by Lie, Vessiot and, less relevantly, other authors. For more than a century, the theory of Lie systems was mainly applied to types of matrix Riccati equations, Ermakov-like systems, problems in control theory, and in time-dependent Schrödinger equations. In 2000, the works by Grundland, Odzijewicz, Marmo, Cariñena, and Grabowski initiated the study of Lie systems through geometric techniques. This approach turned out to be very fruitful and, soon after, the study of Lie systems related to compatible geometric structures commenced.

The authors of this work, along with their collaborators, devised methods to simplify the calculation of superposition rules through different types of geometric structures: symplectic, Poisson, k-symplectic structures, and others. Theoretically, they found new properties of Lie systems with compatible geometric structures. This approach gave rise to a new plethora of applications as illustrated by the analysis of Kummer–Schwarz equations, diffusion models, equations of the Riccati hierarchy, Buchdahl equations, and many more differential equations. The research line of this book is far from being exhausted since other geometric structures can be applied to Lie systems: twisted-Dirac manifolds, Lie algebroids, and multisymplectic structures.

In view of the fast development of the theory of Lie systems with compatible geometric structures, this book provides a comprehensive introduction to the topic starting from the very basic theory of Lie systems from scratch. Moreover, we describe the main authors, works, and trends in the study of Lie systems. We also provide a brief introduction to the main geometric structures employed in the analysis of Lie systems so far. As the main topic, the book details the theory of Lie systems with compatible geometric structures and several of their applications.

We expect that the book will be useful for very different types of readers ranging from final-year students in Mathematics and Physics desiring to study the fundamentals of the theory of Lie systems, or PhD students interested in Lie systems with compatible geometric structures, to researchers looking for a particular reference or technical result on this theory.

About the Authors

Javier de Lucas is an Assistant Professor at the Department of Mathematical Methods in Physics at the Faculty of Physics of the University of Warsaw. He is a specialist on differential geometry, differential equations, and their applications in Physics and Mathematics. His PhD thesis, "Lie Systems and Applications in Quantum Mechanics" was awarded the "Extraordinary Doctoral Prize of the University of Zaragoza". He has authored more than 40 papers on Lie systems and related topics. His research and didactic works have also been awarded several times by the University of Warsaw. He has been a postdoc at the Institute of Mathematics of the Polish Academy of Sciences and he has accomplished several research stays in the Centre de Recherches Mathématiques of the University of Montreal and in the École Normale Supérieure Paris-Saclay.

Cristina Sardón Muñoz is currently a visiting scholar at UC Berkeley and a postdoc at Instituto de Ciencias Matemáticas at the Spanish Research Council. Her research interests are focused on the geometric structures and their applications in Physics, nonlinear phenomena and integrability and dynamics of dynamical systems. More particularly, her most recent works are based on Lie systems and the geometric Hamilton–Jacobi theory, and symplectic reduction.

Contents

Preface	vii
About the Authors	ix

1 Introduction   1

1.1	General Aspects on Lie Systems and Geometric Structures		1
1.2	Historical Introduction to Lie Systems		4
	1.2.1	Vessiot, Lie, and their pioneering works	5
	1.2.2	The next boost: Winternitz and the CRM school	9
	1.2.3	A geometric approach to Lie systems	12
	1.2.4	Lie systems and geometric structures	16
	1.2.5	Modern trends on Lie systems	18
1.3	General Description of the Book		23

2 Geometric Fundamentals   29

2.1	Introduction	29
2.2	Tangent Spaces, Vector Fields, and Integral Curves	31
2.3	t-Dependent Vector Fields	37
2.4	Higher-order Tangent Spaces and Related Notions	41
2.5	Jet Bundles	42

2.6	Poisson Algebras	48
2.7	Symplectic, Presymplectic, and Poisson Manifolds	50
2.8	Dirac Manifolds	54
2.9	Jacobi Manifolds	58
2.10	k-Symplectic Forms	60

3 Basics on Lie Systems and Superposition Rules 65

3.1	Introduction	65
3.2	Vessiot–Guldberg Lie Algebras	65
3.3	Superposition Rules	67
3.4	The Lie System Notion and the Lie–Scheffers Theorem	74
3.5	Lie Systems and Lie Groups	79
3.6	Diagonal Prolongations	83
3.7	Geometric Approach to Superposition Rules	85
3.8	Geometric Lie–Scheffers Theorem	90
3.9	Determination of Superposition Rules	92
3.10	Superposition Rules for Second- and Higher-Order ODEs	96
3.11	New Applications of Lie Systems	99
	3.11.1 The second-order Riccati equation	100
	3.11.2 Kummer–Schwarz equations	104
	3.11.3 Second-order Kummer–Schwarz equations	106
	3.11.4 Third-order Kummer–Schwarz equation	111
	3.11.5 The Schwarzian derivative and third-order Kummer–Schwarz equations	114
3.12	Superposition Rules for PDEs	118

4 Lie–Hamilton Systems 123

4.1	Introduction	123
4.2	The Necessity of Lie–Hamilton Systems	124
	4.2.1 Lie–Hamiltonian structures	127
	4.2.2 t-independent constants of motion	133

	4.2.3	Symmetries, linearization, and comomentum maps	137
	4.2.4	On t-dependent constants of motion	141
	4.2.5	Lie integrals	142
	4.2.6	Polynomial Lie integrals	144
	4.2.7	The coalgebra method, constants of motion and superposition rules	150
4.3		Applications of the Geometric Theory of Lie–Hamilton Systems	155
	4.3.1	The Ermakov system	155
	4.3.2	A superposition rule for Riccati equations	157
	4.3.3	Kummer–Schwarz equations in Hamilton form	159
	4.3.4	The n-dimensional Smorodinsky–Winternitz systems	162
	4.3.5	A trigonometric system	165
4.4		Classification of Lie–Hamilton Systems on the Plane	167
	4.4.1	General definitions and properties	167
	4.4.2	Lie–Hamilton algebras	176
	4.4.3	Local classification	179
4.5		Applications of Lie–Hamilton Systems on the Plane	191
	4.5.1	$\mathfrak{sl}(2, \mathbb{R})$-Lie–Hamilton systems	192
	4.5.2	Lie–Hamilton biological models	199
	4.5.3	Other Lie–Hamilton systems on the plane	209
	4.5.4	Two-photon Lie–Hamilton systems	218
	4.5.5	$\mathfrak{h}_2$-Lie–Hamilton systems	222

5 Dirac–Lie Systems 227

5.1	Introduction	227
5.2	Motivation for Dirac–Lie Systems	227
5.3	Dirac–Lie Hamiltonians	231

5.4	Diagonal Prolongations	237
5.5	Superposition Rules and t-Independent Constants of Motion	240
5.6	Bi-Dirac–Lie Systems	245
5.7	Dirac–Lie Systems and Schwarzian–KdV Equations	248

6 Jacobi–Lie Systems   251

6.1	Introduction	251
6.2	Jacobi–Lie Systems	252
6.3	Jacobi–Lie Hamiltonians	254
6.4	Jacobi–Lie Systems on Low-Dimensional Manifolds	256

7 k-Symplectic Lie Systems   261

7.1	Introduction	261
7.2	The Need of k-Symplectic Lie Systems	263
	7.2.1 Schwarzian equation	263
	7.2.2 Riccati systems	267
	7.2.3 Control problems	269
	7.2.4 Riccati diffusion system	272
	7.2.5 Lotka–Volterra systems	274
	7.2.6 A k-symplectic Lie system on $SL(2,\mathbb{R})$	276
	7.2.7 Definition of k-symplectic Lie systems	278
7.3	A No-Go Theorem for k-Symplectic Lie Systems	280
7.4	Ω-Hamiltonian Functions	283
7.5	Derived Poisson Algebras	289
7.6	k-Symplectic Lie–Hamiltonian Structures	293
7.7	General Properties of k-Symplectic Lie Systems	297
7.8	Diagonal Prolongations of k-Symplectic Lie Systems and Superposition Rules	301

8 Conformal and Killing Lie Algebras on the Plane 307

8.1 Introduction 307
8.2 VG Lie Algebras and Their Distributions 309
8.3 Conformal Geometry and Lie Algebras of Vector Fields on the Plane 311
8.4 VG Lie Algebras of Conformal and Killing Vector Fields on $\mathbb{R}^2$ 316
8.5 Invariant Distributions for VG Lie Algebras on $\mathbb{R}^2$ 332
8.6 Applications in Physics 336
    8.6.1 Milne–Pinney equations 337
    8.6.2 Schrödinger equation on $\mathbb{C}^2$ 338

9 Lie Symmetry for Differential Equations 341

9.1 Introduction 341
9.2 Lie Symmetries for Lie Systems 342
    9.2.1 Certain Lie symmetries for Lie systems . . . 342
    9.2.2 Lie algebras of Lie symmetries for Lie systems 347
    9.2.3 Applications to systems of ODEs and HODEs 350
9.3 Lie Symmetries for Aff($\mathbb{R}$)-Lie Systems 361
9.4 Lie Symmetries for PDE Lie Systems 363
    9.4.1 Lie symmetries for $\mathfrak{sl}(2, \mathbb{R})$-PDE Lie systems 368

Appendix A   371

Bibliography   377

Author Index   399

Subject Index   401

Chapter 1

Introduction

1.1. General Aspects on Lie Systems and Geometric Structures

A *Lie system* is a non-autonomous system of first-order ordinary differential equations whose general solution can be written as a function, the so-called *superposition rule*, of a generic family of particular solutions and a set of constants related to initial conditions [269, 273, 276, 364]. For instance, first-order linear systems of ordinary differential equations, Riccati equations [154, 165], and matrix Riccati equations [164, 184, 239] are Lie systems that appear very frequently in the literature [171, 207, 249, 338, 343]. Since its very first introduction by Lie [269], several generalizations of the notion of Lie system have appeared to encompass systems of partial differential equations (PDEs) [89, 93, 296], superdifferential equations [32, 33, 174], and other types of differential equations [25, 86, 93, 116, 130, 233, 323, 344].

The interest of Lie systems and their generalizations is mainly twofold. On the one hand, their study involves the analysis of connections [93], geometric structures [23, 121, 283], momentum maps [121], local classifications of finite-dimensional Lie algebras of vector fields on manifolds [23, 263], discrete differential equations [329, 356], superequations [32, 33], systems of PDEs [296], conditional symmetries [198–200], stochastic models [253], Bäcklund transformations [89, 196, 197], Poisson–Hopf algebras [25, 26], and others [85, 93, 94].

On the other hand, Lie systems appear in many problems of classical mechanics, quantum mechanics, biology, etc. (see [23, 108] and references therein). Some relevant differential equations that can be studied through Lie systems are dissipative Milne–Pinney equations [103, 116], Caldirola–Kanai oscillators [117], t-dependent frequency harmonic oscillators [118], and second-order Riccati equations [110, 365]. It is surprising that although being a Lie system is more of an exception than a rule [108, 233], Lie systems can be applied to the investigation of a large variety of mathematical and physical problems [12, 32, 108, 171, 253, 329, 343].

As Lie systems are far from being completely explored [99, 283], their applications to old and modern problems give rise to approaches that are still novelties (cf. [283]). Remarkably, there has been a fast development of applications and generalizations of the techniques used in the theory of Lie systems for the last years [23, 27, 50, 197, 283].

In view of the above remarks, this book presents an up-to-date approach to Lie systems presenting the main achievements and notions from the pioneering works of Vessiot, Guldberg, and Lie [202, 267, 272, 362, 363] to the most recent trends on Lie systems with compatible geometric structures [90, 121, 283] and their applications to systems of PDEs [197, 198]. Our comprehensive overview details investigations on Lie systems whose description cannot easily be found elsewhere because some works are only published in research repositories or databases like BnF Gallica (e.g., [243, 362]). Moreover, our aim here is to provide an introductory guide allowing the reader to understand the theory of Lie systems from its fundamentals to the latest works concerning the application of geometric structures.

This handbook starts with a historical review of the development of the theory of Lie systems and their generalizations. We aim at providing a general overview of the subject, the main authors on the topic, trends and open questions, and the main works devoted to describing the fundamental results in this theme. Furthermore, we have detailed a full account on the works published by the main contributors to the theory of Lie systems: Lie [267, 269, 273–276], Vessiot [362–368], Winternitz [12–15, 33, 191, 207, 301, 376, 377, 380], the research group formed by Grabowski, Cariñena, and Marmo [91–93, 98, 108], and others [27, 108, 221]. Additionally, we present further results spread out in works which are not originally written in English, e.g., [269, 362, 363, 365].

After our brief historical survey on Lie systems, we introduce the fundamental notions of the theory and other related topics. More specifically, together with the aforementioned recent modern geometric approach [93], we show results concerning the application of Lie systems to Quantum Mechanics, PDEs, and systems of second- and higher-order ordinary differential equations, etc.

As part of the new trends on Lie systems, this work focuses on modern geometric and algebraic structures as main tools for developing our theory, as well as on the description of the numerous applications of Lie systems accomplished by the authors of the present book and their colleagues over recent years, e.g., [23, 27, 90, 121, 197, 198, 283]. It is important to note that there were just very few and rather simple applications of symplectic forms and Drinfel'd doubles in the theory of Lie systems until 2009 (see e.g., [91, 120]). This illustrated that the authors of this book took initiative in the use of geometric structures to study Lie systems. As a consequence, it has been proven that geometric structures are of utmost relevance in this research field. Indeed, we have applied Poisson, Dirac, Jacobi and k-symplectic structures to study Lie systems, which gave rise to techniques to derive superposition rules and other invariants, like Ermakov–Ince invariants [121] or, more generally, *Casimir tensor fields* for differential equations [23, 189, 263].

As a first remarkable result, we have come to retrieve known invariants and to understand new ones that are geometrically related to Lie systems [27]. For instance, the well-known superposition rule for Riccati equations [276] was derived through a symplectic invariant constructed by means of a Casimir element of $\mathfrak{sl}(2,\mathbb{R})$ [27]. Subsequently, as a second result, this gave rise to techniques simplifying the previous methods to obtain superposition rules and constants of motion, while endowing them with a geometric interpretation [23, 27]. Third, this line of research has led to new theoretical results and applications of modern geometric theories [283]. In particular, we have proven that k-symplectic structures can be related to Poisson algebras of functions, something that had been previously dismissed as impossible and/or useless [283]. Here, we show that these Poisson algebras enable us to simplify the calculation of superposition rules, and we display their usefulness in the study of non-autonomous systems of first-order ordinary differential equations, control systems, and physical models. This has opened a new field of applications for k-symplectic structures, which are mainly applied nowadays to systems of PDEs appearing in field theories [259]. As a byproduct of the previous research, there exist many findings on the existence, properties, and applications of different types of Lie algebras of vector fields on the plane.

The use of geometric structures in the research of Lie systems is far from being concluded, since more geometric structures can also be employed in the study of Lie systems, as noted in [189], where Lie systems were investigated via multisymplectic structures. Currently, there is an open line of study concerning the reduction of multisymplectic Lie systems. Multisymplectic reduction is indeed a relevant problem which has remained unapproachable for several decades now [161, 219, 287, 342].

1.2. Historical Introduction to Lie Systems

This section provides a carefully detailed historical introduction to the theory of Lie systems containing the main authors, the main works, the most important results, and current research trends.

1.2.1. *Vessiot, Lie, and their pioneering works*

It is difficult to establish the origins of the theory of Lie systems and superposition rules, since its pioneering works were published towards the end of the 19th century and many of them are not easily accessible, e.g., [244, 273, 274, 363]. In fact, some of the papers cited above can only be reached in bibliographic repositories such as BnF Gallica, e.g., [202, 273, 363].

Anyhow, it seems that, in an oral communication, Bellman appointed Abel as the founder of this field due to his generalization of the standard linear superposition rule of linear differential equations to differential equations related to nonlinear operators (cf. [233]). Unfortunately, the work in which Abel would have initiated the topic has not been precisely identified.

In 1883, Königsberger proved that the non-autonomous first-order ordinary differential equations in the real line that admit a superposition rule that depends algebraically on particular solutions are, up to an autonomous diffeomorphism, Riccati, linear, or affine differential equations [244]. In 1885, Lie proposed a class of non-autonomous systems of first-order ordinary differential equations [269] whose general solutions can be obtained from certain finite families of particular solutions and sets of constants [52, 357].

But these previous pioneering works did not draw much attention until 1893, when Vessiot and Guldberg proposed, separately, a generalization of Königsberger's result [202, 362]. They showed that all differential equations on the real line that admit a superposition rule can be considered, up to a time-independent change of variables, as particular cases of Riccati, linear, or affine differential equations. Moreover, the problem of determining which ordinary differential equations admit a superposition principle was first proposed by Vessiot [362, 363] for first- and second-order ordinary differential equation with one independent variable and extended by Guldberg [202] to first-order systems of ordinary differential equations with more variables (see [14, p. 164]).

» Si je ne me trompe, *je suis parvenu à démontrer, d'une manière rigou-reuse et bien simple, que mes systèmes (2) sont les seuls qui possèdent la pro priété demandée.* »

Figure 1.1. Lie's statement from [273], where he states that he will provide the characterization of non-autonomous first-order systems of differential equations admitting a superposition rule.

In [273], just some pages after Guldberg's [202], Lie slightly dismissed the value of Vessiot and Guldberg's discovery, and claimed that their contributions are just a consequence of his previous findings appearing in [269] (see [233] for comments on Lie's criticism). More specifically, he stated that the systems of differential equations that admit a superposition rule are the ones that he had previously defined in [274] and ensured that he would prove it. His literal words can be seen in Figure 1.1.

Lie's remarks in [273] gave rise to the most important result in the theory of Lie systems: the one frequently referred to as *Lie–Scheffers Theorem* [276, Theorem 44] (see Figure 1.2). The Lie–Scheffers theorem characterizes non-autonomous first-order systems of ordinary differential equations in normal form that admit a local superposition rule. Its proof also provides a method to obtain a superposition rule in terms of a Lie algebra of vector fields related to the system of differential equations admitting the superposition rule, the so-called *Vessiot–Guldberg (VG) Lie algebra*. Moreover, the Lie–Scheffers theorem additionally relates the dimension of the VG Lie algebra with the number of particular solutions of the superposition rule and the dimension of the manifold in which the Lie system is defined. This latter result is sometimes called the *Lie condition* (see [108] and [276, p. 799]).

In [276], Lie and Scheffers also presented the first detailed discussion on Lie systems. In recognition to Lie and Scheffers' work, some authors also refer to Lie systems as *Lie–Scheffers systems* [91]. Nevertheless, the latter denotation seems quite inappropriate to us, since Scheffers' contribution to this theorem is rather doubtful. As illustrated in Figure 1.1, it is very likely that Lie knew how to prove

Theorem 44: *Damit das System von n simultanen Differentialgleichungen:*

$$\frac{dx_i}{dz} = \eta_i(x_1 \ldots x_n) \quad (i = 1, 2 \ldots n)$$

in $x_1 \ldots x_n$, z *die Eigenschaft besitze, dass das allgemeine Lösungensystem* $x_1 \ldots x_n$ *aus einer gewissen Anzahl* m *von allgemein gewählten particularen Lösungensystemen*

$$x_1 = x_1^{(k)}, \ldots x_n = x_n^{(k)} \quad (k = 1, 2 \ldots m)$$

durch ein Formelsystem:

$$x_i = \varphi_i(x_1^{(1)} \ldots x_n^{1}, \ldots, x_1^{(m)} \ldots x_n^{(m)}, a_1 \ldots a_n)$$
$$(i = 1, 2 \ldots n)$$

mit n *willkürlichen Constanten* $a_1 \ldots a_n$ *darstellbar sei, ist nothwendig und hinreichend, dass es die besondere Form habe:*

$$\frac{dx_i}{dz} = Z_1(z)\xi_{1i}(x) + \cdots + Z_r(z)\xi_{ri}(x)$$
$$(i = 1, 2 \ldots n),$$

in der $Z_1 \ldots Z_r$ *Functionen von* z *allein und die* ξ_{ji} *solche Functionen von* $x_1 \ldots x_n$ *allein sind, dass die infinitesimalen Transformationen*

$$X_j f = \sum_{1}^{r} \xi_{ji} \frac{\partial f}{\partial x_i} \quad (j = 1, 2 \ldots r)$$

eine r-*gliedrige Gruppe erzeugen.*

Auch haben wir gesehen, dass die Zahl m die Ungleichung

$$nm \geq r$$

erfüllt.

Figure 1.2. The celebrated Lie–Scheffers theorem characterizing systems of first-order differential equations admitting a superposition rule as in [276].

the Lie–Scheffers theorem before his work with Scheffers. In addition, as far as we know, Scheffers did not publish any relevant result on Lie systems after his work with Lie [276]. Moreover, Scheffers was one of Lie's students, and the work [276] was written according to Lie's Leipzig lectures (see [345]). Hence, it seems quite possible that Scheffers only helped Lie in writing down his own ideas on Lie systems. This claim can also be supported by the fact that Lie had problems to communicate his findings because he was not fluent in German [345], while Scheffers was known for his dissemination skills. Summarizing, to us, the term *Lie–Scheffers systems* exaggerates Scheffers' contribution in the authorship of this remarkable theorem. Nonetheless, since there exist several Lie theorems in the literature [204], we will keep the term Lie–Scheffers theorem to refer to the most important result in the theory of Lie systems.

Lie's criticism triggered Vessiot's recognition for the relevance of Lie's work and suggested adopting the term *Lie systems* to refer to non-autonomous first-order systems of ordinary differential equations admitting a superposition rule [364, p. H.2]. The above reasons justify the use of the term *Lie systems* all throughout this work. It is interesting that Vessiot posteriorly furnished many new contributions to the theory of Lie systems [363, 364, 366, 368] and proposed various generalizations [365, 367, 368]. For instance, Vessiot proved that some classes of second-order Riccati equations admit solutions in terms of superposition-like expressions depending on four particular solutions, the derivatives of these particular solutions, and two real constants (see [107, 365] for details). As far as we know, this starts the study of superposition rules for nonlinear second-order differential equations. Furthermore, Vessiot also studied the reduction of Lie systems [364], related Lie systems to Galois theory [364], and applied the theory of Lie systems to more practical problems, like the description of orthogonal trajectories to spheres [364, H.15]. Hence, it is also appropriate to refer to Lie systems as *Lie–Vessiot systems* (this convention was adopted in part of the literature [51, 357]).

Vessiot was the first author who noted that a VG Lie algebra of a Lie system allows for the construction of an associated Lie group action relating the solutions of the Lie system to another Lie system on a Lie group, the so-called *automorphic Lie system* (see [364, H.7], [243] and Chapter 3 for further information). In particular, the general solution of a Lie system can be computed by using a unique particular solution of an associated automorphic Lie system and the Lie group action. We recommend to read [364, H.7–H.9] for the original result and [91] for a modern approach.

1.2.2. *The next boost: Winternitz and the CRM school*

After a bright beginning in the thorough study of superposition rules and Lie systems towards the end of the 19th century, the topic was forgotten for almost a century. Most works dealing with Riccati equations and other Lie systems do not make reference to superposition rules [139, 144–146, 262, 322]. In the 1970s, Inselberg [231–233, 237, 303, 352] and other researchers worked on a certain generalization of superposition rules for nonlinear operators [8].

Later on, the topic was brought back and many authors restarted the investigation of Lie systems concerning their generalizations and applications to Mathematics, Physics, and Control Theory [231, 237, 303]. This revival was initiated with an immense number of contributions by Winternitz and his collaborators, mainly from the Centre de Recherches Mathématiques of the University of Montreal (CRM) [12, 14, 33, 58, 208, 249, 376, 380]. One must also highlight the works by Brockett in Lie systems and their role in Control Theory [62, 63]. The application of Lie systems to Control Theory is still an active research field [129, 143, 220, 314, 316, 327, 343].

Let us discuss some of Winternitz's contributions and some results achieved by the hereafter called *CRM School*: Harnad, Hussin, Grundland, and others. By using the diverse results derived by Lie [275, 276], the CRM school developed and applied methods to derive superposition rules of Lie systems [329, 340, 341]. In particular, [375, p. 107] and [14, Section 3] summarize three different methods. One of them, the invariant or distributional method, is based on Lie's proof of the Lie–Scheffers theorem in [276, Theorem 44]. Meanwhile, another one relies on the fact that the dynamics of a Lie system can be described by means of a Lie group action and a curve of a Lie group. The third method is a modified version of the second and implements certain linearizations (see [14, Section 3.2]).

Along with Shnider, Winternitz studied the classification of complex (and some real) Lie systems through transitive primitive Lie algebras [340, 341], a concept that also appeared in some of his posterior works on the integrability of Lie systems [58, 59]. More exactly, Winternitz analyzed *undecoupling Lie systems*, i.e., Lie systems that do not admit a change of dependent variables separating a subset of differential equations involving a smaller number of variables having a superposition formula on their own [340, p. 3164]. This provided methods to obtain superposition rules for such Lie systems [340]. In particular, they classified complex (and some real) transitive primitive Lie algebras in three main categories given in Theorems 1–3 in [340]. Next, they obtained superposition rules for many of such undecoupling Lie systems, in particular, they focused on superposition rules for Lie systems admitting complex primitive transitive Lie algebras isomorphic to semi-direct sums of quotients of complex classical Lie algebras [340, 341]. Many other subcases were studied in other papers [35, 36, 300].

Although the problem of classifying superposition rules for Lie systems seemed almost complete at that time, the most difficult case [340, p. 3161] turned out to be harder to solve than expected and it remains open so far. This hard case seems to be related to classifications of homogeneous spaces. Moreover, the classification of real Lie systems is still an open problem too. As in the theory of classification of Lie algebras, the study of their classification over the reals is much more complicated than over the complex numbers. While studying the classification of Lie systems related to transitive primitive VG Lie algebras, Winternitz also approached the classification of Lie algebras of vector fields on $\mathbb{C}^n$ [290].

It is worth noting that, posteriorly, other works treated the classification of general VG Lie algebras on the plane. Previous findings on the topic by Lie [267] were presented in a modern way by González-López, Kamran, and Olver in [183]. This led to the so-called GKO classification of Lie algebras of vector fields on the plane [23]. Moreover, the description of Lie systems on two-dimensional spaces with superposition rules can also be found in [243].

Winternitz and the CRM school also paid attention to the analysis of discrete problems and numerical approximations of solutions of

Lie systems and related differential equations by means of superposition rules, e.g., [308, 317, 356]. In [191], Winternitz and collaborators accomplished the discretization of Lie systems related to a classical simple complex Lie algebra with a maximal parabolic Lie subalgebra. This allowed the construction of discretizations satisfying the so-called *singular confinement*. Singular confinement can be considered as the discrete analog of the Painlevé integrability property for standard differential equations [191]. Hence, it is a desirable property for discretizations. Among other Lie systems, the CRM school also studied the discretization of matrix Riccati equations [191] and Milne–Pinney equations [329].

Furthermore, solutions to superequations were also retrieved using superposition rules, the *super-superposition rules*, studied by Winternitz and the CRM school as well [32, 33, 174]. In particular, [32] introduced the necessary theory, while [33, 174] extended the results of the previous work to derive super-superposition rules for superequations generated by the action of the orthosymplectic supergroup $OSP(m, 2n)$ on a homogeneous superspace.

Besides the previously mentioned theoretical achievements, it is worth noting that the CRM school applied its methods to many discrete and differential equations with remarkable applications. For instance, many superposition rules were derived for Matrix Riccati equations [12, 207, 249, 300, 317] — which play an important role in Control Theory and Bäcklund transformations — as well as for diverse Lie systems as projective Riccati equations [58], various superequations [32, 33, 174], and others [14, 34, 36, 175, 209]. Finally, it is also worth mentioning Winternitz's research on Milne–Pinney equations [329], which provides one of the first papers (along with Vessiot's) devoted to analyzing second-order differential equations through Lie systems.

Grundland, one of the CRM school members, devoted his research to applications of Lie systems to PDEs and Bäcklund transformations [199, 200, 296]. In particular, Grundland and Odzijewicz developed a generalization of the Lie–Scheffers theorem for systems of PDEs [296]. The systems they describe are nowadays called *PDE*

Lie systems [89, 108]. Grundland *et al.* also pointed out that Lie systems and PDE Lie systems can be employed to determine conditional symmetries [199, 200]. Out of these works, Grundland and Odzijewicz can be considered as the founders of PDE Lie systems and their genuine applications. Winternitz also accomplished the applications of Lie systems to PDEs, e.g., Bäcklund transformations [208, 209, 379], without using PDE Lie systems. Grundland additionally found some applications of standard Lie systems on the theory of Bäcklund transformations [196].

Remarkably, Grundland and Levi developed a work on Riccati equations, where they extended the applications of such equations to other members of the so-called *Riccati hierarchy* [196]. Years later, Grundland and de Lucas achieved other similar extensions in [197]. In the latter work, Grundland provided a method to obtain a superposition rule for second-order Riccati equations using projective Riccati equations. Meanwhile, de Lucas extended it to the whole Riccati hierarchy and gave a geometric interpretation using projective geometry. The paper [197] also provided a full answer to certain partial results concerning the application of Lie systems to members of the Riccati hierarchy given in [110, 111, 188] and other projective properties of the Riccati hierarchy [97]. Finally, [197] also showed that different types of matrix Riccati equations can be used to study Bäcklund transformations so extending results in [196].

1.2.3. *A geometric approach to Lie systems*

After several years of working on Lie systems, the CRM school moved into other topics. Nevertheless, the Lie systems theory continued its development at the beginning of the 21st century in the hands of new authors such as Cariñena *et al.* [91]. After writing the book [91] and motivated by Marmo's initial ideas on the topic, Cariñena started a thorough geometric analysis of the theory and physical applications of Lie systems, as well as spread his interest amidst his PhD students, namely Nasarre [122, 124], Clemente [83, 84, 143], Ramos [113, 114, 129, 130, 314–316], Lázaro-Camí [253], Jover [83], and de Lucas (see references in

Introduction 13

[108, 335] and the bibliography of this book). De Lucas has devoted much of his research to the study of geometric structures that are compatible with Lie systems. As Cariñena, he has also involved his PhD, MSc, and bachelor students in his research [251, 252, 263, 282].

Moreover, other researches started analyzing this topic as well. Let us appoint them along with some of their works: Blázquez and Morales [51–53], Berkovich [44], Grabowski [89, 91, 93], Ibragimov [221–229], Marmo [91–93, 98], Grundland [197, 198], Rañada [115, 116, 118], Flores-Espinoza and Vorobiev [169–172], Campoamor-Stursberg [72–74, 76], Ballesteros and Herranz [23, 25, 26, 215], García-Estévez [166], Vilariño [189, 281], Blasco [23], Gràcia and Muñoz-Lecanda [189], Ibort [98], and others [19, 99, 168, 180, 247, 248, 288, 324, 336, 337]. As a result of their contributions, multiple interesting results about fundamentals, applications, and generalizations of the theory of Lie systems were furnished.

Let us describe the main ideas of some of the above-mentioned works [85, 91, 93, 98]. The book [91] is an introductory manual describing the basic modern differential geometric machinery used in the theory of Lie systems. More specifically, [91] provides a rigorous geometric description of the previous results on Lie systems appearing in the literature, e.g., the Lie group approach to Lie systems [364], along with several new physical applications. It is worth noting that [91] is concerned, although very briefly, with the use of geometric structures (Drinfel'd doubles and symplectic forms) to study Lie systems. The work [91] also presents a new generalization of Lie systems, the so-called *foliated Lie systems*, being compelling how slightly they have been studied so far. Meanwhile, the book [98] treats a broader spectrum of topics containing the themes developed in [91] and a few newer findings of the research carried out by Cariñena and de Lucas.

The celebrated paper [93] provides many relevant contributions to the comprehension of the theory of Lie systems. First, it completes a standard gap in the proofs of the Lie–Scheffers theorem appearing in the literature. This gap concerns the part of the proof given a procedure for the construction of a superposition rule from a VG Lie algebra.

Additionally, [93] establishes that the concept of a superposition rule is equivalent to a certain flat connection. This fact substantially clarifies its properties as shown in [276] and other works by Winternitz [14, 375]. The geometric proof of the Lie–Scheffers theorem in [93] clearly shows that the Lie system notion can naturally be extended to PDEs, which also complements the results on PDE Lie systems described in [296]. This proof also leads naturally to the definition of *partial superposition rules*, namely functions allowing us to generate a certain subfamily of particular solutions of a system of PDEs in terms of a finite family of its particular solutions and some parameters. These partial superposition rules have not been further analyzed so far.

Reference [93] also contains, implicity, clues to the characterization of families of first-order systems of differential equations admitting a t-dependent superposition rule [86]. Further, it provides the definition of the mixed superposition rule notion [93, 116], which allows one to describe the solutions of a Lie system in terms of particular solutions of different Lie systems and sets of constants. These extensions were applied to dissipative Milne–Pinney equations [85, 86], Endem–Fowler equations [99], second-order Riccati equations [95], nonlinear oscillators [85], Bäcklund transformations of modified KdV equations [89], etc.

It is worth noting that the geometric Lie–Scheffers theorem discussed in [93] characterizes non-autonomous first-order systems of differential equations in normal form admitting a local superposition rule [52]. This implies that the superposition rule only retrieves those particular solutions whose initial conditions are contained in an open subset. Moreover, the set of particular solutions and the set of constants of the superposition rule must obey similar restrictions [104]. In most applications, this is sufficiently general to describe the general solution of Lie systems (see [108] and examples in this book).

Theoretically, one may want to know when the superposition rule is globally defined. This is a much more restrictive case, as illustrated in the literature on Lie systems, but it is nevertheless interesting

Introduction 15

and the conditions for such a global definition to take place were given by Blázquez and Morales-Ruiz in [52]. Mostly, only automorphic Lie systems, a Riccati equation on a compactification of the real line, and other particular Lie systems over normal algebras seem to share this property [91, 108, 282]. This topic will be further discussed in our introduction to the geometric framework of Lie systems in Chapter 3.

After [93], the so-called distributional approach to Lie systems was developed. This was fully established in [108], a publication summarizing de Lucas' PhD thesis, where new physical applications of Lie systems, mainly via the distributional approach and their generalizations, are shown. Some displayed systems are second-order Riccati equations, dissipative Milne–Pinney equations, quantum systems, etc. In [108], one can find the state of the art of the theory of Lie systems back in 2009. It is worth noting that this work aims to provide a much detailed up to date description of the subject.

Let us now turn to discuss some contributions that gave rise to [108]. On the one hand, Cariñena, de Lucas, and their collaborators investigated the integrability of Lie systems [94, 108, 112, 113, 117, 125], the study of the applications of the Wei–Norman method [108, 122], the application of Lie systems techniques to analyzing systems of second-order differential equations [106, 108–110], and other topics like the analysis of certain Schrödinger equations [108, 126]. In this way, they provided a continuation of diverse previous articles dedicated to some of these themes [140, 296, 329, 365], they opened several research lines [126], and also gave rise to new applications of Lie systems [108, 136].

Besides the above contributions, the work [108] details numerous applications of Lie systems to classical physics, quantum mechanics, financial mathematics, and control theory.

Apart from the aforementioned generalizations of the Lie system notion [11, 296, 329, 365], a new approach to the generalization of the Lie system and superposition rule notions was carried out by Cariñena, Grabowski, and de Lucas: the theory of *quasi-Lie schemes* [85]. On the one hand, this approach provides a method to transform differential equations of a certain type into equations of the same type, e.g., Abel equations into Abel equations [119]. This can be used to transform differential equations into Lie systems [85], which leads to the *quasi-Lie system* notion. Such systems inherit some properties from Lie systems and, for instance, they admit superposition rules, showing an explicit dependence on the independent variable of the system [85, 110].

Quasi-Lie schemes have multiple applications: they can be used not only to analyze properties of Lie and quasi-Lie systems but also to investigate many other systems, e.g., nonlinear oscillators [85], Emden–Fowler equations [99], Mathews–Lakshmanan oscillators [85], dissipative and non-dissipative Milne–Pinney equations [103], and Abel equations [119], among others. As a consequence, various results about the integrability properties of such differential equations have been obtained, and many others are currently being analyzed. Furthermore, the appearance of t-dependent superposition rules led to the examination of the so-called *Lie families*, which cover, as particular cases, Lie systems and quasi-Lie schemes. Additionally, Lie families can be used to analyze exact solutions of very general families of differential equations [86].

1.2.4. *Lie systems and geometric structures*

From 2009, the theory of Lie systems has been greatly developed. The authors of this book, in collaboration with Grabowski, Winternitz, Grundland, Cariñena, Ballesteros, Herranz, and many others, have led to many generalizations and new research lines.

To understand the evolution of Lie systems from 2009 and the relevance of geometric structures in the theory, one has to recall that, until that year, Lie systems were mainly applied to matrix Riccati-type equations, second-order Riccati equations, Milne–Pinney equations, Ermakov-like systems, a PDE Lie system of Riccati type, and some Lie systems appearing in classical mechanics, control theory, and quantum mechanics found by Vessiot, Cariñena, Marmo, Winternitz, de Lucas, and their coworkers [91–93, 108, 376]. Although Winternitz studied far many more Lie systems during

Introduction 17

his classification theory [340, 341], these were mostly theoretically exploited, while applications were fewer.

Meanwhile, the work accomplished by de Lucas *et al.* gives a larger number of applications, as illustrated by the analysis of viral systems, second- and third-order Kummer–Schwarz equations, diffusion models, all the members of the Riccati hierarchy, Buchdahl equations, non-trigonometric Hamiltonian systems, and many, many more (see e.g., [23, 120], Table A.3, and references there in). It is quite interesting that Lie systems admitting compatible geometric structures seem to have more applications than Lie systems without compatible ones.

Almost all new examples found by de Lucas and Sardón are Lie systems with a VG Lie algebra of Hamiltonian vector fields relative to some geometric structure. This contrasts with what had been achieved until 2009, when there were only a few applications of symplectic structures and Drinfel'd doubles in Lie systems (see e.g., [91, 130]). Then, one can claim that geometric structures can be of utmost relevance in this research field [121].

Subsequently, we applied Poisson, Dirac, Jacobi, Riemann, and k-symplectic structures to construct superposition rules and other invariants of Lie systems appearing in other scientific fields, e.g., Ermakov–Ince invariants [27], viral models [23], or Casimir tensor fields [23, 189, 263].

As a first consequence, this research led to the geometric understanding of some known and new invariants related to Lie systems [27, 281]. Second, this gave rise to techniques simplifying previous methods to obtain superposition rules, constants of motion, etc. [27]. Third, our research helped derive new results and applications of modern geometric theories [283]. This is also the point of view of other authors working on Lie systems (see [98, Introduction]). An important outcome is the understanding of the existence, properties, and applications of Lie algebras of vector fields on the plane [23].

In particular, the well-known superposition rule for Riccati equations [276] was derived through a symplectic invariant that has been constructed by means of a Casimir element of $\mathfrak{sl}(2, \mathbb{R})$ [27]. This led us to devise geometric methods to simplify the calculation of superposition rules for *Lie–Hamilton systems*, i.e., Lie systems admitting a VG Lie algebra of Hamiltonian vector fields relative to a Poisson structure [23, 27]. We proved that k-symplectic structures can be related to Poisson algebras of functions. This was previously considered as impossible and/or useless [283]. Nevertheless, we found that these Poisson algebras enable us to simplify the calculation of superposition rules for physically and mathematically relevant problems [283] while showing new lines of research [259].

1.2.5. *Modern trends on Lie systems*

Out of all the listed available results in the literature, a vast collection of new methods and applications have emerged [98, 189, 281]. There exist nowadays several research groups working on Lie systems and their generalizations. Let us comment on the general lines of current research. In general terms, these concern the following topics:

- properties of low-dimensional VG Lie algebras on low-dimensional manifolds [72, 75, 227, 228];
- applications of quantum groups to Lie systems [25, 26];
- theory and applications of Lie systems to PDEs [89, 197, 198];
- Grassmann-valued differential equations and Lie systems;
- multisymplectic forms, multisymplectic reduction, Riemann structures, and Lie systems [189, 197, 263];
- applications of stochastic Lie systems [253];
- Lie systems and quantum mechanics on infinite-dimensional manifolds [83, 251];
- geometric properties of difference equations coming from Lie systems;
- the generalization of the notion of Lie system [88, 89, 105, 189].

Let us now explain briefly each one of the above research lines. The first works concerning the theory of Lie systems on low-dimensional manifolds can be traced back to Lie [267]. More recent works on the topic are [23, 183, 214]. As previously seen, the interest in Lie systems defined on low-dimensional manifolds that admit a compatible low-dimensional VG Lie algebra is in vogue. There are two distinguished lines of research, one is carried out by Campoamor-Stursberg, and the second is by Ibragimov and Gainetdinova.

Campoamor-Stursberg analyzes second-order differential equations whose related first-order systems of differential equations are Lie systems. Henceforth, we will refer to them as *SODE Lie systems* [108]. This can be considered as an evolution of the Vessiot work [364], where Vessiot studied SODEs admitting a superposition rule.

Campoamor-Stursberg mainly studies SODE Lie systems on $T\mathbb{R}$ related to particular types of VG Lie algebras, e.g., two- and three-dimensional ones in [72] and those isomorphic to the Lie algebra $\mathfrak{sl}_3$, of real 3×3 traceless matrices in [73]. The classification of VG Lie algebras related to SODE-Lie systems on $T\mathbb{R}$ consists in determining which VG Lie algebras of the GKO classification [23, 183] are related to second-order differential equations with one unknown dependent variable.

Campoamor-Stursberg additionally studied other SODE Lie systems related to second-order differential equations with more than one dependent variable [72]. In a closely related line of research, he also developed a method for the construction of VG Lie algebras based upon the reduction of invariants and projections of representations of Lie algebras (see [75]).

On the other hand, Ibragimov and Gainetdinova focused on the study of canonical forms for VG Lie algebras and obtaining their superposition rules [227, 228]. Their works can be considered as complementary to the classification of transitive primitive VG Lie algebras initiated by Winternitz and Shnider in [340, 341].

Another modern trend worthy of mention concerns the study of deformations of Lie–Hamilton systems [25, 26]. In this procedure, a Lie–Hamilton system is endowed with a Kirillov–Kostant–Souriau Poisson algebra, $C^\infty(\mathfrak{g}^*)$, and a universal enveloping algebra, $U(\mathfrak{g})$, as standardly done in [27]. In particular, a t-dependent Hamilton function for the Lie–Hamilton system is determined by a Poisson algebra realization of a curve in $C^\infty(\mathfrak{g}^*)$. Then, quantum deformations of the universal enveloping Lie algebra are performed [135] and their duals give rise to deformations $C_z^\infty(\mathfrak{g}^*)$ of the Konstant–Souriau–Kirillov Poisson algebra $C^\infty(\mathfrak{g}^*)$. The initial curve in $C^\infty(\mathfrak{g}^*)$ gives rise to a new curve in $C_z^\infty(\mathfrak{g}^*)$, which, in turn, allows us to determine an ϵ-deformation of the original Lie–Hamilton system.

The deformation of a Lie–Hamilton system is no longer a Lie system, but its properties can still be studied as if it were through the induced quantum deformations [25]. Particularly, the properties of quantum groups, like Casimir operators, are related to constants of motion and properties of the deformed Lie–Hamilton systems. This suggests the use of the quantum group theory to study Hamiltonian systems. Indeed, this procedure was employed to prove that the deformation of a Milne–Pinney equation is an oscillator with a position-dependent mass and a Winternitz-Rosechatius potential [25], apart from other deformations of other Lie–Hamilton systems.

From these first deformations of Lie–Hamilton systems, we can suggest some continuation since, first, this has only been applied to Lie–Hamilton systems on the plane related to a VG Lie algebra isomorphic to $\mathfrak{sl}_2$, and the deformation was the one related to the so-called non-standard Lie bialgebras [135]. Hence, there exist many other possible deformations and their related discoveries. The formalism was only sketched in [25], thus, a more detailed complementary theory is still necessary. It would also be interesting to investigate the generalization of this formalism to study Lie systems compatible with other structures, e.g., Dirac–Lie systems [90] or k-symplectic Lie systems [281].

As remarked previously, Grundland may be considered as one of the pioneers in the theoretical study of PDE Lie systems [296] in relation with conditional symmetries [199, 200], whose results [296] were posteriorly improved with a more general Lie–Scheffers theorem [93]. Cariñena and Ramos also developed techniques to study PDE Lie systems (see, for example, the reduction for PDE Lie systems in [314]). These results are quite theoretical, but there was a turnout in applications as in [89, 197, 198], where the role of PDE Lie systems in conditional symmetries is devised, transformations for studying Navier-Stokes equations, Toda lattices, Bäcklund transformations to study heat equations, modified and standard KdV equations, Burgers' equations, etc. All these applications fill the void of applications of PDE Lie systems present in [93, 108].

The next point is the study of superposition rules for Grassmann-valued differential equations. This study, initiated by the CRM school [32, 33], was devoted to determining super-superposition rules for certain Grassmann-valued differential equations on Lie supergroups. These works showed that such Grassmann-valued differential equations could be studied through Lie systems. Nevertheless, they did not show outstanding properties coming from their supermanifold structure. It is still an open question how to classify Grassmann-valued differentiable equations with superposition rules in the same way as Shnider and Winternitz did in the case of Lie systems.

To finish, we enumerate Lie systems admitting compatible geometric structures. Apart from the previously known structures, there are two main cases of Lie systems with geometric structures that have been studied lately: Lie systems with compatible Riemann-like [83, 215] and multisymplectic structures [189].

On the one hand, Herranz and his coworkers pioneered the analysis of Lie systems on two-dimensional manifolds with a VG Lie algebra of Killing vector fields relative to Riemann, pseudo-Riemann, and semi-Riemannian structures. In particular, they focused on Lie–Hamilton systems on Riemannian spaces (sphere, Euclidean and hyperbolic plane), on semi-Riemannian spaces (Newtonian spacetimes), as well as on pseudo-Riemannian spaces (anti-de Sitter, de Sitter, and Minkowski spacetimes). As standard in the theory of Lie systems, they obtained constants of motion and superposition rules for these structures. The reader can find the classification of Lie systems on the plane admitting a VG Lie algebra of projective or Killing vector fields relative to pseudo-Riemannian metrics in [197, 263]. As for applications, the first cited work analyzes the properties of the Riccati hierarchy, while the second uses the Casimir elements of universal enveloping Lie algebras and VG Lie algebras of Killing vector fields to study the geometric properties of Lie systems.

Finally, the study of *multisymplectic Lie systems*, namely Lie systems admitting a VG Lie algebra of Hamiltonian vector fields related to a multisymplectic structure, was accomplished in [189]. A new coalgebra approach generalizing [27] was devised in this work so as to study multisymplectic Lie systems. This new approach is more powerful than the one for Lie–Hamilton systems [27]. On the one hand, it allows for the construction of tensor field invariants, which are much more general than the invariant functions found in [27]. On the other hand, these invariants are constructed by means of elements of tensor algebras, which retrieves as a particular case the Casimir elements of the coalgebra approach in [27].

In the conclusions of [189], one can read that multisymplectic Lie systems suggest a proper framework for a Marsden–Weinstein reduction of multisymplectic structures. The study of such a generalization of the symplectic Marsden–Weinstein reduction [78] can be of much interest in Differential Geometry.

Lie systems are also helpful in the study of stochastic differential equations [297]. An extension of the theory of Lie systems to stochastic differential equations appeared in [253]. Reference [253] is rather theoretical, but applications to Brownian motions were also accomplished. It is therefore interesting to find more uses of stochastic Lie systems in physical and mathematical problems.

We close this chapter with the last trend on Lie systems, which is the case of difference Lie systems. Winternitz and the CRM school started studying certain difference equations coming from Lie systems in [191, 308, 329, 356, 378]. Unfortunately, despite the large number of examples, there exists no general/unified theory relative to difference equations coming from Lie systems. For instance, there are many possible ways to discretize Lie systems, as illustrated by the different approaches to the discretization of Milne–Pinney equations [147, 217]. On the other hand, previous works seem to indicate

Introduction 23

that the discretization of a Lie system could be achieved relying on the associated Lie group of one of its VG Lie algebras [91]. In fact, [147] uses a relation between Lie systems related to the same Lie group to carry out their discretization. This topic is still unexplored and needs further inspection.

1.3. General Description of the Book

Let us provide a more careful description of the remaining chapters of this book:

- **Chapter 2 — Fundamentals:** This chapter provides an introduction to Differential Geometry and the geometric structures to be used throughout this work.
- **Chapter 3 — Lie Systems:** We define Lie systems, superposition rules, and VG Lie algebras. We present and prove the Lie–Scheffers theorem. Subsequently, several examples of Lie systems and superposition rules are contemplated. We also review a relevant class of Lie systems on Lie groups, the so-called *automorphic Lie systems*, and their superposition rules. We show how to derive superposition rules by using the so-called diagonal prolongations of the vector fields of a VG Lie algebra. As applications, we study three Lie systems: the second-order Riccati equation and the Kummer–Schwarz equations of second and third order [43, 280]. Relevantly, we study the link between the Schwarzian derivative and the third-order Kummer–Schwarz equation, and we find a superposition rule for the latter.
- **Chapter 4 — Lie–Hamilton Systems:** We introduce Lie–Hamilton systems, i.e., Lie systems with a VG Lie algebra of Hamiltonian vector fields relative to a Poisson structure. We show that planar Riccati equations are Lie–Hamilton systems and we study their properties. Some new examples of Lie–Hamilton systems are given: the coupled Riccati equations, the Hamilton equations of the second-order Riccati equations, the Hamilton equations of the second-order Kummer–Schwarz equations, and the Hamilton equations of the Smorodinsky–Winternitz oscillator.

We study t-dependent Hamiltonians for Lie–Hamilton systems. We analyze the properties of constants of motion for Lie–Hamilton systems. We use Poisson coalgebras to devise an algebraic and geometric method for the construction of superposition rules. We will explicitly construct superposition rules for certain systems of physical and mathematical relevance: the classical Ermakov system, four coupled Riccati equations, the Hamilton equations of the second-order Kummer–Schwarz equations, the Hamilton equations of the Smorodinsky–Winternitz oscillator and a system with trigonometric nonlinearities.

Next, we classify Lie algebras of Hamiltonian vector fields on the plane. First, we introduce the notion generic points of Lie algebras, primitive and imprimitive Lie algebras, integrating factors and modular generating systems. From these concepts, a number of new results arise and we classify VG Lie algebras of Hamiltonian vector fields on the plane (with respect to a Poisson structure). To do so, we rely on the GKO classification for finite-dimensional Lie algebras on the plane. We obtain that, out of the initial non-diffeomorphic 28 classes of finite-dimensional Lie algebras of vector fields on $\mathbb{R}^2$ given by the GKO classification, only 12 of them consist of Hamiltonian vector fields. We will inspect all of them and describe their properties.

The classification of Lie–Hamilton systems on the plane enables us to classify Lie–Hamilton systems on the plane with physical applications. In particular, we describe several examples of Lie–Hamilton systems on the plane of physical and mathematical interest.

We start by Lie–Hamilton systems with Lie algebras isomorphic to $\mathfrak{sl}(2, \mathbb{R})$. Some examples are the Milne–Pinney equations, the second-order Kummer–Schwarz equations, and the complex Riccati equation with t-dependent real coefficients.

Then, we study Lie–Hamilton systems with biological and physical applications: the generalized Buchdahl equations, time-dependent Lotka-Volterra systems, quadratic polynomial models and viral infection models. More Lie–Hamilton systems

Introduction 25

on the plane with relevance are displayed: the Cayley–Klein Riccati equations which is a compendium of Double-Clifford or split-complex Riccati equations and the Dual-Study Riccati equation. Next, we present other Lie–Hamilton systems with VG Lie algebras isomorphic to $\mathfrak{sl}(2,\mathbb{R})$ and we study the equivalence between them. More specifically, we analyze coupled Riccati equations, the Milne–Pinney equations, the second-order Kummer–Schwarz equation and a planar diffusion Riccati system. Later, we present Lie–Hamilton systems of two-photon type, which encompasses the dissipative harmonic oscillator and the second-order Riccati equation. To conclude, we study Lie systems with VG Lie algebras isomorphic to $\mathfrak{h}_2$. This covers the complex Bernoulli equation with t-dependent real coefficients, generalized Buchdahl equations t-dependent Lotka–Volterra systems, etc.

- **Chapter 5 — Dirac–Lie Systems:** We motivate the study of Dirac–Lie systems by showing that third-order Kummer–Schwarz equations and a Riccati system cannot be studied through Lie–Hamilton systems. Nevertheless, they admit a VG Lie algebra of Hamiltonian vector fields relative to a presymplectic form. This has led us to make use of Dirac structures and its corresponding geometry. The non-Lie–Hamiltonian nature of the mentioned two examples also suggests to enunciate the so-called *no-go* theorem, which gives conditions ensuring that a Lie system cannot be a Lie–Hamilton one.

  We state how a Dirac–Lie system covers Lie systems with Hamiltonian VG Lie algebras with respect to presymplectic forms and Poisson structures as particular cases. In this way, we generalize Lie–Hamilton systems to a more general class of Lie systems covering the aforementioned structures.

  In similar fashion as for Lie–Hamilton systems, we derive Dirac–Lie Hamiltonians and all their corresponding concepts are reformulated: the concept of diagonal prolongation is redefined to work out their superposition rules. Certain properties for their constants of motion are reconsidered. To conclude, we show a superposition rule for third-order Kummer–Schwarz equations treated with Dirac–Lie systems theory. It is also obvious that the previously studied coalgebra method to derive superposition rules can also be extrapolated to Dirac–Lie systems. To conclude, we introduce *bi-Dirac–Lie system*, or Lie systems admitting two compatible Dirac structures. A method for generating bi-Lie–Dirac systems is given. We find out that our techniques can be applied to Schwarzian Korteweg–de Vries (SKdV) equations [18, 37]. This provides a new approach to the study of these equations.

Furthermore, we derive soliton-type solutions for Schwarzian–KdV equations, namely shape-preserving traveling wave solutions. We show how Lie systems and our methods can be applied to provide Bäcklund transformations for certain solutions of these equations. This can be considered as one of the first applications of Dirac–Lie systems in the study of PDEs of physical and mathematical interest from the point of view of the theory of Lie systems.

- **Chapter 6 — Jacobi–Lie Systems:** In this chapter, we search for VG Lie algebras of Hamiltonian vector fields relative to a Jacobi structure on the plane.

  As in the previous chapters, we introduce the concept of a Jacobi–Lie system and we study its fundamental properties. An explicit example is given: the continuous Heisenberg group is understood as a Jacobi manifold and a Jacobi–Lie system is constructed out of its left-invariant vector fields.

  Then, it is our purpose to classify all Jacobi–Lie systems on the plane (up to diffeomorphisms) via the GKO classification. Our results are displayed in Table A.4 in Appendix A.

- **Chapter 7 — k-Symplectic Lie Systems:** In this chapter, the theory of Lie systems admitting a VG Lie algebra of Hamiltonian vector fields relative to a symplectic structure is presented. As a consequence of the extension to this new realm of the results obtained in the previous chapters, we obtain new applications of k-symplectic structures and we find that they can be endowed with families of Poisson algebras of functions.

- **Chapter 8 — Lie Systems and Conformal Forms:** This chapter is devoted to the classification of Lie systems, admitting Lie algebras of conformal or Killing vector fields relative to a Riemannian or pseudo-Riemannian metric. In this way, this chapter provides a review of the works [198, 263], which simplifies and generalizes the proofs and results in detailed works.
- **Chapter 9 — Lie Systems and Lie Symmetries:** This final chapter concerns the study of Lie symmetries for Lie systems and PDE Lie systems. In particular, it shows that the VG Lie algebras related to these systems allow us to simplify the determination of classical Lie symmetries.

Chapter 2

Geometric Fundamentals

2.1. Introduction

In short, Differential Geometry is the mathematical discipline generalizing the standard differential and integral calculus on vector spaces to more general spaces, the manifolds, in order to study the geometric properties of the structures defined on them.

During the 18th and 19th centuries, Differential Geometry was mostly concerned with the study of the intrinsic properties of curves and surfaces in $\mathbb{R}^n$, e.g., *Gauss's Theorem Egregium*, and the notion of parallelism [177–179].

The first mature notion of manifold appears in Riemann's habilitation thesis in 1854 [326]. Riemann defined the term *manifold*, coming from the german *Mannigfaltigkeit*, as a topological space that resembles the Euclidean space near each point, but not necessarily globally. Since then, manifolds became a central concept for Geometry and modern Mathematical Physics. Later, manifolds arose as solution sets of systems of equations, graphs of functions, surfaces, etc. They allow us to describe locally very complicated structures in terms of an Euclidean space and techniques from linear algebra.

Riemann settled metrics on manifolds as a way to allow us measure distances and angles. This led, in a natural way, to the existence of non-Euclidean geometries. It is worth noting that by Riemann's time, it was fundamental to understand the existence of such non-Euclidean geometries that had just appeared [39, 68].

Moreover, Gauss's Theorem Egregium is based on the notion of a metric: *the curvature of a surface can be determined entirely by measuring angles and distances on the surface itself*, without further reference on how the surface is embedded in the three-dimensional Euclidean space. In more modern geometrical terms, the Gaussian curvature of a surface is invariant under local isometry. These were ideas fully established by Riemann: *Relevant properties of geometric structures are independent of the spaces in which they might be embedded.*

Since the 19th century, Differential Geometry began to be involved with geometric aspects of the theory of differential equations and geometric structures on manifolds. Indeed, Lie pioneered this line of research by developing a theory of invariance of differential equations under groups of continuous transformations as a way to achieve conserved quantities [267, 271, 276].

Historically, the formulation of Differential Geometry and Symmetry Groups unmodified from Euler's theoretical description of the plane sound waves in fluids until Birkhoff retook the subject during World War II, working on Fluid Dynamics including bazooka changes and problems of air-launched missiles entering water [48, 49]. He emphasized Group Theory for handling symmetries in Hydrodynamics and urged innovative numerical methods relying more heavily on computing. Indeed, he started working on the solution of partial differential equations by successive approximations.

By the end of the 20th century, a big explosion in research of geometrical aspects of differential equations had taken place. In order to study the Hamiltonian formalism of Classical Mechanics, the introduction of symplectic manifolds served as the mathematical description of the phase space. More relevant manifolds were henceforth introduced, such as the Lorentzian manifold model for spacetime in General Relativity [39], etc.

Properties of nonlinear systems of partial differential equations (PDEs) with geometrical origin and natural description in the

Geometric Fundamentals 31

language of infinite-dimensional Differential Geometry were settled. This helped the appearance of their Lagrangian and Hamiltonian formalisms, and the obtention of low-dimensional conservation laws [386]. Many problems in Physics, Medicine, Finance, and industry required descriptions by means of nonlinear PDEs. So, their investigation through Differential Geometry led to an independent field of research with many different directions: optimal transport problems, free boundary problems, nonlinear diffusive systems, singular perturbations, stochastic partial differential equations, regularity issues, and so on [108, 126, 211].

For all the above reasons, we find it important to formulate the content of this book in geometrical terms. This introductory chapter provides an overview of the geometrical structures we shall use in the forthcoming chapters. It provides a guide from the very basic concepts appearing in differential geometry, e.g., tangent vectors, vector fields, and integral curves, to very sophisticated structures appearing in modern differential geometry, for instance, symplectic, presymplectic, Poisson, Dirac, and Jacobi manifolds.

While focusing on the main properties of geometric structures, technical details and more subtle topics have merely been commented on and references for further explanations are detailed. For simplicity, we shall assume every geometric structure to be real, smooth and well defined globally. Minor technical details will be specified when strictly needed.

In the following chapters, we will illustrate how these structures allow us to describe constants of motion, symmetries, superposition rules, and other properties of differential equations.

2.2. Tangent Spaces, Vector Fields, and Integral Curves

All manifolds are assumed to be finite-dimensional. In particular, N is assumed to be an n-dimensional manifold and $\{y_i \,|\, i = 1, \ldots, n\}$ is a coordinate system on N. This allows us to denote each point $y \in N$ by an n-tuple $(y_1, \ldots, y_n) \in \mathbb{R}^n$. We assume ϵ to be a coordinate system on $\mathbb{R}$.

We write curves in N in the form $\gamma : \epsilon \in \mathbb{R} \mapsto \gamma(\epsilon) = (\gamma_1(\epsilon), \dots, \gamma_n(\epsilon)) \in N$. The space of curves in N can be endowed with an equivalence relation R given by

$$\gamma \, R \, \bar{\gamma} \iff \gamma(0) = \bar{\gamma}(0) \quad \text{and} \quad \frac{d\gamma_i}{d\epsilon}(0) = \frac{d\bar{\gamma}_i}{d\epsilon}(0), \quad i = 1, \dots, n. \tag{2.1}$$

In other words, two curves in N are related if their first-order Taylor polynomials around $\epsilon = 0$ coincide. Observe that the above equivalence relation is independent of the chosen coordinate system. That is, if $\gamma \, R \, \bar{\gamma}$ for the coordinate systems ϵ on $\mathbb{R}$ and $\{y_1, \dots, y_n\}$ on N, then $\gamma \, R \, \bar{\gamma}$ also for any other coordinate systems on $\mathbb{R}$ and N. Hence, the equivalence relation (2.1) is independent of the chosen coordinate system and it therefore has a geometrical meaning.

Every *equivalence class* of R is called a *tangent vector*. We write v_y for an equivalence class of curves passing through $y \in N$ for $\epsilon = 0$. We call *tangent space at* $y \in N$ the space $T_y N$ of equivalence classes of the form v_y. In view of (2.1), each tangent vector $v_y \in T_y N$ can be identified with a set of numbers $(v_1, \dots, v_n)$ and vice versa. Hence, the tangent space to each point $y \in N$ can be written as

$$T_y N = \{(v_1, \dots, v_n) | (v_1, \dots, v_n) \in \mathbb{R}^n\}. \tag{2.2}$$

The space $T_y N$ is naturally endowed with an n-dimensional vectorial space structure and there exists an isomorphism $T_y N \simeq \mathbb{R}^n$. Consequently, given a local coordinate system $\{y_1, \dots, y_n\}$ on N around $y \in N$, we can identify each vector $(v_1, \dots, v_n) \in \mathbb{R}^n$ with the tangent vector v_y associated with those curves $\gamma : \epsilon \in \mathbb{R} \mapsto (\gamma_1(\epsilon), \dots, \gamma_n(\epsilon)) \in N$ satisfying that that $d\gamma_i/d\epsilon(0) = v_i$ for $i = 1, \dots, n$. It is worth noting that the above identification does not depend on the chosen coordinate system: if a curve is related to a tangent vector on a coordinate system, then it can be proved that they will be so in any other coordinate system.

Alternatively, tangent vectors can be understood as derivations on a certain $\mathbb{R}$-algebra, i.e., a real vector space along with an $\mathbb{R}$-bilinear commutative multiplication. More specifically, let $C^\infty(U)$ be the space of differentiable functions defined on an open subset $U \subset N$ containing y. The space $C^\infty(U)$ admits an equivalence relation R_g given by

$$f \, R_g \, g \quad \Longleftrightarrow \quad f, g \text{ coincide on an open subset containing } y.$$

The equivalence classes of this relation, let us say $[f]$ with $f \in C^\infty(U)$, are called *germs*. It is easy to verify that the quotient space, $C_g^\infty(U)$, related to the equivalence relation R_g, becomes a vectorial space with the sum $[f] + [g] = [f + g]$ and multiplication by scalars $\lambda \in \mathbb{R}$ given by $\lambda[f] = [\lambda f]$. Additionally, $C^\infty(U)$ forms an $\mathbb{R}$-algebra with the $\mathbb{R}$-bilinear commutative multiplication $[f][g] = [fg]$.

Each v_y induces a mapping $D_{v_y} : [f] \in C_g^\infty(U) \mapsto D_{v_y}[f] \in \mathbb{R}$ given by

$$D_{v_y}[f] = \frac{d}{d\epsilon}\bigg|_{\epsilon=0} f \circ \gamma, \quad \gamma \in v_y. \tag{2.3}$$

Since the first-order Taylor polynomials around 0 of each $\gamma \in v_y$ coincide by definition of v_y, the above mapping is well defined and it does not depend on the chosen $\gamma \in v_y$. It can be proven that each D_{v_y} is a derivation on $C_g^\infty(U)$, i.e., D_{v_y} is linear and satisfies the Leibniz rule $[D_{v_y}([f][g])](y) = D_{v_y}([f])g(y) + f(y)D_{v_y}([g])$ for each $[f], [g] \in C_g^\infty(U)$. Conversely, it can be proven that every derivation on $C_g^\infty(U)$ is of the form (2.3) for a certain $v_y \in T_yN$. This motivates to identify each v_y with a derivation D_{v_y} and vice versa. In particular, the equivalence class v_{v_y} related to the vector $(0, \ldots, 1\,(i - \text{position}), \ldots, 0) \in \mathbb{R}^n$ induces the derivation $(\partial/\partial y_i)_y : [f] \in C_g^\infty(U) \mapsto \frac{\partial f}{\partial y_i}(y) \in \mathbb{R}$.

We call *tangent manifold* to N the space $TN := \bigcup_{y \in N} T_yN$. We can endow TN with a differentiable structure. To do so, we related each coordinate system $\{y_1, \ldots, y_n\}$ on an open U of N to a coordinate system on TN of the form

$$y_i(v_y) := y_i(y), \quad v_i(v_y) := \frac{d\gamma_i}{d\epsilon}(0), \quad i = 1, \ldots, n, \tag{2.4}$$

where $\epsilon \mapsto (\gamma_1(\epsilon), \ldots, \gamma_n(\epsilon))$ is a curve belonging to the class v_y. This gives rise to a coordinate system $\{y_1, \ldots, y_n, v_1, \ldots, v_n\}$ on TN. This coordinate system is called an *adapted coordinate system* to TN. It can be proven that adapted coordinates generate an atlas of TN, which becomes a manifold. In consequence, if N is an n-dimensional manifold, then TN becomes a $2n$-dimensional manifold. We call *tangent bundle* the triple $(TN, N, \tau_N : v_y \in TN \mapsto y \in N)$.

Similarly, we call cotangent manifold the space $T^*N :=$ $\sum_{y \in N} T_y^* N$, where $T_y^* N$ is the dual space to $T_y N$. Let us endow T^*N with a differentiable structure. To do so, we related each coordinate system $\{y_1, \ldots, y_n\}$ on an open U on N to a coordinate system on T^*N given by

$$y_i(\theta_y) := y_i(y), \quad p_i(\theta_y) := \theta_y \left(\frac{d\gamma_i}{d\epsilon}(0) \right), \qquad i = 1, \ldots, n, \quad (2.5)$$

where $\epsilon \mapsto (\gamma_1(\epsilon), \ldots, \gamma_n(\epsilon))$ is a curve belonging to the class v_y. This gives rise to a coordinate system $\{y_1, \ldots, y_n, p_1, \ldots, p_n\}$ on T^*N. This coordinate system is called an *adapted coordinate system* to T^*N. It can be proven that adapted coordinates generate an atlas of T^*N, which becomes a manifold. In consequence, if N is an n-dimensional manifold, then T^*N becomes a $2n$-dimensional manifold. We call *cotangent bundle* the triple $(T^*N, N, \pi_N : v_y \in T^*N \mapsto y \in N)$. With a little abuse of notation, we will hereafter call TN and T^*N the tangent and cotangent bundles, respectively.

Similar to the above bundles, one can also define many other bundles with a base space given by N. For instance, $\Lambda^2 N = \bigcup_{y \in N} \Lambda^2 T_y N$, where $\Lambda^2 T_y N$ is the space generated by anti-symmetric tensor products of elements of $T_y N$, is a bundle over N relative to the natural projection of $\Lambda^2 N$ onto N.

A *vector field* on N is a section of the tangent bundle TN, i.e., a mapping $X : N \to TN$ such that $\tau_N \circ X = \mathrm{Id}_N$, with Id_N being the identity on N. In other words, a vector field is a mapping assigning to every $y \in N$ a tangent vector $v_y \in T_y N$. In the coordinate system $\{y_1, \ldots, y_n\}$, a vector field can be brought into the form

$$X(y) = \sum_{i=1}^{n} \eta_i(y) \frac{\partial}{\partial y_i}, \qquad (2.6)$$

for certain functions $\eta_i \in C^\infty(N)$ with $i = 1, \ldots, n$.

An *integral curve* $\gamma : \epsilon \in \mathbb{R} \mapsto \gamma(\epsilon) \in N$ for X is a curve $\gamma : \epsilon \in \mathbb{R} \mapsto \gamma(\epsilon) \in N$ such that

$$\frac{d\gamma}{dt} = X(\gamma(t)).$$

In other words, the tangent vector to γ at each $t \in \mathbb{R}$ matches the value of X at $\gamma(t)$. In coordinates, one has

$$\begin{cases} \gamma(0) = y_0, \\ \dfrac{d\gamma_i}{d\epsilon}(\epsilon) = \eta_i(\gamma(\epsilon)), \end{cases} \quad \forall \epsilon \in \mathbb{R}, \quad i = 1, \ldots, n. \quad (2.7)$$

Meanwhile, an integral curve of X with initial condition $y_0 \in N$ is the integral curve of X satisfying that $\gamma(0) = y_0$. In view of the theorem of existence and uniqueness of solutions of systems of first-order ordinary differential equations, the solution to the above Cauchy problem exists and is unique. So, there exists a unique integral curve of a vector field for each initial condition.

Observe that if $\gamma : \epsilon \in \mathbb{R} \mapsto (\gamma_1(\epsilon), \ldots, \gamma_n(\epsilon)) \in N$ is a particular solution to system (2.7), then the coordinate expression for γ in another coordinate system is a particular solution for (2.7) in the new coordinate system. This justifies the writing of system (2.7) in an intrinsic form as

$$\begin{cases} \gamma(0) = y_0, \\ \dfrac{d\gamma}{d\epsilon}(\epsilon) = X(\gamma(\epsilon)), \end{cases} \quad (2.8)$$

where $d\gamma/d\epsilon(\epsilon_0)$ is understood as the tangent vector associated with γ at $\gamma(\epsilon_0)$, i.e., the equivalence class related to the curve $\epsilon \mapsto \bar{\gamma}(\epsilon) := \gamma(\epsilon + \epsilon_0)$. Therefore, an integral curve γ of a vector field X on N is a curve in N whose tangent vector at each point $y \in \operatorname{Im}\gamma$ coincides with $X(y)$. We write γ_{y_0} for the integral curve of X with initial condition $\gamma_{y_0}(0) = y_0 \in N$.

By solving system (2.7) for each $y_0 \in N$, we obtain a family of integral curves $\{\gamma_{y_0}\}_{y_0 \in N}$. This gives rise to the so-called *flow* of X, namely the mapping $\Phi : (\epsilon, y) \in \mathbb{R} \times N \mapsto \gamma_y(\epsilon) \in N$. Since

$X(y) = d\gamma_y/d\epsilon(0)$ for each $y \in N$, the flow uniquely determines the vector field X.

The flow allows us to define an ϵ-parametric set of diffeomorphisms $\Phi_\epsilon : y \in N \mapsto \Phi(\epsilon, y) := \gamma_y(\epsilon) \in N$. Observe that $\gamma_{\gamma_y(\bar{\epsilon})}(\epsilon) = \gamma_y(\epsilon + \bar{\epsilon}), \forall \epsilon, \bar{\epsilon} \in \mathbb{R}, \forall y \in N$. Indeed, with fixed $\bar{\epsilon}$, the curve $\epsilon \mapsto \gamma_y(\epsilon + \bar{\epsilon})$ takes the value $\gamma_y(\bar{\epsilon})$ for $\epsilon = 0$ and, defining $\tilde{\epsilon} := \epsilon + \bar{\epsilon}$, we obtain

$$\frac{d\gamma_y(\epsilon + \bar{\epsilon})}{d\epsilon} = \frac{d\gamma_y(\tilde{\epsilon})}{d\tilde{\epsilon}} = X(\gamma_y(\tilde{\epsilon}))$$

$$= X(\gamma_y(\epsilon + \bar{\epsilon})), \quad \gamma_y(0 + \bar{\epsilon}) = \gamma_y(\bar{\epsilon}). \qquad (2.9)$$

Therefore, the curve $\epsilon \mapsto \gamma_y(\epsilon + \bar{\epsilon})$ is an integral curve of X taking the value $\gamma_y(\bar{\epsilon})$ for $\epsilon = 0$. So, it must coincide with $\gamma_{\gamma_y(\bar{\epsilon})}(\epsilon)$ and $\gamma_{\gamma_y(\bar{\epsilon})}(\epsilon) = \gamma_y(\epsilon + \bar{\epsilon})$. In consequence, we have

$$\Phi_\epsilon(\Phi_{\bar{\epsilon}}(y)) = \Phi_\epsilon(\gamma_y(\bar{\epsilon})) = \gamma_{\gamma_y(\bar{\epsilon})}(\epsilon) = \gamma_y(\epsilon + \bar{\epsilon})$$

$$= \Phi_{\epsilon + \bar{\epsilon}}(y), \quad \forall y \in N, \quad \epsilon, \bar{\epsilon} \in \mathbb{R}. \qquad (2.10)$$

On the other hand, $\Phi_0(y) = \gamma_y(0) = y$ for all $y \in N$. Thus,

$$\Phi_\epsilon \circ \Phi_{\bar{\epsilon}} = \Phi_{\epsilon + \bar{\epsilon}}, \quad \Phi_0 = \mathrm{Id}_N, \quad \forall \epsilon, \bar{\epsilon} \in \mathbb{R}, \qquad (2.11)$$

where Id_N is the identity on N. The properties (2.11) tell us that $\{\Phi_\epsilon\}_{\epsilon \in \mathbb{R}}$ is an ϵ-*parametric group of diffeomorphisms* on N. Indeed, the composition of elements of that family belongs to the family in virtue of (2.11). The element $\Phi_0 = \mathrm{Id}_N$ is the neutral element and $\Phi_{-\epsilon} \circ \Phi_\epsilon = \mathrm{Id}_N$. In this way, Φ_ϵ has inverse $\Phi_{-\epsilon}$, which makes each Φ_ϵ, with $\epsilon \in \mathbb{R}$, into a diffeomorphism. We can also understand the uniparametric group of diffeomorphisms $\{\Phi_\epsilon\}_{\epsilon \in \mathbb{R}}$ as a *Lie group action* $\Phi : (\epsilon, y) \in \mathbb{R} \times N \mapsto \Phi_\epsilon(y) \in N$.

Alternatively, every ϵ-parametric group of diffeomorphisms $\{\Phi_\epsilon\}_{\epsilon \in \mathbb{R}}$ on N enables us to define a vector field on N of the form

$$X_\Phi(y) := \frac{d}{d\epsilon}\Big|_{\epsilon=0} \Phi_\epsilon(y), \quad \forall y \in N, \qquad (2.12)$$

whose integral curves are $\gamma_y(\epsilon) = \Phi_\epsilon(y)$, with $y \in N$. If $\{\Phi_\epsilon\}_{\epsilon \in \mathbb{R}}$ is coming from the flow of a vector field X, then $X_\Phi = X$. This shows that each vector field is equivalent to an ϵ-parametric group of diffeomorphisms.

Assuming X to be given by (2.6), the corresponding flow gives rise to a set of ϵ-parametric diffeomorphisms which, for infinitesimal ϵ, takes the form

$$y_i \mapsto y_i + \epsilon\, \eta_i(y) + O(\epsilon^2), \quad i = 1, \ldots, n, \qquad (2.13)$$

where $O(\epsilon^2)$ stands, as usual, for a function of ϵ satisfying that $\lim_{\epsilon \to 0} O(\epsilon^2)/\epsilon = 0$. This transformation is invertible for each fixed value of the parameter ϵ. The set of all transformations for different values of ϵ forms a *group*.

2.3. *t*-Dependent Vector Fields

To study systems of non-autonomous first-order ordinary differential equations from a geometrical viewpoint, we introduce *t-dependent vector fields*. In this section, we present some of their basic properties.

Let $\pi_2 : (t, y) \in \mathbb{R} \times N \mapsto y \in N$ be the projection onto the second factor and let t be the natural coordinate system on $\mathbb{R}$. We write t for the variable in $\mathbb{R}$ because this variable mostly stands for the physical time in the applications to be described hereafter. Recall that $\tau_N : TN \to N$ stands for the tangent bundle projection. A *t-dependent vector field* X on N is a map $X : (t, y) \in \mathbb{R} \times N \mapsto X(t, y) \in TN$ satisfying the commutative diagram

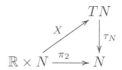

that is, $\pi_N \circ X = \pi_2$. Hence, $X(t, y) \in \tau_N^{-1}(y) = T_y N$ and $X_t : y \in N \mapsto X_t(y) := X(t, y) \in T_y N \subset TN$ is a vector field on N for every $t \in \mathbb{R}$. Conversely, every *t*-parametric family $\{X_t\}_{t \in \mathbb{R}}$ of vector fields on N gives rise to a unique *t*-dependent vector field $X : (t, y) \in \mathbb{R} \times N \mapsto X_t(y) \in TN$.

Thus, each t-dependent vector field X is equivalent to a family $\{X_t\}_{t\in\mathbb{R}}$ of vector fields on N.

The t-dependent vector fields enable us to describe, as a particular instance, standard vector fields. Indeed, a vector field $X : y \in N \mapsto X(y) \in TN$ can naturally be regarded as a t-dependent vector field $X : (t, y) \in \mathbb{R} \times N \mapsto X(y) \in TN$. Conversely, a "constant" t-dependent vector field X on N, i.e., one satisfying that $X_t = X_{t'}$ for every $t, t' \in \mathbb{R}$, can be understood as a vector field $Y = X_0$ on N.

Each t-dependent vector field X gives rise to a linear morphism $X : f \in C^\infty(N) \mapsto Xf \in C^\infty(\mathbb{R} \times N)$, with $(Xf)(t, y) := (X_t f)(y)$ for every $(t, y) \in \mathbb{R} \times N$, satisfying a *Leibniz rule*, namely $[X(fg)](t, y) = (Xf)(t, y)g(y) + f(y)(Xg)(t, y)$ for arbitrary $f, g \in C^\infty(N)$ and $(t, y) \in \mathbb{R} \times N$. Conversely, one can prove that any morphism $X : f \in C^\infty(N) \mapsto Xf \in C^\infty(\mathbb{R} \times N)$ satisfying the Leibniz rule can be understood as a t-dependent vector field.

Similar to vector fields, t-dependent vector fields also admit local integral curves. To define them, we make use of the so-called autonomization of X. The *autonomization* of a vector field X on N is the only vector field $\bar{X}$ on $\mathbb{R} \times N$ satisfying $\iota_{\bar{X}} dt = 1$ and $(\bar{X}\pi_2^* f)(t, y) = (Xf)(t, y)$ for an arbitrary function $f \in C^\infty(N)$ and $(t, y) \in \mathbb{R} \times N$. In coordinates, if $X = \sum_{i=1}^n X_i(t, y)\partial/\partial y_i$, then

$$\bar{X}(t, y) = \frac{\partial}{\partial t} + \sum_{i=1}^n X_i(t, y)\frac{\partial}{\partial y_i}. \tag{2.14}$$

An *integral curve of a t-dependent vector field* X is a map $\gamma : \mathbb{R} \to \mathbb{R} \times N$ satisfying $\pi_1 \circ \gamma = \mathrm{Id}_\mathbb{R}$ and γ an integral curve of $\bar{X}$. Consequently, one has γ of the form $\gamma : t \in \mathbb{R} \mapsto (t, y(t)) \in \mathbb{R} \times N$ and satisfies the differential equation

$$\frac{dy_i}{dt} = X_i(t, y), \quad i = 1, \ldots, n. \tag{2.15}$$

The above system is called the *associated system* of X. Conversely, a first-order system in normal form (2.2) gives rise to a t-dependent vector field of the form $X(t, y) = \sum_{i=1}^n X_i(t, y)\partial/\partial y_i$ whose integral curves $t \to (t, y(t))$ are the particular solutions to (2.2). This motivates to use X to represent a non-autonomous first-order system of ordinary differential equations in normal form and its related t-dependent vector field.

Every t-dependent vector field X on N gives rise to its *generalized flow* g^X, i.e., the map $g^X : \mathbb{R} \times N \to N$ such that $g^X(t, y) := g_t^X(y) = \gamma_y(t)$, where γ_y is the particular solution to X such that $\gamma_y(0) = y$. We already showed that if X is a standard vector field, its particular solutions satisfy that $\gamma_{\gamma_y(s)}(t) = \gamma_y(t + s)$ for each $t, s \in \mathbb{R}$ and $y \in N$. As a consequence, the generalized flow for the t-dependent vector field X associated with an autonomous vector field leads to a uni-parametric group of diffeomorphisms $\{g_t^X\}_{t \in \mathbb{R}}$. If X is not related to an autonomous vector field, this property is no longer valid.

Given a vector bundle $\mathrm{pr} : P \to N$, we denote by $\Gamma(\mathrm{pr})$ the $C^\infty(N)$-module of its smooth sections. Hence, if $\tau_N : TN \to N$ and $\pi_N : T^*N \to N$ are the canonical projections associated with the tangent and cotangent bundle to N, respectively, then $\Gamma(\tau_N)$ and $\Gamma(\pi_N)$ designate the $C^\infty(N)$-modules of vector fields and 1-forms on N, correspondingly.

We call *generalized distribution* $\mathcal{D}$ on N, a correspondence relating each $y \in N$ to a linear subspace $\mathcal{D}_y \subset \mathrm{T}_y N$. A generalized distribution is said to be *regular* at $y' \in N$ when the function $r : y \in N \mapsto \dim \mathcal{D}_y \in \mathbb{N} \cup \{0\}$ is locally constant around y'. Similarly, $\mathcal{D}$ is regular on an open $U \subset N$ when r is constant on U. Finally, a vector field $Y \in \Gamma(\tau_N)$ is said to take values in $\mathcal{D}$, in short $Y \in \mathcal{D}$, when $Y_y \in \mathcal{D}_y$ for all $y \in N$. Likewise, similar notions can be defined for a *generalized codistribution*, namely a correspondence mapping relating every $y \in N$ to a linear subspace of T_y^*N.

It will be very important to our purposes to relate t-dependent vector fields to the so-called Lie algebras as follows. A *Lie algebra* is a pair $(V, [\cdot, \cdot])$, where V stands for a real linear space endowed with a Lie bracket $[\cdot, \cdot] : V \times V \to V$, namely an $\mathbb{R}$-bilinear antisymmetric mapping satisfying the Jacobi identity. Given two subsets $\mathcal{A}, \mathcal{B} \subset V$, we write $[\mathcal{A}, \mathcal{B}]$ for the real vector space spanned by the Lie brackets between elements of $\mathcal{A}$ and $\mathcal{B}$, and we define $\mathrm{Lie}(\mathcal{B}, V, [\cdot, \cdot])$ to be the smallest Lie subalgebra of V containing $\mathcal{B}$. Note that $\mathrm{Lie}(\mathcal{B}, V, [\cdot, \cdot])$

is expanded by

$$\mathcal{B}, [\mathcal{B}, \mathcal{B}], [\mathcal{B}, [\mathcal{B}, \mathcal{B}]], [\mathcal{B}, [\mathcal{B}, [\mathcal{B}, \mathcal{B}]]], [[\mathcal{B}, \mathcal{B}], [\mathcal{B}, \mathcal{B}]], \dots. \qquad (2.16)$$

From now on, we use $\mathrm{Lie}(\mathcal{B})$ and V to represent $\mathrm{Lie}(\mathcal{B}, V, [\cdot, \cdot])$ and $(V, [\cdot, \cdot])$, correspondingly, when their meaning is clear from the context.

Given a t-dependent vector field X, we call *minimal Lie algebra* of X the smallest Lie algebra, V^X, of vector fields (relative to the Lie bracket of vector fields) containing all the vector fields $\{X_t\}_{t \in \mathbb{R}}$, namely $V^X = \mathrm{Lie}(\{X_t\}_{t \in \mathbb{R}})$. It is worth noting that V^X is also called *smallest Lie algebra* or *irreducible Lie algebra* in the literature [194, 197].

Given a t-dependent vector field X on N, its *associated distribution*, $\mathcal{D}^X$, is the generalized distribution on N spanned by the vector fields of V^X, i.e.,

$$\mathcal{D}_x^X = \{Y_x \mid Y \in V^X\} \subset T_x N, \qquad (2.17)$$

and its *associated codistribution*, $\mathcal{V}^X$, is the generalized codistribution on N of the form

$$\mathcal{V}_x^X = \{\vartheta \in T_x^* N \mid \vartheta(Z_x) = 0, \forall\, Z_x \in \mathcal{D}_x^X\} = (\mathcal{D}_x^X)^\circ \subset T_x^* N, \quad (2.18)$$

where $(\mathcal{D}_x^X)^\circ$ is the *annihilator* of $\mathcal{D}_x^X$.

Proposition 2.1. *A function $f : U \to \mathbb{R}$ is a local t-independent constant of motion for a system X if and only if $df \in \mathcal{V}^X|_U$.*

Proof. If f is assumed to be a t-independent constant of motion, then $X_t f|_U = df(X_t)|_U = 0$ for all $t \in \mathbb{R}$. Consequently, df also vanishes on the successive Lie brackets of elements from $\{X_t\}_{t \in \mathbb{R}}$ and hence

$$df(Y)|_U = Y f|_U = 0, \quad \forall Y \in \mathrm{Lie}(\{X_t\}_{t \in \mathbb{R}}). \qquad (2.19)$$

Since the elements of V^X span the generalized distribution $\mathcal{D}^X$, then $df_x(Z_x) = 0$ for all $x \in U$ and $Z_x \in \mathcal{D}_x^X$, i.e., $df \in \mathcal{V}^X|_U$. The converse directly follows from the above considerations. $\qquad \square$

The following lemma can easily be proven [121].

Lemma 2.2. *Given a system X, its associated codistribution $\mathcal{V}^X$ admits a local basis around every $x \in U^X$ of the form $df_1, \dots, df_{p(x)}$,* *with $p(x) = r^X(x)$ and $f_1, \ldots, f_{p(x)} : U \subset U^X \to \mathbb{R}$ being a family of (local) t-independent constant of motion for X. Furthermore, the $\mathbb{R}$-linear space $\mathcal{I}^X|_U$ of t-independent constants of motion of X on U can be written as*

$$\mathcal{I}^X|_U = \{g \in C^\infty(U) \mid \exists F : U \subset \mathbb{R}^{p(x)} \to \mathbb{R},$$
$$g = F(f_1, \ldots, f_{p(x)})\}. \tag{2.20}$$

2.4. Higher-order Tangent Spaces and Related Notions

Given two curves $\rho, \sigma : \mathbb{R} \mapsto N$ such that $\rho(0) = \sigma(0) = y \in N$, we say that they have a *contact of order p at y*, with $p \in \mathbb{N} \cup \{0\}$, if they satisfy

$$\frac{d^p(f \circ \rho)}{dt^p}(0) = \frac{d^p(f \circ \sigma)}{dt^p}(0), \tag{2.21}$$

for every function $f \in C^\infty(N)$. The relation "to have a contact of order p at y" is an equivalence relation. Observe that this relation is purely geometrical, i.e., if two curves are related with respect to a certain coordinate system, then they are so in any other coordinate system. Note also that the above relation amounts to saying that the two curves have the same Taylor expansion around 0 up to order p.

Each equivalence class of the previous equivalence relation, let us say $\mathbf{t}_y^p$, is called a *p-tangent vector* at y. More specifically, $\mathbf{t}_y^p$ stands for an equivalence class of contacts of order p with a curve $\sigma(t)$ with $\sigma(0) = y$. We write $T_y^p N$ for the space of all p-tangent vectors at p and we define

$$T^p N := \bigcup_{y \in N} T_y^p N.$$

It can be proven that $T^p N$ can be endowed with a differential structure turning it into a differential manifold. Additionally, $(T^p N, \pi^p, N)$, with $\pi^p : \mathbf{t}_y^p \in T^p N \mapsto y \in N$, is a fiber bundle. Let us briefly analyze these facts.

Every coordinate system t on $\mathbb{R}$ and $\{y_1, \ldots, y_n\}$ on N induces a natural coordinate system on the space $T^p N$. Indeed, consider again a curve $\rho : t \in \mathbb{R} \mapsto \rho(t) \in N$ with coordinates $\rho_1(t), \ldots, \rho_n(t)$. The p-tangent vector, $\mathbf{t}^p_{\rho(0)}$, associated with this curve admits a representative

$$\rho_i(0) + \frac{t}{1!}\frac{d\rho_i}{dt}(0) + \cdots + \frac{t^p}{p!}\frac{d^p\rho_i}{dt^p}(0), \quad i = 1, \ldots, n, \qquad (2.22)$$

which can be characterized by the coefficients

$$y_i(0) = \rho_i(0), \qquad y_i^{1)}(0) = \frac{d\rho_i}{dt}(0), \ldots,$$

$$y_i^{p)}(0) = \frac{d^p\rho_i}{dt^p}(0), \quad i = 1, \ldots, n. \qquad (2.23)$$

In consequence, the mapping $\varphi : \mathbf{t}^p_{y(0)} \in T^p N \mapsto (y_i(0), y_i^{1)}(0), \ldots, y_i^{p)}(0)) \in \mathbb{R}^{n(p+1)}$ gives a coordinate system for $T^p N$. Obviously, the map π^p becomes a smooth submersion which makes $T^p N$ into a fiber bundle with base N. We hereby denote each element of $T^p N$ by $\mathbf{t}^p_y = (y, y^{1)}, \ldots, y^{p)})$.

Now, given a curve $c : t \in \mathbb{R} \mapsto c(t) \in N$, we call *prolongation to $T^p N$ of c* the curve $\mathbf{t}^p c : t \in \mathbb{R} \mapsto \mathbf{t}^p c(t) \in T^p N$, associating with every t_0 the corresponding equivalence class of $c(t + t_0)$ given in coordinates by

$$\mathbf{t}^p c(t_0) = (c(t_0), c^{1)}(t_0), \ldots, c^{p)}(t_0)). \qquad (2.24)$$

2.5. Jet Bundles

Jet bundles are certain types of bundles constructed out of the sections of a fiber bundle. Our interest in them is due to the geometrical description of systems of higher-order ordinary and PDEs and their Lie symmetries as structures on an appropriate jet bundle.

For simplicity, consider a projection $\pi : (x, u) \in \mathbb{R}^n \times N := N_{\mathbb{R}^n} \mapsto x \in \mathbb{R}^n$ giving rise to a trivial bundle $(N_{\mathbb{R}^n}, \mathbb{R}^n, \pi)$. Let us, hereafter, assume N to be a k-dimensional manifold and let $\{x_1, \ldots, x_n\}$ be a global coordinate system on $\mathbb{R}^n$.

We say that two sections $\sigma_1, \sigma_2 : \mathbb{R}^n \to N_{\mathbb{R}^n}$ are *p-equivalent at a point $x \in \mathbb{R}^n$* or they have a *contact of order p at x* if they have the same Taylor expansion of order p at $x \in \mathbb{R}^n$. Equivalently,

$$\sigma_1(x) = \sigma_2(x), \quad \frac{\partial^{|J|}(\sigma_1)_i}{\partial x_1^{j_1} \ldots \partial x_n^{j_n}}(x) = \frac{\partial^{|J|}(\sigma_2)_i}{\partial x_1^{j_1} \ldots \partial x_n^{j_n}}(x), \quad (2.25)$$

for every multi-index $J = (j_1, \ldots, j_n)$ such that $0 < |J| := j_1 + \cdots + j_n \leq p$ and $i = 1, \ldots, k$. Being p-equivalent induces an equivalence relation in the space $\Gamma(\pi)$ of sections of the bundle $(N_{\mathbb{R}^n}, \mathbb{R}^n, \pi)$. Observe that if two sections have a contact of order p at a point x, then they do have a contact at that point of the same type for any other coordinate systems on $\mathbb{R}^n$ and N, i.e., this equivalence relation is geometric.

We write $j_x^p \sigma$ for the equivalence class of sections that have a *contact of p-order* at $x \in \mathbb{R}^n$ with a section σ. Every such an equivalence class is called a *p-jet*. We write $J_x^p \pi$ for the space of all jets of order p of sections at x. We will denote by $J^p \pi$ the space of all jets of order p. Alternatively, we will write $J^p(\mathbb{R}^n, \mathbb{R}^k)$ for the space of p-jets of sections of the bundle $\pi : (x, u) \in \mathbb{R}^n \times \mathbb{R}^k \mapsto x \in \mathbb{R}^n$.

Given a section $\sigma : \mathbb{R}^n \to J^p \pi$, we can define the functions

$$(u_j)_J(j_x^p \sigma) = \frac{\partial^{|J|}\sigma_j}{\partial x_1^{j_1} \ldots \partial x_n^{j_n}}(x), \quad \forall j, \ |J| \leq p. \quad (2.26)$$

For $|J| = 0$, we define $u_J(x) := u(x)$. Coordinate systems on $\mathbb{R}^n$ and N along with the previous functions give rise to a local coordinate system on $J^p \pi$. We will also hereafter denote the n-tuple and k-tuple, respectively, by $x = (x_1, \ldots, x_n)$, $u = (u_1, \ldots, u_k)$, then

$$(u_j)_J = u_{x_{i_1}^{j_1} \ldots x_{i_n}^{j_n}} = \frac{\partial^{|J|}u_j}{\partial x_{i_1}^{j_1} \ldots \partial x_{i_n}^{j_n}}, \quad \forall j, \ |J| \leq 0. \quad (2.27)$$

All such local coordinate systems give rise to a manifold structure on $J^p \pi$. In this way, every point of $J^p \pi$ can be written as

$$\left(x_i, u_j, (u_j)_{x_i}, (u_j)_{x_{i_1}^{j_1} x_{i_2}^{2-j_1}}, (u_j)_{x_{i_1}^{j_1} x_{i_2}^{j_2} x_{i_3}^{3-j_1-j_2}}, \ldots, \right.$$

$$\left. (u_j)_{x_{i_1}^{j_1} x_{i_2}^{j_2} \ldots x_{i_n}^{p-\sum_{i=1}^{n-1} j_i}} \right), \quad (2.28)$$

where the numb indices run $i_1, \ldots, i_p = 1, \ldots, n$, $j = 1, \ldots, k$, $j_1 + \cdots + j_n \leq p$.

For small values of p, jet bundles have simple descriptions: $J^0\pi = N_{\mathbb{R}^n}$ and $J^1\pi \simeq \mathbb{R}^n \times TN$.

The projections $\pi_{p,l} : j_x^p\sigma \in J^p\pi \mapsto j_x^l\sigma \in J^l\pi$ with $l < p$ lead to define the smooth bundles $(J^p\pi, J^l\pi, \pi_{p,l})$. Conversely, for each section $\sigma : \mathbb{R}^n \to N_{\mathbb{R}^n}$, we have a natural embedding $j^p\sigma : \mathbb{R}^n \ni x \mapsto j_x^p\sigma \in J^p\pi$.

Consider now a vector field on $N_{\mathbb{R}^n}$ of the form

$$X(x, u) = \sum_{i=1}^{n} \xi_i(x, u)\frac{\partial}{\partial x_i} + \sum_{j=1}^{k} \eta_j(x, u)\frac{\partial}{\partial u_j}. \qquad (2.29)$$

This gives rise to an ϵ-parametric group of transformations

$$\begin{cases} x_i \mapsto x_i + \epsilon\xi_i(x, u) + O\left(\epsilon^2\right), \\ u_j \mapsto u_j + \epsilon\eta_j(x, u) + O\left(\epsilon^2\right), \end{cases} \quad i = 1, \ldots, n, \qquad j = 1, \ldots, k.$$

$$(2.30)$$

We can extend this infinitesimal transformation up to p-order derivatives

$$\begin{cases} \bar{x}_i = x_i + \epsilon\xi_i(x, u) + O\left(\epsilon^2\right), \\ \bar{u}_j = u_j + \epsilon\eta_j(x, u) + O\left(\epsilon^2\right), \\ (\bar{u}_j)_J = (u_j)_J + \epsilon(\eta_j)_J(x, u) + O\left(\epsilon^2\right), \end{cases} \qquad (2.31)$$

for $i = 1, \ldots, n$, $j = 1, \ldots, k$, $0 < |J| \leq p$ and $(u_j)_J$ represents (2.3) and the $(\eta_j)_J$ are given in the forthcoming steps.

In coordinates, the above ϵ-parametric group of diffeomorphisms has the associated vector field

$$X^p = \sum_{i=1}^{n} \xi_i \frac{\partial}{\partial x_i} + \sum_{j=1}^{k} \eta_j \frac{\partial}{\partial u_j} + \sum_{0 < |J| \leq p} \sum_{j=1}^{k} \mathfrak{Pr}_J\eta_j \frac{\partial}{\partial(u_j)_J}, \qquad (2.32)$$

where $\mathfrak{Pr}_J\eta_j$ denotes the prolongations of η_j for the multi-index J.

$$\text{Geometric Fundamentals} \qquad 45$$

Lemma 2.3. *Let* $\{\Phi_\epsilon : (x, u) \in N_{\mathbb{R}^n} \mapsto (\bar{x}, \bar{u}) \in N_{\mathbb{R}^n}\}_{\epsilon \in \mathbb{R}}$ *be the one-parametric group of transformations induced by the vector field* $X = \sum_{i=1}^{n} \xi_i(t, u)\partial/\partial x_i + \sum_{j=1}^{k} \eta_j(t, u)\partial/\partial u_j$ *and let* $\sigma : x \in \mathbb{R}^n \mapsto (x, u(x)) \in N_{\mathbb{R}^n}$ *be a section of the bundle* $\pi : N_{\mathbb{R}^n} \to \mathbb{R}^n$. *The section* $\bar{\sigma}_\epsilon = \Phi_\epsilon \circ \sigma$ *has slopes* $\partial \bar{u}_j / \partial \bar{x}_i = \partial u_j / \partial x_i + \epsilon(\eta_j)_{x_i} + O(\epsilon^2)$, *for any two fixed values of* $1 \leq i \leq n$ *and* $1 \leq j \leq k$, *where*

$$\mathfrak{Pr}_{J:=x_i}\eta_j = (\eta_j)_{x_i} = \frac{D\eta_j}{Dx_i} - \sum_{q=1}^{n}(u_j)_{x_q}\frac{D\xi_q}{Dx_i}, \qquad (2.33)$$

and D/Dx_i *is the total derivative with respect to* x_i, *namely*

$$\frac{D}{Dx_i} = \frac{\partial}{\partial x_i} + \sum_{l=1}^{k}(u_l)_{x_i}\frac{\partial}{\partial u_l}. \qquad (2.34)$$

Proof. Observe that

$$\delta \bar{u}_j = \delta u_j + \epsilon \delta \eta_j + O\left(\epsilon^2\right)$$

$$= \sum_{i=1}^{n}\left[\frac{\partial u_j}{\partial x_i} + \epsilon\left(\frac{\partial \eta_j}{\partial x_i} + \sum_{l=1}^{k}\frac{\partial \eta_j}{\partial u_l}\frac{\partial u_l}{\partial x_i}\right)\right]\delta x_i + O\left(\epsilon^2\right), \quad (2.35)$$

and

$$\delta \bar{x}_i = \delta x_i + \epsilon \delta \xi_i + O\left(\epsilon^2\right)$$

$$= \sum_{m=1}^{n}\left[\delta_m^i + \epsilon\left(\frac{\partial \xi_i}{\partial x_m} + \sum_{l=1}^{k}\frac{\partial \xi_i}{\partial u_l}\frac{\partial u_l}{\partial x_m}\right)\right]\delta x_m + O\left(\epsilon^2\right). \quad (2.36)$$

Using the operator defined in (2.10), we can rewrite these two expressions as

$$\delta \bar{u}_j = \sum_{l=1}^{k}\left((u_j)_{x_l} + \epsilon\frac{D\eta_j}{Dx_l}\right)\delta x_l + O\left(\epsilon^2\right),$$

$$\delta \bar{x}_i = \sum_{m=1}^{n}\left(\delta_m^i + \epsilon\frac{D\xi_i}{Dx_m}\right)\delta x_m + O\left(\epsilon^2\right). \qquad (2.37)$$

Hence,

$$
\begin{aligned}
(\bar{u}_j)_{\bar{x}_i} &= \frac{\sum_{l=1}^n \left((u_j)_{x_l} + \epsilon \frac{D\eta_j}{Dx_l} \right)}{\sum_{m=1}^n \left(\delta_m^i + \epsilon \frac{D\xi_i}{Dx_m} \right)} \delta_m^l + O\left(\epsilon^2\right) \\
&= \sum_{l=1}^n \left((u_j)_{x_l} + \epsilon \frac{D\eta_j}{Dx_l} + O\left(\epsilon^2\right) \right) \cdot \left(\delta_i^l - \epsilon \frac{D\xi_l}{Dx_i} + O\left(\epsilon^2\right) \right) \\
&= (u_j)_{x_i} + \epsilon \left(\frac{D\eta_j}{Dx_i} - \sum_{q=1}^n (u_j)_{x_q} \frac{D\xi_q}{Dx_i} \right) + O\left(\epsilon^2\right), \quad (2.38)
\end{aligned}
$$

which finishes the proof. $\qquad\square$

In a similar way as we deduced the expression for the first-order prolongation, we can deduce it for higher-order prolongations. We will deduce the expression by generalization of results in lower-dimensional cases.

Lemma 2.4. *Let* $\sigma : x \in \mathbb{R}^n \mapsto (x, u(x)) \in N_{\mathbb{R}^n}$ *and let* $\Phi_\epsilon : (x, u) \in N_{\mathbb{R}^n} \mapsto (\bar{x}, \bar{u}) \in N_{\mathbb{R}^n}$ *be a one-parametric group of transformations induced by the vector field* $X = \sum_{i=1}^n \xi_i(t, u)\partial/\partial x_i + \sum_{j=1}^k \eta_j(t, u)\partial/\partial u_j$. *We obtain that the section* $\bar{\sigma}_\epsilon = \Phi_\epsilon \circ \sigma$ *of* $\pi : N_{\mathbb{R}^n} \to \mathbb{R}^n$ *has slope* $\partial^2 \bar{u}_j / \partial \bar{x}_{i_1} \partial \bar{x}_{i_2} = \partial^2 u_j / \partial x_{i_1} \partial x_{i_2} + \epsilon(\eta_j)_{x_{i_1} x_{i_2}} + O(\epsilon^2)$, *where*

$$
\begin{aligned}
(\eta_j)_{x_{i_1} x_{i_2}} &= \frac{D^2 \eta_j}{Dx_{i_1} Dx_{i_2}} - \sum_{l=1}^n (u_j)_{x_{i_1}, x_l} \frac{D\xi_l}{Dx_{i_2}} \\
&\quad - \sum_{l=1}^n (u_j)_{x_{i_2}, x_l} \frac{D\xi_l}{Dx_{i_1}} - \sum_{l=1}^n (u_j)_{x_l} \frac{D\xi_l}{Dx_{i_1} Dx_{i_2}}, \quad (2.39)
\end{aligned}
$$

and D/Dx_i, *when acting on functions of* $J^1\pi$, *stands for*

$$
\frac{Df}{Dx_i} = \frac{\partial f}{\partial x_i} + \sum_{j=1}^k \left[(u_j)_{x_i} \frac{\partial f}{\partial u_j} + \sum_{l=1}^n (u_j)_{x_l x_i} \frac{\partial f}{\partial (u_j)_{x_l}} \right]. \quad (2.40)
$$

$$\text{Geometric Fundamentals} \qquad 47$$

Proof. We have $(\bar{u}_j)_{\bar{x}_{i_1}} = (u_j)_{x_{i_1}} + \epsilon(\eta_j)_{x_{i_1}}$. Therefore,

$$\delta[(\bar{u}_j)_{\bar{x}_{i_1}}] = \delta[(u_j)_{x_{i_1}}] + \epsilon\delta[(\eta_j)_{x_{i_1}}]$$

$$= \sum_{l=1}^{n}\left(\frac{D[(u_j)_{x_{i_1}}]}{Dx_l} + \epsilon\frac{D[(\eta_j)_{x_{i_1}}]}{Dx_l}\right)\delta x_l, \qquad (2.41)$$

and

$$\delta\bar{x}_{i_2} = \delta x_{i_2} + \epsilon\delta\xi_{i_2} = \sum_{q=1}^{n}\left(\delta_q^{i_2} + \epsilon\frac{D\xi_{i_2}}{Dx_q}\right)\delta x_q. \qquad (2.42)$$

Consequently, proceeding in similar fashion as for the first-order prolongation,

$$(\bar{u}_j)_{\bar{x}_{i_1}\bar{x}_{i_2}} = \frac{\delta[(\bar{u}_j)_{\bar{x}_{i_1}}]}{\delta\bar{x}_{i_2}} = \frac{\sum_{l=1}^{n}\left(\frac{D[(u_j)_{x_{i_1}}]}{Dx_l} + \epsilon\frac{D[(\eta_j)_{x_{i_1}}]}{Dx_l}\right)}{\sum_{q=1}^{n}\left(\delta_q^{i_2} + \epsilon\frac{D\xi_{i_2}}{Dx_q}\right)}\delta_q^l + O\left(\epsilon^2\right)$$

$$= \sum_{l=1}^{n}\left(\frac{D[(u_j)_{x_{i_1}}]}{Dx_l} + \epsilon\frac{D[(\eta_j)_{x_{i_1}}]}{Dx_l}\right)\cdot\left(\delta_l^{i_2} - \epsilon\frac{D\xi_l}{Dx_{i_2}}\right) + O\left(\epsilon^2\right)$$

$$= (u_j)_{x_{i_1}x_{i_2}} + \epsilon\left(\frac{D[(\eta_j)_{x_{i_1}}]}{Dx_{i_2}} - \sum_{l=1}^{n}(u_j)_{x_{i_1}x_l}\frac{D\xi_l}{Dx_{i_2}}\right) + O\left(\epsilon^2\right).$$

$$(2.43)$$

Introducing the value of $(\eta_j)_{x_{i_1}}$ in (2.43), using (2.9), we arrive at (2.39). $\qquad\square$

Applying this process recursively, we obtain the *p-order prolongations*

$$(\eta_j)_{x_{i_1}^{j_1}x_{i_2}^{j_2}...x_{i_n}^{j_n}} = \frac{D\left((\eta_j)_{x_{i_1}^{j_1}...x_{i_{n-1}}^{j_{n-1}}x_{i_n}^{(j_n)-1}}\right)}{Dx_{i_n}}$$

$$- \sum_{l=1}^{n}(u_j)_{x_{i_1}^{j_1}x_{i_2}^{j_2}...x_{i_{n-1}}^{j_{n-1}}x_{i_n}^{(j_n)-1}x_l}\frac{D\xi_l}{Dx_{i_n}}, \qquad (2.44)$$

or equivalently,

$$\mathfrak{Pr}_J \eta_j = (\eta_j)_{x_{i_1}^{j_1} x_{i_2}^{j_2} \dots x_{i_n}^{j_n}}$$

$$= \frac{D^{|J|}\eta_j}{Dx_{i_1}^{j_1} Dx_{i_2}^{j_2} \dots Dx_{i_n}^{j_n}} - \sum_{l=1}^{n} (u_j)_{x_l} \frac{D^{|J|}\eta_j}{Dx_{i_1}^{j_1} Dx_{i_2}^{j_2} \dots Dx_{i_n}^{j_n}} - \cdots$$

$$- \sum_{i,i'=1}^{n} \sum_{l=1}^{n} (u_j)_{x_l \dots x_{i_{n-2}}^{j_{n-2}}} \frac{D^2 \xi_l}{Dx_i Dx_{i'}} - \sum_{i=1}^{n} \sum_{l=1}^{n} (u_j)_{x_l \dots x_{i_{n-1}}^{j_{n-1}}} \frac{D\xi_l}{Dx_i},$$

$$\tag{2.45}$$

where $x_1 \leq x_{i_1}, \dots, x_{i_n} \leq x_n$ and $j_1 + \cdots + j_n \leq p$. The associated vector field for this transformation reads

$$X^p = \xi \frac{\partial}{\partial t} + \sum_{j=1}^{n} \eta_j \frac{\partial}{\partial u_j} + \sum_{j=1}^{n} \sum_{|J|=p} \mathfrak{Pr}_J \eta_j \frac{\partial}{\partial (u_j)_J}. \tag{2.46}$$

2.6. Poisson Algebras

A *Poisson algebra* is a triple $(A, \star, \{\cdot, \cdot\})$, where A is a real vector space endowed with two bilinear maps, namely '$\star$' and $\{\cdot, \cdot\}$, such that '$\star$' is a commutative and associative real algebra and $(A, \{\cdot, \cdot\})$ is a real Lie algebra whose Lie bracket, the *Poisson bracket*, satisfies the *Leibniz rule* with respect to '$\star$', namely

$$\{f \star g, h\} = f \star \{g, h\} + \{f, h\} \star g, \quad \forall f, g, h \in A.$$

In other words, $\{\cdot, h\}$ is a derivation.

A *Poisson algebra morphism* is a morphism $\mathcal{T} : (A, \star_A, \{\cdot, \cdot\}_A) \to (B, \star_B, \{\cdot, \cdot\}_B)$ of $\mathbb{R}$-algebras $\mathcal{T} : (A, \star_A) \to (B, \star_B)$ that satisfies that $\mathcal{T}(\{a, b\}_A) = \{\mathcal{T}(a), \mathcal{T}(b)\}_B$ for every $a, b \in A$.

The two types of Poisson algebras to be used are the symmetric and universal Poisson algebras. Let us describe their main characteristics. Given a finite-dimensional real Lie algebra $(\mathfrak{g}, [\cdot, \cdot]_\mathfrak{g})$, its *universal enveloping algebra*, $U_\mathfrak{g}$, is obtained from the quotient $T_\mathfrak{g}/\mathcal{R}$ of the tensor algebra $(T_\mathfrak{g}, \otimes)$ of $\mathfrak{g}$ by the bilateral ideal $\mathcal{R}$ spanned by the elements $v \otimes w - w \otimes v - [v, w]$, with $v, w \in \mathfrak{g}$. Given the quotient map $\pi : T_\mathfrak{g} \to U_\mathfrak{g}$, the space $U_\mathfrak{g}$ becomes an $\mathbb{R}$-algebra $(U_\mathfrak{g}, \widetilde{\otimes})$ when endowed with the product $\widetilde{\otimes} : U_\mathfrak{g} \times U_\mathfrak{g} \to U_\mathfrak{g}$ given by $\pi(P) \widetilde{\otimes} \pi(Q) := \pi(P \otimes Q)$, for every $P, Q \in T_\mathfrak{g}$. The Lie bracket on $\mathfrak{g}$ can be extended to a Lie bracket $\{\cdot, \cdot\}_{U_\mathfrak{g}}$ on $U_\mathfrak{g}$ by imposing it to be a derivation of $(U_\mathfrak{g}, \widetilde{\otimes})$ on each factor. This turns $U_\mathfrak{g}$ into a Poisson algebra $(U_\mathfrak{g}, \widetilde{\otimes}, \{\cdot, \cdot\}_{U_\mathfrak{g}})$ [100]. The elements of its Casimir subalgebra are henceforth dubbed as *Casimir elements* of $\mathfrak{g}$ [40].

If we set $\mathcal{R}$ to be the bilateral ideal spanned by the elements $v \otimes w - w \otimes v$ in the above procedure, we obtain a new commutative Poisson algebra $S_\mathfrak{g}$ called the *symmetric algebra* of $(\mathfrak{g}, [\cdot, \cdot]_\mathfrak{g})$. The elements of $S_\mathfrak{g}$ are polynomials on the elements of $\mathfrak{g}$. Via the isomorphism $\mathfrak{g} \simeq (\mathfrak{g}^*)^*$, they can naturally be understood as polynomial functions on $\mathfrak{g}^*$ [1, 100]. The Casimir elements of this Poisson algebra are called *Casimir functions* of $\mathfrak{g}$.

The Poisson algebras $U_\mathfrak{g}$ and $S_\mathfrak{g}$ are related by the *symmetrizer map* [1, 40, 359], i.e., the linear isomorphism $\lambda : S_\mathfrak{g} \to U_\mathfrak{g}$ of the form

$$\lambda(v_{i_1}) := \pi(v_{i_1}),$$

$$\lambda(v_{i_1} v_{i_2} \ldots v_{i_l}) := \frac{1}{l!} \sum_{s \in \Pi_l} \lambda(v_{s(i_1)}) \widetilde{\otimes} \ldots \widetilde{\otimes} \lambda(v_{s(i_l)}), \qquad (2.47)$$

for all $v_{i_1}, \ldots, v_{i_l} \in \mathfrak{g}$ and with Π_l being the set of permutations of l elements. Moreover,

$$\lambda^{-1}\left(\{v, P\}_{U_\mathfrak{g}}\right) = \{v, \lambda^{-1}(P)\}_{S_\mathfrak{g}}, \quad \forall P \in U_\mathfrak{g}, \qquad \forall v \in \mathfrak{g}. \qquad (2.48)$$

So, λ^{-1} maps the Casimir elements of $\mathfrak{g}$ into Casimir elements of $S_\mathfrak{g}$. If $(A, \star_A, \{\cdot, \cdot\}_A)$ and $(B, \star_B, \{\cdot, \cdot\}_B)$ are Poisson algebras and operations: $\star_A$ and $\star_B$ are commutative, then $A \otimes B$ becomes a Poisson algebra $(A \otimes B, \star_{A \otimes B}, \{\cdot, \cdot\}_{A \otimes B})$ by defining

$$\begin{aligned} (a \otimes b) \star_{A \otimes B} (c \otimes d) &:= (a \star_A c) \otimes (b \star_B d), \\ \{a \otimes b, c \otimes d\}_{A \otimes B} &:= \{a, c\}_A \otimes b \star_B d + a \star_A c \otimes \{b, d\}_B, \end{aligned} \qquad (2.49)$$

for all $a, c \in A, \quad \forall b, d \in B$.

Similarly, a Poisson structure on $A^{(m)} := \overbrace{A \otimes \ldots \otimes A}^{m-\text{times}}$ can be constructed by induction.

We say that $(A, \star_A, \{\cdot, \cdot\}_A, \Delta)$ is a *Poisson coalgebra* if $(A, \star_A, \{\cdot, \cdot\}_A)$ is a Poisson algebra and $\Delta \colon (A, \star_A, \{\cdot, \cdot\}_A) \to (A \otimes A, \star_{A \otimes A}, \{\cdot, \cdot\}_{A \otimes A})$, the so-called *coproduct*, is a Poisson algebra morphism which is *coassociative* [135], i.e., $(\Delta \otimes \mathrm{Id}) \circ \Delta = (\mathrm{Id} \otimes \Delta) \circ \Delta$. Then, the mth coproduct map $\Delta^{(m)} \colon A \to A^{(m)}$ can be defined recursively as follows:

$$\Delta^{(m)} := (\overbrace{\mathrm{Id} \otimes \cdots \otimes \mathrm{Id}}^{(m-2)\text{-times}} \otimes \Delta^{(2)}) \circ \Delta^{(m-1)}, \quad m > 2, \qquad (2.50)$$

and $\Delta := \Delta^{(2)}$. Such an induction ensures that $\Delta^{(m)}$ is also a Poisson map.

In particular, $S_{\mathfrak{g}}$ is a Poisson coalgebra with a *coproduct map* given by $\Delta(v) = v \otimes 1 + 1 \otimes v$, for all $v \in \mathfrak{g} \subset S_{\mathfrak{g}}$. The coassociativity of Δ is straightforward, and its mth generalization reads

$$\Delta^{(m)}(v) := v \otimes \overbrace{1 \otimes \cdots \otimes 1}^{(m-1)\text{-times}} + 1 \otimes v \otimes \overbrace{1 \otimes \cdots \otimes 1}^{(m-2)\text{-times}} + \cdots$$

$$+ \overbrace{1 \otimes \cdots \otimes 1}^{(m-1)\text{-times}} \otimes v, \qquad (2.51)$$

for all $v \in \mathfrak{g} \subset S_{\mathfrak{g}}$.

2.7. Symplectic, Presymplectic, and Poisson Manifolds

A *symplectic manifold* is a pair (N, ω), where N stands for a manifold and ω is a non-degenerate closed 2-form on N. We say that a vector field X on N is Hamiltonian with respect to (N, ω) if there exists a function $f \in C^{\infty}(N)$ such that[a]

$$\iota_X \omega = -df. \qquad (2.52)$$

In this case, we say that f is a *Hamiltonian function* for X. Conversely, given a function f, there exists a unique vector field X_f on N,

---

[a]In Geometric Mechanics, the convention $\iota_X \omega = df$ is frequently used.

the so-called *Hamiltonian vector field* of f, satisfying (2.1). This allows us to define a bracket $\{\cdot,\cdot\} : C^\infty(N) \times C^\infty(N) \to C^\infty(N)$ given by[b]

$$\{f,g\} := \omega(X_f, X_g) = X_f(g). \tag{2.53}$$

This bracket turns $C^\infty(N)$ into a *Poisson algebra* $(C^\infty(N), \cdot, \{\cdot,\cdot\})$, i.e., $\{\cdot,\cdot\}$ is a Lie bracket on $C^\infty(N)$ which additionally holds the *Leibniz rule* with respect to the standard product "·" of functions

$$\{fg,h\} = \{f,h\}g + f\{g,h\} \quad \forall f,g,h \in C^\infty(N). \tag{2.54}$$

The Leibniz rule can be rephrased by saying that $\{f,\cdot\}$ is a derivation of the associative algebra $(C^\infty(N), \cdot)$ for each $f \in C^\infty(N)$. Actually, this derivation is represented by the Hamiltonian vector field X_f. The bracket $\{\cdot,\cdot\}$ is called the *Poisson bracket* of $(C^\infty(N), \cdot, \{\cdot,\cdot\})$. Note that if (N, ω) is a symplectic manifold, the non-degeneracy condition for ω implies that N is even dimensional [3].

The above observations lead to the concept of a Poisson manifold which is a natural generalization of the symplectic one. A *Poisson manifold* is a pair $(N, \{\cdot,\cdot\})$, where $\{\cdot,\cdot\} : C^\infty(N) \times C^\infty(N) \to C^\infty(N)$ is the Poisson bracket of $(C^\infty(N), \cdot, \{\cdot,\cdot\})$ which is also referred to as a *Poisson structure* on N. In view of this and (2.2), every symplectic manifold is a particular type of Poisson manifold. Moreover, by noting that $\{f,\cdot\}$ is a derivation on $(C^\infty(N), \cdot)$ for every $f \in C^\infty(N)$, we can associate with every function f a single vector field X_f, called the *Hamiltonian vector field* of f, such that $\{f,g\} = X_f g$ for all $g \in C^\infty(N)$, like in the symplectic case.

As the Poisson structure is a derivation in each entry, it gives rise to a bivector field Λ, i.e., an element of $\Gamma(\wedge^2 TN)$, referred to as *Poisson bivector*, such that $\{f,g\} = \Lambda(df, dg)$. It is known that the Jacobi identity for $\{\cdot,\cdot\}$ amounts to $[\Lambda, \Lambda]_{SN} = 0$, with $[\cdot,\cdot]_{SN}$ being the *Schouten–Nijenhuis bracket* [358]. Conversely, a bivector Λ satisfying $[\Lambda, \Lambda]_{SN} = 0$ gives rise to a Poisson bracket on $C^\infty(N)$

---

[b]In Geometric Mechanics $\{f,g\} := X_g f$. Hence, the mapping $f \to X_f$ becomes a Lie algebra antihomomorphism.

by setting $\{f, g\} = \Lambda(df, dg)$. Hence, a Poisson manifold can be considered, equivalently, as $(N, \{\cdot, \cdot\})$ or (N, Λ). It is remarkable that Λ induces a bundle morphism $\widehat{\Lambda} : \alpha_x \in T^*N \mapsto \widehat{\Lambda}(\alpha_x) \in TN$, where $\bar{\alpha}_x(\widehat{\Lambda}(\alpha_x)) = \Lambda_x(\alpha_x, \bar{\alpha}_x)$ for all $\bar{\alpha}_x \in T_x^*N$, which enables us to write $X_f = \widehat{\Lambda}(df)$ for every $f \in C^\infty(N)$.[c]

Proposition 2.5. *The Casimir codistribution of a Poisson manifold is involutive.*

Proof. Given two sections $\omega, \omega' \in \mathcal{C}^\Lambda$, we have that $\widehat{\Lambda}(\omega) = \widehat{\Lambda}(\omega') = 0$ and then $\Lambda(\omega, \omega') = 0$. In consequence,

$$[\omega, \omega']_\Lambda = \mathcal{L}_{\widehat{\Lambda}(\omega)}\omega' - \mathcal{L}_{\widehat{\Lambda}(\omega')}\omega - d\Lambda(\omega, \omega') = 0. \qquad (2.55)$$

$\square$

Another way of generalizing a symplectic structure is to consider a 2-form ω which is merely closed (not necessarily of constant rank), forgetting the non-degeneracy assumption. In this case, we are led to the theory of presymplectic manifolds [264].

A *presymplectic manifold* is a pair (N, ω), where ω is a closed 2-form on N. A vector field X on N is *Hamiltonian* with respect to (N, ω) if there exists an $h \in C^\infty(N)$ such that[d]

$$\iota_X \omega = -dh.$$

In this case, we call h is a *Hamiltonian function* for X and we say that h is an *admissible function* (relative to (N, ω)). We write $\mathrm{Adm}(\omega)$ for the space of *Hamiltonian functions* relative to (N, ω). We hereafter denote by X_h, with $h \in C^\infty(N)$, a Hamiltonian vector field of h relative to ω. Since ω may be degenerate, there may exist Hamiltonian vector fields related to a zero function. We call these vector fields *gauge vector fields* of ω and we write $G(\omega)$ for the space of such vector fields.

Since $\ker \omega$ may be degenerate, every admissible function h may have different Hamiltonian vector fields. Since the linear combinations and multiplications of admissible functions are also admissible

---

[c] In Geometric Mechanics, one has $X_f = -\widehat{\Lambda}(df)$.
[d] In Geometric Mechanics, one defines $\iota_X \omega = dh$.

functions, the space $\mathrm{Adm}(N, \omega)$ is a real associative algebra. Moreover, $\mathrm{Adm}(\omega)$ becomes a Poisson algebra when endowed with the Poisson bracket $\{\cdot, \cdot\} \colon \mathrm{Adm}(\omega) \times \mathrm{Adm}(\omega) \to \mathrm{Adm}(\omega)$ of the form

$$\{f, g\} = -X_g f, \tag{2.56}$$

where X_g is any Hamiltonian vector field of g. It can be proven that this definition is independent of the chosen X_g [358]. Indeed, every $f \in \mathrm{Adm}(N, \omega)$ is associated with a family of Hamiltonian vector fields of the form $X_f + Z$, where Z is a gauge vector field. Hence, if X_g' is a Hamiltonian vector field related to g, then $X_g' = X_g + Z$ and

$$(X_g + Z)f = df(X_g + Z) = \iota_{X_f}\omega(X_g + Z)$$
$$= \omega(X_f, X_g + Z) = \omega(X_f, X_g) = X_g f,$$

and (2.56) does not depend on the representatives X_f and X_g. Thus, it becomes a Poisson bracket on the space $\mathrm{Adm}(N, \omega)$ making it into a Poisson algebra.

On the other hand,[e]

$$\iota_{[X_f, X_g]}\omega = \mathcal{L}_{X_f}\iota_{X_g}\omega - \iota_{X_g}\mathcal{L}_{X_f}\omega = -\mathcal{L}_{X_f}dg = -d\{f, g\}. \tag{2.57}$$

Consequently, $[X_f, X_g]$ is a Hamiltonian vector field with a Hamiltonian function $\{f, g\}$.

It is immediate that $G(\omega)$ is an ideal of $\mathrm{Ham}(\omega)$. Hence, the space $\mathrm{Ham}(\omega)/G(\omega)$ is also a Lie algebra and the quotient projection $\pi \colon X \in \mathrm{Ham}(\omega) \mapsto [X] \in \mathrm{Ham}(\omega)/G(\omega)$ is a Lie algebra morphism.

Moreover, one has the following exact sequence of Lie algebras

$$0 \hookrightarrow \mathrm{H}^0_{\mathrm{dR}}(N) \hookrightarrow \mathrm{Adm}(\omega) \xrightarrow{\Lambda} \frac{\mathrm{Ham}(\omega)}{G(\omega)} \to 0,$$

where $\mathrm{H}^0_{\mathrm{dR}}(N)$ is the zero cohomology de Rham group of N and $\Lambda \colon f \in \mathrm{Adm}(\omega) \mapsto [X_f] \in \mathrm{Ham}(\omega)/G(\omega)$ (see [90] for details).[f]

---

[e]In Geometric Mechanics $\iota_{[X_f, X_g]}\omega = \mathcal{L}_{X_f}dg = -d\{f, g\}$.

[f]In Geometric Mechanics, one defines $\Lambda \colon f \in \mathrm{Adm}(\omega) \mapsto [-X_f] \in \mathrm{Ham}(\omega)/G(\omega)$.

2.8. Dirac Manifolds

Dirac structures were proposed by Dorfman [158] in the Hamiltonian framework of integrable evolution equations and defined in [150] as a subbundle of the *Whitney sum*, $TN \oplus_N T^*N$, called the *extended tangent* or *Pontryagin bundle*. It was thought of as a common generalization of Poisson and presymplectic structures. It was also designed to deal with constrained systems, including constraints induced by degenerate Lagrangians investigated by Dirac [157], hence their name. The main aim of the section is to present the main properties of Dirac structures (see [66, 150–152, 358] for details).

Definition 2.6. We call *Pontryagin bundle* $\mathcal{P}N$ a vector bundle $TN \oplus_N T^*N$ on N.

Definition 2.7. An *almost-Dirac manifold* is a pair (N, L), where L is a maximally isotropic subbundle of $\mathcal{P}N$ with respect to the pairing

$$\langle X_x + \alpha_x, \bar{X}_x + \bar{\alpha}_x \rangle_+ := \frac{1}{2}(\bar{\alpha}_x(X_x) + \alpha_x(\bar{X}_x)), \qquad (2.58)$$

where $X_x + \alpha_x, \bar{X}_x + \bar{\alpha}_x \in T_x N \oplus T_x^* N = \mathcal{P}_x N$. In other words, L is isotropic and has rank $n = \dim N$.

Definition 2.8. A *Dirac manifold* is an almost-Dirac manifold (N, L) whose subbundle L, its *Dirac structure*, is involutive relative to the *Courant–Dorfman bracket* [150, 158, 187, 238], namely

$$[[X + \alpha, \bar{X} + \bar{\alpha}]]_C := [X, \bar{X}] + \mathcal{L}_X \bar{\alpha} - \iota_{\bar{X}} d\alpha, \qquad (2.59)$$

where $X + \alpha, \bar{X} + \bar{\alpha} \in \Gamma(TN \oplus_N T^*N)$.

The Courant–Dorfman bracket satisfies the Jacobi identity, i.e.,

$$[[\, [[e_1, e_2]]_C, e_3]]_C$$
$$= [[e_1, [[e_2, e_3]]_C]]_C - [[e_2, [[e_1, e_3]]_C]]_C, \ \forall e_1, e_2, e_3 \in \Gamma(\mathcal{P}N), \qquad (2.60)$$

but is not skew-symmetric. It is, however, skew-symmetric on sections of the Dirac subbundle L, defining a *Lie algebroid* structure

$(L, [[\cdot, \cdot]]_C, \rho)$, where $\rho : L \ni X_x + \alpha_x \mapsto X_x \in TN$. This means that $(\Gamma(L), [[\cdot, \cdot]]_C)$ is a Lie algebra and the vector bundle morphism $\rho : L \to TN$, the *anchor*, satisfies

$$[[e_1, f e_2]]_C = (\rho(e_1)f)e_2 + f[[e_1, e_2]]_C \qquad (2.61)$$

for all $e_1, e_2 \in \Gamma(L)$ and $f \in C^\infty(N)$ [150]. One can prove that, automatically, ρ induces a Lie algebra morphism of $(\Gamma(L), [[\cdot, \cdot]]_C)$ into the Lie algebra of vector fields on N. The generalized distribution $\rho(L)$, called the *characteristic distribution* of the Dirac structure, is therefore integrable in the sense of Stefan–Sussmann [349].

Definition 2.9. A vector field X on N is said to be an *L-Hamiltonian vector field* (or simply a *Hamiltonian vector field* if L is fixed) if there exists an $f \in C^\infty(N)$ such that $X + df \in \Gamma(L)$. In this case, f is an *L-Hamiltonian function* for X and an *admissible function* of (N, L). Let us denote by $\mathrm{Ham}(N, L)$ and $\mathrm{Adm}(N, L)$ the spaces of Hamiltonian vector fields and admissible functions of (N, L), respectively.

The space $\mathrm{Adm}(N, L)$ becomes a Poisson algebra $(\mathrm{Adm}(N, L), \cdot, \{\cdot, \cdot\}_L)$ relative to the standard product of functions and the Lie bracket given by $\{f, \bar{f}\}_L = X\bar{f}$ where X is an L-Hamiltonian vector field for f. Since L is isotropic, $\{f, \bar{f}\}_L$ is well defined, i.e., its value is independent on the choice of the L-Hamiltonian vector field associated with f. The elements $f \in \mathrm{Adm}(N, L)$ possessing trivial Hamiltonian vector fields are called the *Casimir functions* of (N, L) [387]. We write $\mathrm{Cas}(N, L)$ for the set of Casimir functions of (N, L). We can also distinguish the space $G(N, L)$ of L-Hamiltonian vector fields which admit zero (or, equivalently, any constant) as an L-Hamiltonian function. We call them *gauge vector fields* of the Dirac structure.

If X and $\bar{X}$ are L-Hamiltonian vector fields with Hamiltonian functions f and $\bar{f}$, then $\{f, \bar{f}\}_L$ is a Hamiltonian for $[X, \bar{X}]$

$$[[X + df, \bar{X} + d\bar{f}]]_C = [X, \bar{X}] + \mathcal{L}_X d\bar{f} - \iota_{\bar{X}} d^2 f$$
$$= [X, \bar{X}] + d\{f, \bar{f}\}_L. \qquad (2.62)$$

56 *A Guide to Lie Systems with Compatible Geometric Structures*

This implies that $(\mathrm{Ham}(N, L), [\cdot, \cdot])$ is a Lie algebra in which $G(N, L)$ is a Lie ideal. Denote the quotient Lie algebra $\mathrm{Ham}(N, L)/G(N, L)$ by $\widehat{\mathrm{Ham}}(N, L)$.

Proposition 2.10. *If (N, L) is a Dirac manifold, then $\{\mathrm{Cas}(N, L), \mathrm{Adm}(N, L)\}_L = 0$, i.e., $\mathrm{Cas}(N, L)$ is an ideal of the Lie algebra $(\mathrm{Adm}(N, L), \{\cdot, \cdot\}_L)$. Moreover, we have the following exact sequence of Lie algebra homomorphisms*

$$0 \hookrightarrow \mathrm{Cas}(N, L) \hookrightarrow \mathrm{Adm}(N, L) \xrightarrow{B_L} \widehat{\mathrm{Ham}}(N, L) \to 0, \qquad (2.63)$$

with $B_L(f) = \pi_H(X_f)$, where the vector field X_f is an L-Hamiltonian vector field of f, and π is the canonical projection $\pi_H : \mathrm{Ham}(N, L) \to \widehat{\mathrm{Ham}}(N, L)$.

Every Dirac manifold (N, L) is endowed with a canonical linear map $\Omega_x^L : \rho(L)_x \subset T_x N \to \rho(L)_x^* \subset T_x^* N$ given by

$$[\Omega_x^L(X_x)](\bar{X}_x) = -\alpha_x(\bar{X}_x), \quad X_x, \bar{X}_x \in \rho(L), \qquad (2.64)$$

where $\alpha_x \in T_x^* N$ is such that As L is isotropic, Ω_x^L is well defined, i.e., the value of

$$\Omega_x^L(X_x, \bar{X}_x) = [\Omega_x^L(X_x)](\bar{X}_x) \qquad (2.65)$$

is independent of the particular α_x and defines a skew-symmetric bilinear form Ω^L on the (generalized) distribution $\rho(L)$. Indeed, given $X_x + \bar{\bar{\alpha}}_x \in L$, we have that $\alpha_x - \bar{\bar{\alpha}}_x \in L$. Since L is isotropic, $\langle \alpha_x - \bar{\alpha}_x, \bar{X}_x + \bar{\alpha}_x \rangle_+ = (\alpha_x - \bar{\alpha}_x)\bar{X}_x/2 = 0$ for all $\bar{X}_x + \bar{\alpha}_x \in L$. Then, $[\Omega_x^L(X_x)](\bar{X}_x) = -\bar{\bar{\alpha}}_x(\bar{X}_x) = -\alpha_x(\bar{X}_x)$ for all $\bar{X}_x \in \rho(L)$ and Ω^L is well defined.

It is easy to see that gauge vector fields generate the *gauge distribution* $\ker \Omega^L$. Moreover, the involutivity of L ensures that $\rho(L)$ is an integrable generalized distribution in the sense of Stefan–Sussmann [349]. Therefore, it induces a (generalized) foliation $\mathfrak{F}^L$. We hereafter call $\mathfrak{F}_x^L$ the unique leaf of $\mathfrak{F}^L$ passing through the point $x \in N$.

Since $\rho(L_x) = T_x \mathfrak{F}_x^L$, if the elements $X_x + \alpha_x$ and $X_x + \bar{\alpha}_x$, with $X_x \in T_x \mathfrak{F}_x^L$, are in $L_x \subset \mathcal{P}_x N = T_x N \oplus T_x^* N$, then $\alpha_x - \bar{\alpha}_x$ is in the annihilator of $T_x \mathfrak{F}_x^L$, so the image of α_x under the canonical restriction $\sigma : \alpha_x \in T_x^* N \mapsto \alpha_x|_{T_x \mathfrak{F}_x^L} \in T_x^* \mathfrak{F}_x^L$ is uniquely determined.

One can verify that $\sigma(\alpha_x) = -\Omega_x^L(X_x)$. The 2-form Ω^L restricted to $\mathfrak{F}_x^L$ turns out to be closed, so that $\mathfrak{F}_x^L$ is canonically a presymplectic manifold, and the canonical restriction of L to $\mathfrak{F}_x^L$ is the graph of this form [150].

As particular instances, Poisson and presymplectic manifolds are particular cases of Dirac manifolds. On one hand, consider a presymplectic manifold (N, ω) and define L^ω to be the graph of (minus) the fiber bundle morphism $\widehat{\omega} : X_x \in TN \mapsto \omega_x(X_x, \cdot) \in T^*N$. The generalized distribution L^ω is isotropic, as

$$\langle X_x - \widehat{\omega}(X_x), \bar{X}_x - \widehat{\omega}(\bar{X}_x) \rangle_+$$
$$= -(\omega_x(X_x, \bar{X}_x) + \omega_x(\bar{X}_x, X_x))/2 = 0. \tag{2.66}$$

As L^ω is the graph of $-\widehat{\omega}$, then $\dim L_x^\omega = \dim N$ and L^ω is a maximally isotropic subbundle of $\mathcal{P}N$. In addition, its integrability relative to the Courant–Dorfman bracket comes from the fact that $d\omega = 0$. Indeed, for arbitrary $X, X' \in \Gamma(TN)$, we have

$$[[X - \iota_X\omega, X' - \iota_{X'}\omega]]_C = [X, X'] - \mathcal{L}_X\iota_{X'}\omega + \iota_{X'}d\iota_X\omega$$
$$= [X, X'] - \iota_{[X,X']}\omega, \tag{2.67}$$

since

$$\mathcal{L}_X\iota_{X'}\omega - \iota_{X'}d\iota_X\omega = \mathcal{L}_X\iota_{X'}\omega - \iota_{X'}\mathcal{L}_X\omega = \iota_{[X,X']}\omega. \tag{2.68}$$

In this case, $\rho : L^\omega \to TN$ is a bundle isomorphism. Conversely, given a Dirac manifold whose $\rho : L \to TN$ is a bundle isomorphism, its characteristic distribution satisfies $\rho(L) = TN$ and it admits a unique integral leaf, namely N, on which Ω^L is a closed 2-form, i.e., (N, Ω^L) is a presymplectic manifold.

On the other hand, every Poisson manifold (N, Λ) induces a subbundle L^Λ given by the graph of $\widehat{\Lambda}$. It is isotropic,

$$\langle \widehat{\Lambda}(\alpha_x) + \alpha_x, \widehat{\Lambda}(\bar{\alpha}_x) + \bar{\alpha}_x \rangle_+ = (\Lambda_x(\bar{\alpha}_x, \alpha_x) + \Lambda_x(\alpha_x, \bar{\alpha}_x))/2 = 0, \tag{2.69}$$

for all $\alpha_x, \bar{\alpha}_x \in T_x^*N$ and $x \in N$, and of rank $\dim N$ as the graph of $\widehat{\Lambda}$ is a map from T^*N. Additionally, L^Λ is integrable. Indeed, as

$\widehat{\Lambda}(d\{f,g\}) = [\widehat{\Lambda}(df), \widehat{\Lambda}(dg)]$ for every $f, g \in C^\infty(N)$ [358], we have

$$[[\widehat{\Lambda}(df) + df, \widehat{\Lambda}(dg) + dg]]_C = [\widehat{\Lambda}(df), \widehat{\Lambda}(dg)] + \mathcal{L}_{\widehat{\Lambda}(df)} dg - \iota_{\widehat{\Lambda}(dg)} d^2 f$$
$$= \widehat{\Lambda}(d\{f,g\}) + d\{f,g\}, \qquad (2.70)$$

and the involutivity follows from the fact that the module of 1-forms is generated locally by exact 1-forms.

Conversely, every Dirac manifold (N, L) such that $\rho^* : L \to T^*N$ is a bundle isomorphism is the graph of $\widehat{\Lambda}$ of a Poisson bivector.

Let us motivate our terminology. We call $\rho(L)$ the characteristic distribution of (N, L), which follows the terminology of [387] instead of the original one by Courant [150]. This is done because when L comes from a Poisson manifold, $\rho(L)$ coincides with the characteristic distribution of the Poisson structure [358]. Meanwhile, the vector fields taking values in $\ker \Omega^L$ are called *gauge vector fields*. In this way, when L is the graph of a presymplectic form, such vector fields are its gauge vector fields [162].

From here, we see that the Dirac structure incorporates the presymplectic forms and Poisson structures as particular cases. Courant [150, 151] provided the theory to this statement.

Recall that every presymplectic manifold (N, ω) gives rise to a Dirac manifold (N, L^ω) whose distribution L^ω is spanned by elements of $\Gamma(TN \oplus_N T^*N)$ of the form $X - \iota_X \omega$ with $X \in \Gamma(TN)$. Obviously, this shows that the Hamiltonian vector fields for (N, ω) are L-Hamiltonian vector fields relative to (N, L).

2.9. Jacobi Manifolds

Jacobi manifolds were independently introduced by Kirillov and Lichnerowicz [241, 265]. We now briefly survey their most fundamental properties.

Definition 2.11. A *Jacobi manifold* is a triple (N, Λ, R), where Λ is a bivector field on N and R is a vector field, referred to as *Reeb vector field*, satisfying

$$[\Lambda, \Lambda]_{SN} = 2R \wedge \Lambda, \quad [R, \Lambda]_{SN} = 0. \qquad (2.71)$$

Example 2.1. Every Poisson manifold (N, Λ) can be considered as a Jacobi manifold $(N, \Lambda, R = 0)$.

Example 2.2. The continuous Heisenberg group [370] can be described as matrix group

$$\mathbb{H} = \left\{ \begin{pmatrix} 1 & x & z \\ 0 & 1 & y \\ 0 & 0 & 1 \end{pmatrix} \middle| x, y, z \in \mathbb{R} \right\}. \tag{2.72}$$

Then, $\{x, y, z\}$ is a natural coordinate system on $\mathbb{H}$ induced by (2.2). Consider the bivector field on $\mathbb{H}$ given by

$$\Lambda_{\mathbb{H}} := -y \frac{\partial}{\partial y} \wedge \frac{\partial}{\partial z} + \frac{\partial}{\partial x} \wedge \frac{\partial}{\partial y}, \tag{2.73}$$

and the vector field $R_{\mathbb{H}} := \partial/\partial z$. After a simple calculation, one obtains

$$[\Lambda_{\mathbb{H}}, \Lambda_{\mathbb{H}}]_{SN} = 2 \frac{\partial}{\partial x} \wedge \frac{\partial}{\partial y} \wedge \frac{\partial}{\partial z} = 2 R_{\mathbb{H}} \wedge \Lambda_{\mathbb{H}}, \qquad [R_{\mathbb{H}}, \Lambda_{\mathbb{H}}]_{SN} = 0, \tag{2.74}$$

and $(\mathbb{H}, \Lambda_{\mathbb{H}}, R_{\mathbb{H}})$ becomes a Jacobi manifold.

Definition 2.12. A vector field X on N is *Hamiltonian vector field* with respect to the Jacobi manifold (N, Λ, R) if there exists an $f \in C^\infty(N)$ such that

$$X = [\Lambda, f]_{SN} + f R = \widehat{\Lambda}(\mathrm{d}f) + f R. \tag{2.75}$$

In this case, f is said to be a *Hamiltonian function* of X and we write $X = X_f$. If f is also a first integral of R, we say that f is a *good Hamiltonian function* and we call X_f a *good Hamiltonian vector field*.

Example 2.3. Given the Jacobi manifold $(\mathbb{H}, \Lambda_{\mathbb{H}}, R_{\mathbb{H}})$ and the vector field $X_1^L := \partial/\partial x$, one has

$$X_1^L = [\Lambda_{\mathbb{H}}, -y]_{SN} - y R_{\mathbb{H}} = \widehat{\Lambda}_{\mathbb{H}}(-\mathrm{d}y) - y R_{\mathbb{H}}. \tag{2.76}$$

Hence, X_1^L is a Hamiltonian vector field with a Hamiltonian function $h_1^L = -y$ with respect to $(\mathbb{H}, \Lambda_{\mathbb{H}}, R_{\mathbb{H}})$.

Each function on N gives rise to a unique Hamiltonian vector field. Meanwhile, each vector field may admit several Hamiltonian functions. This last result will be illustrated afterward within relevant examples concerning Jacobi–Lie systems.

We write $\mathrm{Ham}(\Lambda, R)$ for the space of Hamiltonian vector fields with respect to (N, Λ, R). It can be proven that $\mathrm{Ham}(\Lambda, R)$ is a Lie algebra with respect to the standard Lie bracket of vector fields. Additionally, a Jacobi manifold allow us to define a Lie bracket on $C^\infty(N)$ given by

$$\{f, g\}_{\Lambda, R} = \Lambda(\mathrm{d}f, \mathrm{d}g) + fRg - gRf. \tag{2.77}$$

This Lie bracket becomes a Poisson bracket if and only if $R = 0$. Moreover, the morphism $\phi_{\Lambda, R} : f \in C^\infty(N) \mapsto X_f \in \mathrm{Ham}(\Lambda, R)$ is a Lie algebra morphism. It is important to emphasize that it may not be injective.

2.10. k-Symplectic Forms

Let us now review the main properties of k-symplectic structures. We refer the reader to [259] for further details.

Definition 2.13. Let N be an $n(k+1)$-dimensional manifold and $\omega_1, \ldots, \omega_k$ a set of k presymplectic forms on N. We say that $(\omega_1, \ldots, \omega_k)$ is a k-symplectic structure if

$$\bigcap_{i=1}^k \ker \omega_i(x) = \{0\}, \tag{2.78}$$

for all $x \in N$. We call $(N, \omega_1, \ldots, \omega_k)$ a k-symplectic manifold.

The notion of a k-symplectic structure on a $n(k+1)$-dimensional manifold can be formulated in a equivalent manner as follows. This second approach will be more useful in the study of certain problems.

Definition 2.14. A k-polysymplectic structure (or k-polysymplectic form) on an $n(k+1)$-dimensional manifold N is an $\mathbb{R}^k$-valued closed non-degenerate 2-form on N of the form

$$\Omega = \sum_{i=1}^{k} \eta_i \otimes e^i,$$

where $\{e^1, \ldots, e^k\}$ is any basis for $\mathbb{R}^k$. The pair (N, Ω) is called a k-*polysymplectic manifold.*

Polysymplectic structures (see Definition 2.14) were introduced by Günther in [203]. Meanwhile, k-symplectic manifolds were introduced by Awane [20] and, independently, de León *et al* [257, 258] under the name of k-*cotangent structures.* The notion of k-symplectic structure considered in this book does not exactly match Awane's definition. Indeed, his definition relies on a family of k-closed 2-forms such that (2.8) holds and there exists also an integrable distribution V of dimension nk such that $\omega_r|_{V \times V} = 0$ for all $r = 1, \ldots, k$.

By taking a basis $\{e^1, \ldots, e^k\}$ of $\mathbb{R}^k$, every k-symplectic manifold $(N, \omega_1, \ldots, \omega_k)$ gives rise to a polysymplectic manifold $(N, \Omega = \sum_{i=1}^{k} \omega_i \otimes e^i)$. As Ω depends on the chosen basis, the polysymplectic manifold (N, Ω) is not canonically constructed. Nevertheless, two polysymplectic forms Ω_1 and Ω_2 induced by the same k-symplectic manifold and different bases for $\mathbb{R}^k$ are the same up to a change of basis on $\mathbb{R}^k$. In this case, Ω_1 and Ω_2 are called *gauge equivalent.* Similarly, $(N, \omega_1, \ldots, \omega_k)$ and $(N, \omega_1', \ldots, \omega_k')$ are gauge equivalent if they give rise to gauge equivalent polysymplectic forms. We can summarize these results as follows.

Proposition 2.15. *Let* $\mathrm{Sym}_k(N)$ *and* $\mathrm{Pol}_k(N)$ *be the spaces of* k-*symplectic and* k-*polysymplectic structures on* N, *correspondingly. The relation*

$$(N, \omega_1, \ldots, \omega_k)\mathcal{R}_1(N, \omega_1', \ldots, \omega_k'),$$

if and only if the k-*symplectic structures* (k-*polysymplectic manifolds*) *are gauge equivalent is an equivalence relation. Moreover,*

$$\phi : [(\omega_1, \ldots, \omega_k)] \in \mathrm{Sym}_k(N)/\mathcal{R}_1 \mapsto \left[\sum_{i=1}^{k} \omega_i \otimes e^i \right] \in \mathrm{Pol}_k(N)/\mathcal{R}_2$$

is a bijection.

62 A Guide to Lie Systems with Compatible Geometric Structures

Thus, we can say that k-symplectic and k-polysymplectic manifolds are essentially the same up to gauge equivalence.

Corollary 2.16. *Two k-symplectic manifolds $(N, \omega_1, \ldots, \omega_k)$ and $(N, \omega'_1, \ldots, \omega'_k)$ are equivalent if and only if $\langle \omega_1, \ldots, \omega_k \rangle = \langle \omega'_1, \ldots, \omega'_k \rangle$.*

Definition 2.17. Given a k-symplectic manifold $(N, \omega_1, \ldots, \omega_k)$, a submanifold $S \subset N$ is an l-symplectic submanifold relative to $(N, \omega_1, \ldots, \omega_k)$, $(l \leq k)$ if $\dim S = n_l(l+1)$ for an integer n_l and

$$(T_pS)^{\perp,l} \cap T_pS = \{0\}, \quad \forall p \in S, \tag{2.79}$$

where $(T_pS)^{\perp,l}$ is the l-th orthogonal complement of T_pS with respect to the k-symplectic structure $(N, \omega_1, \ldots, \omega_k)$, i.e., $T_pS^{\perp,l} = \{v \in T_pN : \omega_1(v, w) = \ldots = \omega_l(v, w) = 0, \forall w \in T_pS\}$ [261].

The condition (2.9) is equivalent to

$$\bigcap_{i=1}^{l} (T_pS)^{\perp_i} \cap T_pS = \{0\}, \quad \forall p \in S, \tag{2.80}$$

where $(T_pS)^{\perp_i}$ is the presymplectic annihilator of T_pS, i.e., $T_pS^{\perp_i} = \{v \in T_pN : \omega_i(v, w) = 0, \forall w \in T_pS\}$.

It is easy to prove the following.

Lemma 2.18. *If $(N, \omega_1, \ldots, \omega_k)$ and $(N, \omega'_1, \ldots, \omega'_k)$ are gauge equivalent and $S \subset N$ is a submanifold then*

$$(T_pS)^{\perp,k} = (T_pS)^{\perp',k}, \quad \forall p \in S,$$

where $(T_pS)^{\perp,k}$ and $(T_pS)^{\perp',k}$ are the k-th orthogonal k-symplectic to T_pS with respect to $(N, \omega_1, \ldots, \omega_k)$ and $(N, \omega'_1, \ldots, \omega'_k)$, respectively.

One has that $(T_pS)^{\perp,l} \neq (T_pS)^{\perp',l}$ in general for $l < k$. For instance, consider the linear example given by $N = \mathbb{R}^3$ with the gauge equivalent two-symplectic linear structures ($\omega_1 = e^1 \wedge e^3, \omega_2 = e^2 \wedge e^3$)

and $(\omega_1' = e^2 \wedge e^3, \omega_2' = e^1 \wedge e^3)$, where $\{e^1, e^2, e^3\}$ its the dual of the canonical basis of $\mathbb{R}^3$. Then if $S = \text{span}\{e^1\}$, we obtain

$$S^{\perp,1} = \text{span}\{e_1, e_2\} \quad \text{and} \quad S^{\perp',1} = \mathbb{R}^3.$$

Therefore, $S^{\perp,1} \neq S^{\perp',1}$.

Lemma 2.19. *Given a k-symplectic manifold $(N, \omega_1, \ldots, \omega_k)$ and a submanifold $S \subset N$, with $\iota : S \hookrightarrow N$ a natural embedding, $(\iota^*\omega_1, \ldots, \iota^*\omega_l)$ is an l-symplectic structure on S if and only if S is an l-symplectic submanifold of $(N, \omega_1, \ldots, \omega_k)$.*

Proof. It is a direct consequence of the following relation

$$\bigcap_{i=1}^{l} \ker(\iota^*\omega_i(p)) = \bigcap_{i=1}^{l} \ker(\omega_i(p)) \cap T_pS = (T_pS)^{\perp,l} \cap T_pS, \quad \forall p \in S.$$

$\square$

If a submanifold $S \subset M$ is endowed with an l-symplectic structure $(\iota^*\omega_1, \ldots, \iota^*\omega_l)$ with $l < k$, then for all l' such that $l \leq l' \leq k$ (it is necessary that there exists $n_{l'}$ such that $\dim S = n_{l'}(l' + 1)$), $(\iota^*\omega_1, \ldots, \iota^*\omega_{l'})$ is an l'-symplectic structure on S.

Chapter 3

Basics on Lie Systems and Superposition Rules

3.1. Introduction

This chapter surveys the fundamental theory of Lie systems that is used and analyzed throughout our essay.

3.2. Vessiot–Guldberg Lie Algebras

The Lie algebras associated with t-dependent vector fields, or equivalently, systems of first-order differential equations in normal form, can be divided into two main classes: the finite- and the infinite-dimensional ones. The finite-dimensional ones will be more important in this work, as they are frequently studied in the Lie systems literature [183, 221, 225]. This motivates the following definition.

Definition 3.1. A *Vessiot–Guldberg (VG) Lie algebra* is a finite-dimensional Lie algebra of vector fields.

Example 3.1. The Lie algebra $V_{\mathrm{sl}} = \langle \partial_x, x\partial_x, x^2\partial_x \rangle$ is a VG Lie algebra of vector fields on the real line. Lie proved that all VG Lie algebras on $\mathbb{R}$ whose distribution is $T\mathbb{R}$ are locally diffeomorphic at a generic point to V_{sl} (see [183, 276] for details).

Example 3.2. The vector fields on $\mathbb{R}$ of the form

$$X_\alpha(x) = f_\alpha(x)\frac{\partial}{\partial x}, \quad \alpha = 1, \ldots, n,$$

where each function f_α is different from zero only on an open subset of $[\alpha - 1, \alpha]$, satisfy $[X_\alpha, X_\beta] = 0$ for all $\alpha, \beta = 1, \ldots, n$. Hence, the Lie algebra $V_n := \langle X_1, \ldots, X_n \rangle$ is isomorphic to an Abelian n-dimensional Lie algebra. This example shows that not every VG Lie algebra on the real line is isomorphic to a Lie subalgebra of $\mathfrak{sl}(2, \mathbb{R})$.

Note that the restriction of the vector fields of V_n to $]0, m[$, with m natural and $0 < m \leq n$, is isomorphic to an m-dimensional abelian Lie algebra. Hence, the restriction of a VG Lie algebra to an open submanifold does not need to be isomorphic to the initial Lie algebra.

Lie classified VG Lie algebras on $\mathbb{R}$ and $\mathbb{R}^2$, as well as complex VG Lie algebras on $\mathbb{C}$ and $\mathbb{C}^2$ [183, 276]. It is commented in the literature [183] that Lie also accomplished a partial classification of Lie algebras of vector fields on $\mathbb{C}^3$, but this result is, as far as we know, lost, and whether it even exists is doubtful. The classification of VG Lie algebras on the plane around generic points up to diffeomorphism is given in Table A.1. Further details on this classification will be provided in following chapters.

Throughout this work, two different notions of linear independence of vector fields are used frequently. In order to state the clear meaning of each, we provide the following definitions.

Let us denote by $\mathfrak{X}(N)$ the space of vector fields on N. The vector fields $X_1, \ldots, X_r$, on N are *linearly independent over* $\mathbb{R}$ if they are linearly independent relative to the $\mathbb{R}$-linear space structure of $\mathfrak{X}(N)$, i.e., whenever

$$\sum_{\alpha=1}^r \lambda_\alpha X_\alpha = 0, \quad \lambda_1, \ldots, \lambda_r \in \mathbb{R} \quad \Longleftrightarrow \quad \lambda_1 = \cdots = \lambda_r = 0.$$

On the other hand, the vector fields $X_1, \ldots, X_r$ are *linearly independent at a generic point* if they are linearly independent as elements of $\mathfrak{X}(N)$ when regarded as a $C^\infty(N)$−module. That is, if one has

$$\sum_{\alpha=1}^{r} f_\alpha X_\alpha = 0, \quad f_1,\ldots,f_r \in C^\infty(N) \quad \Longleftrightarrow \quad f_1 = \cdots = f_r = 0.$$

In this essay, we frequently deal with manifolds of the form N^{m+1}. Each point of N^{m+1} is denoted by $(x_{(0)},\ldots,x_{(m)})$, where $x_{(j)}$ stands for a point of the jth copy of the manifold N within N^{m+1}.

The manifold N^{m+1} is related to the group of permutations S_{m+1} whose elements, S_{ij}, with $i \le j = 0, 1, \ldots, m$, act on N^{m+1} by permuting the variables $x_{(i)}$ and $x_{(j)}$. Finally, let us define the projections

$$\mathrm{pr} : (x_{(0)},\ldots,x_{(m)}) \in N^{m+1} \mapsto (x_{(1)},\ldots,x_{(m)}) \in N^m, \qquad (3.1)$$

and

$$\mathrm{pr}_0 : (x_{(0)},\ldots,x_{(m)}) \in N^{m+1} \mapsto x_{(0)} \in N, \qquad (3.2)$$

to be employed hereafter.

3.3. Superposition Rules

Let us introduce the superposition rule concept employed in the theory of Lie systems. In this sense, superposition rules are also known as *superposition principles, superposition formulas*, or *nonlinear superposition rules* in the literature [288, 298, 317, 356]. It is worth noting that there are other less used mathematical structures in the literature known under the same or similar names [185, 186, 233, 237].

To understand the superposition rule notion, let us start by the simplest case of differential equation admitting one. Consider a first-order system of ordinary homogeneous linear differential equations on $\mathbb{R}^n$ of the form

$$\frac{dy^i}{dt} = \sum_{j=1}^{n} A^i_j(t)y^j, \quad i = 1,\ldots,n, \qquad (3.3)$$

where the $A^i_j(t)$, with $i,j = 1,\ldots,n$, form an arbitrary family of t-dependent functions. The general solution, $y(t)$, of the system (3.3) can be written as

$$y(t) = \sum_{j=1}^{n} k_j y_{(j)}(t), \qquad (3.4)$$

where $y_{(1)}(t), \ldots, y_{(n)}(t)$ is any family of n linearly independent particular solutions, and $k_1, \ldots, k_n$ are arbitrary real constants. The above expression is called a *linear superposition rule* for system (3.3). Note that, once a family $y_{(1)}(t), \ldots, y_{(n)}(t)$ is fixed, there is a one-to-one relation between the constants $k_1, \ldots, k_n$ and the initial conditions for (3.3).

Linear superposition rules allow us to reduce the search for the general solution of a linear system to determining a finite set of particular solutions. This fact is relevant, for instance, in numeric techniques for the study of (3.3) (see [376]). Linear superposition rules also find applications in the study of the existence of periodic solutions (cf. [171, 172, 282]).

But not only linear first-order systems of differential equations do admit their general solutions to be written as functions of a family of particular solutions and some constants. If we study the first-order system

$$\frac{dy^i}{dt} = \sum_{j=1}^{n} A_j^i(t) y^j + B^i(t), \quad i = 1, \ldots, n, \qquad (3.5)$$

where $A_j^i(t), B^i(t)$, with $i, j = 1, \ldots, n$, are arbitrary families of t-dependent functions, a general solution $y(t)$ to (3.5) can be written as

$$y(t) = \sum_{j=1}^{n} k_j (y_{(j)}(t) - y_{(0)}(t)) + y_{(0)}(t), \qquad (3.6)$$

where $y_{(0)}(t), \ldots, y_{(n)}(t)$ form a family of $n + 1$ particular solutions such that $y_{(j)}(t) - y_{(0)}(t)$, with $j = 1, \ldots, n$, are linearly independent solutions of the homogeneous problem associated with (3.5), and $k_1, \ldots, k_n$ are arbitrary constants.

As in (3.3), once one has fixed a set $y_{(1)}(t), \ldots, y_{(n)}(t)$ of particular solutions, there exists a one-to-one relation between initial conditions of (3.5) and the constants $k_1, \ldots, k_n$. This new example shows that the number of particular solutions needed to express a general solution does not necessarily match the dimension of the manifold on which the system of ordinary differential equations is defined.

Previous examples suggest us to introduce the following definition.

Definition 3.2. A system of first-order ordinary differential equations on N of the form

$$\frac{dx^i}{dt} = X^i(t, x), \quad i = 1, \ldots, n, \tag{3.7}$$

admits a *global superposition rule* if there exists a t-independent map $\Phi : N^m \times N \to N$ of the form

$$x = \Phi(x_{(1)}, \ldots, x_{(m)}; k), \tag{3.8}$$

such that the general solution, $x(t)$, of (3.7) can be brought into the form

$$x(t) = \Phi(x_{(1)}(t), \ldots, x_{(m)}(t); k), \tag{3.9}$$

where $x_{(1)}(t), \ldots, x_{(m)}(t)$ is any generic family of particular solutions of system (3.7) and k is point of N to be related to the initial conditions of (3.7).

Example 3.3. A relevant family of different equations admitting a superposition rule is given by considering the so-called Riccati equations [230, 318, 343] on $\bar{\mathbb{R}} := \mathbb{R} \cup \{\infty\}$. A Riccati equation on $\bar{\mathbb{R}} := \mathbb{R} \cup \{\infty\}$ is a first-order differential equation of the form

$$\frac{dx}{dt} = b_1(t) + b_2(t)x + b_3(t)x^2, \quad x \in \bar{\mathbb{R}} := \mathbb{R} \cup \{\infty\}, \tag{3.10}$$

for arbitrary t-dependent functions $b_1(t), b_2(t)$, and $b_3(t)$. It steams from the works by Köningsber [244] first, and then by Vessiot [362], Guldberg [202], and Lie [276] that if $x_1(t), x_2(t), x_3(t)$, are three different particular solutions of (3.10), then

$$x(t) = \frac{x_1(t)(x_3(t) - x_2(t)) - kx_2(t)(x_3(t) - x_1(t))}{(x_3(t) - x_2(t)) - k(x_3(t) - x_1(t))}, \qquad (3.11)$$

is a particular solution for (3.10) for every $k \in \mathbb{R}$.

It is worth noting that, given a fixed family of three different particular solutions with initial conditions within $\mathbb{R}$, if we only choose k in $\mathbb{R}$, then the expression (3.11) does not recover the whole general solution of the Riccati equation on $\mathbb{R}$, as $x_2(t)$ cannot be recovered. Nevertheless, if $k \in \bar{\mathbb{R}}$, then (3.11) becomes a global superposition rule for Riccati equations on $\bar{\mathbb{R}}$.

Example 3.4. Let us consider a final relevant example of global superposition rule. Consider a first-order system of differential equations on a Lie group G of the form

$$\frac{dg}{dt} = \sum_{\alpha=1}^{r} b_\alpha(t) X_\alpha^R(g), \qquad g \in G, \qquad (3.12)$$

where the functions $b_1(t), \ldots, b_r(t)$ are arbitrary and the vector fields $X_1^R, \ldots, X_r^R$ form a basis of right-invariant vector fields on G. Owing to the invariance of (3.12) relative to $R_{\bar{g}*}$, where $R_{\bar{g}} : h \in G \mapsto h\bar{g} \in G$, one obtains that if $g_0(t)$ is the particular solution of (3.12) with initial condition $g_0(t) = e$, then $\bar{g}(t) = R_{\bar{g}}g_0(t)$, is a solution of (3.12) for every $\bar{g} \in G$. Let us prove this fact.

Applying $R_{g^{-1}*g}$ to both sides of (3.12), we see that its general solution, $g(t)$, satisfies that

$$R_{g^{-1}(t)*g(t)} \frac{dg}{dt} \dot{g}(t) = \sum_{\alpha=1}^{r} b_\alpha(t) X_\alpha^R(e) \in T_e G. \qquad (3.13)$$

Let us prove that $g'(t) = R_{\bar{g}}g_0(t)$ for a given $\bar{g} \in G$, is another solution of (3.12). In fact,

$$\frac{dg'}{dt}(t) = R_{\bar{g}*g_0(t)} \left(\frac{dg_0}{dt}(t) \right) \Longleftrightarrow \frac{dg'}{dt}(t)$$

$$= R_{\bar{g}*g_0(t)} \left(\sum_{\alpha=1}^{r} b_\alpha(t) X_\alpha^R(g_0(t)) \right).$$

Since $R_{\bar{g}*g_0(t)}X_\alpha^{\mathrm{R}}(g_0(t)) = X_\alpha^{\mathrm{R}}(g_0(t)\bar{g})$, one has that

$$\frac{dg'}{dt}(t) = \sum_{\alpha=1}^{r} b_\alpha(t)X_\alpha^{\mathrm{R}}(R_{\bar{g}}g_0(t)) = \sum_{\alpha=1}^{r} b_\alpha(t)X_\alpha^{\mathrm{R}}(g'(t)),$$

and $g'(t)$ is another particular solution of (3.28) with initial condition $g'(0) = R_{\bar{g}}g_0$. In consequence, the general solution $g(t)$ for Equation (3.13) can be written as $g(t) = R_{\bar{g}}g_0(t)$ for an arbitrary $\bar{g} \in G$ and

$$\Phi : (g; k) \in G \times G \mapsto R_k g \in G \tag{3.14}$$

becomes a global superposition rule for (3.12). Lie systems of the form (3.12) are called *automorphic* [53, 189, 243].

To fully understand Definition 3.2, it is necessary to get a grasp of what the term "generic" means. More precisely, expression (3.9) is valid for any generic family of m particular solutions if there exists an open dense subset $U \subset N^m$, such that expression (3.9) is satisfied for every set of particular solutions $x_{(1)}(t), \ldots, x_{(m)}(t)$, such that $(x_{(1)}(0), \ldots, x_{(m)}(0))$ lies in U.

Let us now show in detail that the systems (3.3) and (3.5) admit global superposition rules. Consider the function $\Phi : N^n \times N \to N$ of the form

$$\Phi(x_{(1)}, \ldots, x_{(n)}; k_1, \ldots, k_n) = \sum_{j=1}^{n} k_j x_{(j)}. \tag{3.15}$$

This mapping is a global superposition rule for the linear system (3.3) since, for each set of particular solutions $x_{(1)}(t), \ldots, x_{(n)}(t)$, of (3.3) such that

$$(x_{(0)}(0), \ldots, x_{(n)}(0)) \in U := \left\{ (x_{(1)}, \ldots, x_{(n)}) \in N^n \,\middle|\, x_{(1)} \wedge \cdots \wedge x_{(n)} \right.$$

$$\left. \subset N^n \neq 0 \right\},$$

the general solution $x(t)$ of (3.3) can be written as (3.4) and U is dense in N^n.

On the other hand, the function $\Phi : N^{n+1} \times N \to N$ of the form

$$\Phi(x_{(0)}, \ldots, x_{(n)}; k_1, \ldots, k_n) = \sum_{j=1}^{n} k_j (x_{(j)} - x_{(0)}) + x_{(0)}, \qquad (3.16)$$

is a global superposition rule for system (3.5). In fact, for each set of particular solutions, $x_{(0)}(t), \ldots, x_{(n)}(t)$, of (3.5) such that the point $(x_{(0)}(0), \ldots, x_{(n)}(0))$ belongs to the open dense subset

$$U := \{(x_{(0)}, \ldots, x_{(n)}) \in N^{n+1} \mid (x_{(1)} - x_{(0)}) \wedge \cdots \wedge (x_{(n)} - x_{(0)}) \neq 0\}$$
$$\subset N^{n+1},$$

and the general solution $x(t)$ of (3.5) can be put in the form (3.6).

Finally, let us analyze the case of Riccati equations in $\bar{\mathbb{R}}$. The map $\Phi : \bar{\mathbb{R}}^3 \times \bar{\mathbb{R}} \to \bar{\mathbb{R}}$ of the form

$$\Phi(x_{(1)}, x_{(2)}, x_{(3)}; k) = \frac{x_{(1)}(x_{(3)} - x_{(2)}) - k x_{(2)}(x_{(3)} - x_{(1)})}{(x_{(3)} - x_{(2)}) - k(x_{(3)} - x_{(1)})} \qquad (3.17)$$

is a global superposition rule for Riccati equations in $\bar{\mathbb{R}}$. To verify this, it is sufficient to note that given one of these equations with three particular solutions, $x_{(1)}(t), x_{(2)}(t), x_{(3)}(t)$, such that $(x_{(1)}(0), x_{(2)}(0), x_{(3)}(0)) \in U$, where

$$U = \{(x_{(1)}, x_{(2)}, x_{(3)}) \in \mathbb{R}^3 \mid (x_{(1)} - x_{(2)})(x_{(1)} - x_{(3)})$$
$$\times (x_{(2)} - x_{(3)}) \neq 0\},$$

then its general solution can be cast into (3.11). Moreover, U is dense in $\bar{\mathbb{R}}$.

There are very few differential equations admitting a global superposition rule. Probably, the most relevant examples of global superposition rule are contained in this section. Meanwhile, most differential equations admit a particular type of superposition rule that retrieves only certain particular solutions (cf. [108, 280, 335, 376]). For instance, if the Riccati equation (3.10) is considered as a differential equation on the real line as done by Königsberger, Guldberg, Lie, and Vessiot [7, 127, 181, 207, 291, 318, 343], then the map $\Phi : \mathbb{R}^m \times \mathbb{R} \to \mathbb{R}$ given by restricting (3.17) to $\mathbb{R}$ is not a global superposition rule. As already said, this map will not retrieve $x_{(2)}(t)$ from a set of different particular solutions $x_{(1)}(t), x_{(2)}(t), x_{(3)}(t)$, for any $k \in \mathbb{R}$. Further, the function (3.17) is not globally defined on $\mathbb{R}^3 \times \mathbb{R}$. Nevertheless, such a function is what is known as a *superposition rule* for Riccati equations in the literature [202, 276, 362]. This illustrates the necessity of the following definition [108].

Definition 3.3. A system of first-order ordinary differential equations on N of the form

$$\frac{dx}{dt} = X(t, x), \tag{3.18}$$

for a certain t-dependent vector field on N, admits a *superposition rule* if there exists a t-independent map $\Phi : U \subset N^m \times N \to N$ such that all solutions, $x(t)$, to (3.18) with initial conditions on an open subset of N can be written in the form

$$x(t) = \Phi(x_{(1)}(t), \ldots, x_{(m)}(t); k), \tag{3.19}$$

for a generic set of particular solutions $x_{(1)}(t), \ldots, x_{(m)}(t)$ and $k \in N$.

Most superposition rules appearing in the literature are such that, for a fixed generic set of particular solutions, they retrieve all solutions of the differential equation under study except those with initial conditions in a set of zero measure (cf. [90, 107, 280]). This is evidently the case of Riccati equations on $\mathbb{R}$.

The above remarks lead to an important simplification. As most superposition rules appearing in applications are "almost global" in the sense given above, it is generally assumed in many works that they are defined globally (cf. [108]). This will also be the approach carried out in this book. Anyhow, for more information on the local or global character of superposition rules, we refer the reader to [52].

Let us finally comment a last feature of superposition rules that will be of interest in the following sections. Given a superposition rule $\Phi : N^m \times N \to N$, in general, one has

$$\Phi(x_{(1)}, \ldots, x_{(i)}, \ldots, x_{(j)}, \ldots, x_{(m)}; k)$$
$$\neq \Phi(x_{(1)}, \ldots, x_{(j)}, \ldots, x_{(i)}, \ldots, x_{(m)}; k). \qquad (3.20)$$

This can easily be seen, for instance, in the case of the superposition rules (3.17) for Riccati equations. Nevertheless, it can be proven (cf. [93]) that there exists a map $\varphi : k \in N \mapsto \varphi(k) \in N$ such that

$$\Phi(x_{(1)}, \ldots, x_{(i)}, \ldots, x_{(j)}, \ldots, x_{(m)}; k)$$
$$= \Phi(x_{(1)}, \ldots, x_{(j)}, \ldots, x_{(i)}, \ldots, x_{(m)}; \varphi(k)).$$

3.4. The Lie System Notion and the Lie–Scheffers Theorem

After the presentation of superposition rules, a relevant question now arises: which first-order systems of ordinary differential equations admit a superposition rule? Several works addressed this question towards the end of the 19th century. As a result, Königsberger [244], Vessiot [362], and Guldberg [202] proved, that every system of first-order differential equations defined over the real line admitting a superposition rule is, up to a diffeomorphism, a Riccati equation or a first-order linear differential equation.

Posteriorly, Lie [269, 273, 276] established the conditions ensuring that a system of first-order differential equations of the form (3.7) admits a superposition rule. We here present a modern geometric approach to his result, the today named *Lie–Scheffers Theorem* [276, Theorem 44], and we illustrate its applications and main properties. We leave its proof for later sections.

Theorem 3.4 (The Lie–Scheffers Theorem). *A first-order system of ordinary differential equations X admits a superposition rule if and only if*

$$X(t, x) = \sum_{\alpha=1}^{r} b_\alpha(t) \, X_\alpha(x), \qquad (3.21)$$

for a family $X_1, \ldots, X_r$, of vector fields on N spanning an r-dimensional real Lie algebra of vector fields V.

The Lie algebra spanned by $X_1, \ldots, X_r$ is called a *Vessiot–Guldberg Lie algebra* admitted by X. Immediately after proving the Lie–Scheffers theorem, Lie showed that X admits a superposition rule that depends on m particular solutions if and only if X admits a VG Lie algebra V such that $\dim N \cdot m \geq \dim V$ (see Figure 1.2). This is the so-called *Lie's condition* [87, 89, 108]. Hence, the larger the dimension of the VG Lie algebra is, the more particular solutions will be involved in the superposition rule.

The Lie–Scheffers Theorem solves theoretically the characterization of systems (3.7) admitting a superposition rule, but in practice, it can be hard to apply. We can find two main drawbacks.

First, given a t-dependent vector field X on a manifold N, it may be difficult to determine whether its smallest Lie algebra is finite-dimensional. For instance, this happens when N or the VG Lie algebras of X are high-dimensional, e.g., in the case of second-order Riccati equations the associated VG Lie algebra is eight-dimensional [107]. In such cases one generally has to rely on other methods [23, 188, 197]. In the case of second-order Riccati equations, one can consider such second-order Riccati equations as the projection of a linear system of ordinary differential equations on $\mathbb{R}^3$, which gives the VG Lie algebra in an easier manner (see [188]).

Second, the Lie–Scheffers theorem does not characterize first-order systems of differential equations of the form $F(t, x, \dot{x}) = 0$, for a certain function F taking values in a manifold Q. This is still an open question in the theory of Lie systems.

The relevance of the Lie–Scheffers theorem (published for the first time in [276] in 1893) suggested Vessiot to propose the following definition [364].

Definition 3.5. A *Lie system* is a first-order system of differential equations in normal form admitting a superposition rule.

Owing to the Lie–Scheffers theorem, the previous definition can be recast into the following form.

Definition 3.6. A *Lie system* is a first-order system of ordinary differential equations X on a manifold N of the form $X(t, x) = \sum_{\alpha=1}^{r} b_\alpha(t) X_\alpha(x)$, for a family $X_1, \ldots, X_r$ of vector fields on N spanning an r-dimensional real Lie algebra of vector fields V.

By virtue of previous definitions, the systems of first-order differential equations (3.3), (3.5), (3.10), and (3.12), which admit global superposition rules (3.15), (3.16), (3.17), and (3.14), respectively, are Lie systems. Anyhow, for the sake of completeness, let us prove that they admit the requirements stipulated by the Lie–Scheffers theorem.

Example 3.5. The homogeneous linear system (3.3) describes the integral curves of the t-dependent vector field

$$X(t, x) = \sum_{i,j=1}^{n} A^i{}_j(t) X_{ij}(x), \tag{3.22}$$

where the n^2 vector fields $X_{ij} = x^j \, \partial/\partial x^i$, with $i, j = 1, \ldots, n$, close on the commutation relations

$$[X_{ij}, X_{lm}] = \delta^i_m X_{lj} - \delta^l_j X_{im},$$

where δ^i_m is the Kronecker delta function. Hence, the vector fields X_{ij} span an n^2-dimensional VG Lie algebra of X isomorphic to the Lie algebra $\mathfrak{gl}(n, \mathbb{R})$ (see [128] for details).

By the Lie–Scheffers theorem, the decomposition (3.22) ensures that each system (3.3) admits a superposition rule. Indeed, recall that (3.3) admits the superposition rule (3.15) depending on n particular solutions. Since the superposition rule (3.15) depends on n particular solutions for a system (3.3) defined on $\mathbb{R}^n$, we obtain that $n \dim \mathbb{R}^n \geq \dim V$ and, therefore, Lie's condition is also satisfied.

Example 3.6. The first-order system of inhomogeneous differential equations (3.5) describes the integral curves of the t-dependent vector field

$$X_t = \sum_{i,j=1}^{n} A^i{}_j(t)\, X_{ij} + \sum_{i=1}^{n} B^i(t)\, X_i, \qquad (3.23)$$

of the vector fields (3.5) and $X_i = \partial/\partial x^i$, $i = 1,\ldots,n$. The vector fields X_{ij} and X_i, with $i,j = 1,\ldots,n$, satisfy the commutation relations

$$[X_i, X_j] = 0, \quad i,j = 1,\ldots,n,$$

$$[X_{ij}, X_l] = -\delta^{lj}\, X_i, \quad i,j,l = 1,\ldots,n$$

and, therefore, they span a Lie algebra V_{aff} of vector fields isomorphic to the $(n^2 + n)$-dimensional Lie algebra of the affine group [128]. Then, the systems (3.23) satisfy the conditions stipulated by the Lie–Scheffers theorem. In fact, systems (3.5) admit a superposition rule (3.16) depending on $n + 1$ particular solutions. Since (3.23) are defined on $\mathbb{R}^n$, one has that $(n + 1)\dim\mathbb{R}^n \geq \dim V_{\mathrm{aff}}$ and Lie's condition is also satisfied.

Examples 3.5 and 3.6 allow us to illustrate that a Lie system may admit multiple VG Lie algebras. In fact, a linear system (3.3) is related to a t-dependent vector field taking values in the VG Lie algebra V_{lin}. But (3.3) can also be considered as a particular instance of an affine system of differential equations (3.5), which admits the $n^2 + n$-dimensional VG Lie algebra V_{aff}. In other words, linear systems admit, at least, two non-isomorphic VG Lie algebras V_{lin} and V_{aff}.

Lie's condition evidences that the existence of different VG Lie algebras for a system of differential equations may cause the existence of different superposition rules for the same Lie system. In particular, linear systems of ordinary differential equations admit a VG Lie algebra V_{lin}, which implies the existence of a superposition rule depending, at least, on n particular solutions, which was determined. Nevertheless, such linear systems, as particular cases of affine systems, admit a second VG Lie algebra V_{aff} of dimension $n(n+1)$, which implies the existence of a second superposition rule depending, at least, on $n+1$ particular solutions.

Example 3.7. Let us check that Riccati equations satisfy the conditions shown in the Lie–Scheffers theorem. A Riccati equation (3.10) determines the integral curves of a t-dependent vector field on $\bar{\mathbb{R}}$ of the form $X(t,x) = \sum_{\alpha=1}^{3} b_\alpha(t) X_\alpha$, where

$$X_1 = \frac{\partial}{\partial x}, \quad X_2 = x\frac{\partial}{\partial x}, \quad X_3 = x^2\frac{\partial}{\partial x}, \tag{3.24}$$

span a three-dimensional Lie algebra V_{Ric} with commuting relations

$$[X_1, X_2] = X_1, \quad [X_1, X_3] = 2X_2, \quad [X_2, X_3] = X_3. \tag{3.25}$$

Thus, Riccati equations obey the conditions given by the Lie–Scheffers theorem to admit a superposition rule. Since Riccati equations are associated with a three-dimensional VG Lie algebra V_{Ric} and they admit a superposition rule depending on three particular solutions, it is immediate that Riccati equations satisfy the corresponding Lie's condition.

The existence of different VG Lie algebras for a system of first-order ordinary differential equations is an important question, since it determines the integrability by quadratures of Lie systems [82].

Let us now turn our attention to determining when a first-order system of differential equations in normal form (3.7) is *not* a Lie system. To study this question, it is useful to rewrite the Lie–Scheffers theorem in the following form.

Proposition 3.7 (The Abbreviated Lie–Scheffers Theorem).
A system X on N is a Lie system if and only if its smallest Lie algebra is finite-dimensional.

Hence, determining that a system X is not a Lie system reduces to showing that Lie $(\{X_t\}_{t\in\mathbb{R}})$ is infinite-dimensional. Proving this is equivalent to showing that there exists an infinite chain, $\{Z_j\}_{j\in\mathbb{N}}$ of linearly independent vector fields over $\mathbb{R}$ obtained through successive Lie brackets of elements in $\{X_t\}_{t\in\mathbb{R}}$. The typical example to illustrate this process relies on studying the Abel equations of the first-type [137]

$$\frac{dx}{dt} = x^2 + b(t)x^3, \tag{3.26}$$

where $b(t)$ is a non-constant function, to prove that these differential equations are not Lie systems. In fact, (3.26) describes the integral curves of the t-dependent vector field

$$X = (x^2 + b(t)x^3)\frac{\partial}{\partial x}.$$

Consider the chain of vector fields

$$Z_1 = x^2\frac{\partial}{\partial x}, \quad Z_2 = x^3\frac{\partial}{\partial x}, \quad Z_j = [X_1, X_{j-1}], \quad j = 3, 4, 5, \ldots.$$

Since $Z_j = (j-2)!x^{j+1}\partial/\partial x$ for $j \geq 3$, one has that $\mathrm{Lie}(\{X_t\}_{t\in\mathbb{R}})$ admits the infinite chain of linearly independent vector fields $\{Z_j\}_{j\in\mathbb{N}}$ and, in consequence, the abbreviated Lie–Scheffers theorem shows that Abel equations of the type (3.26) are not Lie systems.

There are many other relevant systems of differential equations that can be studied through Lie systems: Euler systems [128, 171], matrix Riccati equations [207, 249, 300, 317, 343, 376] and their generalizations over normed division algebras [282], Bernoulli equations and planar differential equations [23], equations of the Riccati hierarchy [197], mechanical systems [168, 180], Schwarzian equation [90], etc. [199, 361]. Other important systems of differential equations which can be studied through Lie systems will be detailed in the following sections.

3.5. Lie Systems and Lie Groups

Vessiot was the first to realize that the general solution of any Lie system can be obtained by a unique particular solution of a type of Lie system defined on a Lie group (see [364, H.7]). This section presents a modern geometric explanation of this result (see also [94, 314] for further details) and some other related topics that appeared during the 20th century, e.g., the Wei–Norman method [122, 371, 372].

Consider a Lie system related to a t-dependent vector field $X = \sum_{\alpha=1}^{r} b_\alpha(t) X_\alpha$ on N, where $\{X_1, \ldots, X_r\}$ form a basis of a VG Lie algebra V for X. This gives rise to a (local) Lie group action $\Phi : G \times N \to N$ whose fundamental vector fields are exactly those of V and such that $\mathfrak{g} \simeq T_e G$ is isomorphic to V. In particular, one can choose a basis $\{a_1, \ldots, a_r\}$ of $\mathfrak{g}$ so that

$$\Phi(\exp(-s a_\alpha), x) = g_s^{(\alpha)}(x), \quad \alpha = 1, \ldots, r, \quad s \in \mathbb{R}, \tag{3.27}$$

where $g^{(\alpha)} : (s, x) \in \mathbb{R} \times N \mapsto g^{(\alpha)}(s, x) = g_s^{(\alpha)}(x) \in N$ is the flow of the vector field X_α. Then, each X_α is the fundamental vector field corresponding to a_α and the map $\phi : \mathfrak{g} \to V$ such that $\phi(a_\alpha) = X_\alpha$ for $\alpha = 1, \ldots, r$, becomes a Lie algebra isomorphism. If the minus sign in (3.27) is replaced with a plus sign, then ϕ becomes a Lie algebra anti-isomorphism (cf. [108, 189]).

Let X_α^{R} be the right-invariant vector field on G with $(X_\alpha^{\mathrm{R}})_e = a_\alpha$, namely $(X_\alpha^{\mathrm{R}})_g = R_{g*e} a_\alpha$ where $R_g : g' \in G \mapsto g'g \in G$. Then, the t-dependent right-invariant vector field

$$X^G(t, g) = -\sum_{\alpha=1}^{r} b_\alpha(t) X_\alpha^{\mathrm{R}}(g), \tag{3.28}$$

is, in virtue of Example 3.4, a Lie system on G whose integral curves are the solutions of the system on G given by

$$\frac{dg}{dt} = -\sum_{\alpha=1}^{r} b_\alpha(t) X_\alpha^{\mathrm{R}}(g). \tag{3.29}$$

Let us show how $x(t) = \Phi(g_0(t), x)$ is a particular solution to X. In fact,

$$\frac{dx}{dt}(t) = \Phi_{x*g_0(t)} \left(\frac{dg_0}{dt}(t) \right) = \Phi_{x*g_0(t)} \left(-\sum_{\alpha=1}^{r} b_\alpha(t) X_\alpha^{\mathrm{R}}(g_0(t)) \right)$$

$$= -\sum_{\alpha=1}^{r} b_\alpha(t)\Phi_{x*g_0(t)}R_{g_0(t)*}\mathtt{a}_\alpha = -\sum_{\alpha=1}^{r} b_\alpha(t)\Phi_{x(t)*e}\mathtt{a}_\alpha$$

$$= \sum_{\alpha=1}^{r} b_\alpha(t)X_\alpha(x(t)).$$

The relevance of the Lie system (3.13) relies on the fact that the integral curves of the t-dependent vector field $X(t,x)$ can be obtained from one particular solution of the so-called automorphic Lie system (3.29) associated with X^G. More explicitly, the general solution $x(t)$ of the Lie system X reads $x(t) = \Phi(g_e(t), x_0)$, where $g_e(t)$ is the particular solution of Equation (3.13) with $g_e(0) = e$ and x_0 is the initial condition of $x(t)$ for X.

In turn, let us prove that the integration of every automorphic Lie system can be reduced to solving an homogeneous first-order system of differential equations. In view of the Ado's theorem [6], every finite-dimensional Lie algebra, e.g., $\mathfrak{g}$, can be represented as a matrix Lie subalgebra of $\mathfrak{gl}(n, \mathbb{R})$ for a certain n, i.e., there exists an injective Lie algebra morphism $\phi : \mathfrak{g} \to \mathfrak{gl}(n, \mathbb{R})$. This gives rise to a Lie group morphism $\Phi : G \to GL(\mathbb{R}^n)$ that is injective around the neutral element (see [359]). Roughly speaking, this implies that the elements of G that are close to the neutral element can be written in matrix form. Multiplying (3.29) by $R_{g^{-1}*}$, we obtain

$$R_{g^{-1}*g}\frac{dg}{dt} = -\sum_{\alpha=1}^{r} R_{g^{-1}*g}X_\alpha^R(g) = -\sum_{\alpha=1}^{r} b_\alpha(t)\mathtt{a}_\alpha. \qquad (3.30)$$

Using the above results, we can consider g as a matrix A and the elements of the basis of the Lie algebra of G, e.g., the $\mathtt{a}_\alpha$, can also be considered as $m \times m$ matrices M_α.

In this way, the system (3.30) related to t-dependent vector field (3.21) amounts to

$$\dot{A}(t)A^{-1}(t) = -\sum_{\alpha=1}^{r} b_\alpha(t)M_\alpha \implies \dot{A} = -\sum_{\alpha=1}^{r} b_\alpha(t)M_\alpha A, \qquad (3.31)$$

with $A(t)$ being a curve taking values in $GL(\mathbb{R}^n)$, and $\{M_1, \ldots, M_r\}$ being a basis of a matrix Lie algebra with opposite structure constants than $X_1, \ldots, X_r$. Equation (3.31) is a homogeneous linear differential equation in the t-dependent coefficients of the matrix $A(t)$.

As a consequence of the above, the general solution of X can be described in terms of the Lie group action Φ and a particular solution of X^G with initial condition in the neutral element of G. In turn, this last problem can be reduced to solving a homogeneous first-order system of differential equations.

Although the above procedure is possible from a theoretical point of view, it is frequently difficult to obtain the Lie group action Φ explicitly. This, in turn, makes the above procedure ineffective.

Apart from the interest in automorphic Lie systems for solving differential equations, automorphic Lie systems have also interest of their own, as they are related to many physical problems, e.g., the determination of evolution operators for quantum mechanical problems and control problems [108].

A generalization of the method [122] used by Wei and Norman for linear systems [371, 372] is very useful for solving equations (3.13). Furthermore, there exist reduction techniques that can also be used to solve them [94]. Such techniques show, for instance, that Lie systems related to solvable VG Lie algebras are integrable by quadratures [94]. Finally, as right-invariant vector fields X^R project onto each homogeneous space of G, the solution of Equation (3.13) enables us to find the general solution for Lie systems on their homogeneous spaces, which was extensively applied, for instance, by Winternitz and Shnider [340, 341]. Conversely, the knowledge of particular solutions of the associated system in a homogeneous space gives us a method for reducing the problem to the corresponding isotropy group [94].

Finally, other methods to solve automorphic Lie systems based on algebraic properties of the Lie algebra have been developed by several authors [189].

3.6. Diagonal Prolongations

To give a geometric proof of Lie–Scheffers theorem and to introduce structures related to superposition rules, we hereafter provide the notion of diagonal prolongations of sections of vector bundles (for further details see [90, 108]).

The *diagonal prolongation* of a vector bundle $\tau : E \to N$ to N^m is the *Cartesian product bundle* $E^{[m]} = E \times \cdots \times E$ of m copies of E, viewed as a vector bundle over N^m in the natural way

$$E^{[m]}_{(x_{(1)},\ldots,x_{(m)})} = E_{x_{(1)}} \oplus \cdots \oplus E_{x_{(m)}}. \tag{3.32}$$

Every section $X : N \to E$ of E has a natural *diagonal prolongation* to a section $X^{[m]}$ of $E^{[m]}$ of the form

$$\widetilde{X}^{[m]}(x_{(1)}, \ldots, x_{(m)}) = X(x_{(1)}) + \cdots + X(x_{(m)}). \tag{3.33}$$

Given a function $f : N \to \mathbb{R}$, we call *diagonal prolongation* of f to N^m the function $\widetilde{f}^{[m]}$ on N^m of the form $\widetilde{f}^{[m]}(x_{(1)}, \ldots, x_{(m)}) = f(x_{(1)}) + \cdots + f(x_{(m)})$. If the dimension of N^m is understood from context, then we will frequently skip the superindex $[m]$ appearing in diagonal prolongations of sections.

We can also consider sections $X^{(j)}$ of $E^{[m]}$ given by

$$X^{(j)}(x_{(1)}, \ldots, x_{(m)}) = 0 + \cdots + X(x_{(j)}) + \cdots + 0. \tag{3.34}$$

It is clear that, if $\{X_i \mid i = 1, \ldots, p\}$ is a basis of local sections of E, then $\{X_i^{(j)} \mid i = 1, \ldots, p, j = 1, \ldots, m\}$ is a basis of local sections of $E^{[m]}$.

Since there are obvious canonical isomorphisms

$$(TN)^{[m]} \simeq TN^m \quad \text{and} \quad (T^*N)^{[m]} \simeq T^*N^m, \tag{3.35}$$

we can interpret the diagonal prolongation $X^{[m]}$ of a vector field on N as a vector field $\widetilde{X}^{[m]}$ on N^m, and the diagonal prolongation $\alpha^{[m]}$ of a 1-form on N as a 1-form $\widetilde{\alpha}^{[m]}$ on N^m. In the case when m is fixed, we will simply write $\widetilde{X}$ and $\widetilde{\alpha}$. The proof of the following properties of diagonal prolongations is straightforward.

Proposition 3.8. *The diagonal prolongation to N^m of a vector field X on N is the unique vector field $\widetilde{X}^{[m]}$ on N^m, projectable under the map $\pi : (x_{(1)}, \ldots, x_{(m)}) \in N^m \mapsto x_{(1)} \in N$ onto X and invariant under the permutation of variables $x_{(i)} \leftrightarrow x_{(j)}$, with $i, j = 1, \ldots, m$. The diagonal prolongation to N^m of a 1-form α on N is the unique 1-form $\widetilde{\alpha}^{[m]}$ on N^m such that $\widetilde{\alpha}^{[m]}(\widetilde{X}^{[m]}) = \widetilde{\alpha(X)}^{[m]}$ for every vector field $X \in \Gamma(TN)$. We have $d\widetilde{\alpha} = \widetilde{d\alpha}$ and $\mathcal{L}_{\widetilde{X}^{[m]}}\widetilde{\alpha}^{[m]} = \widetilde{\mathcal{L}_X\alpha}^{[m]}$. In particular, if α is closed (exact), so is its diagonal prolongation $\widetilde{\alpha}^{[m]}$ to N^m.*

Using local coordinates (x^a) in N and the induced system $(x^a_{(i)})$ of coordinates in N^m, we can write, for $X = \sum_a X^a(x)\partial_{x^a}$ and $\alpha = \sum_a \alpha_a(x)dx^a$,

$$\widetilde{X}^{[m]} = \sum_{a,i} X^a(x_{(i)})\partial_{x^a_{(i)}} \quad \text{and} \quad \widetilde{\alpha}^{[m]} = \sum_{a,i} \alpha_a(x_{(i)})dx^a_{(i)}. \quad (3.36)$$

The following property is useful to study Lie systems.

Lemma 3.9. *For every two vector fields $X, Y \in \mathfrak{X}(N)$, one has that $[\widetilde{X}^{[m]}, \widetilde{Y}^{[m]}] = \widetilde{[X, Y]}^{[m]}$. Consequently, given a Lie algebra of vector fields $V \subset \mathfrak{X}(N)$, the prolongations of its elements to N^m span an isomorphic Lie algebra of vector fields.*

A fundamental notion in the geometrical description of Lie systems is the so-called *diagonal prolongation of a t-dependent vector field*. Its definition goes as follows.

Definition 3.10. Given a t-dependent vector field X on N, its diagonal prolongation to N^m is the t-dependent vector field $\widetilde{X}^{[m]}$ whose $\widetilde{X}^{[m]}_t$, for every fixed $t \in \mathbb{R}$, is the diagonal prolongation to N^m of the vector field X_t.

In particular, if

$$X(t, x_{(1)}) = \sum_{i=1}^{n} X^i(t, x_{(1)})\frac{\partial}{\partial x^i_{(1)}},$$

its *diagonal prolongation* to N^m is the t-dependent vector field on N^m given by

$$\widetilde{X}^{[m]}(t, x_{(1)}, \ldots, x_{(m)}) = \sum_{a=1}^{m} \sum_{i=1}^{n} X^i(t, x_{(a)}) \frac{\partial}{\partial x^i_{(a)}}.$$

3.7. Geometric Approach to Superposition Rules

Let us now review the modern geometric approach to the Lie–Scheffers theorem and superposition rules carried out in [93, 108].

Lemma 3.11. *Let $X_1, \ldots, X_r$, be a family of vector fields on N whose diagonal prolongations to N^m are linearly independent at a generic point. Given the diagonal prolongations, $\widetilde{X}_1^{[m+1]}, \ldots, \widetilde{X}_r^{[m+1]}$, of $X_1, \ldots, X_r$ to N^{m+1}, the vector field $\sum_{\alpha=1}^{r} b_\alpha \widetilde{X}_\alpha^{[m+1]}$, with $b_\alpha \in C^\infty(N^{m+1})$, is a diagonal prolongation if and only if $b_1, \ldots, b_r$, are constant.*

Proof. If the vector fields $X_1, \ldots, X_r$ take the form

$$X_\alpha = \sum_{i=1}^{n} A_\alpha^i(x) \frac{\partial}{\partial x^i}, \quad \alpha = 1, \ldots, r,$$

in local coordinates, then

$$\widetilde{X}_\alpha^{[m+1]} = \sum_{i=1}^{n} \sum_{a=0}^{m} A_\alpha^i(x_{(a)}) \frac{\partial}{\partial x^i_{(a)}}, \quad \alpha = 1, \ldots, r,$$

and

$$\sum_{\alpha=1}^{r} b_\alpha(x_{(0)}, \ldots, x_{(m)}) \widetilde{X}_\alpha^{[m+1]}$$

$$= \sum_{\alpha=1}^{r} \sum_{i=1}^{n} \sum_{a=0}^{m} b_\alpha(x_{(0)}, \ldots, x_{(m)}) A_\alpha^i(x_{(a)}) \frac{\partial}{\partial x^i_{(a)}}.$$

The last vector field is a diagonal prolongation if and only if there exist functions $B^1, \ldots, B^n \in C^\infty(N)$ such that

$$\sum_{\alpha=1}^{r} b_\alpha(x_{(0)}, \ldots, x_{(m)}) A^i_\alpha(t, x_{(a)}) = B^i(x_{(a)}),$$

$$a = 0, \ldots, m, \quad i = 1, \ldots, n.$$

The functions $b_1, \ldots, b_r$ give a solution of the subsystem of linear equations in the variables, $u_1, \ldots, u_r$, given by

$$\sum_{\alpha=1}^{r} u_\alpha A^i_\alpha(x_{(a)}) = B^i(x_{(a)}), \quad a = 1, \ldots, m, \quad i = 1, \ldots, n.$$

As the vector fields $\mathrm{pr}_*(\widetilde{X}^{[m]}_\alpha)$ are linearly independent, the coefficient matrix of the above system of $m \cdot n$ equations with r unknowns has rank $r \leq m \cdot n$. Consequently, the solutions $u_1, \ldots, u_r$ are completely determined via the functions $B^i(x_{(a)})$, with $a = 1, \ldots, m$, and $i = 1, \ldots, n$, and do not depend on $x_{(0)}$. Since diagonal prolongations are invariant under the action of the symmetry group S_{m+1}, the functions $u_\alpha = b_\alpha(x_{(0)}, \ldots, x_{(m)})$, with $\alpha = 1, \ldots, r$, satisfy this symmetry and they cannot depend on the variables $x_{(1)}, \ldots, x_{(m)}$. Therefore, they must be constant. $\qquad \square$

The following two lemmas complete the geometric proof of the Lie–Scheffers theorem. These details were not included in the original work [93].

Proposition 3.12. *Let $X_1, \ldots, X_r$ be a family of vector fields on N linearly independent over $\mathbb{R}$ and let $\sigma(q)$ be the rank at a generic point of the distribution generated by the diagonal prolongations of $X_1, \ldots, X_r$ to N^q. Then, if $\sigma(q) < r$, then $\sigma(q) < \sigma(q+1)$ and there exists $m \in \mathbb{N}$ such that $\sigma(m) = r$.*

Proof. Let us use reduction to absurd. Assume then that $p = \sigma(q) = \sigma(q+1) < r$. By the definition of diagonal prolongations it follows that $\sigma(q) \leq \sigma(q+1)$. One can pick up, among the $X_1, \ldots, X_r$, a family of p vector fields that are linearly independent at a generic point of

N^q and N^{q+1}, let these be, with no loss of generality, $X_1, \ldots, X_p$. Consequently,

$$\bar{f}_1 \widetilde{X}_1^{[q+1]} + \cdots + \bar{f}_p \widetilde{X}_p^{[q+1]} = \widetilde{X}_{p+1}^{[q+1]}, \tag{3.37}$$

for certain uniquely defined functions $\bar{f}_1, \ldots, \bar{f}_p \in C^\infty(N^{q+1})$. Then, the left-hand side is a diagonal prolongation, and, since $\widetilde{X}_1^{[q]}, \ldots, \widetilde{X}_p^{[q]}$ are linearly independent at a generic point by assumption, Lemma 3.11 ensures that $\bar{f}_1, \ldots, \bar{f}_p$, are constant. Projecting (3.37) onto N via pr_0, one has that $X_1, \ldots, X_{p+1}$, are linearly dependent over $\mathbb{R}$. This is a contradiction and, if $\sigma(q) < r$, then $\sigma(q+1) > \sigma(q)$. Finally, one obtains $\sigma(q_0) = r$ for a certain q_0. $\qquad\square$

Let us now turn to describing a geometric interpretation of the superposition rule notion.

Consider a t-dependent vector field $X = \sum_{i=1}^n X^i(t, x)\partial/\partial x^i$ on N associated with the system

$$\frac{dx^i}{dt} = X^i(t, x), \quad i = 1, \ldots, n, \tag{3.38}$$

that describes the integral curves of X. Recall that the above system admits a superposition rule if there exists a map $\Phi : N^m \times N \to N$ given by $x = \Phi(x_{(1)}, \ldots, x_{(m)}; k_1, \ldots, k_n)$ such that the general solution, $x(t)$, of (3.38) can be written as

$$x(t) = \Phi(x_{(1)}(t), \ldots, x_{(m)}(t); k),$$

where $x_{(1)}(t), \ldots, x_{(m)}(t)$ are a generic family of particular solutions and k is an element of N associated univocally with a particular solution of X.

The map $\Phi(x_{(1)}, \ldots, x_{(m)}; \cdot) : N \to N$ can be inverted, at least locally around points of an open dense subset of N^m, to give rise to a map $\Psi : N^m \times N \to N$,

$$k = \Psi(x_{(1)}, \ldots, x_{(m)}, x_{(0)}).$$

Note that the map Ψ is defined so that

$$k = \Psi(x_{(1)}, \ldots, x_{(m)}, \Phi(x_{(1)}, \ldots, x_{(m)}; k)).$$

Hence, Ψ defines an n-codimensional foliation on N^{m+1}.

The fundamental property of Ψ ensures that

$$k = \Psi(x_{(1)}(t), \ldots, x_{(m)}(t), x_{(0)}(t)), \tag{3.39}$$

for any $(m+1)$-tuple of generic particular solutions $x_{(0)}(t), \ldots,$ $x_{(m)}(t)$ of the system (3.38). Therefore, the foliation determined by the level sets of Ψ is invariant under permutations of its $(m+1)$ arguments, $x_{(0)}, \ldots, x_{(m)}$.

Furthermore, the differentiation of (3.39) relative to t gives

$$\sum_{a=0}^{m} \sum_{j=1}^{n} X^j(t, x_{(a)}(t)) \frac{\partial \Psi^l}{\partial x_{(a)}^j} (\bar{p}(t)) = \widetilde{X}_t^{[m+1]} \Psi^l(\bar{p}(t)) = 0,$$

$$l = 1, \ldots, n,$$

where $(\Psi^1, \ldots, \Psi^n) = \Psi$ and $\bar{p}(t) = (x_{(0)}(t), \ldots, x_{(m)}(t))$. Thus, $\Psi^1, \ldots, \Psi^n$ are first integrals for the vector fields $\{\widetilde{X}_t^{[m+1]}\}_{t \in \mathbb{R}}$ defining an n-codimensional foliation $\mathfrak{F}$ on N^{m+1} so that the $\{\widetilde{X}_t^{[m+1]}\}_{t \in \mathbb{R}}$ are tangent to its leaves.

Let us prove another property of the foliation $\mathfrak{F}$ given in (3.20). For each leaf $\mathfrak{F}_k = \Psi^{-1}(k)$ and point $(x_{(1)}, \ldots, x_{(m)}) \in N^m$, there exists a unique point $(x_{(0)}, x_{(1)}, \ldots, x_{(m)}) \in \mathfrak{F}_k$, i.e.,

$$(\Phi(x_{(1)}, \ldots, x_{(m)}; k), x_{(1)}, \ldots, x_{(m)}) \in \mathfrak{F}_k.$$

Consequently, the restriction pr of the projection by (3.1) to each $\mathfrak{F}_k$, namely $\mathrm{pr}|_{\mathfrak{F}_k}$, induces a diffeomorphism between N^m and $\mathfrak{F}_k$. In other words, $\mathfrak{F}$ is horizontal relative to pr. Furthermore, $\mathrm{pr}_*|_{T\mathfrak{F}_k}$ leads to a linear isomorphism among vector fields on N^m and the vector fields tangent to a leaf, namely the 'horizontal' vector fields relative to a connection ∇. Therefore, the foliation $\mathfrak{F}$ corresponds to a connection ∇ on the bundle $\mathrm{pr} : N^{m+1} \to N^m$ with zero curvature.

Note that ∇ determines the superposition rule without referring to the map Ψ. If we fix an $x_{(0)}(0) \in N$ and m particular solutions, $x_{(1)}(t), \ldots, x_{(m)}(t)$, then $x_{(0)}(t)$ is the unique point in N such that $(x_{(0)}(t), x_{(1)}(t), \ldots, x_{(m)}(t))$ belongs to the same leaf as

$(x_{(0)}(0), x_{(1)}(0), \ldots, x_{(m)}(0))$. Thus, $\mathfrak{F}$ determines the superposition rule and vice versa.

On the other hand, given a connection ∇ on the bundle

$$\mathrm{pr} : N^{m+1} \to N^m,$$

with zero curvature, i.e., a horizontal distribution ∇ on N^{m+1} that it is involutive and can be integrated to give a foliation on N^{m+1}, such that the vector fields $\widetilde{X}_t^{[m+1]}$ are tangent to the leaves of ∇, one has that the procedure described above determines a superposition rule for the system (3.38). In fact, the leaves $\mathfrak{F}_k$ of the foliation $\mathfrak{F}$ can be enumerated by elements $k \in N$. Hence, $\Phi(x_{(1)}, \ldots, x_{(m)}; k) \in N$ can be defined as the unique point $x_{(0)}$ of N such that

$$(x_{(0)}, x_{(1)}, \ldots, x_{(m)}) \in \mathfrak{F}_k.$$

This gives rise to a superposition rule $\Phi : N^m \times N \to N$ for the system of first-order differential equations (3.38). Indeed, the inverse relation

$$\Psi(x_{(0)}, \ldots, x_{(m)}) = k,$$

amounts to $(x_{(0)}, \ldots, x_{(m)}) \in \mathfrak{F}_k$. If we fix k and take a generic family of particular solutions $x_{(1)}(t), \ldots, x_{(m)}(t)$ of Equation (3.38), then $x_{(0)}(t)$, defined by the condition $\Psi(x_{(0)}(t), \ldots, x_{(m)}(t)) = k$ satisfies (3.38). In fact, let $x'_{(0)}(t)$ be the solution of (3.38) with initial value $x'_{(0)} = x_{(0)}$. Since the t-dependent vector fields $\widetilde{X}(t, x)$ are tangent to $\mathfrak{F}$, the curve $(x_{(0)}(t), x_{(1)}(t), \ldots, x_{(m)}(t))$ that lies entirely within a leaf of $\mathfrak{F}$, so in $\mathfrak{F}_k$. Since a point of a leaf is entirely determined by its projection by pr, one has that $x'_{(0)}(t) = x_{(0)}(t)$ and $x_{(0)}(t)$ is a solution.

Proposition 3.13. *Giving a superposition rule depending on m generic particular solutions for a Lie system described by a t-dependent vector field X is equivalent to a zero curvature connection ∇ on the bundle $\mathrm{pr} : N^{m+1} \to N^m$ for which the vector fields $\{\widetilde{X}_t^{[m+1]}\}_{t \in \mathbb{R}}$ are horizontal vector fields with respect to this connection.*

Although we refuse to investigate in detail the difference between global superposition rules and superposition rules, it is interesting to comment briefly on this issue now. A rigorous inspection of the above discussion shows that a global or "local" superposition rule leads to a zero curvature connection. Meanwhile, a zero curvature connection *only* ensures the existence of a local superposition rule. This is due to the connection, which only guarantees the existence of a series of *local* first integrals that give rise to a superposition rule locally. To ensure the existence of a global superposition rule, some extra conditions on the connection must be required as well to ensure the foliation to be globally defined (see [52]).

3.8. Geometric Lie–Scheffers Theorem

This section provides a geometric proof of the Lie–Scheffers theorem [276, Theorem 44]. A similar approach, with less details, can be found in [93, Theorem 1].

Main Theorem 1 (The Geometric Lie–Scheffers Theorem).
A system (3.38) *on N admits a superposition rule depending on m generic particular solutions if and only if its associated t-dependent vector field X can be brought into the form*

$$X = \sum_{\alpha=1}^{r} b_\alpha(t) X_\alpha, \tag{3.40}$$

where the vector fields, $X_1, \ldots, X_r$, form a basis for an r-dimensional real Lie algebra and $m \cdot \dim N \geq r$.

Proof. Suppose that system (3.38) admits a superposition rule (3.9) and let $\mathfrak{F}$ be its associated foliation on N^{m+1}. Since the vector fields $\{\widetilde{X}_t^{[m+1]}\}_{t\in\mathbb{R}}$ are tangent to the leaves of $\mathfrak{F}$, the vector fields of $\mathrm{Lie}(\{\widetilde{X}_t^{[m+1]}\}_{t\in\mathbb{R}})$ span a generalized involutive distribution

$$\mathcal{D}_p := \left\{ Y_p \,|\, Y \in \mathrm{Lie}(\{\widetilde{X}_t^{[m+1]}\}_{t\in\mathbb{R}}) \right\} \in \mathrm{T}_p N^{m+1},$$

whose elements are also tangent to the leaves of $\mathfrak{F}$. Since the Lie bracket of two diagonal prolongations is a diagonal prolongation, we can choose, among the elements of $\mathrm{Lie}(\{\widetilde{X}_t^{[m+1]}\}_{t\in\mathbb{R}})$, a finite family $\widetilde{X}_1^{[m+1]},\ldots,\widetilde{X}_r^{[m+1]}$ that gives rise to a local basis of diagonal prolongations or the distribution $\mathcal{D}$. Hence,

$$[\widetilde{X}_\alpha^{[m+1]},\widetilde{X}_\beta^{[m+1]}] = \sum_{\gamma=1}^{r} f_{\alpha\beta\gamma}\widetilde{X}_\gamma^{[m+1]}, \quad \alpha,\beta=1,\ldots,r,$$

for certain functions $f_{\alpha\beta\gamma} \in C^\infty(N^{m+1})$. Meanwhile, the map pr projects each leaf of the foliation $\mathfrak{F}$ onto N^m diffeomorphically and, therefore, the vector fields $\mathrm{pr}_*(\widetilde{X}_\alpha^{[m+1]})$, with $\alpha=1,\ldots,r$, are linearly independent at a generic point of N^m. In view of Lemma 3.11, the functions $f_{\alpha\beta\gamma}$ must be constant, namely $f_{\alpha\beta\gamma}=c_{\alpha\beta\gamma}$ for $\alpha,\beta,\gamma=1,\ldots,r$, and, taking into account the properties of diagonal prolongations, $X_1,\ldots,X_r$ are linearly independent vector fields obeying the relations

$$[X_\alpha,X_\beta] = \sum_{\gamma=1}^{r} c_{\alpha\beta\gamma}X_\gamma, \quad \alpha,\beta=1,\ldots,r.$$

In other words, the vector fields $X_1,\ldots,X_r$ span a finite-dimensional Lie algebra. Since each $\widetilde{X}_t^{[m+1]}$ is spanned by the vector fields $\widetilde{X}_1^{[m+1]},\ldots,\widetilde{X}_r^{[m+1]}$, there are t-dependent functions $b_\alpha \in C^\infty(\mathbb{R} \times N^{m+1})$, with $\alpha=1,\ldots,r$, such that

$$\widetilde{X}^{[m+1]} = \sum_{\alpha=1}^{r} b_\alpha(t)\widetilde{X}_\alpha^{[m+1]}.$$

Since $\widetilde{X}_t^{[m+1]}$ is a diagonal prolongation, Lemma 3.11 implies that the functions $b_1,\ldots,b_r$ only depend on time and

$$\widetilde{X}_t^{[m]} = \sum_{\alpha=1}^{r} b_\alpha(t)\widetilde{X}_\alpha^{[m]}. \tag{3.41}$$

From here, it is immediate that (3.40) holds.

To prove the converse part of this theorem, we assume that X can be brought in the form (3.40) and $X_1,\ldots,X_r$ are linearly independent on $\mathbb{R}$. Hence, Proposition 3.12 ensures that there exists a smallest number $m \leq r$, so that the diagonal prolongations of $X_1,\ldots,X_r$

to N^m are linearly independent at a generic point, and thus $r \leq n \cdot m$. Furthermore, the diagonal prolongations, $\widetilde{X}_1^{[m+1]}, \ldots, \widetilde{X}_r^{[m+1]}$, of $X_1, \ldots, X_r$ to N^{m+1} are linearly independent at every point, forming the basis for an involutive distribution $\mathcal{D}$. This distribution gives rise to a $(n(m+1) - r)$-codimensional foliation $\mathfrak{F}_0$ on N^{m+1}. Since $\mathfrak{F}_0$ is at least n-codimensional, one can consider a n-codimensional foliation $\mathfrak{F}$ whose leaves include the ones of $\mathfrak{F}_0$. The leaves of $\mathcal{F}$ project onto the last $m \cdot n$ factors diffeomorphically, and they are at least n-codimensional. By Proposition 3.13, the foliation $\mathfrak{F}$ leads to a superposition rule depending on m particular solutions. $\qquad \square$

It is worth noting that the converse part of the previous proof shows that systems described by t-dependent vector fields of the form (3.41) give rise to the same distribution $\mathcal{D}$ over the same space N^{m+1}. Consequently, they share a common superposition rule.

Note that if $\mathcal{D}$ has rank $m \dim N$, then, the leaves of the distribution $\mathcal{F}_0$ are n-dimensional, which gives a superposition rule. If $\mathcal{D}$ has a rank smaller than mn, then the leaves of $\mathcal{F}_0$ will have a higher dimension than n, and we will have to look for an additional foliation $\mathcal{F}$ whose leaves will be n-dimensional. This can be done in several non-equivalent ways, which will give rise to different superposition rules. Summarizing, it can be proven that the superposition rule for a system X depending on m particular solutions induced by a VG Lie algebra V is unique if and only if $m \dim N = r$.

3.9. Determination of Superposition Rules

There are several methods to obtain superposition rules for Lie systems (see [375, p. 107] and [14, Section 3]). Nevertheless, the geometric proof of the Lie–Scheffers Theorem provides, implicitly, a method to obtain them. As this will be used in this book frequently, we will now detail the procedure.

Let X be a Lie system on N. The Lie–Scheffers theorem shows that

$$X(t, x) = \sum_{i=1}^{n} \sum_{\alpha=1}^{r} b_\alpha(t) X_\alpha^i(x) \frac{\partial}{\partial x^i},$$

for certain vector fields $X_\alpha(x) = \sum_{i=1}^n X_\alpha^i(x)\partial/\partial x^i$, with $\alpha = 1,\ldots,r$, on N spanning an r-dimensional VG Lie algebra V. The geometric proof of Lie–Scheffers theorem allows us to use the previous decomposition to derive a superposition rule depending on m generic particular solutions and that satisfy the Lie's condition, namely $\dim V \leq m \cdot \dim N$. More exactly, the number m coincides with the minimal integer that makes the diagonal prolongations of $X_1,\ldots,X_r$, to N^m linearly independent at a generic point, namely if $f_1,\ldots,f_r \in C^\infty(N^m)$ are such that

$$\sum_{\alpha=1}^r f_\alpha \widetilde{X}_\alpha^{[m]} = 0, \tag{3.42}$$

then $f_1 = \cdots = f_r = 0$.

The procedure to obtain a superposition rule goes as follows (see [93, 107] for details and examples).

1. Choose the smallest positive integer m, so that the diagonal prolongations of $X_1,\ldots,X_r$ to N^m are linearly independent at a generic point.
2. Take local coordinates $x^1,\ldots,x^n$ on N. By defining this coordinate system on each copy of N within N^{m+1}, we get a coordinate system $\{x_{(a)}^i \mid i = 1,\ldots,n,\ a = 0,\ldots,m\}$ on N^{m+1}. Obtain n functionally independent first integrals $F_1,\ldots,F_n$ common to all diagonal prolongations $\widetilde{X}_1^{[m+1]},\ldots,\widetilde{X}_r^{[m+1]}$ of $X_1,\ldots,X_r$ to N^{m+1} such that

$$\frac{\partial(F_1,\ldots,F_n)}{\partial(x_{(0)}^1,\ldots,x_{(0)}^n)} \neq 0. \tag{3.43}$$

This can be performed, for instance, via the well-known *method of characteristics*.
3. Consider the equations $F_i = k_i$, for $i = 1,\ldots,n$. The condition (3.43) allows us to ensure that previous equations enable us to write the expressions of the variables $x_{(0)}^1,\ldots,x_{(0)}^n$ in terms of $x_{(a)}^1,\ldots,x_{(a)}^n$, with $a = 1,\ldots,m$, and $k_1,\ldots,k_n$.
4. The obtained expressions give rise to a superposition rule in terms of any generic family of m particular solutions and the

constants $k_1, \ldots, k_n$. It is worth stressing that if the vector fields, $\widetilde{X}_1^{[m+1]}, \ldots, \widetilde{X}_r^{[m+1]}$, on N^{m+1} span a distribution of codimension larger than n, then the system X admits more than one superposition rule (see [93]).

Let us illustrate our comments through a simple example. Consider the Riccati equation

$$\frac{dx}{dt} = b_1(t) + b_2(t)\,x + b_3(t)x^2,$$

for arbitrary functions $b_1(t), b_2(t), b_3(t)$. This Riccati equation describes the integral curves of the t-dependent vector field

$$X = (b_1(t) + b_2(t)x + b_3(t)x^2)\frac{\partial}{\partial x}.$$

Recall that the vector fields $\{X_t\}_{t\in\mathbb{R}}$ take values in the three-dimensional Lie algebra V spanned by the vector fields

$$X_1 = \frac{\partial}{\partial x}, \quad X_2 = x\frac{\partial}{\partial x}, \quad X_3 = x^2\frac{\partial}{\partial x}.$$

Now, the number of particular solutions of a superposition rule for Riccati equations is given by the smallest m such that system (3.42) only admits the trivial solution. For $m = 2$, this system reads

$$f_1\left(\frac{\partial}{\partial x_{(1)}} + \frac{\partial}{\partial x_{(2)}}\right) + f_2\left(x_{(1)}\frac{\partial}{\partial x_{(1)}} + x_{(2)}\frac{\partial}{\partial x_{(2)}}\right)$$
$$+ f_3\left(x_{(1)}^2\frac{\partial}{\partial x_{(1)}} + x_{(2)}^2\frac{\partial}{\partial x_{(2)}}\right) = 0,$$

and then

$$f_1 + f_2 x_{(1)} + f_3 x_{(1)}^2 = 0, \quad f_1 + f_2 x_{(2)} + f_3 x_{(2)}^2 = 0,$$

and it has non-trivial solutions. Nevertheless, the system for the diagonal prolongations to $\mathbb{R}^3$, that is,

$$f_1 + f_2 x_{(1)} + f_3 x_{(1)}^2 = 0, \quad f_1 + f_2 x_{(2)} + f_3 x_{(2)}^2 = 0,$$
$$f_1 + f_2 x_{(3)} + f_3 x_{(3)}^2 = 0,$$

does not admit any non-trivial solution in the f_1, f_2, f_3 because the determinant of the coefficients, i.e.,

$$\left| \begin{pmatrix} 1 & x_{(1)} & x_{(1)}^2 \\ 1 & x_{(2)} & x_{(2)}^2 \\ 1 & x_{(3)} & x_{(3)}^2 \end{pmatrix} \right| = (x_{(2)} - x_{(1)})(x_{(2)} - x_{(3)})(x_{(1)} - x_{(3)}),$$

is different from zero on the dense subset of $\mathbb{R}^3$ of the points $(x_{(1)}, x_{(2)}, x_{(3)})$ such that $x_{(1)}$, $x_{(2)}$, and $x_{(3)}$ are different. Thus, we get that $m = 3$ and $n = 1$, which implies that we can obtain a superpoition rule for the Riccati equation depending on three particular solutions.

Once m has been determined, the superposition rule for Riccati equations can be derived in terms of first integrals for the diagonal prolongations, $\widetilde{X}_1^{[4]}, \widetilde{X}_2^{[4]}, \widetilde{X}_3^{[4]}$, on $\mathbb{R}^4$.

To obtain the integrals of the diagonal prolongations $\widetilde{X}_1^{[4]}, \widetilde{X}_2^{[4]}, \widetilde{X}_3^{[4]}$, on $\mathbb{R}^4$, we can use the method of characteristics. For instance, given the vector field $\widetilde{X}_1^{[4]}$, the characteristic system for $\widetilde{X}_1^{[4]}$ takes the form

$$dx_{(1)} = dx_{(2)} = dx_{(3)} = dx_{(4)}$$
$$\Longrightarrow \Upsilon_1 = x_{(1)} - x_{(4)}, \Upsilon_2 = x_{(2)} - x_{(4)}, \Upsilon_3 = x_{(3)} - x_{(4)},$$

and $\Upsilon_1, \Upsilon_2, \Upsilon_3$ are first integrals for $\bar{X}_1$. Hence, every common first integral for $\widetilde{X}_1^{[4]}, \widetilde{X}_2^{[4]}, \widetilde{X}_3^{[4]}$, must be a function $F = F(\Upsilon_1, \Upsilon_2, \Upsilon_3)$. To obtain the exact form of F, we have to impose, in particular, $\widetilde{X}_3^{[4]} F = 0$. Writing $\widetilde{X}_3^{[4]}$ in coordinates $\Upsilon_1, \Upsilon_2, \Upsilon_3, x_{(4)}$, we obtain

$$\Upsilon_1 \frac{\partial F}{\partial \Upsilon_1} + \Upsilon_2 \frac{\partial F}{\partial \Upsilon_2} + \Upsilon_3 \frac{\partial F}{\partial \Upsilon_3} + 2x_4 \left(\Upsilon_1^2 \frac{\partial F}{\partial \Upsilon_1} + \Upsilon_2^2 \frac{\partial F}{\partial \Upsilon_2} + \Upsilon_3^2 \frac{\partial F}{\partial \Upsilon_3} \right) = 0,$$

for every x_4. Hence,

$$\Upsilon_1 \frac{\partial F}{\partial \Upsilon_1} + \Upsilon_2 \frac{\partial F}{\partial \Upsilon_2} + \Upsilon_3 \frac{\partial F}{\partial \Upsilon_3} = 0.$$

Using the characteristic method, we obtain that F must be a function of $\Xi_1 = \Upsilon_1/\Upsilon_3$ and $\Xi_2 = \Upsilon_2/\Upsilon_3$. This suggests us to use coordinates

$\Xi_1, \Xi_2, \Upsilon_3, x_{(4)}$ to describe the condition

$$\Upsilon_1^2 \frac{\partial F}{\partial \Upsilon_1} + \Upsilon_2^2 \frac{\partial F}{\partial \Upsilon_2} + \Upsilon_3^2 \frac{\partial F}{\partial \Upsilon_3} = 0,$$

using that $F = F(\Xi_1, \Xi_2)$. Hence, one obtains

$$(\Xi_1^2 - \Xi_1) \frac{\partial}{\partial \Xi_1} + (\Xi_2^2 - \Xi_2) \frac{\partial}{\partial \Xi_2} = 0 \Rightarrow \frac{d\Xi_1}{\Xi_1^2 - \Xi_1} = \frac{d\Xi_2}{\Xi_2^2 - \Xi_2}.$$

Therefore, $F = F(\Xi_2(\Xi_1 - 1)/[\Xi_1(\Xi_2 - 1)])$ and

$$k = \frac{(x_{(2)} - x_{(4)})(x_{(1)} - x_{(3)})}{(x_{(1)} - x_{(4)})(x_{(2)} - x_{(3)})}.$$

Writing x_1 as a function of the remaining variables, we obtain the superposition rule for Riccati equations.

3.10. Superposition Rules for Second- and Higher-Order ODEs

Lie systems can also be applied to investigating systems of second- and higher-order differential equations, e.g., the so-called SODE and HODE Lie systems [87, 108, 329]. Therefore, one can apply the theory of Lie systems to derive t-dependent and t-independent constants of motion, exact solutions, superposition rules, and other properties for these differential equations. This section addresses the theory of SODE and HODE Lie systems.

Vessiot initiated the analysis of systems of second-order differential equations via Lie systems [365]. Much later, Chisholm and Common, and mainly Winternitz, revived the topic [140, 329]. These previous works were mainly concerned with the study of relevant physical systems determined by second-order differential equations, e.g., Milne–Pinney equations. Meanwhile, the theoretical properties of these systems were left aside. Nevertheless, the theoretical study of second- and higher-order differential equations was accomplished in the 21st century (see [108, 314] and references there in). This allowed for retrieving previous *ad hoc* results from a new geometric point of view [210, 320].

The theory of Lie systems studies systems of first-order differential equations admitting its general solution as families of particular solutions and a set of constants. This property applies to more general differential equations than first-order systems of differential equations. For instance, each second-order differential equation of the form

$$\ddot{x} = a(t)x, \quad x \in \mathbb{R}, \tag{3.44}$$

where $a(t)$ is any t-dependent real function, satisfies that its general solution, $x(t)$, can be written as

$$x(t) = k_1 x_{(1)}(t) + k_2 x_{(2)}(t), \tag{3.45}$$

where k_1, k_2 are arbitrary constants and $x_{(1)}(t), x_{(2)}(t)$ are two particular solutions such that, for each fixed valued $t \in \mathbb{R}$, we have that $(x_{(1)}(t), \dot{x}_{(1)}(t)), (x_{(2)}(t), \dot{x}_{(2)}(t))$ are linearly independent when considered as vectors in $\mathbb{R}^2$.

Superposition rules for SODEs do not only apply to linear second-order systems. For instance, a change of variables $y = 1/x$ transforms (3.44) into

$$y\ddot{y} - 2\dot{y}^2 = -a(t)y^2, \tag{3.46}$$

which, as a consequence of the linear superposition rule (3.45), admits its general solution in the form

$$y(t) = \left(k_1 y_1^{-1}(t) + k_2 y_2^{-1}(t)\right)^{-1}, \tag{3.47}$$

in terms of two generic particular solutions, $y_{(1)}(t), y_{(2)}(t)$, of (3.45). More specifically, $y_{(1)}(t), y_{(2)}(t)$, are particular solutions of (3.46) such that $1/y_{(1)}(t), 1/y_{(2)}(t)$ form a family of admissible particular solutions for the superposition rule (3.45).

In view of the previous examples and many other ones that can be found in the literature [85, 101], it is natural to define superposition rules for second-order differential equations as follows.

Definition 3.14. A second-order differential equation on N of the form

$$\ddot{x}^i = F^i(t, x, \dot{x}), \quad i = 1, \ldots, n, \tag{3.48}$$

admits a global superposition rule if there exists a map $\Psi : (TN)^m \times TN \to N$ such that its general solution $x(t)$ can be written as

$$x(t) = \Psi(x_{(1)}(t), \ldots, x_{(m)}(t), \dot{x}_{(1)}(t), \ldots, \dot{x}_{(m)}(t); k), \tag{3.49}$$

in terms of a generic family, $x_{(1)}(t), \ldots, x_{(m)}(t)$, of particular solutions, their derivatives, and a point $k \in TN$.

As in the case of standard superposition rules, "generic" means that expression (3.49) allows us to recover particular solutions, $x(t)$, with initial conditions on an open subset of N for families of particular solutions, $x_1(t), \ldots, x_m(t)$, satisfying that $(x_1(0), \dot{x}_1(0), \ldots, x_m(0), \dot{x}_m(0)) \in U$, where U is an open dense subset of $(TN)^m$.

There exists no characterization of systems of SODEs of the form (3.48) that admits a superposition rule (cf. [107]). Nevertheless, there exists a special class of such systems, the so-called *SODE Lie systems* [116], admitting superposition rules [110]. We next furnish the definition of SODE Lie system with a proof that shows that every SODE Lie system admits a superposition rule. We are going to discuss further properties and the interest of this notion.

Definition 3.15. The system of second-order differential equations (3.48) is a SODE Lie system if the system of first-order differential equations

$$\begin{cases} \dot{x}^i = v^i, \\ \dot{v}^i = F^i(t, x, v), \end{cases} \quad i = 1, \ldots, n, \tag{3.50}$$

obtained from (3.48) by defining the new variables $v^i = \dot{x}^i$, with $i = 1, \ldots, n$, is a Lie system.

Proposition 3.16. *Every SODE Lie system* (3.48) *admits a super-position rule* $\Psi : (TN)^m \times TN \to N$ *of the form* $\Psi = \pi \circ \Phi$, *where* $\Phi : (TN)^m \times TN \to TN$ *is a superposition rule for the system* (3.50) *and* $\pi : TN \to N$ *is the projection associated with the tangent bundle* TN.

Proof. Each SODE Lie system (3.48) is associated with a first-order system of differential equations (3.50) admitting a superposition rule $\Phi : (TN)^m \times TN \to TN$. This allows us to describe the general solution $(x(t), v(t))$ of system (3.50) in terms of a generic set $(x_a(t), v_a(t))$, with $a = 1, \ldots, m$, of particular solutions and a parameter $k \in N$, i.e.,

$$(x(t), v(t)) = \Phi(x_1(t), \ldots, x_m(t), v_1(t), \ldots, v_m(t); k). \qquad (3.51)$$

Each solution, $x_p(t)$, of (3.48) can be associated in a univocally manner with one solution $(x_p(t), v_p(t))$ of the system of first-order differential equations (3.50) and vice versa. Furthermore, since one has that $(x_p(t), v_p(t)) = (x_p(t), \dot{x}_p(t))$, the general solution $x(t)$ of (3.48) can be written as

$$x(t) = \pi \circ \Phi(x_1(t), \ldots, x_m(t), \dot{x}_1(t), \ldots, \dot{x}_m(t); k), \qquad (3.52)$$

in terms of a generic family $x_a(t)$, with $a = 1, \ldots, n$, of particular solutions of (3.48). That is, the map $\Psi = \pi \circ \Phi$ is a superposition rule for the system of SODEs (3.48). $\qquad \square$

The above considerations can be extended *mutatis mutandis* to systems of higher-order systems of differential equations. For a detailed survey on these and related topics, we refer the reader to [87, 110].

3.11. New Applications of Lie Systems

Here, we show a few recent applications of Lie systems [121, 280].

3.11.1. The second-order Riccati equation

The most general class of second-order Riccati equations is given by the family of second-order differential equations of the form

$$\frac{d^2x}{dt^2} + (f_0(t) + f_1(t)x)\frac{dx}{dt} + c_0(t) + c_1(t)x + c_2(t)x^2 + c_3(t)x^3 = 0, \tag{3.53}$$

with

$$f_1(t) = 3\sqrt{c_3(t)}, \quad f_0(t) = \frac{c_2(t)}{\sqrt{c_3(t)}} - \frac{1}{2c_3(t)}\frac{dc_3}{dt}(t), \quad c_3(t) > 0, \tag{3.54}$$

and $c_0(t), c_1(t), c_2(t)$ being t-dependent functions. These equations arise by reducing third-order linear differential equations through a dilation symmetry and a t-reparametrization [131]. Their interest is due to their use in the study of several physical and mathematical problems [107, 131, 140, 196].

It was recently discovered that a quite general family of second-order Riccati equations (3.53) admits a t-dependent non-natural regular Lagrangian of the form

$$L(t, x, v) = \frac{1}{v + U(t, x)}, \tag{3.55}$$

with $U(t, x) = a_0(t) + a_1(t)x + a_2(t)x^2$ and $a_0(t), a_1(t), a_2(t)$ being certain functions related to the t-dependent coefficients of (3.53), see [131]. Therefore,

$$p = \frac{\partial L}{\partial v} = \frac{-1}{(v + U(t, x))^2}, \tag{3.56}$$

and the image of the Legendre transform $\mathbb{F}L : (t, x, v) \in \mathcal{W} \subset \mathbb{R} \times T\mathbb{R} \mapsto (t, x, p) \in \mathbb{R} \times T^*\mathbb{R}$, where $\mathcal{W} = \{(t, x, v) \in \mathbb{R} \times T\mathbb{R} \mid v + U(t, x) \neq 0\}$, is the open submanifold $\mathbb{R} \times \mathcal{O}$ where $\mathcal{O} = \{(x, p) \in T^*\mathbb{R} \mid p < 0\}$. The Legendre transform is not injective, as $(t, x, p) = \mathbb{F}L(t, x, v)$ for $v = \pm 1/\sqrt{-p} - U(t, x)$. Nevertheless, it can become so by restricting it to the open $\mathcal{W}_+ = \{(t, x, v) \in \mathbb{R} \times T\mathbb{R} \mid v + $

$U(t,x) > 0\}$. In such a case, $v = 1/\sqrt{-p} - U(t,x)$ and we can define over $\mathbb{R} \times \mathcal{O}$ the t-dependent Hamiltonian

$$h(t,x,p) = p\left(\frac{1}{\sqrt{-p}} - U(t,x)\right) - \sqrt{-p} = -2\sqrt{-p} - pU(t,x).$$

$$(3.57)$$

Its Hamilton equations read

$$\begin{cases} \dfrac{dx}{dt} = \dfrac{\partial h}{\partial p} = \dfrac{1}{\sqrt{-p}} - U(t,x) = \dfrac{1}{\sqrt{-p}} - a_0(t) - a_1(t)x - a_2(t)x^2, \\ \dfrac{dp}{dt} = -\dfrac{\partial h}{\partial x} = p\dfrac{\partial U}{\partial x}(t,x) = p(a_1(t) + 2a_2(t)x). \end{cases}$$

$$(3.58)$$

Since the general solution $x(t)$ of a second-order Riccati equation (3.53) can be recovered from the general solution $(x(t), p(t))$ of its corresponding system (3.58), the analysis of the latter provides information about general solutions of such second-order Riccati equations.

The important point now is that system (3.58) is a Lie system. Indeed, consider the vector fields on $\mathcal{O}$ of the form

$$X_1 = \frac{1}{\sqrt{-p}}\frac{\partial}{\partial x}, \quad X_2 = \frac{\partial}{\partial x}, \quad X_3 = x\frac{\partial}{\partial x} - p\frac{\partial}{\partial p},$$

$$X_4 = x^2\frac{\partial}{\partial x} - 2xp\frac{\partial}{\partial p}, \quad X_5 = \frac{x}{\sqrt{-p}}\frac{\partial}{\partial x} + 2\sqrt{-p}\frac{\partial}{\partial p}. \quad (3.59)$$

Their non-vanishing commutation relations read

$$[X_1, X_3] = \frac{1}{2}X_1, \quad [X_1, X_4] = X_5, \quad [X_2, X_3] = X_2,$$

$$[X_2, X_4] = 2X_3, \quad [X_2, X_5] = X_1, \quad [X_3, X_4] = X_4, \quad [X_3, X_5] = \frac{1}{2}X_5,$$

$$(3.60)$$

and therefore span a five-dimensional Lie algebra V of vector fields. Additionally, the t-dependent vector field X_t associated with

(3.58) holds

$$X_t = X_1 - a_0(t)X_2 - a_1(t)X_3 - a_2(t)X_4. \tag{3.61}$$

In view of (3.61), system (3.58) is a Lie system. Note also that a similar result would have been obtained by restricting the Legendre transform over the open $\mathcal{W}_- = \{(t, x, v) \in \mathbb{R} \times T\mathbb{R} \mid v + U(t, x) < 0\}$.

Let us use the theory of Lie systems to reduce the integration of (3.58) to solving a Lie system on a Lie group. Using a Levi decomposition of V, we get $V \simeq V_1 \oplus_s V_2$, with $V_2 = \langle X_2, X_3, X_4 \rangle$ being a semisimple Lie algebra isomorphic to $\mathfrak{sl}(2, \mathbb{R})$ and $V_1 = \langle X_1, X_5 \rangle$ being the radical of V. Hence, V is isomorphic to the Lie algebra of a Lie group $G := \mathbb{R}^2 \rtimes SL(2, \mathbb{R})$, where $\rtimes$ denotes a semidirect product of $\mathbb{R}^2$ by $SL(2, \mathbb{R})$, and there exists a local action $\Phi : G \times \mathcal{O} \to \mathcal{O}$ whose fundamental vector fields are those of V. It is a long but simple computation to show that

$$\Phi\left(\left((\lambda_1, \lambda_5), \begin{pmatrix} \alpha & \beta \\ \gamma & \delta \end{pmatrix}\right), (x, p)\right) = \left(\frac{\sqrt{-\bar{p}}\bar{x} - \lambda_1}{\sqrt{-\bar{p}} + \lambda_5}, -(\sqrt{-\bar{p}} + \lambda_5)^2\right), \tag{3.62}$$

where $\bar{x} = (\alpha x + \beta)/(\gamma x + \delta)$, $\bar{p} = p(\gamma x + \delta)^2$ and $\alpha\delta - \beta\gamma = 1$, is one of such actions (for a detailed example of how to derive these actions see [108, Chapter 2] and Section 3.9).

The above action enables us to write the general solution $\xi(t)$ of system (3.58) in the form $\xi(t) = \Phi(g(t), \xi_0)$, where $\xi_0 \in N$ and $g(t)$ is a particular solution of

$$\frac{dg}{dt} = -\left(X_1^R(g) - a_0(t)X_2^R(g) - a_1(t)X_3^R(g) - a_2(t)X_4^R(g)\right),$$
$$g(0) = e, \tag{3.63}$$

on G, with the X_α^R being a family of right-invariant vector fields on G whose vectors $X_\alpha^R(e) \in T_eG$ close on the opposite commutation relations of the vector fields X_α (cf. [108]).

We now turn to apply the theory of reduction for Lie systems. Since $T_eG \simeq \mathbb{R}^2 \oplus_s \mathfrak{sl}(2, \mathbb{R})$, a particular solution of a Lie system of

Basics on Lie Systems and Superposition Rules 103

the form (3.63) but on $SL(2, \mathbb{R})$, which amounts to integrating (first-order) Riccati equations (cf. [12, 108]), provides us with a transformation which maps system (3.63) into an easily integrable Lie system on $\mathbb{R}^2$. In short, the explicit determination of the general solution of a second-order Riccati equation reduces to solving Riccati equations.

Another way of analyzing the solutions of (3.58) is based on the determination of a superposition rule. According to the method sketched in Section 3.9, a superposition rule for a Lie system (3.58), which admits a decomposition of the form (3.61), can be obtained through two common functionally independent first integrals I_1, I_2, for the diagonal prolongations $\widetilde{X}_1, \widetilde{X}_2, \widetilde{X}_3, \widetilde{X}_4, \widetilde{X}_5$ to a certain $\mathrm{T}^* N^{(m+1)}$ provided their prolongations to $\mathrm{T}^* \mathbb{R}^m$ are linearly independent at a generic point and

$$\frac{\partial(I_1, I_2)}{\partial(x_{(0)}, p_{(0)})} \neq 0. \tag{3.64}$$

In our case, it can easily be verified that $m = 3$. The resulting first integrals, derived through a long but easy calculation (see [107] for a similar procedure), read

$$\begin{aligned}
F_0 &= \left(x_{(2)} - x_{(3)}\right)\sqrt{p_{(2)}p_{(3)}} + \left(x_{(3)} - x_{(1)}\right)\sqrt{p_{(3)}p_{(1)}} \\
&\quad + \left(x_{(1)} - x_{(2)}\right)\sqrt{p_{(1)}p_{(2)}}, \\
F_1 &= \left(x_{(1)} - x_{(2)}\right)\sqrt{p_{(1)}p_{(2)}} + \left(x_{(2)} - x_{(0)}\right)\sqrt{p_{(2)}p_{(0)}} \\
&\quad + \left(x_{(0)} - x_{(1)}\right)\sqrt{p_{(0)}p_{(1)}}, \\
F_2 &= \left(x_{(1)} - x_{(3)}\right)\sqrt{p_{(1)}p_{(3)}} + \left(x_{(3)} - x_{(0)}\right)\sqrt{p_{(3)}p_{(0)}} \\
&\quad + \left(x_{(0)} - x_{(1)}\right)\sqrt{p_{(0)}p_{(1)}}. \tag{3.65}
\end{aligned}$$

Note that given a family of solutions $(x_{(i)}(t), p_{(i)}(t))$, with $i = 0, \ldots, 3$, of (3.58), then $d\bar{F}_j/dt = \widetilde{X}_t F_j = 0$ for $j = 0, 1, 2$ and

$$\bar{F}_j = F_j(x_{(0)}(t), p_{(0)}(t), \ldots, x_{(3)}(t), p_{(3)}(t)).$$

In order to derive a superposition rule, we just need to obtain the value of $p_{(0)}$ from the equation $k_1 = F_1$, where k_1 is a real constant, to get

$$\sqrt{-p_{(0)}} = \frac{k_1 + (x_{(2)} - x_{(1)})\sqrt{p_{(1)}p_{(2)}}}{(x_{(2)} - x_{(0)})\sqrt{-p_{(2)}} + (x_{(0)} - x_{(1)})\sqrt{-p_{(1)}}}, \qquad (3.66)$$

and then plug this value into the equation $k_2 = F_2$ to have

$$x_{(0)} = \frac{k_1\Gamma(x_{(1)}, p_{(1)}, x_{(3)}, p_{(3)}) + k_2\Gamma(x_{(2)}, p_{(2)}, x_{(1)}, p_{(1)}) - F_0 x_{(1)}\sqrt{-p_{(1)}}}{k_1(\sqrt{-p_{(1)}} - \sqrt{-p_{(3)}}) + k_2(\sqrt{-p_{(2)}} - \sqrt{-p_{(1)}}) - \sqrt{-p_{(1)}}F_0},$$

$$\begin{aligned} p_{(0)} = -[k_1/F_0(\sqrt{-p_{(3)}} - \sqrt{-p_{(1)}}) \\ + k_2/F_0(\sqrt{-p_{(1)}} - \sqrt{-p_{(2)}}) + \sqrt{-p_{(1)}}]^2, \qquad (3.67) \end{aligned}$$

where $\Gamma(x_{(i)}, p_{(i)}, x_{(j)}, p_{(j)}) = \sqrt{-p_{(i)}}x_{(i)} - \sqrt{-p_{(j)}}x_{(j)}$. The above expressions give us a superposition rule Φ : $(x_{(1)}, p_{(1)}, x_{(2)}, p_{(2)}, x_{(3)}, p_{(3)}; k_1, k_2) \in \mathrm{T}^*\mathbb{R}^3 \times \mathbb{R}^2 \mapsto (x_{(0)}, p_{(0)}) \in \mathrm{T}^*\mathbb{R}$ for system (3.58). Finally, since every $x_{(i)}(t)$ is a particular solution for (3.53), the map $\Upsilon = \tau \circ \Phi$ gives the general solution of second-order Riccati equations in terms of three generic particular solutions $x_{(1)}(t), x_{(2)}(t), x_{(3)}(t)$ of (3.53), the corresponding $p_{(1)}(t), p_{(2)}(t), p_{(3)}(t)$ and two real constants k_1, k_2.

3.11.2. *Kummer–Schwarz equations*

Here, we analyze second- and third-order Kummer–Schwarz equations (KS-2 and KS-3, respectively), [41, 42, 201, 254], with the aid of Lie systems.

The KS-2 equations take the form

$$\frac{d^2x}{dt^2} = \frac{3}{2x}\left(\frac{dx}{dt}\right)^2 - 2c_0x^3 + 2b_1(t)x, \qquad (3.68)$$

with c_0 being a real constant and $b_1(t)$ a t-dependent function. KS-2 equations are a particular case of second-order Gambier equations [87, 201] and appear in the study of cosmological models [294].

Basics on Lie Systems and Superposition Rules 105

In addition, their relations to other differential equations like Milne–Pinney equations [201], make them an alternative approach to the analysis of many physical problems [87, 234, 254].

In the case of KS-3 equations, i.e., we have

$$\frac{d^3 x}{dt^3} = \frac{3}{2} \left(\frac{dx}{dt} \right)^{-1} \left(\frac{d^2 x}{dt^2} \right)^2 - 2c_0(x) \left(\frac{dx}{dt} \right)^3 + 2b_1(t) \frac{dx}{dt}, \qquad (3.69)$$

where $c_0 = c_0(x)$ and $b_1 = b_1(t)$ are arbitrary.

The relevance of KS-3 equations resides in their relation to the Kummer's problem [41, 42, 45], Milne–Pinney [254] and Riccati equations [148, 254]. Such relations can be useful in the interpretation of physical systems through KS-3 equations, e.g., the case of quantum non-equilibrium dynamics of many body systems [192]. Furthermore, KS-3 equations with $c_0 = 0$ can be rewritten as $\{x, t\} = 2b_1(t)$, where $\{x, t\}$ is the *Schwarzian derivative* [255] of the function $x(t)$ relative to t.

KS-2 and KS-3 appear in other related themes [18, 41, 42, 47, 350]. Moreover, there is some interest in studying the particular solutions of KS-2 and KS-3 equations, which have been analyzed in different ways in the literature, e.g., through non-local transformations or in terms of solutions to other differential equations [41, 45, 201, 254]. From a physical viewpoint, KS-2 and KS-3 equations occur in the study of Milne–Pinney equations, Riccati equations, and time-dependent frequency harmonic oscillators [87, 148, 201, 254], which are of certain relevancy in two-body problems [38, 40], Quantum Mechanics [234, 245], Classical Mechanics [294], etc. [310].

We show that KS-2 equations can be studied through two $\mathfrak{sl}(2, \mathbb{R})$-Lie systems [311], i.e., Lie systems that describe the integral curves of a t-dependent vector field taking values in VG Lie algebra isomorphic to $\mathfrak{sl}(2, \mathbb{R})$. This new result slightly generalizes previous findings about these equations [87].

Afterwards, we obtain two Lie group actions whose fundamental vector fields correspond with those of the above-mentioned VG Lie algebras. These actions allow us to prove that the explicit integration

106   *A Guide to Lie Systems with Compatible Geometric Structures*

of KS-2 and KS-3 equations is equivalent to working out a particular solution of a Lie system on $SL(2, \mathbb{R})$. Further, we will see that Riccati and Milne–Pinney equations exhibit similar features.

We show that the knowledge of the general solution of any of the reviewed equations enables us to solve simultaneously any other related one to the same equation on $SL(2, \mathbb{R})$. This fact provides a new powerful and general way of linking solutions of these equations, which were previously known to be related through ad hoc expressions in certain cases [148, 254].

3.11.3. *Second-order Kummer–Schwarz equations*

Consider the first-order system associated with (3.68)

$$\begin{cases} \dfrac{dx}{dt} = v, \\[2mm] \dfrac{dv}{dt} = \dfrac{3}{2}\dfrac{v^2}{x} - 2c_0 x^3 + 2b_1(t)x, \end{cases} \tag{3.70}$$

on $T\mathbb{R}_0$, with $\mathbb{R}_0 = \mathbb{R} - \{0\}$, obtained by adding the new variable $v := dx/dt$ to the KS-2 equation (3.68). This system describes the integral curves of the t-dependent vector field

$$X_t = v\frac{\partial}{\partial x} + \left(\frac{3}{2}\frac{v^2}{x} - 2c_0 x^3 + 2b_1(t)x\right)\frac{\partial}{\partial v} = M_3 + b_1(t)M_1, \tag{3.71}$$

where

$$M_1 = 2x\frac{\partial}{\partial v}, \quad M_2 = x\frac{\partial}{\partial x} + 2v\frac{\partial}{\partial v},$$

$$M_3 = v\frac{\partial}{\partial x} + \left(\frac{3}{2}\frac{v^2}{x} - 2c_0 x^3\right)\frac{\partial}{\partial v}, \tag{3.72}$$

satisfy the commutation relations

$$[M_1, M_3] = 2M_2, \quad [M_1, M_2] = M_1, \quad [M_2, M_3] = M_3. \tag{3.73}$$

These vector fields span a three-dimensional real Lie algebra V of vector fields isomorphic to $\mathfrak{sl}(2, \mathbb{R})$ [87, 108]. Hence, in view of (3.71) and the Lie–Scheffers theorem, X admits a superposition rule and becomes a Lie system associated with a VG Lie algebra isomorphic to $\mathfrak{sl}(2, \mathbb{R})$, i.e., an $\mathfrak{sl}(2, \mathbb{R})$-Lie system.

The knowledge of a Lie group action $\varphi_{2\mathrm{KS}} : G \times T\mathbb{R}_0 \to T\mathbb{R}_0$ whose fundamental vector fields are V and $T_e G \simeq V$ allows us to express the general solution of X, let us say $x(t)$, in the form $x(t) = \Phi(g(t), x_0)$, with x_0 being any element of $\mathcal{O}^2$ in terms of a particular solution of the Lie system

$$\frac{dg}{dt} = -\sum_{\alpha=1}^{3} b_\alpha(t) X_\alpha^R(g), \quad g \in SL(2, \mathbb{R}).$$

For details, we refer the reader to Example 3.4. Let us determine φ_{2KS} in such a way that our procedure can easily be extended to third-order Kummer–Schwarz equations.

Consider the basis of matrices of $\mathfrak{sl}(2, \mathbb{R})$ given by

$$\mathsf{a}_1 = \begin{pmatrix} 0 & 1 \\ 0 & 0 \end{pmatrix}, \quad \mathsf{a}_2 = \frac{1}{2} \begin{pmatrix} -1 & 0 \\ 0 & 1 \end{pmatrix}, \quad \mathsf{a}_3 = \begin{pmatrix} 0 & 0 \\ -1 & 0 \end{pmatrix}, \tag{3.74}$$

satisfying the commutation relations

$$[\mathsf{a}_1, \mathsf{a}_3] = 2\mathsf{a}_2, \quad [\mathsf{a}_1, \mathsf{a}_2] = \mathsf{a}_1, \quad [\mathsf{a}_2, \mathsf{a}_3] = \mathsf{a}_3, \tag{3.75}$$

which match those satisfied by M_1, M_2, and M_3. So, the linear function $\rho : \mathfrak{sl}(2, \mathbb{R}) \to V$ mapping a_α into M_α, with $\alpha = 1, 2, 3$, is a Lie algebra isomorphism. If we consider it as an infinitesimal Lie group action, we can then ensure that there exists a local Lie group action $\varphi_{2\mathrm{KS}} : SL(2, \mathbb{R}) \times T\mathbb{R}_0 \to T\mathbb{R}_0$ obeying the required properties. In particular,

$$\frac{d}{ds} \varphi_{2\mathrm{KS}}(\exp(-s\mathsf{a}_\alpha), \mathbf{t}_x) = M_\alpha(\varphi_{2\mathrm{KS}}(\exp(-s\mathsf{a}_\alpha), \mathbf{t}_x)), \tag{3.76}$$

where $\mathbf{t}_x := (x, v) \in T_x \mathbb{R}_0 \subset T\mathbb{R}_0$, $\alpha = 1, 2, 3$, and $s \in \mathbb{R}$. This condition determines the action on $T\mathbb{R}_0$ of the elements of $SL(2, \mathbb{R})$ of the form $\exp(-s\mathsf{a}_\alpha)$, with $\alpha = 1, 2, 3$ and $s \in \mathbb{R}$. By integrating

M_1 and M_2, we obtain

$$\varphi(\exp_{2KS}(-\lambda_1 a_1), \mathbf{t}_x) = (x, v + 2x\lambda_1),$$

$$\varphi(\exp_{2KS}(-\lambda_2 a_2), \mathbf{t}_x) = \left(xe^{\lambda_2}, ve^{2\lambda_2}\right). \tag{3.77}$$

Observe that M_3 is not defined on $T\mathbb{R}_0$. Hence, its integral curves, let us say $(x(\lambda_3), v(\lambda_3))$, must be fully contained in either $T\mathbb{R}^+$ or $T\mathbb{R}^-$. These integral curves are determined by the system

$$\frac{dx}{d\lambda_3} = v, \quad \frac{dv}{d\lambda_3} = \frac{3}{2}\frac{v^2}{x} - 2c_0 x^3. \tag{3.78}$$

When $v \neq 0$, we obtain

$$\frac{dv^2}{dx} = \frac{3v^2}{x} - 4c_0 x^3 \implies v^2(\lambda_3) = x^3(\lambda_3)\Gamma - 4c_0 x^4(\lambda_3), \tag{3.79}$$

for a real constant Γ. Hence, for each integral curve $(x(\lambda_3), v(\lambda_3))$, we have

$$\Gamma = \frac{v^2(\lambda_3) + 4c_0 x^4(\lambda_3)}{x^3(\lambda_3)}. \tag{3.80}$$

Moreover, $d\Gamma/d\lambda_3 = 0$ not only for solutions of (3.78) with $v(\lambda_3) \neq 0$ for every λ_3, but for any solution of (3.78). Using the above results and (3.78), we see that

$$\frac{dx}{d\lambda_3} = \mathrm{sg}(v)\sqrt{\Gamma x^3 - 4c_0 x^4} \implies x(\lambda_3) = \frac{x(0)}{F_{\lambda_3}(x(0), v(0))}, \tag{3.81}$$

where sg is the well-known *sign function* and

$$F_{\lambda_3}(\mathbf{t}_x) = \left(1 - \frac{v\lambda_3}{2x}\right)^2 + c_0 x^2 \lambda_3^2. \tag{3.82}$$

Now, from (3.81) and taking into account the first equation within (3.78), it immediately follows that

$$\varphi_{2KS}(\exp(-\lambda_3 a_3), \mathbf{t}_x) = \left(\frac{x}{F_{\lambda_3}(\mathbf{t}_x)}, \frac{v - \frac{v^2 + 4c_0 x^4}{2x}\lambda_3}{F_{\lambda_3}^2(\mathbf{t}_x)}\right). \tag{3.83}$$

Let us employ previous results to determine the action on $\mathfrak{sl}(2, \mathbb{R})$ of those elements g close to the neutral element $e \in SL(2, \mathbb{R})$. Using the so-called *canonical coordinates of the second kind* [235], we can write g within an open neighborhood U of e in a unique form as

$$g = \exp(-\lambda_3 \mathtt{a}_3) \exp(-\lambda_2 \mathtt{a}_2) \exp(-\lambda_1 \mathtt{a}_1), \tag{3.84}$$

for real constants λ_1, λ_2 and λ_3. This allows us to obtain the action of every $g \in U$ on $T\mathbb{R}_0$ through the composition of the actions of elements $\exp(-\lambda_\alpha \mathtt{a}_\alpha)$, with $\lambda_\alpha \in \mathbb{R}$ for $\alpha = 1, 2, 3$. To do so, we determine the constants λ_1, λ_2 and λ_3 associated with each $g \in U$ in (3.84).

Considering the standard matrix representation of $SL(2, \mathbb{R})$, we can express every $g \in SL(2, \mathbb{R})$ as

$$g = \begin{pmatrix} \alpha & \beta \\ \gamma & \delta \end{pmatrix}, \quad \alpha\delta - \beta\gamma = 1, \quad \alpha, \beta, \gamma, \delta \in \mathbb{R}. \tag{3.85}$$

In view of (3.74), and comparing (3.84) and (3.85), we obtain

$$\alpha = e^{\lambda_2/2}, \quad \beta = -e^{\lambda_2/2}\lambda_1, \quad \gamma = e^{\lambda_2/2}\lambda_3. \tag{3.86}$$

Consequently,

$$\lambda_1 = -\beta/\alpha, \quad \lambda_2 = 2\log\alpha, \quad \lambda_3 = \gamma/\alpha, \tag{3.87}$$

and, from the basis (3.74), the decomposition (3.84) and expressions (3.77) and (3.83), the action reads

$$\varphi_{2\mathrm{KS}}(g, \mathbf{t}_x) = \left(\frac{x}{F_g(\mathbf{t}_x)}, \frac{1}{F_g^2(\mathbf{t}_x)} \left[(v\alpha - 2x\beta)\left(\delta - \frac{\gamma v}{2x}\right) - 2c_0 x^3 \alpha\gamma \right] \right), \tag{3.88}$$

where

$$F_g(\mathbf{t}_x) = \left(\delta - \frac{\gamma v}{2x}\right)^2 + c_0 x^2 \gamma^2. \tag{3.89}$$

Although this expression has been derived for g being close to e, it can be proven that the action is properly defined at points $(g, \mathbf{t}_x)$ such that $F_g(\mathbf{t}_x) \neq 0$. If $c_0 > 0$, then $F_g(\mathbf{t}_x) > 0$ for all $g \in SL(2, \mathbb{R})$ and $\mathbf{t}_x \in T\mathbb{R}_0$. So, $\varphi_{2\mathrm{KS}}$ becomes globally defined. Otherwise, $F_g(\mathbf{t}_x) > 0$ for g close enough to e. Then, $\varphi_{2\mathrm{KS}}$ is only defined on a neighborhood of e.

The action $\varphi_{2\mathrm{KS}}$ also permits us to write the general solution of system (3.70) in the form $(x(t), v(t)) = \varphi_{2\mathrm{KS}}(g(t), \mathbf{t}_x)$, with $g(t)$ being a particular solution of

$$\frac{dg}{dt} = -X_3^R(g) - b_1(t)X_1^R(g), \tag{3.90}$$

where X_α^R, with $\alpha = 1, 2, 3$, are the right-invariant vector fields on $SL(2, \mathbb{R})$ satisfying $X_\alpha^R(e) = \mathrm{a}_\alpha$ [91, 108]. Additionally, as $x(t)$ is the general solution of KS-2 equation (3.68), we readily see that

$$x(t) = \tau \circ \varphi_{2\mathrm{KS}}(g(t), \mathbf{t}_x), \tag{3.91}$$

with $\tau : (x, v) \in \mathrm{T}\mathbb{R} \mapsto x \in \mathbb{R}$, the natural tangent bundle projection, provides us with the general solution of (3.68) in terms of a particular solution of (3.90).

Conversely, we prove that we can recover a particular solution to (3.90) from the knowledge of the general solution of (3.68). For simplicity, we will determine the particular solution $g_1(t)$ with $g_1(0) = e$. Given two particular solutions $x_1(t)$ and $x_2(t)$ of (3.68) with $dx_1/dt(t) = dx_2/dt(t) = 0$, the expression (3.91) implies that

$$(x_i(t), v_i(t)) = \varphi_{2\mathrm{KS}}(g_1(t), (x_i(0), 0)), \quad i = 1, 2. \tag{3.92}$$

Writing the above expression explicitly, we get

$$-\frac{x_i(0)v_i(t)}{2x_i^2(t)} = \beta(t)\delta(t) + c_0 x_i^2(0)\alpha(t)\gamma(t),$$

$$\frac{x_i(0)}{x_i(t)} = \delta^2(t) + c_0 x_i^2(0)\gamma^2(t), \tag{3.93}$$

for $i = 1, 2$. The first two equations allow us to determine the value of $\beta(t)\delta(t)$ and $\alpha(t)\gamma(t)$. Meanwhile, we can obtain the value of $\delta^2(t)$ and $\gamma^2(t)$ from the other two. As $\delta(0) = 1$, we know that $\delta(t)$ is positive when close to $t = 0$. Taking into account that we have already worked out $\delta^2(t)$, we can determine $\delta(t)$ for small values of t. Since we have already obtained $\beta(t)\delta(t)$, we can also derive $\beta(t)$ for small values of t using $\delta(t)$. Note that $\alpha(0) = 1$. So, $\alpha(t)$ is positive for small values of t, and the sign of $\alpha(t)\gamma(t)$ determines the sign of $\gamma(t)$ around $t = 0$. In view of this, the value of $\gamma(t)$ can be determined

Basics on Lie Systems and Superposition Rules

from $\gamma^2(t)$ in the interval around $t = 0$. Summing up, we can obtain algebraically a particular solution of (3.93) with $g_1(0) = e$ from the general solution of (3.68).

3.11.4. *Third-order Kummer–Schwarz equation*

Let us write the KS-3 equations (3.69) as a first-order system

$$
\begin{cases}
\dfrac{dx}{dt} = v, \\[2mm]
\dfrac{dv}{dt} = a, \\[2mm]
\dfrac{da}{dt} = \dfrac{3}{2}\dfrac{a^2}{v} - 2c_0(x)v^3 + 2b_1(t)v,
\end{cases}
\tag{3.94}
$$

in the open submanifold $\mathcal{O}_2 = \{(x, v, a) \in T^2\mathbb{R} \mid v \neq 0\}$ of $T^2\mathbb{R} \simeq \mathbb{R}^3$, the referred to as *second-order tangent bundle* of $\mathbb{R}$ [3].

Consider now the set of vector fields on $\mathcal{O}_2$ given by

$$
\begin{aligned}
N_1 &= 2v\frac{\partial}{\partial a}, \\[2mm]
N_2 &= v\frac{\partial}{\partial v} + 2a\frac{\partial}{\partial a}, \\[2mm]
N_3 &= v\frac{\partial}{\partial x} + a\frac{\partial}{\partial v} + \left(\frac{3}{2}\frac{a^2}{v} - 2c_0(x)v^3\right)\frac{\partial}{\partial a},
\end{aligned}
\tag{3.95}
$$

which satisfy the commutation relations

$$
[N_1, N_3] = 2N_2, \quad [N_1, N_2] = N_1, \quad [N_2, N_3] = N_3.
\tag{3.96}
$$

Thus, they span a three-dimensional Lie algebra of vector fields V isomorphic to $\mathfrak{sl}(2, \mathbb{R})$. Since (3.94) is determined by the t-dependent vector field

$$
X_t = v\frac{\partial}{\partial x} + a\frac{\partial}{\partial v} + \left(\frac{3}{2}\frac{a^2}{v} - 2c_0(x)v^3 + 2b_1(t)v\right)\frac{\partial}{\partial a},
\tag{3.97}
$$

we can write $X_t = N_3 + b_1(t)N_1$. Consequently, X takes values in the finite-dimensional VG Lie algebra V and becomes an $\mathfrak{sl}(2, \mathbb{R})$-Lie system. This generalizes the result provided in [87] for $c_0(x) = \text{const}$.

We shall now reduce the integration of (3.94) with $c_0(x) = \text{const.}$, and in consequence the integration of the related (3.69), to working out a particular solution of the Lie system (3.90). To do so, we employ the Lie group action $\varphi_{3\text{KS}} : SL(2, \mathbb{R}) \times \mathcal{O}_2 \to \mathcal{O}_2$ whose infinitesimal action is given by the Lie algebra isomorphism $\rho : \mathfrak{sl}(2, \mathbb{R}) \to V$ satisfying that $\rho(\mathsf{a}_\alpha) = N_\alpha$, with $\alpha = 1, 2, 3$. This Lie group action holds that

$$\frac{d}{ds}\varphi_{3\text{KS}}(\exp(-s\mathsf{a}_\alpha), \mathbf{t}_x^2) = N_\alpha(\varphi_{3\text{KS}}(\exp(-s\mathsf{a}_\alpha), \mathbf{t}_x^2)), \qquad (3.98)$$

with $\mathbf{t}_x^2 := (x, v, a) \in \mathcal{O}_2$ and $\alpha = 1, 2, 3$. Integrating N_1 and N_2, we obtain that

$$\varphi_{3\text{KS}}\left(\exp(-\lambda_1\mathsf{a}_1), \mathbf{t}_x^2\right) = \begin{pmatrix} x \\ v \\ a + 2v\lambda_1 \end{pmatrix}, \qquad (3.99)$$

and

$$\varphi_{3\text{KS}}\left(\exp(-\lambda_2\mathsf{a}_2), \mathbf{t}_x^2\right) = \begin{pmatrix} x \\ ve^{\lambda_2} \\ ae^{2\lambda_2} \end{pmatrix}. \qquad (3.100)$$

To integrate N_3, we need to obtain the solutions of

$$\frac{dx}{d\lambda_3} = v, \quad \frac{dv}{d\lambda_3} = a, \quad \frac{da}{d\lambda_3} = \frac{3}{2}\frac{a^2}{v} - 2c_0v^3. \qquad (3.101)$$

Proceeding mutatis mutandis, as in the analysis of system (3.78), we obtain

$$v(\lambda_3) = \frac{v(0)}{F_{\lambda_3}(x(0), v(0), a(0))}, \qquad (3.102)$$

with

$$F_{\lambda_3}(\mathbf{t}_x^2) = \left(1 - \frac{a\lambda_3}{2v}\right)^2 + c_0v^2\lambda_3^2. \qquad (3.103)$$

Taking into account this and the first two equations within (3.101), we see that

$$\varphi_{3KS}\left(e^{-\lambda_3 a_3}, \mathbf{t}_x^2\right) = \begin{pmatrix} x + v \int_0^{\lambda_3} F_{\lambda_3'}^{-1}(\mathbf{t}_x^2) d\lambda_3' \\ F_{\lambda_3}^{-1}(\mathbf{t}_x^2) v \\ v \partial (F_{\lambda_3}^{-1}(\mathbf{t}_x^2)) / \partial \lambda_3 \end{pmatrix}. \tag{3.104}$$

Using decomposition (3.84), we can reconstruct the new action

$$\varphi_{3KS}\left(g, \mathbf{t}_x^2\right) = \begin{pmatrix} x + v \int_0^{\gamma/\alpha} \bar{F}_{\lambda_3,g}^{-1}(\mathbf{t}_x^2) d\lambda_3 \\ \bar{F}_{\gamma/\alpha,g}^{-1}(\mathbf{t}_x^2) v \\ v \frac{\partial (\bar{F}_{\lambda_3,g}^{-1}(\mathbf{t}_x^2))}{\partial \lambda_3} \Big|_{\lambda_3 = \gamma/\alpha} \end{pmatrix}, \tag{3.105}$$

with $\bar{F}_{\lambda_3,g}(\mathbf{t}_x^2) = \alpha^{-2} F_{\lambda_3}(x, v\alpha^2, (a\alpha - 2v\beta)\alpha^3 \lambda_3)$, i.e.,

$$\bar{F}_{\lambda_3,g}(\mathbf{t}_x^2) = \left(\frac{1}{\alpha} - \frac{a\alpha - 2v\beta}{2v}\lambda_3\right)^2 + c_0 v^2 \alpha^2 \lambda_3^2. \tag{3.106}$$

This action enables us to write the general solution of (3.94) as

$$(x(t), v(t), a(t)) = \varphi_{3KS}(g(t), \mathbf{t}_x^2), \tag{3.107}$$

where $\mathbf{t}_x^2 \in \mathcal{O}_2$ and $g(t)$ is a particular solution of the equation on $SL(2, \mathbb{R})$ given by (3.90). Hence, if $\tau^{2)} : (x, v, a) \in \mathrm{T}^2\mathbb{R} \mapsto x \in \mathbb{R}$ is the fiber bundle projection corresponding to the second-order tangent bundle on $\mathbb{R}$, we can write the general solution of (3.69) in the form

$$x(t) = \tau^2 \circ \varphi_{3KS}(g(t), \mathbf{t}_x^2), \tag{3.108}$$

where $g(t)$ is any particular solution of (3.90).

Conversely, given the general solution of (3.69), we can obtain a particular solution of (3.90). As before, we focus on obtaining the particular solution $g_1(t)$, with $g_1(0) = e$. In this case, given two particular solutions $x_1(t), x_2(t)$ of (3.69) with $d^2 x_1/dt^2(0) = d^2 x_2/dt^2(0) = 0$, we obtain that the t-dependent coefficients $\alpha(t)$, $\beta(t)$, $\gamma(t)$ and $\delta(t)$ corresponding to the matrix expression of $g_1(t)$ obey a system that is similar to (3.93) where v and x have been replaced by a and v, respectively.

3.11.5. *The Schwarzian derivative and third-order Kummer–Schwarz equations*

The Schwarzian derivative of a real function $f = f(t)$ is defined by

$$\{f, t\} := \frac{d^3 f}{dt^3} \left(\frac{df}{dt} \right)^{-1} - \frac{3}{2} \left[\frac{d^2 f}{dt^2} \left(\frac{df}{dt} \right)^{-1} \right]^2. \tag{3.109}$$

This derivative is clearly related to KS-3 equations (3.69) with $c_0 = 0$, which can be written as $\{x, t\} = 2b_1(t)$.

Although a superposition rule for studying KS-3 equations was developed in [87], the result provided in there was not valid when $c_0 = 0$, which retrieves the relevant equation $\{x, t\} = 2b_1(t)$. This is why we aim to reconsider this case and its important connection to the Schwarzian derivative.

The vector fields N_1, N_2, N_3 are linearly independent at a generic point of $\mathcal{O}_2 \subset \mathrm{T}^2 \mathbb{R}_0$. Therefore, obtaining a superposition rule for (3.94) amounts to obtaining three functionally independent first integrals F_1, F_2, F_3 common to all diagonal prolongations $\widetilde{N}_1, \widetilde{N}_2, \widetilde{N}_3$ in $(\mathcal{O}_2)^2$, such that

$$\frac{\partial(F_1, F_2, F_3)}{\partial(x_0, v_0, a_0)} \neq 0. \tag{3.110}$$

As $[\widetilde{N}_1, \widetilde{N}_3] = 2\widetilde{N}_2$, it suffices to obtain common first integrals for $\widetilde{N}_1, \widetilde{N}_3$ to describe first integrals common to the integrable distribution $\mathcal{D}$ spanned by $\widetilde{N}_1, \widetilde{N}_2, \widetilde{N}_3$.

Let us start by solving $\widetilde{N}_1 F = 0$, with $F : \mathcal{O}_2 \to \mathbb{R}$, i.e.,

$$v_0 \frac{\partial F}{\partial a_0} + v_1 \frac{\partial F}{\partial a_1} = 0. \tag{3.111}$$

The method of characteristics shows that F must be constant along the solutions of the associated *Lagrange–Charpit equations* [156], namely

$$\frac{da_0}{v_0} = \frac{da_1}{v_1}, \quad dx_0 = dx_1 = dv_0 = dv_1 = 0. \tag{3.112}$$

Such solutions are the curves $(x_0(\lambda), v_0(\lambda), a_0(\lambda), x_1(\lambda), v_1(\lambda), a_1(\lambda))$ within $\mathcal{O}_2$ with $\Delta = v_1(\lambda) a_0(\lambda) - a_1(\lambda) v_0(\lambda)$, for a real constant $\Delta \in$

Basics on Lie Systems and Superposition Rules 115

$\mathbb{R}$, and constant $x_i(\lambda)$ and $v_i(\lambda)$, with $i = 0, 1$. In other words, there exists a function $F_2 : \mathbb{R}^5 \to \mathbb{R}$ such that $F(x_0, v_0, a_0, x_1, v_1, a_1) = F_2(\Delta, x_0, x_1, v_0, v_1)$.

If we now impose $\widetilde{N}_3 F = 0$, we obtain

$$\widetilde{N}_3 F = \widetilde{N}_3 F_2 = \frac{\Delta + a_1 v_0}{v_1} \frac{\partial F_2}{\partial v_0} + a_1 \frac{\partial F_2}{\partial v_1}$$

$$+ v_0 \frac{\partial F_2}{\partial x_0} + v_1 \frac{\partial F_2}{\partial x_1} + \frac{3\Delta^2 + 6\Delta a_1 v_0}{2 v_1 v_0} \frac{\partial F_2}{\partial \Delta} = 0. \qquad (3.113)$$

We can then write that $\widetilde{N}_3 F_2 = (a_1/v_1)\Xi_1 F_2 + \Xi_2 F_2 = 0$, where

$$\Xi_1 = v_0 \frac{\partial}{\partial v_0} + v_1 \frac{\partial}{\partial v_1} + 3\Delta \frac{\partial}{\partial \Delta},$$

$$\Xi_2 = v_0 \frac{\partial}{\partial x_0} + v_1 \frac{\partial}{\partial x_1} + \frac{\Delta}{v_1} \frac{\partial}{\partial v_0} + \frac{3\Delta^2}{2 v_0 v_1} \frac{\partial}{\partial \Delta}. \qquad (3.114)$$

As F_2 does not depend on a_1 in the chosen coordinate system, it follows $\Xi_1 F_2 = \Xi_2 F_2 = 0$. Using the characteristics method again, we obtain that $\Xi_1 F_2 = 0$ implies the existence of a new function $F_3 : \mathbb{R}^4 \to \mathbb{R}$ such that $F_2(\Delta, x_0, x_1, v_0, v_1) = F_3(K_1 := v_1/v_0, K_2 := v_0^3/\Delta, x_0, x_1)$.

The only condition remaining is $\Xi_2 F_3 = 0$. In the local coordinate system $\{K_1, K_2, x_0, x_1\}$, this equation reads

$$v_0 \left(\frac{3}{2K_1} \frac{\partial F_3}{\partial K_2} - \frac{1}{K_2} \frac{\partial F_3}{\partial K_1} + \frac{\partial F_3}{\partial x_0} + K_1 \frac{\partial F_3}{\partial x_1} \right) = 0, \qquad (3.115)$$

and its Lagrange–Charpit equations become

$$-K_2 dK_1 = \frac{2K_1 dK_2}{3} = dx_0 = \frac{dx_1}{K_1}. \qquad (3.116)$$

From the first equality, we obtain that $K_1^3 K_2^2 = \Upsilon_1$ for a certain real constant Υ_1. In view of this and with the aid of the above system, it turns out

$$\frac{2}{3} K_1^2 dK_2 = dx_1 \longrightarrow \frac{2}{3} \Upsilon_1^{2/3} K_2^{-4/3} dK_2 = dx_1. \qquad (3.117)$$

Integrating, we see that $-2K_2 K_1^2 - x_1 = \Upsilon_2$ for a certain real constant Υ_2. Finally, these previous results are used to solve the last part of the Lagrange–Charpit system, i.e.,

$$dx_0 = \frac{dx_1}{K_1} = \frac{4\Upsilon_1 dx_1}{(x_1 + \Upsilon_2)^2} \longrightarrow \Upsilon_3 = x_0 + \frac{4\Upsilon_1}{x_1 + \Upsilon_2}. \tag{3.118}$$

Note that $\partial(\Upsilon_1, \Upsilon_2, \Upsilon_3)/\partial(x_0, v_0, a_0) \neq 0$. Therefore, considering $\Upsilon_1 = k_1$, $\Upsilon_2 = k_2$ and $\Upsilon_3 = k_3$, we can obtain a mixed superposition rule. From these equations, we easily obtain

$$x_0 = \frac{x_1 k_3 + k_2 k_3 - 4k_1}{x_1 + k_2}. \tag{3.119}$$

Multiplying numerator and denominator of the right-hand side by a non-null constant Υ_4, the above expression can be rewritten as

$$x_0 = \frac{\alpha x_1 + \beta}{\gamma x_1 + \delta}, \tag{3.120}$$

with $\alpha = \Upsilon_4 k_3, \beta = \Upsilon_4(k_2 k_3 - 4k_1), \gamma = \Upsilon_4, \delta = k_2 \Upsilon_4$. Observe that

$$\alpha\delta - \gamma\beta = 4\Upsilon_4^2 \Upsilon_1 = \frac{4\Upsilon_4^2 v_0^3 v_1^3}{(v_1 a_0 - a_1 v_0)^2} \neq 0. \tag{3.121}$$

Then, choosing an appropriate Υ_4, we obtain that (3.119) can be rewritten as (3.120) for a family of constants $\alpha, \beta, \gamma, \delta$ such that $\alpha\delta - \gamma\beta = \pm 1$. It is important to recall that the matrices

$$\begin{pmatrix} \alpha & \beta \\ \gamma & \delta \end{pmatrix}, \quad I = \alpha\delta - \beta\gamma = \pm 1, \tag{3.122}$$

are the matrix description of the Lie group $PGL(2, \mathbb{R})$. Recall that we understand each matrix as the equivalence class related to it.

Operating, we also obtain that

$$v_0 = \frac{Iv_1}{(\gamma x_1 + \delta)}, \quad a_0 = I\left[\frac{a_1}{(\gamma x_1 + \delta)^2} - \frac{2v_1^2 \gamma}{(\gamma x_1 + \delta)^3}\right]. \tag{3.123}$$

The above expression together with (3.120) become a superposition rule for KS-3 equations with $c_0 = 0$ (written as a first-order system). In other words, the general solution $(x(t), v(t), a(t))$ of (3.69)

Basics on Lie Systems and Superposition Rules 117

with $c_0 = 0$ can be written as

$$(x(t), v(t), a(t)) = \Phi(A, x_1(t), v_1(t), a_1(t)), \tag{3.124}$$

with $(x_1(t), v_1(t), a_1(t))$ being a particular solution, $A \in PGL(2, \mathbb{R})$ and

$$\Phi(A, x_1, v_1, a_1) = \left(\frac{\alpha x_1 + \beta}{\gamma x_1 + \delta}, \frac{I v_1}{(\gamma x_1 + \delta)^2}, I \left[\frac{a_1(\gamma x_1 + \delta) - 2v_1^2 \gamma}{(\gamma x_1 + \delta)^3} \right] \right). \tag{3.125}$$

Moreover, $x(t)$, which is the general solution of a KS-3 equation with $c_0 = 0$, can be determined out of a particular solution $x_1(t)$ and three constants through

$$x(t) = \tau^{2)} \circ \Phi \left(A, x_1(t), \frac{dx_1}{dt}(t), \frac{d^2 x_1}{dt^2}(t) \right), \tag{3.126}$$

where we see that the right-hand part does merely depend on A and $x_1(t)$. This constitutes a *basic superposition rule* [87] for equations $\{x(t), t\} = 2b_1(t)$, i.e., it is an expression that allows us to describe the general solution of any of these equations in terms of a particular solution (without involving its derivatives) and some constants to be related to initial conditions. We shall now employ this superposition rule to describe some properties of the Schwarzian derivative.

From the equation above, we analyze the relation between two particular solutions $x_1(t)$ and $x_2(t)$ of the same equation $\{x, t\} = 2b_1(t)$, i.e., $\{x_1(t), t\} = \{x_2(t), t\}$. Our basic superposition rule (3.126) tells us that from $x_1(t)$ we can generate every other solution of the equation. In particular, there must exist certain real constants c_1, c_2, c_3, c_4 such that

$$x_2(t) = \frac{c_1 x_1(t) + c_2}{c_3 x_1(t) + c_4}, \quad c_1 c_4 - c_2 c_3 \neq 0. \tag{3.127}$$

In this way, we recover a relevant property of this type of equations [305].

Our superposition rule (3.126) also provides us with information about the Lie symmetries of $\{x(t), t\} = 2b_1(t)$. Indeed, note that

118   A Guide to Lie Systems with Compatible Geometric Structures

(3.126) implies that the local Lie group action $\varphi : PGL(2,\mathbb{R})\times\mathbb{R} \to \mathbb{R}$

$$\varphi(A,x) = \frac{\alpha x + \beta}{\gamma x + \delta},\qquad (3.128)$$

transforms solutions of $\{x(t),t\} = 2b_1(t)$ into solutions of the same equation. The prolongation [60, 302] $\widehat{\varphi} : PGL(2,\mathbb{R}) \times T^2\mathbb{R}_0 \to T^2\mathbb{R}_0$ of φ to $T^2\mathbb{R}_0$, i.e.,

$$\widehat{\varphi}(A,\mathbf{t}_x^2) = \left(\frac{\alpha x + \beta}{\gamma x + \delta}, \frac{Iv}{(\gamma x + \delta)^2}, I\frac{a(\gamma x + \delta) - 2\gamma v^2}{(\gamma x + \delta)^3}\right),\qquad (3.129)$$

gives rise to a group of symmetries $\varphi(A,\cdot)$ of (3.94) when $c_0 = 0$. The fundamental vector fields of this action are spanned by

$$Z_1 = -\frac{\partial}{\partial x},\quad Z_2 = x\frac{\partial}{\partial x} + v\frac{\partial}{\partial v} + a\frac{\partial}{\partial a},$$

$$Z_3 = -\left[x^2\frac{\partial}{\partial x} + 2vx\frac{\partial}{\partial v} + 2(ax + v^2)\frac{\partial}{\partial a}\right],\qquad (3.130)$$

which close on a Lie algebra of vector fields isomorphic to $\mathfrak{sl}(2,\mathbb{R})$ and commute with X_t for every $t \in \mathbb{R}$. In addition, their projections onto $\mathbb{R}$ must be Lie symmetries of $\{x(t),t\} = 2b_1(t)$. Indeed, they read

$$S_1 = -\frac{\partial}{\partial x},\quad S_2 = x\frac{\partial}{\partial x},\quad S_3 = -x^2\frac{\partial}{\partial x},\qquad (3.131)$$

which are the known Lie symmetries for these equations [304].

Consider now the equation $\{x(t),t\} = 0$. Obviously, this equation admits the particular solution $x(t) = t$. This, together with our basic superposition rule, show that the general solution of this equation is

$$x(t) = \frac{\alpha t + \beta}{\gamma t + \delta},\quad \alpha\delta - \gamma\beta \neq 0,\qquad (3.132)$$

recovering another relevant known solution of these equations.

3.12.   Superposition Rules for PDEs

The geometrical formulation of the theory of Lie systems enables one to extend the superposition rule notion to first-order systems of

Basics on Lie Systems and Superposition Rules 119

PDEs in normal form. A first approach to this topic was accomplished in [296], while a characterization of systems of PDEs in normal form admitting a superposition rule was provided in [93]. Finally, numerous properties of such PDEs were detailed in [314].

Consider the system of first-order PDEs in normal form given by

$$\frac{\partial x^i}{\partial t^a} = X_a^i(t, x), \quad x \in \mathbb{R}^n, \quad t = (t^1, \ldots, t^s) \in \mathbb{R}^s, \tag{3.133}$$

whose solutions are maps $x(t) : \mathbb{R}^s \to \mathbb{R}^n$. Since (3.133) is in normal form, each particular solution $x(t)$ is locally determined by its value at a single point $t_0 \in \mathbb{R}^s$. The main difference between these systems is that for $s > 1$ there exists no solution with a given initial condition in general. For a better understanding of this problem, let us put (3.133) in a more general and geometric framework.

Let P_N^s be the trivial fiber bundle

$$P_N^s := \mathbb{R}^s \times N \xrightarrow{\text{pr}} \mathbb{R}^s.$$

A connection ∇ on this bundle is a way of determining an s-dimensional distribution transversal to the fibers, i.e., a so-called *horizontal distribution* HP_N^s on TP_N^s. This gives rise to an isomorphism $T_{(t,x)}\text{pr} : T_{(t,x)}\mathbb{R}^s \times N \to T_x N$ for every element $(t, x) \in \mathbb{R}^s \times N$. Additionally, every vector field on N can be considered as the projection onto N of a unique vector field on TP_N^s, its so-called *horizontal lift*. Consequently, HP_N^s can be determined by the horizontal lifts of the vector fields $\partial/\partial t^a$ on $\mathbb{R}^s$, i.e.,

$$\overline{X}_a(t, x) = \frac{\partial}{\partial t^a} + X_a(t, x), \tag{3.134}$$

where

$$X_a(t, x) = \sum_{i=1}^n X_a^i(t, x) \frac{\partial}{\partial x^i}.$$

Since (3.133) is a system of PDEs in normal form, its solutions, $x(t)$, can be identified with the integral submanifolds $(t, x(t))$ of the distribution spanned by the $\overline{X}_a$, namely

$$(t, X_a(t, x)), \quad t \in \mathbb{R}^s, \ x \in N.$$

Therefore, there exists a solution of (3.133) for every initial condition if and only if the distribution HP_N^s is integrable. This means that

$$[\overline{X}_a, \overline{X}_b] = \sum_{c=1}^{r} f_{abc} \, \overline{X}_c,$$

for some functions f_{abc} in P_N^s. In view of the expressions (3.134), we see that $[\overline{X}_a, \overline{X}_b]$ is vertical. Since the $\overline{X}_c$ are linearly independent horizontal vector fields, one has $f_{abc} = 0$. Hence, the integrability of HP_N^s amounts to the system of equations $[\overline{X}_a, \overline{X}_b] = 0$. In local coordinates,

$$\frac{\partial X_b^i}{\partial t^a}(t, x) - \frac{\partial X_a^i}{\partial t^b}(t, x) + \sum_{j=1}^{n} \left(X_a^j(t, x) \frac{\partial X_b^i}{\partial x^j}(t, x) \right.$$

$$\left. - X_b^j(t, x) \frac{\partial X_a^i}{\partial x^j}(t, x) \right) = 0. \tag{3.135}$$

In this case, we say that the connection HP_N^s has zero curvature.

Let us assume now that the system of first-order PDEs of the form (3.133) satisfies the integrability conditions (3.135). We will hereafter say that (3.133) is *integrable*. In other words, for a given initial condition, there exists a unique solution of system (3.133). Furthermore, it is immediate that the geometrical interpretation of superposition rules for first-order described in Section 3.7 can be generalized straightforwardly to the case of PDEs. In consequence, Proposition 3.13 takes now the following form.

Proposition 3.17. *A superposition rule for an integrable system* (3.133) *amounts to a connection on the bundle* pr $: N^{m+1} \to N^m$ *with a zero curvature such that the family of vector fields* $\{(X_1^{[m+1]})_t, \ldots, (X_s^{[m+1]})_t\}_{t \in \mathbb{R}}$ *consists of horizontal vector fields.*

Similarly, the proof of the Lie–Scheffers theorem remains the same. Consequently, one obtains the following analog of the Lie–Scheffers theorem for PDEs (see [93]).

Theorem 3.18. *An integrable system (3.133) of PDEs defined on N admits a superposition rule if and only if the vector fields $\{(X_a)_t\}_{t\in\mathbb{R}}$, with $a = 1,\ldots s$, on N can be written as*

$$(X_a)_t = \sum_{\alpha=1}^{r} u_a^\alpha(t) X_\alpha, \quad a = 1,\ldots, s, \tag{3.136}$$

for certain t-dependent functions $u_a^\alpha(t)$ and a family of vector fields $X_1,\ldots,X_r$ spanning a finite-dimensional real Lie algebra of vector fields.

In similarity with the theory of Lie systems, we say that the Lie algebra spanned by $X_1,\ldots,X_s$ is a *VG Lie algebra for the PDE Lie system* (3.133).

The integrability condition for a system of PDEs of the form (3.133) whose right-hand side can be written as $X_a = \sum_{\alpha=1}^{r} u_a^\alpha(t) X_\alpha$ and the vector fields X_α have commutation relations $[X_\alpha, X_\beta] = \sum_{\gamma=1}^{r} c_{\alpha\beta}^\gamma X_\gamma$ for $\alpha, \beta = 1,\ldots,r$, can be brought into the form

$$\sum_{\alpha,\beta,\gamma=1}^{r} \left[(u_b^\gamma)'(t) - (u_a^\gamma)'(t) + u_a^\alpha(t) u_b^\beta(t) c_{\alpha\beta}^\gamma \right] X_\gamma = 0.$$

Let us illustrate the above results through a relevant example. Consider the following system of PDEs on $\mathbb{R}^2$ associated with the $SL(2,\mathbb{R})$-action on $\bar{\mathbb{R}}$,

$$\begin{cases} \dfrac{\partial u}{\partial x} = a(x, y)u^2 + b(x, y)u + c(x, y), \\[2mm] \dfrac{\partial u}{\partial y} = d(x, y)u^2 + e(x, y)u + f(x, y). \end{cases} \tag{3.137}$$

Systems of this type appear, for instance, in the study of conditional symmetries of PDEs such as sine-Gordon equations or Bäcklund transactions between Korteweg-de Vries equations and their modified versions [89]. The PDE system (3.137) can be written in the form of a "total differential equation"

$$(a(x,y)u^2 + b(x,y)u + c(x,y))\mathrm{d}x + (d(x,y)u^2$$
$$+ e(x,y)u + f(x,y))\mathrm{d}y = \mathrm{d}u.$$

Then, the integrability condition amounts to saying that the 1-form

$$\omega = (a(x,y)u^2 + b(x,y)u + c(x,y))\mathrm{d}x + (d(x,y)u^2$$
$$+ e(x,y)u + f(x,y))\mathrm{d}y,$$

is closed for an arbitrary function $u = u(x,y)$. If this is the case, there is a unique solution with the initial condition $u(x_0, y_0) = u_0$ and there is a superposition rule giving a general solution as a function of three independent solutions, exactly as in the case of Riccati equations

$$u = \frac{(u_{(1)} - u_{(3)})u_{(2)}k + u_{(1)}(u_{(3)} - u_{(2)})}{(u_{(1)} - u_{(3)})k + (u_{(3)} - u_{(2)})}.$$

This expression can be achieved by applying the methods given in Section 3.9, which follow the same steps as in the case of the Riccati equations. This is why the superposition rule for (3.137) has the same form as the standard superposition rule for Riccati equations.

There are many other types of PDE Lie systems in the literature with relevant applications (cf. [89, 198]). Apart from the cases of PDE Lie system found by Grundland, most new applications can be found in above-mentioned works.

Chapter 4

Lie–Hamilton Systems

4.1. Introduction

We have found that many instances of relevant Lie systems possess VG Lie algebras of Hamiltonian vector fields with respect to a Poisson structure. Such Lie systems are hereafter called *Lie–Hamilton systems.*

Lie–Hamilton systems admit a plethora of geometric properties. For instance, we here prove that every Lie–Hamilton system admits a t-dependent Hamiltonian which can be understood as a curve in finite-dimensional Lie algebra of functions (with respect to a certain Poisson structure). These t-dependent Hamiltonians, the so-called *Lie–Hamiltonian structures*, are the key to understand the properties of Lie–Hamilton systems [16, 91, 121, 170]. Additionally, Lie–Hamilton systems appear in the analysis of relevant physical and mathematical problems, like second-order Kummer–Schwarz equations and t-dependent Smorodinsky–Winternitz oscillators [121].

In this section, we study Lie–Hamilton systems and some of their applications. In particular, our achievements are employed to study superposition rules, Lie symmetries and constants of motion for these systems. It is worth noting that this chapter uses the sign convention appearing in Geometric Mechanics.

4.2. The Necessity of Lie–Hamilton Systems

In this section, we start by showing how several relevant Lie systems admit a VG Lie algebra of Hamilton vector fields with respect to a Poisson structure or symplectic form. This will allow us to justify the interest of defining a new particular type of Lie systems enjoying such a property: the Lie–Hamilton systems.

Example 4.1. Let us show that *planar Riccati equations* (4.206), described geometrically by the t-dependent vector field $X = a_0(t)X_1 + a_1(t)X_2 + a_2(t)X_3$ given by vector fields (4.207) admit a VG Lie algebra consisting of Hamiltonian vector fields with respect to a symplectic form. To do so, we search for a symplectic form, let us say $\omega = f(x, y)\mathrm{d}x \wedge \mathrm{d}y$, turning the VG Lie algebra $V = \langle X_1, X_2, X_3 \rangle$ into a Lie algebra of Hamiltonian vector fields with respect to it. To ensure that the basis for V given by the vector fields X_1, X_2, and X_3, (see commutation relations (4.208)), is locally Hamiltonian relative to ω, we impose $\mathcal{L}_{X_i}\omega = 0$ $(i = 1, 2, 3)$, where $\mathcal{L}_{X_i}\omega$ stands for the Lie derivative of ω with respect to X_i. In coordinates, these conditions read

$$\frac{\partial f}{\partial x} = 0, \quad x\frac{\partial f}{\partial x} + y\frac{\partial f}{\partial y} + 2f = 0, \quad (x^2 - y^2)\frac{\partial f}{\partial x} + 2xy\frac{\partial f}{\partial y} + 4xf = 0. \tag{4.1}$$

From the first equation, we obtain $f = f(y)$. Using this in the second equation, we find that $f = y^{-2}$ is a particular solution of both equations (the third one is therefore automatically fulfilled). This leads to a closed and non-degenerate 2-form on $\mathbb{R}^2_{y\neq 0} := \{(x, y) \in \mathbb{R}^2 \,|\, y \neq 0\}$, namely

$$\omega = \frac{\mathrm{d}x \wedge \mathrm{d}y}{y^2}. \tag{4.2}$$

Using the relation $\iota_X\omega = \mathrm{d}h$ between a Hamiltonian vector field X and one of its corresponding Hamiltonian functions h, we observe that X_1, X_2, and X_3 are Hamiltonian vector fields with

Hamiltonian functions

$$h_1 = -\frac{1}{y}, \quad h_2 = -\frac{x}{y}, \quad h_3 = -\frac{x^2 + y^2}{y}, \tag{4.3}$$

respectively. Obviously, the remaining vector fields of V also become Hamiltonian. Thus, planar Riccati equations (4.206) admit a VG Lie algebra of Hamiltonian vector fields relative to (4.2).

Example 4.2. Let us now focus on the Hamilton equations for an n-dimensional *Smorodinsky–Winternitz oscillator* [381] on $\mathrm{T}^*\mathbb{R}_0^n$, where $\mathbb{R}_0^n = \mathbb{R} - 0$, of the form

$$\begin{cases} \dfrac{dx_i}{dt} = p_i, \\[2mm] \dfrac{dp_i}{dt} = -\omega^2(t)x_i + \dfrac{k}{x_i^3}, \end{cases} \quad i = 1, \dots, n, \tag{4.4}$$

with $\omega(t)$ being any t-dependent function, $\mathbb{R}_0^n := \{(x_1, \dots, x_n) \in \mathbb{R}_0^n \mid x_1, \dots, x_n \neq 0\}$ and $k \in \mathbb{R}$. These oscillators have attracted quite much attention in Classical and Quantum Mechanics for their special properties [194, 213, 384]. In addition, observe that Winternitz–Smorodinsky system oscillators reduce to t-dependent isotropic harmonic oscillators when $k = 0$.

System (4.4) describes the integral curves of the t-dependent vector field

$$X_t = \sum_{i=1}^{n} \left[p_i \frac{\partial}{\partial x_i} + \left(-\omega^2(t)x_i + \frac{k}{x_i^3} \right) \frac{\partial}{\partial p_i} \right], \tag{4.5}$$

on $\mathrm{T}^*\mathbb{R}_0^n$. This cotangent bundle admits a natural Poisson bivector Λ related to the restriction to $\mathrm{T}^*\mathbb{R}_0^n$ of the canonical symplectic structure on $\mathrm{T}^*\mathbb{R}^n$, namely $\Lambda = \sum_{i=1}^{n} \partial/\partial x_i \wedge \partial/\partial p_i$. If we consider the vector fields

$$X_1 = -\sum_{i=1}^{n} x_i \frac{\partial}{\partial p_i}, \quad X_2 = \sum_{i=1}^{n} \frac{1}{2} \left(p_i \frac{\partial}{\partial p_i} - x_i \frac{\partial}{\partial x_i} \right),$$

$$X_3 = \sum_{i=1}^{n} \left(p_i \frac{\partial}{\partial x_i} + \frac{k}{x_i^3} \frac{\partial}{\partial p_i} \right), \tag{4.6}$$

we can write $X_t = X_3 + \omega^2(t)X_1$. Additionally, since

$$[X_1, X_3] = 2X_2, \quad [X_1, X_2] = X_1, \quad [X_2, X_3] = X_3, \quad (4.7)$$

it follows that (4.4) is a Lie system related to a VG Lie algebra isomorphic to $\mathfrak{sl}(2, \mathbb{R})$. In addition, this Lie algebra is again made of Hamiltonian vector fields. In fact, it is easy to check that $X_\alpha = -\widehat{\Lambda}(dh_\alpha)$, with $\alpha = 1, 2, 3$, and

$$h_1 = \frac{1}{2}\sum_{i=1}^{n} x_i^2, \quad h_2 = -\frac{1}{2}\sum_{i=1}^{n} x_i p_i, \quad h_3 = \frac{1}{2}\sum_{i=1}^{n}\left(p_i^2 + \frac{k}{x_i^2}\right). \quad (4.8)$$

Consequently, system (4.4) admits a VG Lie algebra of Hamiltonian vector fields relative to the Poisson structure Λ.

Example 4.3. Finally, let us turn to the system of Riccati equations

$$\frac{dx_i}{dt} = a_0(t) + a_1(t)x_i + a_2(t)x_i^2, \quad i = 1, \ldots, n. \quad (4.9)$$

The relevance of this system is due to the fact that it appears in the calculation of superposition rules for Riccati equations [108].

The system (4.9) admits a VG Lie algebra spanned by the vector fields

$$X_1 = \sum_{i=1}^{n}\frac{\partial}{\partial x_i}, \quad X_2 = \sum_{i=1}^{n} x_i\frac{\partial}{\partial x_i}, \quad X_3 = \sum_{i=1}^{n} x_i^2\frac{\partial}{\partial x_i}. \quad (4.10)$$

Hence, (4.9) becomes a Lie system possessing a VG Lie algebra isomorphic to $\mathfrak{sl}(2, \mathbb{R})$ spanned by the vector fields X_1, X_2, X_3. Let us prove that this VG Lie algebra consists of Hamiltonian vector fields. Indeed, we can consider the Poisson bivector

$$\Lambda_R = (x_1 - x_2)^2\frac{\partial}{\partial x_1}\wedge\frac{\partial}{\partial x_2} + (x_3 - x_4)^2\frac{\partial}{\partial x_3}\wedge\frac{\partial}{\partial x_4}, \quad (4.11)$$

on $\mathcal{O} = \{(x_1, x_2, x_3, x_4)|(x_1 - x_2)(x_2 - x_3)(x_3 - x_4) \neq 0\} \subset \mathbb{R}^4$. We have that $X_i = -\widehat{\Lambda}(dh_i)$ on $(\mathcal{O}, \omega_R)$ for

$$h_1 = \frac{1}{x_1 - x_2} + \frac{1}{x_3 - x_4}, \quad h_2 = \frac{1}{2}\left(\frac{x_1 + x_2}{x_1 - x_2} + \frac{x_3 + x_4}{x_3 - x_4}\right),$$

$$h_3 = \frac{x_1 x_2}{x_1 - x_2} + \frac{x_3 x_4}{x_3 - x_4}. \quad (4.12)$$

Hence, the vector fields X_1, X_2, X_3, given in (4.10) are Hamiltonian with respect to $(\mathcal{O}, \Lambda_R)$ and our system of Riccati equations admits a VG Lie algebra of Hamiltonian vector fields. The relevance of this result is due to the fact that it allowed us to obtain the superposition rule for Riccati equations through a Casimir of $\mathfrak{sl}(2, \mathbb{R})$ [27].

All previous examples lead us to propose the following definition.

Definition 4.1. A system X on N is said to be a *Lie–Hamilton system* if N can be endowed with a Poisson bivector Λ in such a way that V^X becomes a finite-dimensional real Lie algebra of Hamiltonian vector fields relative to Λ.

Although Lie–Hamilton systems form a particular class of Lie systems, it turns out that it is easier to find applications of Lie–Hamilton systems than applications of other types of Lie systems. Moreover, Lie–Hamilton systems admit much more geometrical properties than general Lie systems.

4.2.1. *Lie–Hamiltonian structures*

Standard Hamiltonian systems can be described through a Hamiltonian. Likewise, Lie–Hamilton systems admit a similar structure playing an analogous role: the Lie–Hamiltonian structures.

Definition 4.2. A *Lie–Hamiltonian structure* is a triple (N, Λ, h), where (N, Λ) stands for a Poisson manifold and h represents a t-parametrized family of functions $h_t : N \to \mathbb{R}$ such that $\mathrm{Lie}(\{h_t\}_{t \in \mathbb{R}}, \{\cdot, \cdot\}_\Lambda)$ is a finite-dimensional real Lie algebra.

Definition 4.3. A t-dependent vector field X is said to admit, or to possess, a Lie–Hamiltonian structure (N, Λ, h) if X_t is the Hamiltonian vector field corresponding to h_t for each $t \in \mathbb{R}$. The Lie algebra $\mathrm{Lie}(\{h_t\}_{t \in \mathbb{R}}, \{\cdot, \cdot\}_\Lambda)$ is called a *Lie–Hamilton algebra* for X.

Example 4.4. Recall that the vector fields X_1, X_2, and X_3, for planar Riccati equations (4.206), are Hamiltonian with respect to the symplectic form ω given by (4.2) and form a basis for a VG

Lie algebra $V \simeq \mathfrak{sl}(2,\mathbb{R})$. Assume that the minimal Lie algebra for a planar Riccati equation, X, is V, i.e., $V^X = V$. If $\{\cdot,\cdot\}_\omega : C^\infty(\mathbb{R}^2_{y\neq 0}) \times C^\infty(\mathbb{R}^2_{y\neq 0}) \to C^\infty(\mathbb{R}^2_{y\neq 0})$ stands for the Poisson bracket induced by ω (see [358]), then

$$\{h_1, h_2\}_\omega = -h_1, \quad \{h_1, h_3\}_\omega = -2h_2, \quad \{h_2, h_3\}_\omega = -h_3. \quad (4.13)$$

Hence, the planar Riccati equation X possesses a Lie–Hamiltonian structure of the form $\left(\mathbb{R}^2_{y\neq 0}, \omega, h = a_0(t)h_1 + a_1(t)h_2 + a_2(t)h_3\right)$ and, as $V^X \simeq \mathfrak{sl}(2,\mathbb{R})$, we have that $(\mathcal{H}_\Lambda, \{\cdot,\cdot\}_\omega) := (\langle h_1, h_2, h_3 \rangle, \{\cdot,\cdot\}_\omega)$ is a Lie–Hamilton algebra for X isomorphic to $\mathfrak{sl}(2,\mathbb{R})$.

Proposition 4.4. *If a system X admits a Lie–Hamiltonian structure, then X is a Lie–Hamilton system.*

Proof. Let (N, Λ, h) be a Lie–Hamiltonian structure for X. Thus, $\mathrm{Lie}(\{h_t\}_{t\in\mathbb{R}})$ is a finite-dimensional Lie algebra. Moreover, $\{X_t\}_{t\in\mathbb{R}} \subset \widehat{\Lambda} \circ d[\mathrm{Lie}(\{h_t\}_{t\in\mathbb{R}})]$, and as $\widehat{\Lambda} \circ d$ is a Lie algebra morphism, it follows that $V = \widehat{\Lambda} \circ d[\mathrm{Lie}(\{h_t\}_{t\in\mathbb{R}})]$ is a finite-dimensional Lie algebra of Hamiltonian vector fields containing $\{X_t\}_{t\in\mathbb{R}}$. Therefore, $V^X \subset V$ and X is a Lie–Hamilton system. $\qquad\square$

Example 4.5. Consider again the second-order Riccati equation (3.53) in Hamiltonian form (3.58) and the vector fields (3.59) which span a five-dimensional VG Lie algebra for such equations. All the vector fields of this Lie algebra are Hamiltonian vector fields with respect to the Poisson bivector $\Lambda = \partial/\partial x \wedge \partial/\partial p$ on $\mathcal{O}$. Indeed, note that $X_\alpha = -\widehat{\Lambda}(dh_\alpha)$, with $\alpha = 1,\ldots,5$ and

$$h_1(x,p) = -2\sqrt{-p}, \quad h_2(x,p) = p, \quad h_3(x,p) = xp, \quad h_4(x,p) = x^2 p,$$
$$h_5(x,p) = -2x\sqrt{-p}. \quad (4.14)$$

This system is therefore a Lie–Hamilton system. Moreover, the previous Hamiltonian functions span along with $\tilde{h}_0 = 1$ a Lie algebra of functions isomorphic to the *two-photon Lie algebra* $\mathfrak{h}_6$ [27]

with non-vanishing Lie brackets given by (4.257). Consequently, system (3.53) admits a t-dependent Hamiltonian structure given by $(\mathrm{T}^*\mathbb{R}, \Lambda, h = h_1 - a_0(t)h_2 - a_1(t)h_3 - a_2(t)h_4)$.

Consider X to be a Lie–Hamilton system admitting a Lie–Hamiltonian structure (N, Λ, h) leading to a Lie–Hamilton algebra $\mathrm{Lie}(\{h_t\}_{t\in\mathbb{R}}, \{\cdot,\cdot\}_\Lambda)$. Let us now analyze the relations between all the previous geometric structures.

Lemma 4.5. *Given a system X on N with a Lie–Hamiltonian structure (N, Λ, h), we have that*

$$0 \hookrightarrow \mathrm{Cas}(N, \Lambda) \cap \mathrm{Lie}(\{h_t\}_{t\in\mathbb{R}}) \hookrightarrow \mathrm{Lie}(\{h_t\}_{t\in\mathbb{R}}) \xrightarrow{\mathcal{J}_\Lambda} V^X \to 0,$$

$$(4.15)$$

where $\mathcal{J}_\Lambda : f \in \mathrm{Lie}(\{h_t\}_{t\in\mathbb{R}}) \mapsto \widehat{\Lambda} \circ df \in V^X$, is an exact sequence of Lie algebras.

Proof. Consider the exact sequence of (generally) infinite-dimensional real Lie algebras

$$0 \hookrightarrow \mathrm{Cas}(N, \Lambda) \hookrightarrow C^\infty(N) \xrightarrow{\widehat{\Lambda}\circ d} \mathrm{Ham}(N, \Lambda) \to 0. \qquad (4.16)$$

Since $X_t = -\widehat{\Lambda} \circ dh_t$, we see that $V^X = \mathrm{Lie}(\widehat{\Lambda} \circ d(\{h_t\}_{t\in\mathbb{R}}))$. Using that $\widehat{\Lambda} \circ d$ is a Lie algebra morphism, we have $V^X = \widehat{\Lambda} \circ d[\mathrm{Lie}(\{h_t\}_{t\in\mathbb{R}})] = \mathcal{J}_\Lambda(\mathrm{Lie}(\{h_t\}_{t\in\mathbb{R}}))$. Additionally, as $\mathcal{J}_\Lambda$ is the restriction to $\mathrm{Lie}(\{h_t\}_{t\in\mathbb{R}})$ of $\widehat{\Lambda} \circ d$, we obtain that its kernel consists of Casimir functions belonging to $\mathrm{Lie}(\{h_t\}_{t\in\mathbb{R}})$, i.e., $\ker \mathcal{J}_\Lambda = \mathrm{Lie}(\{h_t\}_{t\in\mathbb{R}}) \cap \mathrm{Cas}(N, \Lambda)$. The exactness of sequence (4.15) easily follows from these results. $\qquad \square$

The above proposition entails that every X that possesses a Lie–Hamiltonian structure (N, Λ, h) is such that $\mathrm{Lie}(\{h_t\}_{t\in\mathbb{R}})$ is a Lie algebra extension of V^X by $\mathrm{Cas}(N, \Lambda) \cap \mathrm{Lie}(\{h_t\}_{t\in\mathbb{R}})$, i.e., the sequence of Lie algebras (4.15) is exact. Note that if $\mathrm{Lie}(\{h_t\}_{t\in\mathbb{R}})$ is finite-dimensional by assumption, all the Lie algebras appearing in such a sequence are finite-dimensional. For instance, the first-order system (3.58) associated with second-order Riccati equations in Hamiltonian form admits a Lie–Hamiltonian structure

$$\left(\mathcal{O}, \frac{\partial}{\partial x} \wedge \frac{\partial}{\partial p}, h_1 - a_0(t)h_2 - a_1(t)h_3 - a_2(t)h_4\right), \qquad (4.17)$$

where h_1, h_2, h_3, h_4 are given by (4.14). Note that $\mathrm{Lie}(\{h_t\}_{t\in\mathbb{R}})$, for generic functions $a_0(t), a_1(t), a_2(t)$, is a six-dimensional Lie algebra of functions $\mathfrak{W} \simeq V^X \oplus \mathbb{R}$. Hence, we see that $\mathfrak{W}$ is an extension of V^X.

Note that every t-dependent vector field admitting a Lie–Hamiltonian structure necessarily possesses many other Lie–Hamiltonian structures. For instance, if a system X admits (N, Λ, h), then it also admits a Lie–Hamiltonian structure (N, Λ, h'), with $h' : (t, x) \in \mathbb{R} \times N \mapsto h(t, x) + f_{\mathcal{C}}(x) \in \mathbb{R}$, where $f_{\mathcal{C}}$ is any Casimir function with respect to Λ. Indeed, it is easy to see that if $h_1, \ldots, h_r$ is a basis for $\mathrm{Lie}(\{h_t\}_{t\in\mathbb{R}})$, then $h_1, \ldots, h_r, f_{\mathcal{C}}$ span $\mathrm{Lie}(\{h'_t\}_{t\in\mathbb{R}})$, which also becomes a finite-dimensional real Lie algebra. As shown later, this has relevant implications for the linearization of Lie–Hamilton systems.

We have already proven that every system X admitting a Lie–Hamiltonian structure must possess several ones. Nevertheless, we have not yet studied the conditions ensuring that a Lie–Hamilton system X possesses a Lie–Hamiltonian structure. Let us answer this question.

Proposition 4.6. *Every Lie–Hamilton system admits a Lie–Hamiltonian structure.*

Proof. Assume X to be a Lie–Hamilton system on a manifold N with respect to a Poisson bivector Λ. Since $V^X \subset \mathrm{Ham}(N, \Lambda)$ is finite-dimensional, there exists a finite-dimensional linear space $\mathfrak{W}_0 \subset C^\infty(N)$ isomorphic to V^X and such that $\widehat{\Lambda} \circ d(\mathfrak{W}_0) = V^X$. Consequently, there exists a curve h_t in $\mathfrak{W}_0$ such that $X_t = -\widehat{\Lambda} \circ d(h_t)$. To ensure that h_t gives rise to a Lie–Hamiltonian structure, we need to demonstrate that $\mathrm{Lie}(\{h_t\}_{t\in\mathbb{R}}, \{\cdot, \cdot\}_\Lambda)$ is finite-dimensional. This will be done by constructing a finite-dimensional Lie algebra of functions containing the curve h_t.

Define the linear isomorphism $T : X_f \in V^X \mapsto -f \in \mathfrak{W}_0 \subset C^\infty(N)$ associating each vector field in V^X with minus its unique Hamiltonian function within $\mathfrak{W}_0$. This can be done by choosing a representative for each element of a basis of V^X and extending the map by linearity.

Note that this mapping need not be a Lie algebra morphism and hence $\operatorname{Im} T = \mathfrak{W}_0$ does not need to be a Lie algebra. Indeed, we can define a bilinear map $\Upsilon : V^X \times V^X \to C^\infty(N)$ of the form

$$\Upsilon(X_f, X_g) = \{f, g\}_\Lambda - T[X_f, X_g],$$

measuring the obstruction for T to be a Lie algebra morphism, i.e., Υ is identically null if and only if T is a Lie algebra morphism. In fact, if $\mathfrak{W}_0$ were a Lie algebra, then $\{f, g\}_\Lambda$ would be the only element of $\mathfrak{W}_0$ with Hamiltonian vector field $-[X_f, X_g]$, i.e., $T[X_f, X_g]$, and Υ would be a zero function.

Note that $\Upsilon(X_f, X_g)$ is the difference between two functions, namely $\{f, g\}_\Lambda$ and $T[X_f, X_g]$, sharing the same Hamiltonian vector field. Consequently, $\operatorname{Im} \Upsilon \subset \operatorname{Cas}(N, \Lambda)$ and it can be injected into a finite-dimensional Lie algebra of Casimir functions of the form

$$\mathfrak{W}_C := \langle \Upsilon(X_i, X_j) \rangle, \quad i, j = 1, \ldots, r, \tag{4.18}$$

where $X_1, \ldots, X_r$ is a basis for V^X. From here, it follows that

$$\{\mathfrak{W}_C, \mathfrak{W}_C\}_\Lambda = 0, \quad \{\mathfrak{W}_C, \mathfrak{W}_0\}_\Lambda = 0, \quad \{\mathfrak{W}_0, \mathfrak{W}_0\}_\Lambda \subset \mathfrak{W}_C + \mathfrak{W}_0. \tag{4.19}$$

Hence, $\mathfrak{W} := \mathfrak{W}_0 + \mathfrak{W}_C$ is a finite-dimensional Lie algebra of functions containing the curve h_t. From here, it readily follows that X admits a Lie–Hamiltonian structure $(N, \Lambda, -TX_t)$. $\qquad \square$

Since every Lie–Hamilton system possesses a Lie–Hamiltonian structure and every Lie–Hamiltonian structure determines a Lie–Hamilton system, we obtain the following theorem.

Theorem 4.7. *A system X admits a Lie–Hamiltonian structure if and only if it is a Lie–Hamilton system.*

Example 4.6. Let us consider the Hamilton equations for the second-order Kummer–Schwarz equations (3.68) in Hamiltonian form [45, 87]. They read

$$\begin{cases} \dfrac{dx}{dt} = \dfrac{px^3}{2}, \\ \dfrac{dp}{dt} = -\dfrac{3p^2x^2}{4} - 4c_0 + \dfrac{4b_1(t)}{x^2}, \end{cases} \tag{4.20}$$

on $T^*\mathbb{R}_0$, where $\mathbb{R}_0 = \mathbb{R} - \{0\}$. Once again, the above system is a Lie system as it describes the integral curves of the t-dependent vector field $X_t = X_3 + b_1(t)X_1$, where

$$X_1 = \frac{4}{x^2}\frac{\partial}{\partial p}, \quad X_2 = x\frac{\partial}{\partial x} - p\frac{\partial}{\partial p},$$

$$X_3 = \frac{px^3}{2}\frac{\partial}{\partial x} - \left(\frac{3p^2x^2}{4} + 4c_0\right)\frac{\partial}{\partial p}, \tag{4.21}$$

span a three-dimensional Lie algebra $V^{2\mathrm{KS}}$ isomorphic to $\mathfrak{sl}(2,\mathbb{R})$. Indeed,

$$[X_1, X_3] = 2X_2, \quad [X_1, X_2] = X_1, \quad [X_2, X_3] = X_3. \tag{4.22}$$

Apart from providing a new approach to Kummer–Schwarz equations (see [87] for a related method), our new description gives an additional relevant property: $V^{2\mathrm{KS}}$ consists of Hamiltonian vector fields with respect to the Poisson bivector $\Lambda = \partial/\partial x \wedge \partial/\partial p$ on $T^*\mathbb{R}_0$. In fact, $X_\alpha = -\widehat{\Lambda}(dh_\alpha)$ with $\alpha = 1, 2, 3$ and

$$h_1 = \frac{4}{x}, \quad h_2 = xp, \quad h_3 = \frac{1}{4}p^2x^3 + 4c_0x. \tag{4.23}$$

Therefore, (4.20) is a Lie–Hamilton system. Moreover, we have that

$$\{h_1, h_2\} = -h_1, \quad \{h_1, h_3\} = -2h_2, \quad \{h_2, h_3\} = -h_3. \tag{4.24}$$

Then, (4.20) admits a Lie–Hamiltonian structure $(T^*\mathbb{R}_0, \Lambda, h = h_3 + b_1(t)h_1)$ as ensured by Theorem 4.7.

4.2.2. t-independent constants of motion

Let us now study the structure of the space of t-independent constants of motion for Lie–Hamilton systems. In particular, we are interested in investigating the use of Poisson structures to study such constants of motion.

Proposition 4.8. *Given a system X with a Lie–Hamiltonian structure (N, Λ, h), then $\mathcal{C}^\Lambda \subset \mathcal{V}^X$, where we recall that $\mathcal{C}^\Lambda$ is the Casimir distribution relative to Λ.*

Proof. Consider a $\theta_x \in \mathcal{C}_x^\Lambda$, with $x \in N$. As X is a Lie–Hamilton system, for every $Y \in V^X$ there exists a function $f \in C^\infty(N)$ such that $Y = -\widehat{\Lambda}(df)$. Then,

$$\theta_x(Y_x) = -\theta_x(\widehat{\Lambda}_x(df_x)) = -\Lambda_x(df_x, \theta_x) = 0, \qquad (4.25)$$

where $\widehat{\Lambda}_x$ is the restriction of $\widehat{\Lambda}$ to T_x^*N. As the vectors Y_x, with $Y \in V^X$, span $\mathcal{D}_x^X$, then $\theta_x \in \mathcal{V}_x^X$ and $\mathcal{C}^\Lambda \subset \mathcal{V}^X$. $\square$

Observe that different Lie–Hamiltonian structures for a Lie–Hamilton system X may lead to different families of Casimir functions, which may determine different constants of motion for X.

Theorem 4.9. *Let X be a system admitting a Lie–Hamiltonian structure (N, Λ, h), the space $\mathcal{I}^X|_U$ of t-independent constants of motion of X on an open $U \subset U^X$ is a Poisson algebra. Additionally, the codistribution $\mathcal{V}^X|_{U_X}$ is involutive with respect to the Lie bracket $[\cdot, \cdot]_\Lambda$ induced by Λ on the space $\Gamma(\pi_N)$ of smooth 1-forms on N.*

Proof. Let $f_1, f_2 : U \to \mathbb{R}$ be two t-independent functions constants of motion for X, i.e., $X_t f_i = 0$, for $i = 1, 2$ and $t \in \mathbb{R}$. As X is a Lie–Hamilton system, all the elements of V^X are Hamiltonian vector fields and we can write $Y\{f, g\}_\Lambda = \{Yf, g\}_\Lambda + \{f, Yg\}_\Lambda$ for every $f, g \in C^\infty(N)$ and $Y \in V^X$. In particular, $X_t(\{f_1, f_2\}_\Lambda) = \{X_t f_1, f_2\}_\Lambda + \{f_1, X_t f_2\}_\Lambda = 0$. Hence, the Poisson bracket of t-independent constants of motion is a new one. As $\lambda f_1 + \mu f_2$ and $f_1 \cdot f_2$ are also t-independent constants of motion for every $\lambda, \mu \in \mathbb{R}$, it easily follows that $\mathcal{I}^X|_U$ is a Poisson algebra.

In view of Lemma 2.2 in Chapter 2, the codistribution $\mathcal{V}^X$ admits a local basis of exact forms $df_1, \ldots, df_{p(x)}$ for every point $x \in U^X$, where $\mathcal{V}^X$ has local constant rank $p(x) := \dim N - \dim \mathcal{D}_x^X$. Now, $[df_i, df_j]_\Lambda = d(\{f_i, f_j\}_\Lambda)$ for $i, j = 1, \ldots, p(x)$. We have already proven that the function $\{f_i, f_j\}_\Lambda$ is another first integral. Therefore, from Lemma 2.2 in Chapter 2, it easily follows that $\{f_i, f_j\}_\Lambda = G(f_1, \ldots, f_{p(x)})$. Thus, $[df_i, df_j]_\Lambda \in \mathcal{V}^X|_{U_X}$. Using the latter and the properties of the Lie bracket $[\cdot, \cdot]_\Lambda$, it directly turns out that the Lie bracket of two 1-forms taking values in $\mathcal{V}^X|_{U_X}$ belongs to $\mathcal{V}^X|_{U_X}$. Hence, $\mathcal{V}^X|_{U_X}$ is involutive. $\qquad\square$

Corollary 4.10. *Given a Lie–Hamilton system X, the space $\mathcal{I}^X|_U$, where $U \subset U^X$ is such that $\mathcal{V}^X$ admits a local basis of exact forms, is a function group, that is*

1. *The space $\mathcal{I}^X|_U$ is a Poisson algebra.*
2. *There exists a family of functions $f_1, \ldots, f_s \in \mathcal{I}^X|_U$ such that every element f of $\mathcal{I}^X|_U$ can be put in the form $f = F(f_1, \ldots, f_s)$ for a certain function $F : \mathbb{R}^s \to \mathbb{R}$.*

Proof. In view of the previous theorem, $\mathcal{I}^X|_U$ is a Poisson algebra with respect to a certain Poisson bracket. Taking into account Proposition 2.1 in Chapter 2 and the form of $\mathcal{I}^X|_U$ given by Lemma 2.2 in Chapter 2, we obtain that this space becomes a function group. $\qquad\square$

Previous properties do not necessarily hold for systems other than Lie–Hamilton systems, as they do not need to admit any *a priori* relation among a Poisson bracket of functions and the t-dependent vector field describing the system. Let us exemplify this.

Example 4.7. Consider the Poisson manifold $(\mathbb{R}^3, \Lambda_{GM})$, where

$$\Lambda_{GM} = \sigma_3 \frac{\partial}{\partial \sigma_2} \wedge \frac{\partial}{\partial \sigma_1} - \sigma_1 \frac{\partial}{\partial \sigma_2} \wedge \frac{\partial}{\partial \sigma_3} + \sigma_2 \frac{\partial}{\partial \sigma_3} \wedge \frac{\partial}{\partial \sigma_1}, \qquad (4.26)$$

and $(\sigma_1, \sigma_2, \sigma_3)$ is a coordinate basis for $\mathbb{R}^3$, appearing in the study of Classical XYZ Gaudin Magnets [31]. The system $X = \partial/\partial\sigma_3$ is not a Lie–Hamilton system with respect to this Poisson structure as

X is not Hamiltonian, namely $\mathcal{L}_X \Lambda_{GM} \neq 0$. In addition, this system admits two first integrals σ_1 and σ_2. Nevertheless, their Lie bracket reads $\{\sigma_1, \sigma_2\} = -\sigma_3$, which is not a first integral for X. On the other hand, consider the system

$$Y = \sigma_3 \frac{\partial}{\partial \sigma_2} + \sigma_2 \frac{\partial}{\partial \sigma_3}, \qquad (4.27)$$

which is a Lie–Hamilton system, as it can be written in the form $Y = -\widehat{\Lambda}_{GM}(d\sigma_1)$, and it possesses two first integrals given by σ_1 and $\sigma_2^2 - \sigma_3^2$. Unsurprisingly, $Y\{\sigma_1, \sigma_2^2 - \sigma_3^2\} = 0$, i.e., the Lie bracket of two t-independent constants of motion is also a constant of motion.

Let us prove some final interesting results about the t-independent constants of motion for Lie–Hamilton systems.

Proposition 4.11. *Let X be a Lie–Hamilton system that admits a Lie–Hamiltonian structure (N, Λ, h). The function $f : N \to \mathbb{R}$ is a constant of motion for X if and only if f Poisson commutes with all elements of* $\mathrm{Lie}(\{h_t\}_{t \in \mathbb{R}}, \{\cdot, \cdot\}_\Lambda)$.

Proof. The function f is a t-independent constant of motion for X if and only if

$$0 = X_t f = \{f, h_t\}_\Lambda, \quad \forall t \in \mathbb{R}. \qquad (4.28)$$

From here,

$$\{f, \{h_t, h_{t'}\}_\Lambda\}_\Lambda = \{\{f, h_t\}_\Lambda, h_{t'}\}_\Lambda$$
$$+ \{h_t, \{f, h_{t'}\}_\Lambda\}_\Lambda = 0, \quad \forall t, t' \in \mathbb{R}, \quad (4.29)$$

and inductively follows that f Poisson commutes with all successive Poisson brackets of elements of $\{h_t\}_{t \in \mathbb{R}}$ and their linear combinations. As these elements span $\mathrm{Lie}(\{h_t\}_{t \in \mathbb{R}})$, we get that f Poisson commutes with $\mathrm{Lie}(\{h_t\}_{t \in \mathbb{R}})$.

Conversely, if f Poisson commutes with $\mathrm{Lie}(\{h_t\}_{t \in \mathbb{R}})$, it commutes with the elements $\{h_t\}_{t \in \mathbb{R}}$, and, in view of (4.28), becomes a constant of motion for X. $\qquad \square$

In order to illustrate the above proposition, let us show an example.

Example 4.8. Consider a Smorodinsky–Winternitz oscillator (4.4) with $n = 2$. Recall that this system admits a Lie–Hamiltonian structure $(T^*\mathbb{R}_0^2, \Lambda, h = h_3 + \omega^2(t)h_1)$, where $\Lambda = \sum_{i=1}^2 \partial/\partial x_i \wedge \partial/\partial p_i$ is a Poisson bivector on $T^*\mathbb{R}_0^2$ and the functions h_1, h_3 are given within (4.8). For non-constant $\omega(t)$, we obtain that $\mathrm{Lie}(\{h_t\}_{t\in\mathbb{R}}, \{\cdot,\cdot\}_\Lambda)$ is a real Lie algebra of functions isomorphic to $\mathfrak{sl}(2,\mathbb{R})$ generated by the functions h_1, h_2 and h_3 detailed in (4.8). When $\omega(t) = \omega_0 \in \mathbb{R}$, the Lie algebra $\mathrm{Lie}(\{h_t\}_{t\in\mathbb{R}}, \{\cdot,\cdot\}_\Lambda)$ becomes a one-dimensional Lie subalgebra of the previous one. In any case, it is known that

$$I = (x_1 p_2 - p_1 x_2)^2 + k\left[\left(\frac{x_1}{x_2}\right)^2 + \left(\frac{x_2}{x_1}\right)^2\right], \qquad (4.30)$$

is a t-independent constant of motion (cf. [116]). A simple calculation shows that

$$\{I, h_\alpha\}_\Lambda = 0, \quad \alpha = 1,2,3. \qquad (4.31)$$

Then, the function I always Poisson commutes with the whole $\mathrm{Lie}(\{h_t\}_{t\in\mathbb{R}}, \{\cdot,\cdot\}_\Lambda)$, as expected.

Obviously, every autonomous Hamiltonian system is a Lie–Hamilton system possessing a Lie–Hamiltonian structure (N, Λ, h), with h being a t-independent Hamiltonian. Consequently, Proposition 4.11 shows that the t-independent first integrals for a Hamiltonian system are those functions that Poisson commutes with its Hamiltonian, recovering as a particular case this widely known result.

Moreover, the above proposition suggests that the role played by autonomous Hamiltonians for Hamiltonian systems is performed by the finite-dimensional Lie algebras of functions associated with Lie–Hamiltonian structures in the case of Lie–Hamilton systems. This can be employed to study t-independent first integrals of Lie–Hamilton systems or analyze the maximal number of such first integrals in involution, which would lead to the interesting analysis of integrability/superintegrability of Lie–Hamilton systems.

4.2.3. Symmetries, linearization, and comomentum maps

Definition 4.12. We say that a Lie system X admitting a Lie–Hamilton structure (N, Λ, h) possesses a compatible *strong comomentum map* with respect to this Lie–Hamilton structure if there exists a Lie algebra morphism $\lambda : V^X \to \mathrm{Lie}(\{h_t\}_{t\in\mathbb{R}}, \{\cdot,\cdot\}_\Lambda)$ such that the following diagram

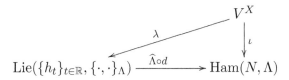

where $\iota : V^X \hookrightarrow \mathrm{Ham}(N, \Lambda)$ is the natural injection of V^X into $\mathrm{Ham}(N, \Lambda)$, is commutative.

Proposition 4.13. *Let X be a Lie system possessing a Lie–Hamiltonian structure (N, Λ, h) compatible with a strong comomentum map λ such that $\dim \mathcal{D}_x^X = \dim N = \dim V^X$ at a generic $x \in N$. Then, there exists a local coordinate system defined on a neighborhood of each x such that X and Λ are simultaneously linearizable and where X possesses a linear superposition rule.*

Proof. As it is assumed that $n := \dim N = \dim V^X = \dim \mathcal{D}_x^X$ at a generic x, every basis $X_1, \ldots, X_n$ of V^X gives rise to a basis for the tangent bundle TN on a neighborhood of x. Since X admits a strong comomentum map compatible with (N, Λ, h), we have $(V^X, [\cdot, \cdot]) \simeq (\lambda(V^X), \{\cdot, \cdot\}_\Lambda)$ and the family of functions, $h_\alpha = \lambda(X_\alpha)$, with $\alpha = 1, \ldots, n$, forms a basis for the Lie subalgebra $\lambda(V^X)$. Moreover, since $\widehat{\Lambda} \circ d \circ \lambda(V^X) = V^X$ and $\dim V^X = \dim \mathcal{D}_{x'}^X$ for x' in a neighborhood of x, then $\widehat{\Lambda}_{x'} \circ d(\lambda(V^X)) \simeq T_{x'}N$ and $dh_1 \wedge \cdots \wedge dh_n \neq 0$ at a generic point. Hence, the set $(h_1, \ldots, h_n)$ is a coordinate system on an open dense subset of N. Now, using again that $(\lambda(V^X), \{\cdot, \cdot\}_\Lambda)$ is a real Lie algebra, the Poisson bivector Λ can be put in the form

$$\Lambda = \frac{1}{2} \sum_{i,j=1}^n \{h_i, h_j\}_\Lambda \frac{\partial}{\partial h_i} \wedge \frac{\partial}{\partial h_j} = \frac{1}{2} \sum_{i,j,k=1}^n c_{ijk} h_k \frac{\partial}{\partial h_i} \wedge \frac{\partial}{\partial h_j}, \quad (4.32)$$

for certain real n^3 constants c_{ijk}. In other words, the Poisson bivector Λ becomes linear in the chosen coordinate system.

Since we can write $X_t = -\widehat{\Lambda}(d\bar{h}_t)$, with $\bar{h}_t = -\lambda(X_t)$ being a curve in the Lie algebra $\lambda(V^X) \subset \text{Lie}(\{h_t\}_{t\in\mathbb{R}})$, expression (4.32) yields

$$X_t = -\widehat{\Lambda}(d\bar{h}_t) = -\widehat{\Lambda} \circ d \left(\sum_{l=1}^{n} b_l(t)h_l \right)$$

$$= -\sum_{l=1}^{n} b_l(t)(\widehat{\Lambda} \circ dh_l) = -\sum_{l,j,k=1}^{n} b_l(t)c_{ljk}h_k \frac{\partial}{\partial h_j}, \quad (4.33)$$

and X_t is linear in this coordinate system. Consequently, as every linear system, X admits a linear superposition rule in the coordinate system $(h_1, \ldots, h_n)$. $\qquad\square$

Let us turn to describing some features of t-independent Lie symmetries for Lie–Hamilton systems. Our exposition will be based upon the properties of the hereafter called *symmetry distribution*.

Definition 4.14. Given a Lie–Hamilton system X that possesses a Lie–Hamiltonian structure (N, Λ, h), we define its *symmetry distribution*, $\mathcal{S}_\Lambda^X$, by

$$(\mathcal{S}_\Lambda^X)_x = \widehat{\Lambda}_x(\mathcal{V}_x^X) \in T_x N, \quad x \in N. \quad (4.34)$$

As its name indicates, the symmetry distribution can be employed to investigate the t-independent Lie symmetries of a Lie–Hamilton system. Let us give some basic examples of how this can be done.

Proposition 4.15. *Given a Lie–Hamilton system X with a Lie–Hamiltonian structure (N, Λ, h), then*

1. *The symmetry distribution $\mathcal{S}_\Lambda^X$ associated with X and Λ is involutive on an open subset of U^X, where U^X is the open dense subset of N where $\mathcal{V}^X$ is differentiable.*
2. *If f is a t-independent constant of motion for X, then $\widehat{\Lambda}(df)$ is a t-independent Lie symmetry of X.*

Lie–Hamilton Systems

3. *The distribution $\mathcal{S}_\Lambda^X$ admits a local basis of t-independent Lie symmetries of X defined around a generic point of N. The elements of such a basis are Hamiltonian vector fields of t-independent constants of motion of X.*

Proof. By definition of $\mathcal{S}_\Lambda^X$ and using that $\mathcal{V}^X$ has constant rank on the connected components of an open $U \subset U_X$, where Λ has locally constant rank, we can ensure that given two vector fields in $Y_1, Y_2 \in \mathcal{S}_\Lambda^X|_U$, there exist 2-forms $\omega, \omega' \in \mathcal{V}^X|_U$ such that $Y_1 = \widehat{\Lambda}(\omega)$, $Y_2 = \widehat{\Lambda}(\omega')$. Since X is a Lie–Hamilton system, $\mathcal{V}^X|_U$ is involutive and $\widehat{\Lambda}$ is an anchor, i.e., a Lie algebra morphism from $(\Gamma(\pi_N), [\cdot, \cdot]_\Lambda)$ to $(\Gamma(\tau_N), [\cdot, \cdot])$, then

$$[Y_1, Y_2] = [\widehat{\Lambda}(w), \widehat{\Lambda}(w')] = \widehat{\Lambda}([w, w']_\Lambda) \in \mathcal{S}_\Lambda^X. \tag{4.35}$$

In other words, since $\mathcal{V}^X$ is involutive on U, then $\mathcal{S}_\Lambda^X$ is so, which proves (1).

To prove (2), note that

$$[X_t, \widehat{\Lambda}(df)] = -[\widehat{\Lambda}(dh_t), \widehat{\Lambda}(df)] = -\widehat{\Lambda}(d\{h_t, f\}_\Lambda) = \widehat{\Lambda}[d(X_t f)] = 0. \tag{4.36}$$

Finally, the proof of (3) is based upon the fact that $\mathcal{V}^X$ admits, around a point $x \in U^X \subset N$, a local basis of 1-forms $df_1, \ldots, df_{p(x)}$, with $f_1, \ldots, f_{p(x)}$ being a family of t-independent constants of motion for X and $p(x) = \dim N - \dim \mathcal{D}_x^X$. From (2), the vector fields $X_{f_1}, \ldots, X_{f_{p(x)}}$ form a family of Lie symmetries of X locally spanning $\mathcal{S}_\Lambda^X$. Hence, we can easily choose among them a local basis for $\mathcal{S}_\Lambda^X$. $\square$

Example 4.9. As a particular example of the usefulness of the above result, let us turn to a two-dimensional Smorodinsky–Winternitz oscillator X given by (4.4) and its known constant of motion (4.30). In view of the previous proposition, $Y = \widehat{\Lambda}(dI)$ must be a Lie symmetry for these systems. A short calculation leads to

$$Y = 2(x_1p_2 - p_1x_2)\left(x_2\frac{\partial}{\partial x_1} - x_1\frac{\partial}{\partial x_2}\right) + 2\left[(x_1p_2 - p_1x_2)p_2\right.$$
$$\left. +k\frac{x_1^4 - x_2^4}{x_1^3x_2^2}\right]\frac{\partial}{\partial p_1} - 2\left[(x_1p_2 - p_1x_2)p_1 + k\frac{x_1^4 - x_2^4}{x_2^3x_1^2}\right]\frac{\partial}{\partial p_2},$$

(4.37)

and it is straightforward to verify that Y commutes with X_1, X_2, X_3, given by (4.6), and therefore with every X_t, with $t \in \mathbb{R}$, i.e., Y is a Lie symmetry for X.

Proposition 4.16. *Let X be a Lie–Hamilton system with a Lie–Hamiltonian structure (N, Λ, h). If $[V^X, V^X] = V^X$ and $Y \in$ Ham(N, Λ) is a Lie symmetry of X, then $Y \in \mathcal{S}_\Lambda^X$.*

Proof. As Y is a t-independent Lie symmetry, then $[Y, X_t] = 0$ for every $t \in \mathbb{R}$. Since Y is a Hamiltonian vector field, then $Y = -\widehat{\Lambda} \circ df$ for a certain $f \in C^\infty(N)$. Using that $X_t = -\widehat{\Lambda}(dh_t)$, we obtain

$$0 = [Y, X_t] = [\widehat{\Lambda}(df), \widehat{\Lambda}(dh_t)] = \widehat{\Lambda}(d\{f, h_t\}_\Lambda) = \widehat{\Lambda}[d(X_t f)]. \quad (4.38)$$

Hence, $X_t f$ is a Casimir function. Therefore, as every $X_{t'}$ is a Hamiltonian vector field for all $t' \in \mathbb{R}$, it turns out that $X_{t'}X_t f = 0$ for every $t, t' \in \mathbb{R}$ and, in consequence, $Z_1 f$ is a Casimir function for every $Z_1 \in V^X$. Moreover, as every $Z_2 \in V^X$ is Hamiltonian, we have

$$Z_2 Z_1 f = Z_1 Z_2 f = 0 \Longrightarrow (Z_2 Z_1 - Z_1 Z_2)f = [Z_2, Z_1]f = 0. \quad (4.39)$$

As $[V^X, V^X] = V^X$, every element Z of V^X can be written as the commutator of two elements of V^X and, in view of the above expression, $Zf = 0$ which shows that f is a t-independent constant of motion for X. Finally, as $Y = -\widehat{\Lambda}(df)$, then $Y \in \mathcal{S}_\Lambda^X$. $\qquad \square$

Note that, roughly speaking, the above proposition ensures that when V^X is *perfect*, i.e., $[V^X, V^X] = V^X$ (see [70]), then $\mathcal{S}_\Lambda^X$ contains all Hamiltonian Lie symmetries of X. This is the case for Smorodinsky–Winternitz systems (4.4) with a non-constant $\omega(t)$, whose V^X was already shown to be isomorphic to $\mathfrak{sl}(2, \mathbb{R})$.

4.2.4. On t-dependent constants of motion

The aim of this section is to present an analysis of the algebraic properties of the t-dependent constants of motion for Lie–Hamilton systems. More specifically, we prove that a Lie–Hamiltonian structure (N, Λ, h) for a Lie–Hamilton system X induces a Poisson bivector on $\mathbb{R} \times N$. This allows us to endow the space of constants of motion for X with a Poisson algebra structure, which can be used to produce new constants of motion from known ones.

Definition 4.17. We call *autonomization* of a vector field X to the explicit t-dependent vector field $\bar{X} = X + d/dt$.

Given a system X on N, a constant of motion for X is a first integral $f \in C^\infty(\mathbb{R} \times N)$ of the autonomization $\bar{X}$.

$$\frac{\partial f}{\partial t} + Xf = \bar{X}f = 0, \tag{4.40}$$

where X is understood as a vector field on $\mathbb{R} \times N$. Using this, we can straightforwardly prove the following proposition.

Proposition 4.18. *The space $\bar{\mathcal{I}}^X$ of t-dependent constants of motion for a system X forms an $\mathbb{R}$-algebra $(\bar{\mathcal{I}}^X, \cdot)$.*

Lemma 4.19. *Every Poisson manifold (N, Λ) induces a Poisson manifold $(\mathbb{R} \times N, \bar{\Lambda})$ with Poisson structure*

$$\{f, g\}_{\bar{\Lambda}}(t, x) := \{f_t, g_t\}_\Lambda(x), \quad (t, x) \in \mathbb{R} \times N. \tag{4.41}$$

Definition 4.20. Given a Poisson manifold (N, Λ), the associated Poisson manifold $(\mathbb{R} \times N, \bar{\Lambda})$ is called the *autonomization* of (N, Λ). Likewise, the Poisson bivector $\bar{\Lambda}$ is called the *autonomization* of Λ.

The following lemma allows us to prove that $(\bar{\mathcal{I}}^X, \cdot, \{\cdot, \cdot\}_{\bar{\Lambda}})$ is a Poisson algebra.

Lemma 4.21. *Let (N, Λ) be a Poisson manifold and X be a Hamiltonian vector field on N relative to Λ. Then, $\mathcal{L}_{\bar{X}}\bar{\Lambda} = 0$.*

Proof. Given a coordinate system $\{x_1, \ldots, x_n\}$ for N and $x_0 := t$ in $\mathbb{R}$, we can naturally define a coordinate system $\{x_0, x_1, \ldots, x_n\}$

on $\mathbb{R} \times N$. Since $(x_0)_t = t$ is constant as a function on N, then $\bar{\Lambda}(\mathrm{d}x_0, \mathrm{d}f) = \{(x_0)_t, f_t\}_\Lambda = 0$ for every $f \in C^\infty(\mathbb{R} \times N)$. Additionally, $(x_i)_t = x_i$ for $i = 1, \dots, n$. Hence, we have that $\bar{\Lambda}(\mathrm{d}x_i, \mathrm{d}x_j) = \{x_i, x_j\}_{\bar{\Lambda}}$ is a x_0-independent function for $i, j = 0, \dots, n$. So,

$$(\mathcal{L}_{\bar{X}}\bar{\Lambda})(t, x) = \left[\mathcal{L}_{\frac{\partial}{\partial x_0} + X} \left(\sum_{i<j=1}^{n} \{x_i, x_j\}_\Lambda \frac{\partial}{\partial x_i} \wedge \frac{\partial}{\partial x_j} \right) \right](x)$$
$$= (\mathcal{L}_X \Lambda)(x). \tag{4.42}$$

Since X is Hamiltonian, we obtain $0 = (\mathcal{L}_X \Lambda)(x) = (\mathcal{L}_{\bar{X}}\bar{\Lambda})(t, x) = 0$. $\qquad\square$

Now, we can establish the main result of this section.

Proposition 4.22. *Let X be a Lie–Hamilton system on N with a Lie–Hamiltonian structure (N, Λ, h), then the space $(\bar{\mathcal{I}}^X, \cdot, \{\cdot, \cdot\}_{\bar{\Lambda}})$ is a Poisson algebra.*

Proof. From Proposition 4.18, we see that $(\bar{\mathcal{I}}^X, \cdot)$ is an $\mathbb{R}$-algebra. To demonstrate that $(\bar{\mathcal{I}}^X, \cdot, \{\cdot, \cdot\}_{\bar{\Lambda}})$ is a Poisson algebra, it remains to prove that the Poisson bracket of any two elements f, g of $\bar{\mathcal{I}}^X$ remains in it, i.e., $X\{f, g\}_{\bar{\Lambda}} = 0$. By taking into account that the vector fields $\{X_t\}_{t \in \mathbb{R}}$ are Hamiltonian relative to (N, Λ) and Lemma 4.21, we find that $\bar{\Lambda}$ is invariant under the autonomization of each vector field $X_{t'}$ with $t' \in \mathbb{R}$, i.e., $\mathcal{L}_{\overline{X_{t'}}}\bar{\Lambda} = 0$. Therefore,

$$\bar{X}\{f, g\}_{\bar{\Lambda}}(t', x) = \overline{X_{t'}}\{f, g\}_{\bar{\Lambda}}(t', x)$$
$$= \{\overline{X_{t'}}f, g\}_{\bar{\Lambda}}(t', x) + \{f, \overline{X_{t'}}g\}_{\bar{\Lambda}}(t', x)$$
$$= \{\bar{X}f, g\}_{\bar{\Lambda}}(t', x) + \{f, \bar{X}g\}_{\bar{\Lambda}}(t', x) = 0. \tag{4.43}$$

That is, $\{f, g\}_{\bar{\Lambda}}$ is a t-dependent constant of motion for X. $\qquad\square$

4.2.5. *Lie integrals*

Definition 4.23. Given a Lie–Hamilton system X on N possessing a Lie–Hamiltonian structure (N, Λ, h), a *Lie integral* of X with respect to (N, Λ, h) is a (generally t-dependent) constant of motion f of X

such that $\{f_t\}_{t\in\mathbb{R}} \subset \mathcal{H}_\Lambda$. In other words, given a basis $h_1, \ldots, h_r$ of the Lie algebra $(\mathcal{H}_\Lambda, \{\cdot, \cdot\}_\Lambda)$, we have that $\bar{X}f = 0$ and $f_t = \sum_{\alpha=1}^{r} f_\alpha(t)h_\alpha$ for every $t \in \mathbb{R}$ and certain t-dependent functions $f_1, \ldots, f_r$.

The Lie integrals of a Lie–Hamilton system X relative to a Lie–Hamiltonian structure (N, Λ, h) are the solutions of the equation

$$0 = \bar{X}f = \frac{\partial f}{\partial t} + Xf = \frac{\partial f}{\partial t} + \{f, h\}_{\bar{\Lambda}} \implies \frac{\partial f}{\partial t} = \{h, f\}_{\bar{\Lambda}}. \quad (4.44)$$

Since f and h can be understood as curves $t \mapsto f_t$ and $t \mapsto g_t$ within $\mathcal{H}_\Lambda$, the above equation can be rewritten as

$$\frac{\mathrm{d}f_t}{\mathrm{d}t} = \{h_t, f_t\}_\Lambda, \quad (4.45)$$

which can be thought of as a Euler equation on the Lie algebra $(\mathcal{H}_\Lambda, \{\cdot, \cdot\}_\Lambda)$. Equations of this type quite frequently appear in the literature such as in the Lewis–Riesenfeld method and works concerning Lie–Hamilton systems [170, 279, 284].

Proposition 4.24. *Given a Lie–Hamilton system X with a Lie–Hamiltonian structure (N, Λ, h), the space $\mathfrak{L}_h^\Lambda$ of Lie integrals relative to (N, Λ, h) gives rise to a Lie algebra $(\mathfrak{L}_h^\Lambda, \{\cdot, \cdot\}_{\bar{\Lambda}})$ isomorphic to $(\mathcal{H}_\Lambda, \{\cdot, \cdot\}_\Lambda)$.*

Proof. Since the Lie integrals of X are the solutions of the system of ODEs (4.45) on $\mathcal{H}_\Lambda$, they span an $\mathbb{R}$-linear space of dimension $\dim \mathcal{H}_\Lambda$. In view of Proposition 4.22, the Poisson bracket $\{f, g\}_{\bar{\Lambda}}$ of two constants of motion f, g for X is another constant of motion. If f and g are Lie integrals, the function $\{f, g\}_{\bar{\Lambda}}$ is then a new constant of motion that can additionally be considered as a curve $t \mapsto \{f_t, g_t\}_\Lambda$ taking values in $\mathcal{H}_\Lambda$, i.e., a new Lie integral.

Consider now the linear morphism $\mathcal{E}_0 : f \in \mathfrak{L}_h^\Lambda \mapsto f_0 \in \mathcal{H}_\Lambda$ relating every Lie integral to its value in $\mathcal{H}_\Lambda$ at $t = 0$. As every initial condition in $\mathcal{H}_\Lambda$ is related to a single solution of (4.45), we can relate every $v \in \mathcal{H}_\Lambda$ to a unique Lie integral f of X such that $f_0 = v$. Therefore, $\mathcal{E}_0$ is an isomorphism. Indeed, it is a Lie algebra isomorphism since $(\{f, g\}_{\bar{\Lambda}})_0 = \{f_0, g_0\}_\Lambda$ for every $f, g \in \mathfrak{L}_h^\Lambda$. $\square$

Proposition 4.25. *Given a Lie–Hamilton system X on N which possesses a Lie–Hamiltonian structure (N, Λ, h), then $\mathfrak{L}_h^\Lambda$ consists of t-independent constants of motion if and only if $\mathcal{H}_\Lambda$ is abelian.*

Proof. If $(\mathcal{H}_\Lambda, \{\cdot, \cdot\}_\Lambda)$ is abelian, then $\{f_t, h_t\}_\Lambda = 0$ and the system (4.45) reduces to $df_t/dt = 0$, whose solutions are of the form $f_t = g \in \mathcal{H}_\Lambda$, i.e., $\mathfrak{L}_h^\Lambda = \mathcal{H}_\Lambda$. Conversely, if $\mathfrak{L}_h^\Lambda = \mathcal{H}_\Lambda$, then every $g \in \mathcal{H}_\Lambda$ is a solution of (4.45) and $\{g, h_t\}_\Lambda = 0 \ \forall t \in \mathbb{R}$. Hence, every $g \in \mathcal{H}_\Lambda$ commutes with the whole $\mathcal{H}_\Lambda$, which becomes Abelian. $\qquad\square$

4.2.6. *Polynomial Lie integrals*

Let us formally define and investigate a remarkable class of constants of motion for Lie–Hamilton systems appearing in the literature [170, 284], hereafter called *Lie integrals*, and a relevant generalization of them, the *polynomial Lie integrals*. We first prove that Lie integrals can be characterized by an Euler equation on a finite-dimensional real Lie algebra of functions, retrieving as a particular case a result given in [170]. Then, we show that Lie integrals form a finite-dimensional real Lie algebra and we devise several methods to determine them. Our results can easily be extended to investigate certain quantum mechanical systems [279]. Finally, we investigate polynomial Lie integrals and the relevance of Casimir functions to derive them.

Definition 4.26. Given a Lie–Hamilton system X on N possessing a Lie–Hamiltonian structure (N, Λ, h), a *Lie integral* of X with respect to (N, Λ, h) is a constant of motion f of X such that $\{f_t\}_{t\in\mathbb{R}} \subset \mathcal{H}_\Lambda$. In other words, given a basis $h_1, \ldots, h_r$ of the Lie algebra $(\mathcal{H}_\Lambda, \{\cdot, \cdot\}_\Lambda)$, we have that $\bar{X}f = 0$ and $f_t = \sum_{\alpha=1}^r f_\alpha(t) h_\alpha$ for every $t \in \mathbb{R}$ and certain t-dependent functions $f_1, \ldots, f_r$.

The Lie integrals of a Lie–Hamilton system X relative to a Lie–Hamiltonian structure (N, Λ, h) are the solutions of the equation

$$0 = \bar{X}f = \frac{\partial f}{\partial t} + Xf = \frac{\partial f}{\partial t} + \{f, h\}_{\bar{\Lambda}} \implies \frac{\partial f}{\partial t} = \{h, f\}_{\bar{\Lambda}}. \quad (4.46)$$

Since f and h can be understood as curves $t \mapsto f_t$ and $t \mapsto g_t$ within $\mathcal{H}_\Lambda$, the above equation can be rewritten as

$$\frac{\mathrm{d}f_t}{\mathrm{d}t} = \{h_t, f_t\}_\Lambda, \tag{4.47}$$

which can be thought of as a Euler equation on the Lie algebra $(\mathcal{H}_\Lambda, \{\cdot, \cdot\}_\Lambda)$ [170]. Equations of this type quite frequently appear in the literature such as in the Lewis–Riesenfeld method and works concerning Lie–Hamilton systems [170, 279, 284].

4.2.6.1. *Polynomial Lie integrals and Casimir functions*

We here investigate a generalization of Lie integrals: the hereafter called *polynomial Lie integrals*. Although we prove that these constants of motion can be determined by Lie integrals, we also show that their determination can be simpler in some cases. In particular, we can obtain polynomial Lie integrals algebraically by means of the Casimir functions related to the Lie algebra of Lie integrals.

Definition 4.27. Let X be a Lie–Hamilton system admitting a compatible Lie–Hamiltonian structure (N, Λ, h). A *polynomial Lie integral* for X with respect to (N, Λ, h) is a constant of motion f for X of the form $f_t = \sum_{I \in M} \lambda_I(t) h^I$, where I's are r-multi-indexes, i.e., sets $(i_1, \dots, i_r)$ of nonnegative integers, set M is a finite family of multi-indexes, $\lambda_I(t)$ are certain t-dependent functions, and $h^I = h_1^{i_1} \cdot \ldots \cdot h_r^{i_r}$ for a fixed basis $\{h_1, \dots, h_r\}$ for $\mathcal{H}_\Lambda$.

The study of polynomial Lie integrals can be approached through the symmetric algebra $S_\mathfrak{g}$, where $\mathfrak{g} \simeq \mathcal{H}_\Lambda$.

Lemma 4.28. *Every Lie algebra isomorphism* $\phi : (\mathfrak{g}, [\cdot, \cdot]_\mathfrak{g}) \to (\mathcal{H}_\Lambda, \{\cdot, \cdot\}_\Lambda)$ *can be extended to a Poisson algebra morphism* $D : (S_\mathfrak{g}, \cdot, \{\cdot, \cdot\}_{S_\mathfrak{g}}) \to (C^\infty(N), \cdot, \{\cdot, \cdot\}_\Lambda)$ *in a unique way. Indeed, if* $\{v_1, \dots, v_r\}$ *is a basis for* $\mathfrak{g}$*, then* $D(P(v_1, \dots, v_r)) = P(\phi(v_1), \dots, \phi(v_r))$*, for every polynomial* $P \in S_\mathfrak{g}$*.*

Proof. The elements of $S_\mathfrak{g}$ given by $v^I := v_1^{i_1} \cdot \ldots \cdot v_r^{i_r}$, where the I's are r-multi-indexes, "·" denotes the\product of elements of $\mathfrak{g}$ as functions on $\mathfrak{g}^*$ and $\{v_1, \ldots, v_r\}$ is a basis for $\mathfrak{g}$, form a basis of $S_{\mathfrak{g}}$. Then, every $P \in S_{\mathfrak{g}}$ can be written in a unique way as $P = \sum_{I \in M} \lambda_I v^I$, where M is a finite family of multi-indexes and each $\lambda_I \in \mathbb{R}$. Hence, the $\mathbb{R}$-algebra morphism $D : (S_{\mathfrak{g}}, \cdot) \to (C^\infty(N), \cdot)$ extending $\phi : \mathfrak{g} \to \mathcal{H}_\Lambda$ is determined by the image of the elements of a basis for $\mathfrak{g}$. Indeed,

$$D(P) = \sum_I \lambda_I D(v^I) = \sum_I \lambda_I \phi(v_1^{i_1}) \cdot \ldots \cdot \phi(v_r^{i_r}). \qquad (4.48)$$

Let us prove that D is also a $\mathbb{R}$-algebra morphism. From (4.48), we see that D is linear. Moreover, $D(PQ) = D(P)D(Q)$ for every $P, Q \in S_{\mathfrak{g}}$. In fact, if we write $Q = \sum_{J \in M} \lambda_J v^J$, we obtain

$$D(PQ) = D\left(\sum_I \lambda_I v^I \sum_J \lambda_J v^J\right) = \sum_K \sum_{I+J=K} \lambda_I \lambda_J D(v^K)$$
$$= \sum_I \lambda_I D(v^I) \sum_J \lambda_J D(v^J) = D(P)D(Q), \qquad (4.49)$$

where $I + J = (i_1 + j_1, \ldots, i_r + j_r)$ with $I = (i_1, \ldots, i_r)$ and $J = (j_1, \ldots, j_r)$.

Let us show that D is also a Poisson morphism. By linearity, this reduces to proving that $D\left(\{v^I, v^J\}_{S_{\mathfrak{g}}}\right) = \{D(v^I), D(v^J)\}_\Lambda$ for arbitrary I and J. Define $|I| = i_1 + \cdots + i_r$. If $|J| = 0$ or $|I| = 0$ this is satisfied, as a Poisson bracket vanishes when any entry is a constant. We now prove by induction the remaining cases. For $|I| + |J| = 2$, we have

$$D\left(\{v_\alpha, v_\beta\}_{S_{\mathfrak{g}}}\right) = \phi([v_\alpha, v_\beta]_{\mathfrak{g}}) = \{\phi(v_\alpha), \phi(v_\beta)\}_\Lambda$$
$$= \{D(v_\alpha), D(v_\beta)\}_\Lambda, \qquad (4.50)$$

for all $\alpha, \beta = 1, \ldots, r$. If D is a Poisson morphism for $|I| + |J| = m > 2$, then for $|I| + |J| = m + 1$ we can set $v^I = v^{\bar{I}} v_\gamma^{i_\gamma}$ for $i_\gamma \neq 0$ and some

γ to obtain

$$
\begin{aligned}
D\left(\{v^I, v^J\}_{S_{\mathfrak{g}}}\right) &= D\left(\{v^{\bar{I}} v_\gamma^{i_\gamma}, v^J\}_{S_{\mathfrak{g}}}\right) \\
&= D\left(\{v^{\bar{I}}, v^J\}_{S_{\mathfrak{g}}} v_\gamma^{i_\gamma} + v^{\bar{I}}\{v_\gamma^{i_\gamma}, v^J\}_{S_{\mathfrak{g}}}\right) \\
&= \{D(v^{\bar{I}}), D(v^J)\}_\Lambda D(v_\gamma^{i_\gamma}) + D(v^{\bar{I}})\{D(v_\gamma^{i_\gamma}), D(v^J)\}_\Lambda \\
&= \{D(v^{\bar{I}})D(v_\gamma^{i_\gamma}), D(v^J)\}_\Lambda = \{D(v^I), D(v^J)\}_\Lambda.
\end{aligned}
$$

$$(4.51)$$

By induction, $D\left(\{v^I, v^J\}_{S_{\mathfrak{g}}}\right) = \{D(v^I), D(v^J)\}_\Lambda$ for any I and J. $\square$

Recall that "·" denotes the standard product of elements of $S_{\mathfrak{g}}$ understood as polynomial functions on $S_{\mathfrak{g}}$. It is remarkable that D does not need to be injective, which causes that $S_{\mathfrak{g}}$ is not, in general, isomorphic to the space of polynomials on the elements of a basis of $\mathcal{H}_\Lambda$. For instance, consider the Lie algebra isomorphism $\phi : (\mathfrak{sl}(2, \mathbb{R}), [\cdot, \cdot]_{\mathfrak{sl}(2,\mathbb{R})}) \to (\mathcal{H}_\Lambda, \{\cdot, \cdot\}_\Lambda)$, with $\{v_1, v_2, v_3\}$ being a basis of $\mathfrak{sl}(2, \mathbb{R})$, of the form $\phi(v_1) = p^2$, $\phi(v_2) = xp$ and $\phi(v_3) = x^2$ and $\{\cdot, \cdot\}_\Lambda$ being the standard Poisson structure on T*$\mathbb{R}$. Then, $D(v_1 v_3 - v_2^2) = \phi(v_1)\phi(v_3) - \phi^2(v_2) = 0$.

The following notion enables us to simplify the statement and proofs of our results.

Definition 4.29. Given a curve P_t in $S_{\mathfrak{g}}$, its *degree*, $\deg(P_t)$, is the highest degree of the polynomials $\{P_t\}_{t\in\mathbb{R}}$. If there exists no finite highest degree, we say that $\deg(P_t) = \infty$.

Proposition 4.30. *A function f is a polynomial Lie integral for a Lie–Hamilton system X with respect to the Lie–Hamiltonian structure (N, Λ, h) if and only if for every $t \in \mathbb{R}$ we have $f_t = D(P_t)$, where D is the Poisson algebra morphism $D : (S_{\mathfrak{g}}, \cdot, \{\cdot, \cdot\}_{S_{\mathfrak{g}}}) \to (C^\infty(N), \cdot, \{\cdot, \cdot\}_\Lambda)$ induced by $\phi : (\mathfrak{g}, [\cdot, \cdot]_{\mathfrak{g}}) \to (\mathcal{H}_\Lambda, \{\cdot, \cdot\}_\Lambda)$, and the curve P_t is a solution of finite degree for*

$$
\frac{\mathrm{d}P}{\mathrm{d}t} + \{P, w_t\}_{S_{\mathfrak{g}}} = 0, \quad P \in S_{\mathfrak{g}}, \tag{4.52}
$$

where w_t stands for a curve in $\mathfrak{g}$ such that $D(w_t) = h_t$ for every $t \in \mathbb{R}$.

Proof. Let P_t be a particular solution of (4.52). Since D is a Poisson algebra morphism and $h_t = D(w_t)$ for every $t \in \mathbb{R}$, we obtain by applying D to (4.52) that

$$\frac{\partial D(P_t)}{\partial t} + \{D(P_t), D(w_t)\}_{\bar{\Lambda}} = \frac{\partial D(P_t)}{\partial t} + \{D(P_t), h_t\}_{\bar{\Lambda}} = 0. \quad (4.53)$$

So, $D(P_t)$ is a Lie integral for X. Note that this does not depend on the chosen curve w_t satisfying $D(w_t) = h_t$.

Conversely, given a polynomial Lie integral f for X, there exists a curve P_t of finite degree such that $D(P_t) = f_t$ for every $t \in \mathbb{R}$. Hence, we see that

$$D\left(\frac{dP_t}{dt} + \{P_t, w_t\}_{S_{\mathfrak{g}}}\right) = \frac{\partial D(P_t)}{\partial t} + \{D(P_t), D(w_t)\}_{\bar{\Lambda}} = 0$$

$$\implies \frac{dP_t}{dt} + \{P_t, w_t\}_{S_{\mathfrak{g}}} = \xi_t, \quad (4.54)$$

where ξ_t is a curve in $\ker D$. As $\deg(dP_t/dt)$ and $\deg(\{P_t, w_t\}_{S_{\mathfrak{g}}})$ are at most $\deg(P_t)$, then $\deg(\xi_t) \leq \deg(P_t)$. Next, consider the equation

$$\frac{d\eta}{dt} + \{\eta, w_t\}_{S_{\mathfrak{g}}} = \xi_t, \quad \deg(\eta) \leq \deg(P) \quad \text{and} \quad \eta \subset \ker D. \quad (4.55)$$

Note that this equation is well defined. Indeed, since $\deg(\eta) \leq \deg(P)$ and $\deg(w_t) \leq 1$ for every $t \in \mathbb{R}$, then $\deg(\{\eta, w_t\}_{S_{\mathfrak{g}}}) \leq \deg(P)$ for all $t \in \mathbb{R}$. In addition, as $D(\eta_t) \subset \ker D$, then $\{\eta, w_t\}_{S_{\mathfrak{g}}} \in \ker D$. Then, the above equation can be restricted to the finite-dimensional space of elements of $\ker D$ with degree at most $\deg(P_t)$. Given a particular solution η_t of this equation, which exists for the theorem of existence and uniqueness, we have that $P_t - \eta_t$ is a solution of (4.52) projecting into f_t. $\qquad \square$

Proposition 4.31. *Every polynomial Lie integral f of a Lie–Hamilton system X admitting a Lie–Hamiltonian structure (N, Λ, h) can be brought into the form $f = \sum_{I \in M} c_I l^I$, where M is a finite set of multi-indexes, the c_I's are certain real constants, and $l^I = f_1^{i_1} \cdot \ldots \cdot f_r^{i_r}$, with $f_1, \ldots, f_r$ being a basis of the space $\mathfrak{L}_h^\Lambda$.*

Proof. From Proposition 4.30, we have that $f_t = D(P_t)$ for a solution P_t of finite degree p for (4.52). So, it is a solution of the restriction of this system to $\mathbb{P}(p, \mathfrak{g})$, i.e., the elements of $S_\mathfrak{g}$ of degree at most p. Given the isomorphism $\phi : \mathfrak{g} \to \mathcal{H}_\Lambda$, define $\phi^{-1}(f_j)$, with $j = 1, \ldots, r$, to be the curve in $\mathfrak{g}$ of the form $t \mapsto \phi^{-1}(f_j)_t$. Note that $v_1 := \phi^{-1}(f_1)_0, \ldots, v_r := \phi^{-1}(f_r)_0$ form a basis of $\mathfrak{g}$. Hence, their polynomials up to order p span a basis for $\mathbb{P}(p, \mathfrak{g})$ and we can write $P_0 = \sum_{I \in M} c_I v^I$, where $v^I = v_1^{i_1} \cdot \ldots \cdot v_r^{i_r}$. As $P'_t = \sum_{I \in M} c_I [\phi^{-1}(f_1)]_t^{i_1} \cdots [\phi^{-1}(f_r)]_t^{i_r}$ and P_t are solutions with the same initial condition of the restriction of (4.52) to $\mathbb{P}(p, \mathfrak{g})$, they must be the same in virtue of the theorem of existence and uniqueness of systems of differential equations. Applying D, we obtain that
$$f_t = D(P_t) = D(\textstyle\sum_{I \in M} c_I [\phi^{-1}(f_1)_t]^{i_1} \cdots [\phi^{-1}(f_r)_t]^{i_r}) = \sum_{I \in M} c_I l_t^I.$$
$\square$

Corollary 4.32. *Let X be a Lie–Hamilton system that possesses a Lie–Hamiltonian structure (N, Λ, h) inducing a Lie algebra isomorphism $\phi : \mathfrak{g} \to \mathcal{H}_\Lambda$ and a Poisson algebra morphism $D : S_\mathfrak{g} \to C^\infty(N)$. The function $F = D(C)$, where C is a Casimir element of $S_\mathfrak{g}$, is a t-independent constant of motion of X. If $\mathcal{C}$ is a Casimir element of $U_\mathfrak{g}$, then $F = D(\lambda^{-1}(\mathcal{C}))$ is t-independent constant of motion for X.*

Note that if C is a constant of motion for X, it is also for any other X' whose $V^{X'} \subset V^X$. From Proposition 4.31 and Corollary 4.32, it follows that $F = D(C) = \sum_{I \in M} c_I l^I$. Therefore, the knowledge of Casimir elements provides not only constants of motion for Lie–Hamilton systems, but also information about the Lie integrals of the system.

As Casimir functions are known for many Lie algebras, we can use them to derive constants of motion for the corresponding Lie–Hamilton systems algebraically instead of applying the usual procedure, i.e., by solving a system of PDEs or ODEs.

In particular, Casimir functions for (semi)simple Lie algebras of arbitrary dimension are known [195, 309]. The same is true for the so-called "quasi-simple" Lie algebras, which can be obtained from

150   *A Guide to Lie Systems with Compatible Geometric Structures*

simple Lie algebras through contraction techniques [216]. Moreover, the Casimir invariants (Casimir elements of the Poisson algebra $(C^\infty(\mathfrak{g}^*), \{\cdot, \cdot\}_{\mathfrak{g}^*})$, being $\{\cdot, \cdot\}_{\mathfrak{g}^*}$ the Poisson structure induced by the Lie bracket for all real Lie algebras with dimension $d \leq 5$ were given in [307] (recall that the Casimir invariants for some of the solvable cases are not polynomial, i.e., they cannot be considered as elements of $S_\mathfrak{g}$), and the literature dealing with Casimir invariants for solvable and nilpotent Lie algebras is not scarce (see, e.g., [9, 69, 71]).

4.2.7. *The coalgebra method, constants of motion and superposition rules*

We here prove that each Lie–Hamiltonian structure (N, Λ, h) for a Lie–Hamilton system X gives rise in a natural way to a Poisson coalgebra $(S_\mathfrak{g}, \cdot, \{\cdot, \cdot\}_{S_\mathfrak{g}}, \Delta)$ where $\mathfrak{g} \simeq (\mathcal{H}_\Lambda, \{\cdot, \cdot\}_\Lambda)$. This allows us to use the coproduct of this coalgebra to construct new Lie–Hamiltonian structures for all the diagonal prolongations of X and to derive algebraically t-independent constants of motion for such diagonal prolongations. In turn, these constants can further be employed to obtain a superposition rule for the initial system. Our findings, which are only applicable to "primitive" Poisson coalgebras, give rigorous proof and generalizations of previous achievements established in [24, 28, 31].

Lemma 4.33. *If X is a Lie–Hamilton system possessing a Lie–Hamiltonian structure (N, Λ, h), then the space $(S_\mathfrak{g}, \cdot, \{\cdot, \cdot\}_{S_\mathfrak{g}}, \Delta)$, with $\mathfrak{g} \simeq (\mathcal{H}_\Lambda, \{\cdot, \cdot\}_\Lambda)$, is a Poisson coalgebra with a coproduct $\Delta : S_\mathfrak{g} \to S_\mathfrak{g} \otimes S_\mathfrak{g}$ satisfying*

$$\Delta(v) = v \otimes 1 + 1 \otimes v, \quad \forall v \in \mathfrak{g} \subset S_\mathfrak{g}. \qquad (4.56)$$

Proof. We know that $(S_\mathfrak{g}, \cdot, \{\cdot, \cdot\}_{S_\mathfrak{g}})$ and $(S_\mathfrak{g} \otimes S_\mathfrak{g}, \cdot_{S_\mathfrak{g} \otimes S_\mathfrak{g}}, \{\cdot, \cdot\}_{S_\mathfrak{g} \otimes S_\mathfrak{g}})$ are Poisson algebras. The coassociativity property for the coproduct map (4.56) is straightforward. Therefore, let us prove that there exists a Poisson algebra morphism $\Delta : (S_\mathfrak{g}, \cdot, \{\cdot, \cdot\}_{S_\mathfrak{g}}) \to (S_\mathfrak{g} \otimes S_\mathfrak{g}, \cdot_{S_\mathfrak{g} \otimes S_\mathfrak{g}}, \{\cdot, \cdot\}_{S_\mathfrak{g} \otimes S_\mathfrak{g}})$ satisfying (4.56), which turns $(S_\mathfrak{g}, \cdot, \{\cdot, \cdot\}_{S_\mathfrak{g}}, \Delta)$ into a Poisson coalgebra.

The elements of $S_\mathfrak{g}$ of the form $v^I := v_1^{i_1} \cdot \ldots \cdot v_r^{i_r}$, where the I's are r-multi-index with $r = \dim \mathfrak{g}$, form a basis for $S_\mathfrak{g}$ (considered as a linear space). Then, every $P \in S_\mathfrak{g}$ can be written in a unique way as $P = \sum_{I \in M} \lambda_I v^I$, where the λ_I are real constants and I runs all the elements of a finite set M. In view of this, an $\mathbb{R}$-algebra morphism $\Delta : S_\mathfrak{g} \to S_\mathfrak{g} \otimes S_\mathfrak{g}$ is determined by the image of the elements of a basis for $\mathfrak{g}$, i.e.,

$$\Delta(P) = \sum_I \lambda_I \Delta(v^I) = \sum_I \lambda_I \Delta(v_1^{i_1}) \cdot \ldots \cdot \Delta(v_r^{i_r}). \qquad (4.57)$$

Therefore, two $\mathbb{R}$-algebra morphisms that coincide on the elements on $\mathfrak{g}$ are necessarily the same. Hence, if there exists such a morphism, it is unique. Let us prove that there exists an $\mathbb{R}$-algebra morphism Δ satisfying (4.56).

From (4.57), we easily see that Δ is $\mathbb{R}$-linear. Let us also prove that $\Delta(PQ) = \Delta(P)\Delta(Q)$ for every $P, Q \in S_\mathfrak{g}$, which shows that Δ is an $\mathbb{R}$-algebra morphism. If we write $Q = \sum_{J \in M} \lambda_J v^J$, we obtain that

$$\Delta(PQ) = \sum_K \left(\sum_{I+J=K} \lambda_I \lambda_J \right) \Delta(v^K) = \sum_I \lambda_I \Delta(v^I) \sum_J \lambda_J \Delta(v^J)$$
$$= \Delta(P)\Delta(Q). \qquad (4.58)$$

Finally, we show that Δ is also a Poisson morphism. By linearity, this reduces to proving that $\Delta(\{v^I, v^J\}_{S_\mathfrak{g}}) = \{\Delta(v^I), \Delta(v^J)\}_{S_\mathfrak{g} \otimes S_\mathfrak{g}}$. If $|I| = 0$ or $|J| = 0$, this result is immediate as the Poisson bracket involving a constant is zero. For the remaining cases and starting by $|I| + |J| = 2$, we have that $\Delta(\{v_\alpha, v_\beta\}_{S_\mathfrak{g}}) = \{\Delta(v_\alpha), \Delta(v_\beta)\}_{S_\mathfrak{g} \otimes S_\mathfrak{g}}$, $\forall \alpha, \beta = 1, \ldots, r$. Proceeding by induction, we prove that this holds for every value of $|I| + |J|$; by writing $v^I = v^{\bar{I}} v_\gamma^{i_\gamma}$ and using induction hypothesis, we get

$$\Delta\left(\{v^I, v^J\}_{S_\mathfrak{g}}\right) = \Delta\left(\{v^{\bar{I}} v_\gamma^{i_\gamma}, v^J\}_{S_\mathfrak{g}}\right)$$
$$= \Delta\left(\{v^{\bar{I}}, v^J\}_{S_\mathfrak{g}} v_\gamma^{i_\gamma} + v^{\bar{I}} \{v_\gamma^{i_\gamma}, v^J\}_{S_\mathfrak{g}}\right)$$

$$= \left\{\Delta(v^{\bar{I}}), \Delta(v^J)\right\}_{S_{\mathfrak{g}} \otimes S_{\mathfrak{g}}} \Delta(v_\gamma^{i\gamma}) + \Delta(v^{\bar{I}}) \left\{\Delta(v_\gamma^{i\gamma}), \Delta(v^J)\right\}_{S_{\mathfrak{g}} \otimes S_{\mathfrak{g}}}$$

$$= \left\{\Delta(v^{\bar{I}})\Delta(v_\gamma^{i\gamma}), \Delta(v^J)\right\}_{S_{\mathfrak{g}} \otimes S_{\mathfrak{g}}} = \left\{\Delta(v^I), \Delta(v^J)\right\}_{S_{\mathfrak{g}} \otimes S_{\mathfrak{g}}}. \qquad (4.59)$$

$\square$

The coproduct defined in the previous lemma gives rise to a new Poisson algebra morphism as stated in the following immediate lemma.

Lemma 4.34. *The map* $\Delta^{(m)} : (S_{\mathfrak{g}}, \cdot, \{\cdot, \cdot\}_{S_{\mathfrak{g}}}) \to (S_{\mathfrak{g}}^{(m)}, \cdot_{S_{\mathfrak{g}}^{(m)}}, \{\cdot, \cdot\}_{S_{\mathfrak{g}}^{(m)}})$, *with* $m > 1$ *and* $\Delta^{(2)} = \Delta$ *given by* (4.56), *is a Poisson algebra morphism.*

The injection $\iota : \mathfrak{g} \to \mathcal{H}_\Lambda \subset C^\infty(N)$ is a Lie algebra morphism that can be extended to a Poisson algebra morphism $D : S_{\mathfrak{g}} \to C^\infty(N)$ given by $D(P(v_1, \ldots, v_r)) = P(\iota(v_1), \ldots, \iota(v_r))$. Recall that this map need not to be injective.

Lemma 4.35. *The Lie algebra morphism* $\mathfrak{g} \hookrightarrow C^\infty(N)$ *gives rise to a family of Poisson algebra morphisms* $D^{(m)} : S_{\mathfrak{g}}^{(m)} \hookrightarrow C^\infty(N)^{(m)} \subset C^\infty(N^m)$ *satisfying, for all* $v_1, \ldots, v_m \in \mathfrak{g} \subset S_{\mathfrak{g}}$, *that*

$$[D^{(m)}(v_1 \otimes \ldots \otimes v_m)](x_{(1)}, \ldots, x_{(m)}) = [D(v_1)](x_{(1)}) \cdot \ldots \cdot$$
$$[D(v_m)](x_{(m)}), \qquad (4.60)$$

where $x_{(i)}$ *is a point of the manifold* N *placed in the* i-*position within the product* $N \times \cdots \times N := N^m$.

From the above results, we can easily demonstrate the following statement which shows that the diagonal prolongations of a Lie–Hamilton system X are also Lie–Hamilton ones admitting a structure induced by that of X.

Proposition 4.36. *If* X *is a Lie–Hamilton system on* N *with a Lie–Hamiltonian structure* (N, Λ, h), *then the diagonal prolongation* $\widetilde{X}$ *to each* N^{m+1} *is also a Lie–Hamilton system endowed with a Lie–Hamiltonian structure* $(N^{m+1}, \Lambda^{m+1}, \widetilde{h})$ *given by* $\Lambda^{m+1}(x_{(0)}, \ldots, x_{(m)}) = \sum_{a=0}^m \Lambda(x_{(a)})$, *where we make use of the* *vector bundle isomorphism* $TN^{m+1} \simeq TN \oplus \cdots \oplus TN$, *and* $\widetilde{h}_t = D^{(m+1)}(\Delta^{(m+1)}(h_t))$, *where* $D^{(m+1)}$ *is the Poisson algebra morphism* (4.60) *induced by the Lie algebra morphism* $\mathfrak{g} \hookrightarrow \mathcal{H}_\Lambda \subset C^\infty(N)$.

The above results enable us to prove the following theorem that provides a method to obtain t-independent constants of motion for the diagonal prolongations of a Lie–Hamilton system. From this theorem, one may obtain superposition rules for Lie–Hamilton systems in an algebraic way. Additionally, this theorem is a generalization, only valid in the case of primitive coproduct maps, of the integrability theorem for coalgebra symmetric systems given in [31].

Theorem 4.37. *If X is a Lie–Hamilton system with a Lie–Hamiltonian structure (N, Λ, h) and C is a Casimir element of $(S_\mathfrak{g}, \cdot, \{,\}_{S_\mathfrak{g}})$, where $\mathfrak{g} \simeq \mathcal{H}_\Lambda$, then*
(i) *The functions defined as*

$$F^{(k)} = D^{(k)}(\Delta^{(k)}(C)), \quad k = 2, \dots, m, \tag{4.61}$$

are t-independent constants of motion for the diagonal prolongation $\widetilde{X}$ to N^m. Furthermore, if all the $F^{(k)}$ are non-constant functions, they form a set of $(m-1)$ functionally independent functions in involution.
(ii) *The functions given by*

$$F_{ij}^{(k)} = S_{ij}(F^{(k)}), \quad 1 \leq i < j \leq k, \quad k = 2, \dots, m, \tag{4.62}$$

where S_{ij} is the permutation of variables $x_{(i)} \leftrightarrow x_{(j)}$, are t-independent constants of motion for the diagonal prolongation $\widetilde{X}$ to N^m.

Proof. Every $P \in S_\mathfrak{g}^{(j)}$ can naturally be considered as an element $P \otimes \overbrace{1 \otimes \ldots \otimes 1}^{(k-j)\text{-times}} \in S_\mathfrak{g}^{(k)}$. Since $j \leq k$, we have that

$$\{\Delta^{(j)}(\bar{v}), \Delta^{(k)}(v)\}_{S_{\mathfrak{g}}^{(k)}} = \{\Delta^{(j)}(\bar{v}), \Delta^{(j)}(v)\}_{S_{\mathfrak{g}}^{(j)}}, \ \forall \bar{v}, v \in \mathfrak{g}. \text{ So,}$$

$$\{\Delta^{(j)}(C), \Delta^{(k)}(v)\}_{S_{\mathfrak{g}}^{(k)}} = \{\Delta^{(j)}(C), \Delta^{(j)}(v)\}_{S_{\mathfrak{g}}^{(j)}}$$

$$= \Delta^{(j)}(\{C, v\}_{S_{\mathfrak{g}}^{(j)}}) = 0. \tag{4.63}$$

Hence, by using that every function $f \in C^{\infty}(N^j)$ can be understood as a function $\pi^* f \in C^{\infty}(N^k)$, being $\pi : N^j \times N^{k-j} \to N^j$ the projection onto the first factor, and by applying the Poisson algebra morphisms introduced in Lemma 4.35, we get

$$\{D^{(j)}(\Delta^{(j)}(C)), D^{(k)}(\Delta^{(k)}(v))\}_{\Lambda^k} = \{F^{(j)}, D^{(k)}(\Delta^{(k)}(v))\}_{\Lambda^k}$$

$$= 0, \quad \forall v \in \mathfrak{g}, \tag{4.64}$$

which leads to $\{F^{(j)}, F^{(k)}\}_{\Lambda^k} = 0$, that is, the functions (4.61) are in involution. By construction (see Lemma 4.34), if these are non-constant, then they are functionally independent functions since $F^{(j)}$ lives in N^j, meanwhile $F^{(k)}$ is defined on N^k.

Let us prove now that all the functions (4.61) and (4.62) are t-independent constants of motion for $\tilde{X}$. Using that $\mathcal{H}_{\Lambda} \simeq \mathfrak{g}$ and $X_t = -\hat{\Lambda} \circ dh_t$, we see that X can be brought into the form $X_t = -\hat{\Lambda} \circ d \circ D(v_t)$ for a unique curve $t \to v_t$ in $\mathfrak{g}$. From this and Proposition 4.36, it follows that

$$\tilde{X}_t = -\Lambda^m \circ dD^{(m)}(\Delta^{(m)}(v_t))$$

$$\implies \tilde{X}_t(F^{(k)}) = \left\{ D^{(k)}(\Delta^{(k)}(C)), D^{(m)}(\Delta^{(m)}(v_t)) \right\}_{\Lambda^m}$$

$$= 0. \tag{4.65}$$

Then, $F^{(k)}$ is a common first integral for every $\tilde{X}_t$. Finally, consider the permutation operators S_{ij}, with $1 \leq i < j \leq k$ for $k = 2, \ldots, m$. Note that

$$0 = S_{ij} \left\{ F^{(k)}, D^{(m)}(\Delta^{(m)}(v_t)) \right\}_{\Lambda^m}$$

$$= \left\{ S_{ij}(F^{(k)}), S_{ij}\left(D^{(m)}(\Delta^{(m)}(v_t)) \right) \right\}_{\Lambda^m}$$

$$= \left\{ F_{ij}^{(k)}, D^{(m)}(\Delta^{(m)}(v_t)) \right\}_{\Lambda^m} = \tilde{X}_t(F_{ij}^{(k)}). \tag{4.66}$$

Consequently, the functions $F_{ij}^{(k)}$ are t-independent constants of motion for $\widetilde{X}$. $\qquad\square$

Note that the "omitted" case with $k = 1$ in the set of constants (4.61) is, precisely, the one provided by Corollary 4.32 as $F^{(1)} := F = D(C)$. Depending on the system X, or more specifically, on the associated $\mathcal{H}_\Lambda$, the function F can be either a useless trivial constant or a relevant function. It is also worth noting that constants (4.62) need not be functionally independent, but we can always choose those fulfilling such a property. Finally, observe that if X' is such that $V^{X'} \subset V^X$, then the functions (4.61) and (4.62) are also constants of motion for the diagonal prolongation of X' to N^m.

4.3. Applications of the Geometric Theory of Lie–Hamilton Systems

In this section, we aim to achieve superposition rules and constants of motion through the coalgebra formalism. We find of special interest our achievement of finding a superposition rule for Riccati equations with the aid of the coalgebra method.

4.3.1. The Ermakov system

Let us consider the classical Ermakov system [108]

$$
\begin{cases}
\dfrac{d^2 x}{dt^2} = -\omega^2(t)x + \dfrac{b}{x^3}, \\[2mm]
\dfrac{d^2 y}{dt^2} = -\omega^2(t)y,
\end{cases}
\tag{4.67}
$$

with a non-constant t-dependent frequency $\omega(t)$, being b a real constant. This system appears in a number of applications related to problems in Quantum and Classical Mechanics [254]. By writing this system as a first-order one,

$$
\begin{cases}
\dfrac{dx}{dt} = v_x, \quad \dfrac{dv_x}{dt} = -\omega^2(t)x + \dfrac{b}{x^3}, \\[2mm]
\dfrac{dy}{dt} = v_y, \quad \dfrac{dv_y}{dt} = -\omega^2(t)y,
\end{cases}
\tag{4.68}
$$

156   *A Guide to Lie Systems with Compatible Geometric Structures*

we can apply the theory of Lie systems. Indeed, this is a Lie system related to a VG Lie algebra V isomorphic to $\mathfrak{sl}(2,\mathbb{R})$ [102]. In fact, system (4.2) describes the integral curves of the t-dependent vector field $X = X_3 + \omega^2(t)X_1$, where the vector fields

$$X_1 = -x\frac{\partial}{\partial v_x} - y\frac{\partial}{\partial v_y}, \quad X_2 = \frac{1}{2}\left(v_x\frac{\partial}{\partial v_x} + v_y\frac{\partial}{\partial v_y} - x\frac{\partial}{\partial x} - y\frac{\partial}{\partial y}\right),$$

$$X_3 = v_x\frac{\partial}{\partial x} + v_y\frac{\partial}{\partial y} + \frac{b}{x^3}\frac{\partial}{\partial v_x}, \tag{4.69}$$

satisfy the commutation relations

$$[X_1, X_2] = X_1, \quad [X_1, X_3] = 2X_2, \quad [X_2, X_3] = X_3. \tag{4.70}$$

As a first new result, we show that this is a Lie–Hamilton system. The vector fields are Hamiltonian with respect to the Poisson bivector $\Lambda = \partial/\partial x \wedge \partial/\partial v_x + \partial/\partial y \wedge \partial/\partial v_y$ provided that $X_\alpha = -\widehat{\Lambda}(dh_\alpha)$ for $\alpha = 1, 2, 3$. Thus, we find the following Hamiltonian functions which form a basis for $(\mathcal{H}_\Lambda, \{\cdot, \cdot\}_\Lambda) \simeq (\mathfrak{sl}(2,\mathbb{R}), [\cdot, \cdot])$:

$$h_1 = \frac{1}{2}(x^2 + y^2), \quad h_2 = -\frac{1}{2}(xv_x + yv_y), \quad h_3 = \frac{1}{2}\left(v_x^2 + v_y^2 + \frac{b}{x^2}\right),$$

$$\tag{4.71}$$

as they fulfill (4.13).

Since $X = X_3 + \omega^2(t)X_1$ and $\omega(t)$ is not a constant, every t-independent constant of motion f for X is a common first integral for X_1, X_2, X_3. Instead of searching an f by solving the system of PDEs given by $X_1 f = X_2 f = X_3 f = 0$, we use Corollary 4.32. This easily provides such a first integral through the Casimir element of the symmetric algebra of $\mathfrak{sl}(2,\mathbb{R})$. Explicitly, given a basis $\{v_1, v_2, v_3\}$ for $\mathfrak{sl}(2,\mathbb{R})$ satisfying

$$[v_1, v_2] = -v_1, \quad [v_1, v_3] = -2v_2, \quad [v_2, v_3] = -v_3, \tag{4.72}$$

the Casimir element of $\mathfrak{sl}(2,\mathbb{R})$ reads $\mathcal{C} = \frac{1}{2}(v_1\widetilde{\otimes}v_3 + v_3\widetilde{\otimes}v_1) - v_2\widetilde{\otimes}v_2 \in U_{\mathfrak{sl}(2,\mathbb{R})}$. Then, the inverse of symmetrizer map (2.1), $\lambda^{-1} : U_{\mathfrak{sl}(2,\mathbb{R})} \to$

$S_{\mathfrak{sl}(2,\mathbb{R})}$, gives rise to the Casimir element of $S_{\mathfrak{sl}(2,\mathbb{R})}$

$$C = \lambda^{-1}(\mathcal{C}) = v_1 v_3 - v_2^2. \tag{4.73}$$

According to Lemma 4.28, we consider the Poisson algebra morphism D induced by the isomorphism $\phi : \mathfrak{sl}(2,\mathbb{R}) \to \mathcal{H}_\Lambda$ defined by $\phi(v_\alpha) = h_\alpha$ for $\alpha = 1, 2, 3$. Subsequently, via Corollary 4.32, we obtain

$$F = D(C) = \phi(v_1)\phi(v_3) - \phi^2(v_2) = h_1 h_3 - h_2^2$$

$$= (v_y x - v_x y)^2 + b\left(1 + \frac{y^2}{x^2}\right). \tag{4.74}$$

In this way, we recover, up to an additive and multiplicative constant, the well-known Lewis–Riesenfeld invariant [254]. Note that when $\omega(t)$ is a constant, then $V^X \subset V$ and the function F is also a constant of motion for X (4.2).

4.3.2. A superposition rule for Riccati equations

Let us turn to the system of Riccati equations on $\mathcal{O} = \{(x_1, x_2, x_3, x_4) \,|\, x_i \neq x_j, i \neq j = 1, \ldots, 4\} \subset \mathbb{R}^4$, given by

$$\frac{dx_i}{dt} = a_0(t) + a_1(t)x_i + a_2(t)x_i^2, \quad i = 1, \ldots, 4, \tag{4.75}$$

where $a_0(t), a_1(t), a_2(t)$ are arbitrary t-dependent functions. The knowledge of a non-constant t-independent constant of motion for any system of this type leads to obtaining a superposition rule for Riccati equations [93]. Usually, this requires the integration of a system of PDEs [93] or ODEs [376]. We here obtain such a t-independent constant of motion through algebraic methods by showing that (4.9) is a Lie–Hamilton system with a given Lie–Hamiltonian structure and obtaining an associated polynomial Lie integral.

Observe that (4.9) is a Lie system related to a t-dependent vector field $X_R = a_0(t)X_1 + a_1(t)X_2 + a_2(t)X_3$, where

$$X_1 = \sum_{i=1}^{4} \frac{\partial}{\partial x_i}, \quad X_2 = \sum_{i=1}^{4} x_i \frac{\partial}{\partial x_i}, \quad X_3 = \sum_{i=1}^{4} x_i^2 \frac{\partial}{\partial x_i} \tag{4.76}$$

span a VG Lie algebra V for (4.9) isomorphic to $\mathfrak{sl}(2, \mathbb{R})$ satisfying the same commutation relations (4.4). For simplicity, we assume $V^X = V$. Nevertheless, our final results are valid for any other case.

To show that (4.9) is a Lie–Hamilton system for arbitrary functions $a_0(t)$, $a_1(t)$, $a_2(t)$, we need to search for a symplectic form ω such that V consists of Hamiltonian vector fields. By imposing $\mathcal{L}_{X_\alpha}\omega = 0$, for $\alpha = 1, 2, 3$, we obtain the 2-form

$$\omega_R = \frac{\mathrm{d}x_1 \wedge \mathrm{d}x_2}{(x_1 - x_2)^2} + \frac{\mathrm{d}x_3 \wedge \mathrm{d}x_4}{(x_3 - x_4)^2}, \tag{4.77}$$

which is closed and non-degenerate on $\mathcal{O}$. Now, observe that $\iota_{X_\alpha}\omega = \mathrm{d}h_\alpha$, with $\alpha = 1, 2, 3$, and

$$h_1 = \frac{1}{x_1 - x_2} + \frac{1}{x_3 - x_4}, \quad h_2 = \frac{1}{2}\left(\frac{x_1 + x_2}{x_1 - x_2} + \frac{x_3 + x_4}{x_3 - x_4}\right),$$

$$h_3 = \frac{x_1 x_2}{x_1 - x_2} + \frac{x_3 x_4}{x_3 - x_4}.$$

So, h_1, h_2, and h_3 are Hamiltonian functions for X_1, X_2, and X_3, correspondingly. Using the Poisson bracket $\{\cdot, \cdot\}_\omega$ induced by ω, we obtain that h_1, h_2 and h_3 satisfy the commutation relations (4.13), and $(\langle h_1, h_2, h_3 \rangle, \{\cdot, \cdot\}_\omega) \simeq \mathfrak{sl}(2, \mathbb{R})$. Next, we again express $\mathfrak{sl}(2, \mathbb{R})$ in the basis $\{v_1, v_2, v_3\}$ with Lie brackets (4.6) and Casimir function (4.7), and we consider the Poisson algebra morphism $D : S_{\mathfrak{sl}(2,\mathbb{R})} \to C^\infty(\mathcal{O})$ given by the isomorphism $\phi(v_\alpha) = h_\alpha$ for $\alpha = 1, 2, 3$. As $(\mathcal{O}, \{\cdot, \cdot\}_\omega, h_t = a_0(t)h_1 + a_1(t)h_2 + a_2(t)h_3)$ is a Lie–Hamiltonian structure for X and applying Corollary 4.32, we obtain the t-independent constant of motion for X

$$F = D(C) = h_1 h_3 - h_2^2 = \frac{(x_1 - x_4)(x_2 - x_3)}{(x_1 - x_2)(x_3 - x_4)}. \tag{4.78}$$

As in the previous example, if $V^X \subset V$, then F is also a constant of motion for X. It is worth noting that F is the known constant of motion obtained for deriving a superposition rule for Riccati equations [93, 376], which is here deduced through a simple algebraic calculation.

It is also interesting that V also becomes a Lie algebra of Hamiltonian vector fields with respect to a second symplectic structure given by $\omega = \sum_{1 \leq i < j}^{4} \frac{\mathrm{d}x_i \wedge \mathrm{d}x_j}{(x_i - x_j)^2}$. Consequently, the system (4.9) can be considered, in fact, as a *bi-Lie–Hamilton system*.

4.3.3. *Kummer–Schwarz equations in Hamilton form*

It was proven in [121] that the second-order Kummer–Schwarz equations [46, 87] admit a t-dependent Hamiltonian which can be used to work out their Hamilton's equations, namely

$$
\begin{cases}
\dfrac{\mathrm{d}x}{\mathrm{d}t} = \dfrac{px^3}{2}, \\[2mm]
\dfrac{\mathrm{d}p}{\mathrm{d}t} = -\dfrac{3p^2 x^2}{4} - \dfrac{b_0}{4} + \dfrac{4b_1(t)}{x^2},
\end{cases}
\tag{4.79}
$$

where $b_1(t)$ is assumed to be a non-constant t-dependent function, $(x, p) \in \mathrm{T}^* \mathbb{R}_0$ with $\mathbb{R}_0 := \mathbb{R} - \{0\}$, and b_0 is a real constant. This is a Lie system associated with the t-dependent vector field $X = X_3 + b_1(t)X_1$ [121], where the vector fields

$$
X_1 = \frac{4}{x^2} \frac{\partial}{\partial p}, \quad X_2 = x \frac{\partial}{\partial x} - p \frac{\partial}{\partial p},
$$
$$
X_3 = \frac{px^3}{2} \frac{\partial}{\partial x} - \frac{1}{4} \left(3p^2 x^2 + b_0 \right) \frac{\partial}{\partial p},
\tag{4.80}
$$

span a VG Lie algebra V isomorphic to $\mathfrak{sl}(2, \mathbb{R})$ fulfilling (4.4). Moreover, X is a Lie–Hamilton system as V consists of Hamiltonian vector fields with respect to the Poisson bivector $\Lambda = \partial/\partial x \wedge \partial/\partial p$ on $\mathrm{T}^* \mathbb{R}_0$. Indeed, $X_\alpha = -\widehat{\Lambda}(dh_\alpha)$, with $\alpha = 1, 2, 3$, and (4.23) forms a basis of a Lie algebra isomorphic to $\mathfrak{sl}(2, \mathbb{R})$ satisfying the commutation relations (4.13). Therefore, (4.13) is a Lie–Hamilton system possessing a Lie–Hamiltonian structure $(\mathrm{T}^* \mathbb{R}_0, \Lambda, h)$, where $h_t = h_3 + b_1(t)h_1$.

To obtain a superposition rule for X, we need to determine an integer m so that the diagonal prolongations of X_α to $\mathrm{T}^* \mathbb{R}_0^m$ ($\alpha = 1, 2, 3$) become linearly independent at a generic point (see [93]). This happens for $m = 2$. We consider a coordinate system in $\mathrm{T} \mathbb{R}_0^3$,

160   *A Guide to Lie Systems with Compatible Geometric Structures*

namely $\{x_{(1)}, p_{(1)}, x_{(2)}, p_{(2)}, x_{(3)}, p_{(3)}\}$. A superposition rule for X can be obtained by determining two common first integrals for the diagonal prolongations $\widetilde{X}_\alpha$ to $T^*\mathbb{R}^3_0$ satisfying

$$\frac{\partial(F_1, F_2)}{\partial(x_{(1)}, p_{(1)})} \neq 0. \tag{4.81}$$

Instead of searching F_1, F_2 in the standard way, i.e., by solving the system of PDEs given by $\widetilde{X}_\alpha f = 0$, we make use of Theorem 4.37. This provides such first integrals through the Casimir element C (4.7) of the symmetric algebra of $\mathcal{H}_\Lambda \simeq \mathfrak{sl}(2, \mathbb{R})$. Indeed, the coproduct (4.56) enables us to define the elements

$$\begin{aligned}
\Delta(C) &= \Delta(v_1)\Delta(v_3) - \Delta(v_2)^2 \\
&= (v_1 \otimes 1 + 1 \otimes v_1)(v_3 \otimes 1 + 1 \otimes v_3) - (v_2 \otimes 1 + 1 \otimes v_2)^2, \\
\Delta^{(3)}(C) &= \Delta^{(3)}(v_1)\Delta^{(3)}(v_3) - \Delta^{(3)}(v_2)^2 \\
&= (v_1 \otimes 1 \otimes 1 + 1 \otimes v_1 \otimes 1 + 1 \otimes 1 \otimes v_1)(v_3 \otimes 1 \otimes 1 \\
&\quad + 1 \otimes v_3 \otimes 1 + 1 \otimes 1 \otimes v_3) - (v_2 \otimes 1 \otimes 1 + 1 \otimes v_2 \otimes 1 \\
&\quad + 1 \otimes 1 \otimes v_2)^2, \tag{4.82}
\end{aligned}$$

for $S^{(2)}_{\mathfrak{sl}(2,\mathbb{R})}$ and $S^{(3)}_{\mathfrak{sl}(2,\mathbb{R})}$, respectively. By applying D, $D^{(2)}$ and $D^{(3)}$ coming from the isomorphism $\phi(v_\alpha) = h_\alpha$ for the Hamiltonian functions (4.23), we obtain, via Theorem 4.37, the following constants of motion of the type (4.61):

$$\begin{aligned}
F &= D(C) = h_1(x_1, p_1)h_3(x_1, p_1) - h_2^2(x_1, p_1) = b_0, \\
F^{(2)} &= D^{(2)}(\Delta(C)) \\
&= [h_1(x_1, p_1) + h_1(x_2, p_2)] [h_3(x_1, p_1) + h_3(x_2, p_2)] \\
&\quad - [h_2(x_1, p_1) + h_2(x_2, p_2)]^2 \\
&= \frac{b_0 (x_1 + x_2)^2 + \left(p_1 x_1^2 - p_2 x_2^2\right)^2}{x_1 x_2} \\
&= \frac{b_0(x_1^2 + x_2^2) + \left(p_1 x_1^2 - p_2 x_2^2\right)^2}{x_1 x_2} + 2b_0,
\end{aligned}$$

$$F^{(3)} = D^{(3)}(\Delta(C)) = \sum_{i=1}^{3} h_1(x_i, p_i) \sum_{j=1}^{3} h_3(x_j, p_j) - \left(\sum_{i=1}^{3} h_2(x_i, p_i) \right)^2$$

$$= \sum_{1 \leq i < j}^{3} \frac{b_0(x_i + x_j)^2 + (p_i x_i^2 - p_j x_j^2)^2}{x_i x_j} - 3b_0, \tag{4.83}$$

where, for simplicity, hereafter we denote by (x_i, p_i) the coordinates $(x_{(i)}, p_{(i)})$. Thus, F simply gives rise to the constant b_0, while $F^{(2)}$ and $F^{(3)}$ are, by construction, two functionally independent constants of motion for $\widetilde{X}$ fulfilling (4.15) which, in turn, allows us to to derive a superposition rule for X. Furthermore, the function $F^{(2)} := F_{12}^{(2)}$ provides two other constants of the type (4.62) given by $F_{13}^{(2)} = S_{13}(F^{(2)})$ and $F_{23}^{(2)} = S_{23}(F^{(2)})$ that verify $F^{(3)} = F^{(2)} + F_{13}^{(2)} + F_{23}^{(2)} - 3b_0$. Since it is simpler to work with $F_{23}^{(2)}$ than with $F^{(3)}$, we choose the pair $F^{(2)}, F_{23}^{(2)}$ as the two functionally independent first integrals to obtain a superposition rule. We set

$$F^{(2)} = k_1 + 2b_0, \quad F_{23}^{(2)} = k_2 + 2b_0, \tag{4.84}$$

and compute x_1, p_1 in terms of the other variables and k_1, k_2. From (4.18), we have

$$p_1 = p_1(x_1, x_2, p_2, x_3, p_3, k_1) = \frac{p_2 x_2^2 \pm \sqrt{k_1 x_1 x_2 - b_0(x_1^2 + x_2^2)}}{x_1^2}. \tag{4.85}$$

Substituting in the second relation within (4.18), we obtain

$$x_1 = x_1(x_2, p_2, x_3, p_3, k_1, k_2) = \frac{A^2 B_+ + b_0 B_- (x_2^2 - x_3^2) \pm 2A\sqrt{\Upsilon}}{B_-^2 + 4b_0 A^2}, \tag{4.86}$$

provided that the functions $A, B_\pm, \Upsilon$ are defined by

$$A = p_2 x_2^2 - p_3 x_3^2, \quad B_\pm = k_1 x_2 \pm k_2 x_3,$$

$$\Upsilon = A^2 \left[k_1 k_2 x_2 x_3 - 2b_0^2 \left(x_2^2 + x_3^2 \right) - b_0 A^2 \right]$$

$$+ b_0 x_2 x_3 B_- (k_2 x_2 - k_1 x_3) - b_0^3 \left(x_2^2 - x_3^2 \right)^2. \tag{4.87}$$

162 *A Guide to Lie Systems with Compatible Geometric Structures*

By introducing this result into (4.19), we obtain $p_1 = p_1(x_2, p_2, x_3, p_3, k_1, k_2)$ which, along with $x_1 = x_1(x_2, p_2, x_3, p_3, k_1, k_2)$, provides a superposition rule for X.

In particular for (4.13) with $b_0 = 0$, it results

$$
\begin{aligned}
x_1 &= \frac{A^2 \left(B_+ \pm 2\sqrt{k_1 k_2 x_2 x_3} \right)}{B_-^2}, \\
p_1 &= B_-^3 \frac{\left[B_- p_2 x_2^2 \pm A \sqrt{k_1 x_2 \left(B_+ \pm 2\sqrt{k_1 k_2 x_2 x_3} \right)} \right]}{A^4 \left(B_+ \pm 2\sqrt{k_1 k_2 x_2 x_3} \right)^2},
\end{aligned}
\tag{4.88}
$$

where the functions $A, B_\pm$ remain in the above same form. As the constants of motion were derived for non-constant $b_1(t)$, when $b_1(t)$ is constant we have $V^X \subset V$. As a consequence, the functions F, $F^{(2)}$, $F^{(3)}$ and so on are still constants of motion for the diagonal prolongation $\widetilde{X}$ and the superposition rules are still valid for any system (4.13).

4.3.4. *The n-dimensional Smorodinsky–Winternitz systems*

Let us focus on the n-dimensional Smorodinsky–Winternitz systems [173, 381] with unit mass and a non-constant t-dependent frequency $\omega(t)$ whose Hamiltonian is given by

$$
h = \frac{1}{2} \sum_{i=1}^n p_i^2 + \frac{1}{2}\omega^2(t) \sum_{i=1}^n x_i^2 + \frac{1}{2} \sum_{i=1}^n \frac{b_i}{x_i^2},
\tag{4.89}
$$

where the b_i's are n real constants. The corresponding Hamilton's equations read

$$
\begin{cases}
\dfrac{\mathrm{d}x_i}{\mathrm{d}t} = p_i, \\
\dfrac{\mathrm{d}p_i}{\mathrm{d}t} = -\omega^2(t)x_i + \dfrac{b_i}{x_i^3},
\end{cases}
\quad i = 1, \ldots, n.
\tag{4.90}
$$

These systems have been recently attracting quite much attention in Classical and Quantum Mechanics for their special properties [121, 194, 212, 384]. Observe that Ermakov systems (4.2) arise as the particular case of (4.24) for $n = 2$ and $b_2 = 0$. For $n = 1$ the above system maps into the Milne–Pinney equations, which are of interest in the study of several cosmological models [108, 312], through the diffeomorphism $(x, p) \in T^*\mathbb{R}_0 \mapsto (x, v = p) \in T\mathbb{R}_0$.

Let us show that the system (4.24) can be endowed with a Lie–Hamiltonian structure. This system describes the integral curves of the t-dependent vector field on $T^*\mathbb{R}_0^n$ given by $X = X_3 + \omega^2(t)X_1$, where the vector fields

$$
X_1 = -\sum_{i=1}^n x_i \frac{\partial}{\partial p_i}, \quad X_2 = \frac{1}{2}\sum_{i=1}^n \left(p_i \frac{\partial}{\partial p_i} - x_i \frac{\partial}{\partial x_i} \right),
$$

$$
X_3 = \sum_{i=1}^n \left(p_i \frac{\partial}{\partial x_i} + \frac{b_i}{x_i^3} \frac{\partial}{\partial p_i} \right), \tag{4.91}
$$

fulfill the commutation rules (4.4). Hence, (4.24) is a Lie system. The space $T^*\mathbb{R}_0^n$ admits a natural Poisson bivector $\Lambda = \sum_{i=1}^n \partial/\partial x_i \wedge \partial/\partial p_i$ related to the restriction to this space of the canonical symplectic structure on $T^*\mathbb{R}^n$. Moreover, the preceding vector fields are Hamiltonian vector fields with Hamiltonian functions

$$
h_1 = \frac{1}{2}\sum_{i=1}^n x_i^2, \quad h_2 = -\frac{1}{2}\sum_{i=1}^n x_i p_i, \quad h_3 = \frac{1}{2}\sum_{i=1}^n \left(p_i^2 + \frac{b_i}{x_i^2} \right), \tag{4.92}
$$

which satisfy the commutation relations (4.13), so that $\mathcal{H}_\Lambda \simeq \mathfrak{sl}(2, \mathbb{R})$. Consequently, every curve h_t that takes values in the Lie algebra spanned by h_1, h_2 and h_3 gives rise to a Lie–Hamiltonian structure $(T^*\mathbb{R}_0^n, \Lambda, h)$. Then, the system (4.24), described by the t-dependent vector field $X = X_3 + \omega^2(t)X_1 = -\hat{\Lambda}(dh_3 + \omega^2(t)dh_1)$, is a Lie–Hamilton system with a Lie–Hamiltonian structure $(T^*\mathbb{R}_0^n, \Lambda, h_t = h_3 + \omega^2(t)h_1)$.

Subsequently, we derive an explicit superposition rule for the simplest case of the system (4.24) corresponding to $n = 1$, and proceed as in the previous subsection. The prolongations of X_α ($\alpha = 1, 2, 3$) again become linearly independent for $m = 2$ and we need to obtain two first integrals for the diagonal prolongations $\widetilde{X}_\alpha$ of $T^*\mathbb{R}_0^3$ fulfilling (4.15) for the coordinate system $\{x_{(1)}, p_{(1)}, x_{(2)}, p_{(2)}, x_{(3)}, p_{(3)}\}$

of $T^*\mathbb{R}_0^3$. Similarly to the previous example, we have an injection $D : \mathfrak{sl}(2, \mathbb{R}) \to C^\infty(T^*\mathbb{R}_0)$ which leads to the morphisms $D^{(2)}$ and $D^{(3)}$.

Then, by taking into account the Casimir function (4.7) and the Hamiltonians (4.26), we apply Theorem 4.37 obtaining the following first integrals

$$F^{(2)} = D^{(2)}(\Delta(C)) = \frac{1}{4}(x_1 p_2 - x_2 p_1)^2 + \frac{b(x_1^2 + x_2^2)^2}{4 x_1^2 x_2^2},$$

$$F^{(3)} = D^{(3)}(\Delta(C)) = \frac{1}{4} \sum_{1 \leq i < j}^{3} \left[(x_i p_j - x_j p_i)^2 + \frac{b(x_i^2 + x_j^2)^2}{x_i^2 x_j^2} \right] - \frac{3}{4}b,$$

$$F_{13}^{(2)} = S_{13}(F^{(2)}),$$

$$F_{23}^{(2)} = S_{23}(F^{(2)}),$$

$$F^{(3)} = F^{(2)} + F_{13}^{(2)} + F_{23}^{(2)} - 3b/4, \tag{4.93}$$

where (x_i, p_i) denote the coordinates $(x_{(i)}, p_{(i)})$; notice that $F = D(C) = b/4$. We choose $F^{(2)}$ and $F_{23}^{(2)}$ as the two functionally independent constants of motion and we shall use $F_{13}^{(2)}$ in order to simplify the results. Recall that these functions are exactly the first integrals obtained in other works, e.g., [102], for describing superposition rules of dissipative Milne–Pinney equations (up to the diffeomorphism $\varphi : (x, p) \in T^*\mathbb{R}_0 \mapsto (x, v) = (x, p) \in T\mathbb{R}_0$ to system (4.24) with $n = 1$), and lead straightforwardly to deriving a superposition rule for these equations [104].

Indeed, we set

$$F^{(2)} = \frac{k_1}{4} + \frac{b}{2}, \quad F_{23}^{(2)} = \frac{k_2}{4} + \frac{b}{2}, \quad F_{13}^{(2)} = \frac{k_3}{4} + \frac{b}{2}, \tag{4.94}$$

and from the first equation we obtain p_1 in terms of the remaining variables and k_1

$$p_1 = p_1(x_1, x_2, p_2, x_3, p_3, k_1) = \frac{p_2 x_1^2 x_2 \pm \sqrt{k_1 x_1^2 x_2^2 - b(x_1^4 + x_2^4)}}{x_1 x_2^2}. \tag{4.95}$$

By introducing this value in the second expression of (4.28), one can determine the expression of x_1 as a function of x_2, p_2, x_3, p_3 and the constants k_1, k_2. Such a result is rather simplified when the third constant of (4.28) enters, yielding

$$x_1 = x_1(x_2, p_2, x_3, p_3, k_1, k_2) = x_1(x_2, x_3, k_1, k_2, k_3)$$

$$= \left\{ \mu_1 x_2^2 + \mu_2 x_3^2 \pm \sqrt{\mu \left[k_3 x_2^2 x_3^2 - b(x_2^4 + x_3^4) \right]} \right\}^{1/2}, \quad (4.96)$$

where the constants μ_1, μ_2, μ are defined in terms of k_1, k_2, k_3 and b as

$$\mu_1 = \frac{2bk_1 - k_2 k_3}{4b^2 - k_3^2}, \quad \mu_2 = \frac{2bk_2 - k_1 k_3}{4b^2 - k_3^2},$$

$$\mu = \frac{4 \left[4b^3 + k_1 k_2 k_3 - b(k_1^2 + k_2^2 + k_3^2) \right]}{(4b^2 - k_3^2)^2}. \quad (4.97)$$

And by introducing (4.30) into (4.29), we obtain

$$p_1 = p_1(x_2, p_2, x_3, p_3, k_1, k_2) = p_1(x_2, p_2, x_3, k_1, k_2, k_3), \quad (4.98)$$

which together with (4.30) provide a superposition rule for (4.24) with $n = 1$. These expressions constitute the known superposition rule for Milne–Pinney equations [104]. Observe that, instead of solving systems of PDEs for obtaining the first integrals as in [102, 104], we have obtained them algebraically in a simpler way. When $b = 0$ we recover, as expected, the superposition rule for the harmonic oscillator with a t–dependent frequency. Similarly to previous examples, the above superposition rule is also valid when $\omega(t)$ is constant.

4.3.5. *A trigonometric system*

Let us study a final example appearing in the study of integrability of classical systems [16, 17]. Consider the system

$$\begin{cases} \dfrac{\mathrm{d}x}{\mathrm{d}t} = \sqrt{1 - x^2} \left(B_x(t) \sin p - B_y(t) \cos p \right), \\[2mm] \dfrac{\mathrm{d}p}{\mathrm{d}t} = -(B_x(t) \cos p + B_y(t) \sin p) \dfrac{x}{\sqrt{1 - x^2}} - B_z(t), \end{cases} \quad (4.99)$$

where $B_x(t), B_y(t), B_z(t)$ are arbitrary t-dependent functions and $(x, p) \in T^*I$, with $I = (-1, 1)$. This system describes the integral

166   *A Guide to Lie Systems with Compatible Geometric Structures*

curves of the t-dependent vector field

$$X = \sqrt{1-x^2}(B_x(t)\sin p - B_y(t)\cos p)\frac{\partial}{\partial x}$$
$$- \left[\frac{(B_x(t)\cos p + B_y(t)\sin p)x}{\sqrt{1-x^2}} + B_z(t)\right]\frac{\partial}{\partial p}, \qquad (4.100)$$

which can be brought into the form $X = B_x(t)X_1 + B_y(t)X_2 + B_z(t)X_3$, where

$$X_1 = \sqrt{1-x^2}\sin p\frac{\partial}{\partial x} - \frac{x}{\sqrt{1-x^2}}\cos p\frac{\partial}{\partial p},$$
$$X_2 = -\sqrt{1-x^2}\cos p\frac{\partial}{\partial x} - \frac{x}{\sqrt{1-x^2}}\sin p\frac{\partial}{\partial p}, \qquad (4.101)$$

and $X_3 = -\partial/\partial p$ satisfy the commutation relations

$$[X_1, X_2] = X_3, \quad [X_3, X_1] = X_2, \quad [X_2, X_3] = X_1. \qquad (4.102)$$

In other words, X describes a Lie system associated with a VG Lie algebra isomorphic to $\mathfrak{su}(2)$. As in the previous examples, we assume $V^X = V$. Now, the vector fields X_α ($\alpha = 1, 2, 3$) are Hamiltonian ones with Hamiltonian functions given by

$$h_1 = -\sqrt{1-x^2}\cos p, \quad h_2 = -\sqrt{1-x^2}\sin p, \quad h_3 = x, \qquad (4.103)$$

thus spanning a real Lie algebra isomorphic to $\mathfrak{su}(2)$. Indeed,

$$\{h_1, h_2\} = -h_3, \quad \{h_3, h_1\} = -h_2, \quad \{h_2, h_3\} = -h_1. \qquad (4.104)$$

Next, we consider a basis $\{v_1, v_2, v_3\}$ for $\mathfrak{su}(2)$ satisfying

$$[v_1, v_2] = -v_3, \quad [v_3, v_1] = -v_2, \quad [v_2, v_3] = -v_1, \qquad (4.105)$$

so that $\mathfrak{su}(2)$ admits the Casimir $\mathcal{C} = v_1\widetilde{\otimes}v_1 + v_2\widetilde{\otimes}v_2 + v_3\widetilde{\otimes}v_3 \in U_{\mathfrak{su}(2)}$. Then, the Casimir element of $S_{\mathfrak{su}(2)}$ reads $C = \lambda^{-1}(\mathcal{C}) = v_1^2 + v_2^2 + v_3^2$.

The diagonal prolongations of X_1, X_2, X_3 are linearly independent at a generic point for $m = 2$ and we have to derive two first integrals for the diagonal prolongations $\widetilde{X}_1, \widetilde{X}_2, \widetilde{X}_3$ on $T^*\mathbb{I}^3$ satisfying (4.15) working with the coordinate system $\{x_{(1)}, p_{(1)}, x_{(2)}, p_{(2)}, x_{(3)}, p_{(3)}\}$ of $T^*\mathbb{I}^3$. Then, by taking into account the Casimir function C, the

Hamiltonians (4.37), the isomorphism $\phi(v_\alpha) = h_\alpha$ and the injection $D : \mathfrak{sl}(2, \mathbb{R}) \to C^\infty(T^*I)$, we apply Theorem 4.37 obtaining the following first integrals:

$$F^{(2)} = 2\left[\sqrt{1 - x_1^2}\sqrt{1 - x_2^2}\cos(p_1 - p_2) + x_1 x_2 + 1\right],$$

$$F^{(3)} = 2\sum_{1 \leq i < j}^{3}\left[\sqrt{1 - x_i^2}\sqrt{1 - x_j^2}\cos(p_i - p_j) + x_i x_j\right] + 3,$$

$$F_{13}^{(2)} = S_{13}(F^{(2)}),$$

$$F_{23}^{(2)} = S_{23}(F^{(2)}),$$

$$F^{(3)} = F^{(2)} + F_{13}^{(2)} + F_{23}^{(2)} - 3,$$

$$(4.106)$$

and $F = D(C) = 1$. We again choose $F^{(2)}$ and $F_{23}^{(2)}$ as the two functionally independent constants of motion, which provide us, after cumbersome but straightforward computations, with a superposition rule for these systems. This leads to a quartic equation, whose solution can be obtained through known methods. All our results are also valid for the case when $V^X \subset V$.

4.4. Classification of Lie–Hamilton Systems on the Plane

In this section, we classify finite-dimensional real Lie algebras of Hamiltonian vector fields on the plane. In view of the definition of a Lie–Hamilton system, our classification of finite-dimensional Lie algebras on the plane implies a classification of Lie–Hamilton systems on the plane. This classification will be of primordial importance, given the number of physical system which can be interpreted in terms of Lie–Hamilton systems. This will be illustrated in the following sections.

4.4.1. General definitions and properties

A natural problem in the theory of Lie systems is the classification of Lie systems on a fixed manifold, which amounts to classifying finite-dimensional Lie algebras of vector fields on it. Lie accomplished the local classification of finite-dimensional real Lie algebras of vector fields on the real line. More precisely, he showed that each such a Lie algebra is locally diffeomorphic to a Lie subalgebra of $\langle \partial_x, x\partial_x, x^2\partial_x \rangle \simeq \mathfrak{sl}(2, \mathbb{R})$ on a neighborhood of each *generic point* x_0 of the Lie algebra [183]. He also performed the local classification of finite-dimensional real Lie algebras of planar vector fields and started the study of the analogous problem on $\mathbb{C}^3$ [5].

Lie's local classification on the plane presented some unclear points which were misunderstood by several authors during the following decades. To solve this, González-López, Kamran and Olver retook the problem and provided a clearer insight in [183]. They proved that every non-zero Lie algebra of vector fields on the plane is locally diffeomorphic around each generic point to one of the finite-dimensional real Lie algebras (the GKO classification) given in Table A.1 in Appendix A.

As every VG Lie algebra on the plane has to be locally diffeomorphic to one of the Lie algebras of the GKO classification, every Lie system on the plane is locally diffeomorphic to a Lie system taking values in a VG Lie algebra within the GKO classification. So, the local properties of all Lie systems on the plane can be studied through the Lie systems related to the GKO classification. As a consequence, we say that the GKO classification gives the local classification of Lie systems on the plane. Their classification divides finite-dimensional Lie algebras of vector fields on $\mathbb{R}^2$ into 28 non-diffeomorphic classes, which, in fact, can be regarded as a local classification of Lie systems on the plane [23, 183].

For instance, we see from the matrix Riccati equations on $\mathbb{R}^2$ [23, 197, 376]

$$\begin{cases} \dfrac{dx}{dt} = a_0(t) + a_1(t)x + a_2(t)y + c_1(t)x^2 + c_2(t)xy, \\ \dfrac{dy}{dt} = b_0(t) + b_1(t)x + b_2(t)y + c_2(t)y^2 + c_1(t)xy \end{cases} \qquad (4.107)$$

that such a system is related to a non-solvable eight-dimensional VG Lie algebra

$$X_1 = \frac{\partial}{\partial x}, \ X_2 = x\frac{\partial}{\partial x}, \ X_3 = y\frac{\partial}{\partial x}, \ X_4 = x^2\frac{\partial}{\partial x} + xy\frac{\partial}{\partial y},$$
$$X_5 = \frac{\partial}{\partial y}, \ X_6 = x\frac{\partial}{\partial y}, \ X_7 = y\frac{\partial}{\partial y}, \ X_8 = y^2\frac{\partial}{\partial y} + xy\frac{\partial}{\partial x}. \tag{4.108}$$

In view of the GKO classification in Table A.1 (cf. [23, 183]), all Lie algebras of this type are locally diffeomorphic to $P_8 \simeq \mathfrak{sl}(3, \mathbb{R})$. That is why we say that the system (4.1) is *a Lie system of class* P_8.

The GKO classification of finite-dimensional Lie algebras of vector fields X_i $(i = 1, \ldots, n)$ on the plane covers two subclasses called *primitive* (8 cases P_x) and *imprimitive* (20 cases I_x) [183].

Definition 4.38. A finite-dimensional real Lie algebra V of vector fields on an open subset $U \subset \mathbb{R}^2$ is *imprimitive* when there exists a one-dimensional distribution $\mathcal{D}$ on $\mathbb{R}^2$ invariant under the action of V by Lie brackets, i.e., for every $X \in V$ and every vector field Y taking values in $\mathcal{D}$, we have that $[X, Y]$ takes values in $\mathcal{D}$. Otherwise, V is called *primitive*.

Example 4.10. Consider the Lie algebra I_4 which is spanned by the vector fields X_1, X_2, and X_3 given in Table A.1 in Appendix A. If we define $\mathcal{D}$ to be the distribution on $\mathbb{R}^2$ generated by $Y = \partial_x$, we see that

$$[X_1, Y] = 0, \quad [X_2, Y] = -Y, \quad [X_3, Y] = -2xY. \tag{4.109}$$

We infer from this that $\mathcal{D}$ is a one-dimensional distribution invariant under the action of I_4. Hence, I_4 is an imprimitive Lie algebra of vector fields.

To determine which of the 28 classes can be considered as VG Lie algebras of Hamiltonian vector fields, a symplectic form

$$\omega = f(x, y)\mathrm{d}x \wedge \mathrm{d}y \tag{4.110}$$

must be found so that each X_i, belonging to the basis of the considered Lie algebra, becomes Hamiltonian (see [23] for details). In particular, X_i are Hamiltonian vector fields with respect to ω whenever the Lie derivative of ω relative to X_i vanishes, that is, $\mathcal{L}_{X_i}\omega = 0$. If ω exists, then the vector fields X_i become Hamiltonian and their corresponding Hamiltonian functions h_i are obtained by using the relation $\iota_{X_i}\omega = \mathrm{d}h_i$. The functions $\langle h_1, \ldots, h_n \rangle$ and their successive brackets with respect to the Lie bracket

$$\{\cdot, \cdot\}_\omega \; : \; C^\infty\left(\mathbb{R}^2\right) \times C^\infty\left(\mathbb{R}^2\right) \to C^\infty\left(\mathbb{R}^2\right) \qquad (4.111)$$

induced by ω span a finite-dimensional Lie algebra of functions $(\mathfrak{W}, \{\cdot, \cdot\}_\omega)$. We will call this Lie algebra a *Lie–Hamilton algebra* (LH algebra, in short).

In the case of the VG Lie algebra V^{MR} for matrix Riccati equations, we see that there exists no such structure. Indeed, if $X_0 = \partial_x$ and $X_2 = x\partial_x$, which belong to V^{MR}, were Hamiltonian relative to a symplectic form, let us say $\omega = f(x, y)\mathrm{d}x \wedge \mathrm{d}y$, we would have $\mathcal{L}_{X_0}\omega = \mathcal{L}_{X_2}\omega = 0$. This implies that $\partial_x f = x\partial_x f + f = 0$. Hence, $f = 0$ and ω cannot be symplectic. In other words, V^{MR} is not a Lie algebra of Hamiltonian vector fields.

This example illustrates that not every Lie system admits a VG Lie algebra of Hamiltonian vector fields with respect to a Poisson structure [358]. When a Lie system does, we call it a *Lie–Hamilton system* (LH system) [121].

The problem of which of the Lie algebras within the GKO classification consist of Hamiltonian vector fields with respect to some Poisson structure has recently been solved by the present writer and collaborators in [23], showing that, among the 28 classes of the GKO classification, only 12 remain as Lie algebras of Hamiltonian vector fields. Furthermore, examples of LH systems having applications in the above-mentioned fields have also been worked out in [23]. We recall that one advantage of LH systems is that their superposition rules might be obtained straightforwardly by applying a coalgebra approach as it has been formulated in the previous section [27].

It has recently been proven that the initial $8 + 20$ cases of the GKO classification are reduced to $4 + 8$ classes of finite-dimensional

Lie algebras of Hamiltonian vector fields [23]. The final result is summarized in Table A.1 in Appendix A, where we indicate the Lie algebra $\mathfrak{g}$ of Hamiltonian vector fields X_i, a family of their corresponding Hamiltonian functions h_i and an associated symplectic form ω.

4.4.1.1. Minimal Lie algebras of Lie–Hamilton systems on the plane

In this section, we study the local structure of the minimal Lie algebras of Lie–Hamilton systems on the plane around their generic points.

Our main result, Theorem 4.43, and the remaining findings of this section enable us to give the local classification of Lie–Hamilton systems on the plane in the following subsections. To simplify the notation, U will hereafter stand for a contractible open subset of $\mathbb{R}^2$.

Definition 4.39. Given a finite-dimensional Lie algebra V of vector fields on a manifold N, we say that $\xi_0 \in N$ is a *generic point* of V when the rank of the generalized distribution

$$\mathcal{D}_\xi^V := \{X(\xi) \mid X \in V\} \subset T_\xi N, \quad \xi \in N, \tag{4.112}$$

i.e., the function $r^V(\xi) := \dim \mathcal{D}_\xi^V$, is locally constant around ξ_0. We call *generic domain* or simply *domain* of V the set of generic points of V.

Example 4.11. Consider the Lie algebra $I_4 = \langle X_1, X_2, X_3 \rangle$ of vector fields on $\mathbb{R}^2$ detailed in Table A.1. By using the expressions of X_1, X_2 and X_3 in coordinates, we see that $r^{I_4}(x, y)$ equals the rank of the matrix

$$\begin{pmatrix} 1 & x & x^2 \\ 1 & y & y^2 \end{pmatrix}, \tag{4.113}$$

which is two for every $(x, y) \in \mathbb{R}^2$ except for points with $y - x = 0$, where the rank is one. So, the domain of I_4 is $\mathbb{R}^2_{x \neq y} := \{(x, y) \mid x \neq y\} \subset \mathbb{R}^2$.

Lemma 4.40. *Let V be a finite-dimensional real Lie algebra of Hamiltonian vector fields on $\mathbb{R}^2$ with respect to a Poisson structure* and let $\xi_0 \in \mathbb{R}^2$ be a generic point of V. There exists a $U \ni \xi_0$ such that $V|_U$ consists of Hamiltonian vector fields relative to a symplectic structure.

Proof. If $\dim \mathcal{D}^V_{\xi_0} = 0$, then $\dim \mathcal{D}^V_\xi = 0$ for every ξ in a $U \ni \xi_0$ because the rank of $\mathcal{D}^V$ is locally constant around generic points. Consequently, $V|_U = 0$ and its unique element become Hamiltonian relative to the restriction of $\omega = \mathrm{d}x \wedge \mathrm{d}y$ to U. Let us assume now $\dim \mathcal{D}^V_{\xi_0} \neq 0$. By assumption, the elements of V are Hamiltonian vector fields with respect to a Poisson bivector $\Lambda \in \Gamma(\Lambda^2 \mathrm{T}\mathbb{R}^2)$. Hence, $\mathcal{D}^V_\xi \subset \mathcal{D}^\Lambda_\xi$ for every $\xi \in \mathbb{R}^2$, with $\mathcal{D}^\Lambda$ being the *characteristic distribution* of Λ [358]. Since $\dim \mathcal{D}^V_{\xi_0} \neq 0$ and r^V is locally constant at ξ_0, then $\dim \mathcal{D}^V_\xi > 0$ for every ξ in a $U \ni \xi_0$. Since the rank of $\mathcal{D}^\Lambda$ is even at every point of $\mathbb{R}^2$ and $\mathcal{D}^V_\xi \subset \mathcal{D}^\Lambda_\xi$ for every $\xi \in U$, the rank of $\mathcal{D}^\Lambda$ is two on U. So, Λ comes from a symplectic structure on U and $V|_U$ is a Lie algebra of Hamiltonian vector fields relative to it. $\qquad\square$

Roughly speaking, the previous lemma establishes that any Lie–Hamilton system X on $\mathbb{R}^2$ can be considered around each generic point of V^X as a Lie–Hamilton system admitting a minimal Lie algebra of Hamiltonian vector fields with respect to a symplectic structure. As our study of such systems is local, we hereafter focus on analysing minimal Lie algebras of this type.

A *volume form* Ω on an n-dimensional manifold N is a non-vanishing n-form on N. The divergence of a vector field X on N with respect to Ω is the unique function $\mathrm{div}X : N \to \mathbb{R}$ satisfying $\mathcal{L}_X \Omega = (\mathrm{div}\, X)\, \Omega$. An *integrating factor* for X on $U \subset N$ is a function $f : U \to \mathbb{R}$ such that $\mathcal{L}_{fX} \Omega = 0$ on U. Next, we have the following result [277].

Lemma 4.41. *Consider the volume form $\Omega = \mathrm{d}x \wedge \mathrm{d}y$ on a $U \subset \mathbb{R}^2$ and a vector field X on U. Then, X is Hamiltonian with respect to a symplectic form $\omega = f\Omega$ on U if and only if $f : U \to \mathbb{R}$ is a non-vanishing integrating factor of X with respect to Ω, i.e., $Xf = -f\mathrm{div}X$ on U.*

Proof. Since ω is a symplectic form on U, then f must be non-vanishing. As

$$\mathcal{L}_X\omega = \mathcal{L}_X(f\Omega) = (Xf)\Omega + f\mathcal{L}_X\Omega = (Xf + f\operatorname{div}X)\Omega = \mathcal{L}_{fX}\Omega,$$
$$(4.114)$$

then X is locally Hamiltonian with respect to ω, i.e., $\mathcal{L}_X\omega = 0$, if and only if f is a non-vanishing integrating factor for X on U. As U is a contractible open subset, the Poincaré Lemma ensures that X is a local Hamiltonian vector field if and only if it is a Hamiltonian vector field. Consequently, the lemma follows. $\qquad\square$

Definition 4.42. Given a vector space V of vector fields on U, we say that V admits a *modular generating system* $(U_1, X_1, \ldots, X_p)$ if U_1 is a dense open subset of U such that every $X \in V|_{U_1}$ can be brought into the form $X|_{U_1} = \sum_{i=1}^{p} g_i X_i|_{U_1}$ for certain functions $g_1, \ldots, g_p \in C^\infty(U_1)$ and vector fields $X_1, \ldots, X_p \in V$.

Example 4.12. Given the Lie algebra $P_3 \simeq \mathfrak{so}(3)$ on $\mathbb{R}^2$ of Table A.1 of Appendix A, the vector fields

$$X_1 = y\frac{\partial}{\partial x} - x\frac{\partial}{\partial y}, \quad X_2 = (1 + x^2 - y^2)\frac{\partial}{\partial x} + 2xy\frac{\partial}{\partial y}, \quad (4.115)$$

of P_3 satisfy that $X_3 = g_1 X_1 + g_2 X_2$ on $U_1 = \{(x, y) \in \mathbb{R}^2 \mid x \neq 0\}$ for the functions $g_1, g_2 \in C^\infty(U_1)$

$$g_1 = \frac{x^2 + y^2 - 1}{x}, \quad g_2 = \frac{y}{x}. \quad (4.116)$$

Obviously, U_1 is an open dense subset of $\mathbb{R}^2$. As every element of V is a linear combination of X_1, X_2 and $X_3 = g_1 X_1 + g_2 X_2$, then every $X \in V|_{U_1}$ can be written as a linear combination with smooth functions on U_1 of X_1 and X_2. So, (U_1, X_1, X_2) form a generating modular system for P_3.

In Table A.1 in Appendix A, we detail a modular generating system, which is indicated by the first one or two vector fields written

174 *A Guide to Lie Systems with Compatible Geometric Structures*

between brackets in the list of the X_i's, for every finite-dimensional Lie algebra of vector fields of the GKO classification.

Theorem 4.43. *Let V be a Lie algebra of vector fields on $U \subset \mathbb{R}^2$ admitting a modular generating system $(U_1, X_1, \ldots, X_p)$. We have that*

(1) The space V consists of Hamiltonian vector fields relative to a symplectic form on U if and only if

- *(i) Let $g_1, \ldots, g_p$ be certain smooth functions on $U_1 \subset U$. Then,*

$$X|_{U_1} = \sum_{i=1}^p g_i X_i|_{U_1} \in V|_{U_1} \implies \mathrm{div} X|_{U_1} = \sum_{i=1}^p g_i \mathrm{div} X_i|_{U_1}.$$

$$(4.117)$$

- *(ii) The elements $X_1, \ldots, X_p$ admit a common non-vanishing integrating factor on U.*

(2) If the rank of $\mathcal{D}^V$ is two on U, the symplectic form is unique up to a multiplicative non-zero constant.

Proof. Let us prove the direct part of 1). Since $(U_1, X_1, \ldots, X_p)$ form a modular generating system for V, we have that every $X|_{U_1} \in V|_{U_1}$ can be brought into the form $X|_{U_1} = \sum_{i=1}^p g_i X_i|_{U_1}$ for certain $g_1, \ldots, g_p \in C^\infty(U_1)$. As V is a Lie algebra of Hamiltonian vector fields with respect to a symplectic structure on U, let us say

$$\omega = f(x, y)\mathrm{d}x \wedge \mathrm{d}y, \qquad (4.118)$$

then Lemma 4.41 ensures that $Yf = -f\mathrm{div}Y$ for every $Y \in V$. Then,

$$f\mathrm{div}X = -Xf = -\sum_{i=1}^p g_i X_i f = f \sum_{i=1}^p g_i \mathrm{div} X_i$$

$$\iff f\left(\mathrm{div}X - \sum_{i=1}^p g_i \mathrm{div} X_i\right) = 0$$

$$(4.119)$$

on U_1. As ω is non-degenerate, then f is non-vanishing and (i) follows. Since all the vector fields of V are Hamiltonian with respect to ω, they share a common non-vanishing integrating factor, namely f. From this, (ii) holds.

Conversely, if (ii) is satisfied, then Lemma 4.41 ensures that $X_1, \ldots, X_p$ are Hamiltonian with respect to (4.12) on U, with f being a non-vanishing integrating factor. As $(U_1, X_1, \ldots, X_p)$ form a generating modular system for V, every $X \in V$ can be written as $\sum_{i=1}^{p} g_i X_i$ on U_1 for certain functions $g_1, \ldots, g_p \in C^{\infty}(U_1)$. From (i), we obtain div $X = \sum_{i=1}^{p} g_i \mathrm{div} X_i$ on U_1. Then,

$$Xf = \sum_{i=1}^{p} g_i X_i f = -f \sum_{i=1}^{p} g_i \mathrm{div} X_i = -f \mathrm{div} X, \qquad (4.120)$$

on U_1 and, since the elements of V are smooth and U_1 is dense on U, the above expression also holds on U. Hence, f is a non-vanishing integrating factor for X, which becomes a Hamiltonian vector field with respect to ω on U in virtue of Lemma 4.41. Hence, part (1) is proven.

As far as part (2) of the theorem is concerned, if the vector fields of V are Hamiltonian with respect to two different symplectic structures on U, they admit two different non-vanishing integrating factors f_1 and f_2. Hence,

$$X(f_1/f_2) = (f_2 X f_1 - f_1 X f_2)/f_2^2$$
$$= (f_2 f_1 \mathrm{div} X - f_1 f_2 \mathrm{div} X)/f_2^2 = 0, \qquad (4.121)$$

and f_1/f_2 is a common constant of motion for all the elements of V. Hence, it is a constant of motion for every vector field taking values in the distribution $\mathcal{D}^V$. Then rank of $\mathcal{D}^V$ on U is two by assumption. So, $\mathcal{D}^V$ is generated by the vector fields ∂_x and ∂_y on U. Thus, the only constants of motion on U common to all the vector fields taking values in $\mathcal{D}^V$, and consequently common to the elements of V, are constants. Since f_1 and f_2 are non-vanishing, then $f_1 = \lambda f_2$ for a $\lambda \in \mathbb{R} \backslash \{0\}$ and the associated symplectic structures are the same up to an irrelevant non-zero proportionality constant. $\qquad \square$

Using Theorem 4.43, we can immediately prove the following result.

Corollary 4.44. *If a Lie algebra of vector fields V on a $U \subset \mathbb{R}^2$ consists of Hamiltonian vector fields with respect to a symplectic form and admits a modular generating system whose elements are divergence free, then every element of V is divergence free.*

4.4.2. Lie–Hamilton algebras

It is known that given a Lie–Hamilton system X, its Lie–Hamilton algebras are not uniquely defined in general. Moreover, the existence of different types of Lie–Hamilton algebras for the same Lie–Hamilton system is important in its linearization and the use of certain methods [121]. For instance, if a Lie–Hamilton system X on N admits a Lie–Hamilton algebra isomorphic to V^X and $\dim V^X = \dim N$, then X can be linearized together with its associated Poisson structure [121].

Example 4.13. Consider again the Lie–Hamilton system X given by the complex Riccati equations with t-dependent real coefficients (4.206) and assume $V^X \simeq \mathfrak{sl}(2, \mathbb{R})$. Recall that X admits a Lie–Hamilton algebra $(\mathcal{H}_\Lambda, \{\cdot, \cdot\}_\omega) \simeq \mathfrak{sl}(2, \mathbb{R})$ spanned by the Hamiltonian functions h_1, h_2, h_3 given by (4.3) relative to the symplectic structure ω detailed in (4.2). We can also construct a second (non-isomorphic) Lie–Hamilton algebra for X with respect to ω. The vector fields X_i, with $i = 1, 2, 3$, spanning V^X (see (4.208)) have also Hamiltonian functions $\bar{h}_i = h_i + 1$, for $i = 1, 2, 3$, respectively. Hence, $(\mathbb{R}^2_{y \neq 0}, \omega, h = a_0(t)\bar{h}_1 + a_1(t)\bar{h}_2 + a_2(t)\bar{h}_3)$ is a Lie–Hamiltonian structure for X giving rise to a Lie–Hamilton algebra $(\overline{\mathcal{H}}_\Lambda, \{\cdot, \cdot\}_\omega) := (\langle \bar{h}_1, \bar{h}_2, \bar{h}_3, 1 \rangle, \{\cdot, \cdot\}_\omega) \simeq \mathfrak{sl}(2, \mathbb{R}) \oplus \mathbb{R}$ for X.

Proposition 4.45. *A Lie–Hamilton system X on a symplectic connected manifold (N, ω) possesses an associated Lie–Hamilton algebra $(\mathcal{H}_\Lambda, \{\cdot, \cdot\}_\omega)$ isomorphic to V^X if and only if every Lie–Hamilton algebra non-isomorphic to V^X is isomorphic to $V^X \oplus \mathbb{R}$.*

Proof. Let $(\overline{\mathcal{H}}_\Lambda, \{\cdot, \cdot\}_\omega)$ be an arbitrary Lie–Hamilton algebra for X. As X is defined on a connected manifold, the sequence of Lie algebras

$$0 \hookrightarrow (\overline{\mathcal{H}}_\Lambda, \{\cdot, \cdot\}_\omega) \cap \langle 1 \rangle \hookrightarrow (\overline{\mathcal{H}}_\Lambda, \{\cdot, \cdot\}_\omega) \xrightarrow{\varphi} V^X \to 0, \qquad (4.122)$$

where $\varphi : \overline{\mathcal{H}}_\Lambda \to V^X$ maps every function of $\overline{\mathcal{H}}_\Lambda$ to minus its Hamiltonian vector field, is always exact (cf. [121]). Hence, $(\overline{\mathcal{H}}_\Lambda, \{\cdot, \cdot\}_\omega)$ can be isomorphic either to V^X or to a Lie algebra extension of V^X of dimension $\dim V^X + 1$.

If $(\mathcal{H}_\Lambda, \{\cdot, \cdot\}_\omega)$ is isomorphic to V^X and there exists a second Lie–Hamilton algebra $(\overline{\mathcal{H}}_\Lambda, \{\cdot, \cdot\}_\omega)$ for X non-isomorphic to V^X, we see from (4.16) that $1 \in \overline{\mathcal{H}}_\Lambda$ and $1 \notin \mathcal{H}_\Lambda$. Given a basis $X_1, \ldots, X_r$ of V^X, each element X_i, with $i = 1, \ldots, r$, has a Hamiltonian function $\overline{h}_i \in \overline{\mathcal{H}}_\Lambda$ and another $h_i \in \mathcal{H}_\Lambda$. As V^X is defined on a connected manifold, then $h_i = \overline{h}_i - \lambda_i \in \overline{\mathcal{H}}_\Lambda$ with $\lambda_i \in \mathbb{R}$ for every $i = 1, \ldots, r$. From this and using again that $1 \in \overline{\mathcal{H}}_\Lambda \backslash \mathcal{H}_\Lambda$, we obtain that $\{h_1, \ldots, h_r, 1\}$ is a basis for $\overline{\mathcal{H}}_\Lambda$ and $(\overline{\mathcal{H}}_\Lambda, \{\cdot, \cdot\}_\omega) \simeq (\mathcal{H}_\Lambda \oplus \mathbb{R}, \{\cdot, \cdot\}_\omega)$.

Let us assume now that every Lie–Hamilton algebra $(\overline{\mathcal{H}}_\Lambda, \{\cdot, \cdot\}_\omega)$ non-isomorphic to V^X is isomorphic to $V^X \oplus \mathbb{R}$. We can define a Lie algebra anti-isomorphism $\mu : V^X \to \overline{\mathcal{H}}_\Lambda$ mapping each element of V^X to a Hamiltonian function belonging to a Lie subalgebra of $(\overline{\mathcal{H}}_\Lambda, \{\cdot, \cdot\}_\omega)$ isomorphic to V^X. Hence, $(N, \omega, h = \mu(X))$, where $h_t = \mu(X_t)$ for each $t \in \mathbb{R}$, is a Lie–Hamiltonian structure for X and $(\mu(V^X), \{\cdot, \cdot\}_\omega)$ is a Lie–Hamilton algebra for X isomorphic to V^X. $\square$

Proposition 4.46. *If a Lie–Hamilton system X on a symplectic connected manifold (N, ω) admits an associated Lie–Hamilton algebra $(\mathcal{H}_\Lambda, \{\cdot, \cdot\}_\omega)$ isomorphic to V^X, then it admits a Lie–Hamilton algebra isomorphic to $V^X \oplus \mathbb{R}$.*

Proof. Let (N, ω, h) be a Lie–Hamiltonian structure for X giving rise to the Lie–Hamilton algebra $(\mathcal{H}_\Lambda, \{\cdot, \cdot\}_\omega)$. Consider the linear space L_h spanned by linear combinations of the functions $\{h_t\}_{t \in \mathbb{R}}$. Since we assume $\mathcal{H}_\Lambda \simeq V^X$, the exact sequence (4.16) involves that $1 \notin L_h$. Moreover, we can write $h = \sum_{i=1}^p b_i(t) h_{t_i}$, where h_{t_i} are the values of h at certain times $t_1, \ldots, t_p$ such that $\{h_{t_1}, \ldots, h_{t_p}\}$ are linearly independent and $b_1, \ldots, b_p$ are certain t-dependent functions. Observe that the vector fields $(b_1(t), \ldots, b_p(t))$, with $t \in \mathbb{R}$, span a p-dimensional linear space. If we choose a t-dependent Hamiltonian $\bar{h} = \sum_{i=1}^{p} b_i(t) h_{t_i} + b_{p+1}(t)$, where $b_{p+1}(t)$ is not a linear combination of $b_1(t), \ldots, b_p(t)$, and we recall that $1, h_{t_1}, \ldots, h_{t_p}$ are linearly independent over $\mathbb{R}$, we obtain that the linear hull of the functions $\{\bar{h}_t\}_{t \in \mathbb{R}}$ has dimension $\dim L_h + 1$. Moreover, $(N, \{\cdot, \cdot\}_\omega, \bar{h})$ is a Lie–Hamiltonian structure for X. Hence, they span, along with their successive Lie brackets, a Lie–Hamilton algebra isomorphic to $\mathcal{H}_\Lambda \oplus \mathbb{R}$. $\qquad \square$

Corollary 4.47. *If X is a Lie–Hamilton system with respect to a symplectic connected manifold (N, ω) admitting a Lie–Hamilton algebra $(\mathcal{H}_\Lambda, \{\cdot, \cdot\}_\omega)$ satisfying that $1 \in \{\mathcal{H}_\Lambda, \mathcal{H}_\Lambda\}_\omega$, then X does not possess any Lie–Hamilton algebra isomorphic to V^X.*

Proof. If $1 \in \{\mathcal{H}_\Lambda, \mathcal{H}_\Lambda\}_\omega$, then $\mathcal{H}_\Lambda$ cannot be isomorphic to $V^X \oplus \mathbb{R}$ because the derived Lie algebra of $\mathcal{H}_\Lambda$, i.e., $\{\mathcal{H}_\Lambda, \mathcal{H}_\Lambda\}_\omega$, contains the constant function 1 and the derived Lie algebra of a $\mathcal{H}_\Lambda$ isomorphic to $V^X \oplus \mathbb{R}$ does not. In view of Proposition 4.45, system X does not admit any Lie–Hamilton algebra isomorphic to V^X. $\qquad \square$

Proposition 4.48. *If X is a Lie–Hamilton system on a connected manifold N whose V^X consists of Hamiltonian vector fields with respect to a symplectic form ω that does not possess any Lie–Hamilton algebra $(\mathcal{H}_\Lambda, \{\cdot, \cdot\}_\omega)$ isomorphic to V^X, then all its Lie–Hamilton algebras (with respect to $\{\cdot, \cdot\}_\omega$) are isomorphic.*

Proof. Let $(\mathcal{H}_\Lambda, \{\cdot, \cdot\}_\omega)$ and $(\overline{\mathcal{H}}_\Lambda, \{\cdot, \cdot\}_\omega)$ be two Lie–Hamilton algebras for X. Since they are not isomorphic to V^X and in view of the exact sequence (4.16), then $1 \in \mathcal{H}_\Lambda \cap \overline{\mathcal{H}}_\Lambda$. Let $X_1, \ldots, X_r$ be a basis of V^X. Every vector field X_i admits a Hamiltonian function $h_i \in \mathcal{H}_\Lambda$ and another $\bar{h}_i \in \overline{\mathcal{H}}_\Lambda$. The functions $h_1, \ldots, h_r$ are linearly independent and the same applies to $\bar{h}_1, \ldots, \bar{h}_r$. Then, $\{h_1, \ldots, h_r, 1\}$ is a basis for $\mathcal{H}_\Lambda$ and $\{\bar{h}_1, \ldots, \bar{h}_r, 1\}$ is a basis for $\overline{\mathcal{H}}_\Lambda$. As N is connected, then $h_i = \bar{h}_i - \lambda_i$ with $\lambda_i \in \mathbb{R}$ for each $i \in \mathbb{R}$. Hence, the functions h_i belong to $\overline{\mathcal{H}}_\Lambda$ and the functions $\bar{h}_i$ belong to $\mathcal{H}_\Lambda$. Thus $\mathcal{H}_\Lambda = \overline{\mathcal{H}}_\Lambda$.

$\square$

For instance, the case P_1 from Table A.1 in Appendix A corresponds to the two-dimensional Euclidean algebra $\mathfrak{iso}(2) \simeq \langle X_1, X_2, X_3 \rangle$, but the Hamiltonian functions $\langle h_1, h_2, h_3, h_0 = 1 \rangle$ span the centrally extended Euclidean algebra $\overline{\mathfrak{iso}}(2)$. A similar fact arises in classes $\mathrm{P}_3 \simeq \mathfrak{so}(3)$, $\mathrm{P}_5 \simeq \mathfrak{sl}(2, \mathbb{R}) \ltimes \mathbb{R}^2$, $\mathrm{I}_8 \simeq \mathfrak{iso}(1,1)$ (the $(1+1)$-dimensional Poincaré algebra), $\mathrm{I}_{14B} \simeq \mathbb{R} \ltimes \mathbb{R}^r$ and $\mathrm{I}_{16} \simeq \mathfrak{h}_2 \ltimes \mathbb{R}^{r+1}$. Among them, only the family $\mathrm{P}_3 \simeq \mathfrak{so}(3)$ is a simple Lie algebra in such a manner that $h_0 = 1$ gives rise to a trivial central extension which means that the LH algebra is isomorphic to $\mathfrak{so}(3) \oplus \mathbb{R}$; otherwise, the central extension is a non-trivial one and it cannot be "removed".

In this respect, notice that the appearance of a non-trivial central extension is the difference between the family I_{14B} with respect to I_{14A}. We also recall that the LH algebra corresponding to the class P_5, that is $\overline{\mathfrak{sl}}(2, \mathbb{R}) \ltimes \mathbb{R}^2$, is isomorphic to the two-photon Lie algebra $\mathfrak{h}_6$ (see [22, 385] and references therein) and, therefore, also to the $(1+1)$-dimensional centrally extended Schrödinger Lie algebra [30].

We stress that the Lie algebra $\mathfrak{sl}(2, \mathbb{R})$ appears four times (classes $\mathrm{P}_2, \mathrm{I}_3, \mathrm{I}_4$, and I_5 in the GKO classification) which means that there may be different LH systems sharing isomorphic VG Lie algebras that are non-diffeomorphic, namely there exists no diffeomorphism mapping the elements of one into the other. In other words, only LH systems belonging to each class can be related through a change of variables. We shall explicitly apply this property throughout the paper.

4.4.3. *Local classification*

In this section, we describe the local structure of Lie–Hamilton systems on the plane, i.e., given the minimal Lie algebra of a Lie–Hamilton system X on the plane, we prove that V^X is locally diffeomorphic around a generic point of V^X to one of the Lie algebras given in Table A.2 in Appendix A. We also prove that, around a generic point of V^X, the Lie–Hamilton algebras of X must have one of the algebraic structures described in Table A.2, Appendix A.

If X is a Lie–Hamilton system, its minimal Lie algebra must be locally diffeomorphic to one of the Lie algebras of the GKO classification that consists of Hamiltonian vector fields with respect to a Poisson structure. As we are concerned with generic points of minimal Lie algebras, Lemma 4.40 ensures that V^X is locally diffeomorphic around generic points to a Lie algebra of Hamiltonian vector fields with respect to a symplectic structure. So, its minimal Lie algebra is locally diffeomorphic to one of the Lie algebras of the GKO classification consisting of Hamiltonian vector fields with respect to a symplectic structure on a certain open contractible subset of its domain. By determining which of the Lie algebras of the GKO classification admit such a property, we can classify the local structure of all Lie–Hamilton systems on the plane.

Proposition 4.49. *The primitive Lie algebras* $\mathrm{P}_1^{\alpha \neq 0}$, P_4, P_6–P_8 *and the imprimitive ones* I_2, I_3, I_6, I_7, $\mathrm{I}_8^{(\alpha \neq -1)}$, I_9–I_{11}, I_{13}, I_{15}, $\mathrm{I}_{16}^{(\alpha \neq -1)}$, I_{17}–I_{20} *are not Lie algebras of Hamiltonian vector fields on any* $U \subset \mathbb{R}^2$.

Proof. Apart from I_{15}, the remaining Lie algebras detailed in this statement admit a modular generating system whose elements are divergence free on the whole $\mathbb{R}^2$ (see the elements written between brackets in Table A.1). At the same time, we also observe in Table A.1 that these Lie algebras admit a vector field with non-zero divergence on any U. In view of Corollary 4.44, they cannot be Lie algebras of Hamiltonian vector fields with respect to any symplectic structure on any $U \subset \mathbb{R}^2$.

In the case of the Lie algebra I_{15}, we have that $(\mathbb{R}^2_{y \neq 0}, X_1 = \partial_x, X_2 = y\partial_y)$ form a generating modular system of I_{15}. Observe that $X_2 = y\partial_y$ and $X_3 = \eta_1(x)\partial_y$, where η_1 is a non-null function — it forms with $\eta_2(x), \ldots, \eta_r(x)$ a basis of solutions of a system of r first-order linear homogeneous differential equations in normal form with constant coefficients (cf. [183, 267]) — satisfy $\mathrm{div} X_2 = 1$ and $\mathrm{div} X_3 = 0$. Obviously, $\mathrm{div} X_3 \neq \eta_1 \mathrm{div} X_2 / y$ on any U. So, I_{15} does not satisfy Theorem 4.43 on any U and it is not a Lie algebra of Hamiltonian vector fields on any $U \subset \mathbb{R}^2$. $\square$

To simplify the notation, we assume in this section that all objects are defined on a contractible $U \subset \mathbb{R}^2$ of the domain of the Lie algebra under study. Additionally, U_1 stands for a dense open subset of U. In the following two subsections, we explicitly show that *all* of the Lie algebras of the GKO classification *not* listed in Proposition 4.49 consist of Hamiltonian vector fields on any U of their domains. For each Lie algebra, we compute the integrating factor f of ω given by (4.12) turning the elements of a basis of the Lie algebra into Hamiltonian vector fields and we work out their Hamiltonian functions. Additionally, we obtain the algebraic structure of all the Lie–Hamilton algebras of the Lie–Hamilton systems admitting such minimal Lie algebras.

We stress that the main results covering the resulting Hamiltonian functions h_i, the symplectic form ω and the Lie–Hamilton algebra are summarized in Table A.2 accordingly to the GKO classification of Table A.1. We point out that the Lie algebras of the class I_{14} give rise to two non-isomorphic Lie–Hamilton algebras: I_{14A} whenever $1 \notin \langle \eta_1, \ldots, \eta_r \rangle$ and I_{14B} otherwise. Consequently, we obtain twelve finite-dimensional real Lie algebras of Hamiltonian vector fields on the plane.

In order to shorten the presentation of the following results, we remark that for some of such Lie–Hamilton algebras their symplectic form is just the standard one.

Proposition 4.50. *The Lie algebras* $\mathrm{P}_1^{(\alpha=0)}$, P_5, $\mathrm{I}_8^{(\alpha=-1)}$, I_{14B} *and* $\mathrm{I}_{16}^{(\alpha=-1)}$ *are Lie algebras of Hamiltonian vector fields with respect to the standard symplectic form* $\omega = \mathrm{d}x \wedge \mathrm{d}y$, *that is,* $f := 1$.

Proof. We see in Table A.1 that all the aforementioned Lie algebras admit a modular generating system $(U, X_1 = \partial_x, X_2 = \partial_y)$ and all their elements have zero divergence. So, they satisfy condition (4.11). The vector fields X_1, X_2 are Hamiltonian with respect to the symplectic structure $\omega = \mathrm{d}x \wedge \mathrm{d}y$. In view of Theorem 4.43, the whole Lie algebra consists of Hamiltonian vector fields relative to ω. $\square$

4.4.3.1. *Primitive Lie algebras*

Lie algebra $P_1^{(\alpha=0)}$: $A_0 \simeq \mathfrak{iso}(2)$

Proposition 4.50 states that A_0 is a Lie algebra of Hamiltonian vector fields with respect to the symplectic form $\omega = \mathrm{d}x \wedge \mathrm{d}y$. The basis of vector fields X_1, X_2, X_3 of A_0 given in Table A.1 in Appendix A satisfies the commutation relations

$$[X_1, X_2] = 0, \quad [X_1, X_3] = -X_2, \quad [X_2, X_3] = X_1. \tag{4.123}$$

So, A_0 is isomorphic to the two-dimensional Euclidean algebra $\mathfrak{iso}(2)$. Using the relation $\iota_X \omega = \mathrm{d}h$ between a Hamiltonian vector field and one of its Hamiltonian functions, we get that the Hamiltonian functions for X_1, X_2, X_3 read

$$h_1 = y, \quad h_2 = -x, \quad h_3 = \tfrac{1}{2}(x^2 + y^2), \tag{4.124}$$

correspondingly. Along with $h_0 = 1$, these functions fulfill

$$\{h_1, h_2\}_\omega = h_0, \quad \{h_1, h_3\}_\omega = h_2, \quad \{h_2, h_3\}_\omega = -h_1, \quad \{h_0, \cdot\}_\omega = 0. \tag{4.125}$$

Consequently, if X is a Lie–Hamilton system admitting a minimal Lie algebra A_0, i.e., $X = \sum_{i=1}^{3} b_i(t) X_i$ for certain t-dependent functions $b_1(t), b_2(t), b_3(t)$ such that $V^X \simeq A_0$, then it admits a Lie–Hamiltonian structure $(U, \omega, h = \sum_{i=1}^{3} b_i(t) h_i)$ and a Lie–Hamilton algebra $(\mathcal{H}_\Lambda, \{\cdot, \cdot\}_\omega)$ generated by the functions $\langle h_1, h_2, h_3, h_0 \rangle$. Hence, $(\mathcal{H}_\Lambda, \{\cdot, \cdot\}_\omega)$ is a finite-dimensional real Lie algebra of Hamiltonian functions isomorphic to the *centrally extended* Euclidean algebra $\overline{\mathfrak{iso}}(2)$ [21]. Indeed, note that $1 \in \{\mathcal{H}_\Lambda, \mathcal{H}_\Lambda\}_\omega$. In virtue of Corollary 4.47, system X does not admit any Lie–Hamilton algebra isomorphic to V^X. Moreover, Proposition 4.48 ensures that all Lie–Hamilton algebras for X are isomorphic to $\overline{\mathfrak{iso}}(2)$.

Lie algebra P_2: $\mathfrak{sl}(2, \mathbb{R})$

We have already proven that the Lie algebra of vector fields P_2, which is spanned by the vector fields (4.208), is a Lie algebra of Hamiltonian vector fields with respect to the symplectic structure (4.2).

The Hamiltonian functions h_1, h_2, h_3 for X_1, X_2 and X_3 were found to be (4.3), correspondingly. Then, a Lie system X with minimal Lie algebra P_2, i.e., a system of the form $X = \sum_{i=1}^{3} b_i(t)X_i$ for certain t-dependent functions $b_1(t), b_2(t), b_3(t)$ such that $V^X = P_2$, is a Lie–Hamilton system with respect to the Poisson bracket induced by (4.2). Then, X admits a Lie–Hamiltonian structure $(U, \omega, h = \sum_{i=1}^{3} b_i(t)h_i)$ and a Lie–Hamilton algebra isomorphic to $\mathfrak{sl}(2, \mathbb{R})$ with commutation relations (4.3). In view of Proposition 4.46, any Lie–Hamilton system associated with P_2 also admits a Lie–Hamilton algebra isomorphic to $\mathfrak{sl}(2, \mathbb{R}) \oplus \mathbb{R}$. In view of Proposition 4.45, these are the only algebraic structures of the Lie–Hamilton algebras for such Lie–Hamilton systems.

Lie algebra P_3: $\mathfrak{so}(3)$

In this case, we must determine a symplectic structure ω turning the elements of the modular generating system (U_1, X_1, X_2) of P_3 into locally Hamiltonian vector fields with respect to a symplectic structure ω (4.12). In view of Theorem 4.43, this ensures that every element of P_3 is Hamiltonian with respect to ω. The condition $\mathcal{L}_{X_1}\omega = 0$ gives

$$y\frac{\partial f}{\partial x} - x\frac{\partial f}{\partial y} = 0. \tag{4.126}$$

Applying the characteristics method, we find that f must be constant along the integral curves of the system $x\,dx + y\,dy = 0$, namely curves with $x^2 + y^2 = k \in \mathbb{R}$. So, $f = f(x^2 + y^2)$. If we now require $\mathcal{L}_{X_2}\omega = 0$, we obtain that

$$(1 + x^2 - y^2)\frac{\partial f}{\partial x} + 2xy\frac{\partial f}{\partial y} + 4xf = 0. \tag{4.127}$$

Using that $f = f(x^2 + y^2)$, we have

$$\frac{f'}{f} = -\frac{2}{1 + x^2 + y^2} \Rightarrow f(x^2 + y^2) = (1 + x^2 + y^2)^{-2}. \tag{4.128}$$

184   A Guide to Lie Systems with Compatible Geometric Structures

Then,

$$\omega = \frac{\mathrm{d}x \wedge \mathrm{d}y}{(1 + x^2 + y^2)^2}. \tag{4.129}$$

So, P_3 becomes a Lie algebra of Hamiltonian vector fields relative to ω. The vector fields X_1, X_2, and X_3 admit the Hamiltonian functions

$$h_1 = -\frac{1}{2(1 + x^2 + y^2)}, \quad h_2 = \frac{y}{1 + x^2 + y^2}, \quad h_3 = -\frac{x}{1 + x^2 + y^2}, \tag{4.130}$$

which along with $h_0 = 1$ satisfy the commutation relations

$$\{h_1, h_2\}_\omega = -h_3, \quad \{h_1, h_3\}_\omega = h_2,$$
$$\{h_2, h_3\}_\omega = -4h_1 - h_0, \quad \{h_0, \cdot\}_\omega = 0, \tag{4.131}$$

with respect to the Poisson bracket induced by ω. Then, $\langle h_1, h_2, h_3, h_0 \rangle$ span a Lie algebra of Hamiltonian functions isomorphic to a *central extension* of $\mathfrak{so}(3)$, denoted $\overline{\mathfrak{so}}(3)$. It is well known [21] that the central extension associated with h_0 is a trivial one; if we define $\bar{h}_1 = h_1 + h_0/4$, then $\langle \bar{h}_1, h_2, h_3 \rangle$ span a Lie algebra isomorphic to $\mathfrak{so}(3)$ and $\overline{\mathfrak{so}}(3) \simeq \mathfrak{so}(3) \oplus \mathbb{R}$. In view of this and using Propositions 4.45 and 4.46, a Lie system admitting a minimal Lie algebra P_3 admits Lie–Hamiltonian structure isomorphic to $\mathfrak{so}(3) \oplus \mathbb{R}$ and $\mathfrak{so}(3)$.

Lie algebra P_5: $\mathfrak{sl}(2, \mathbb{R}) \ltimes \mathbb{R}^2$

From Proposition 4.50, this Lie algebra consists of Hamiltonian vector fields with respect to the symplectic form $\omega = \mathrm{d}x \wedge \mathrm{d}y$. The vector fields of the basis $X_1, \ldots, X_5$ for P_5 given in Table A.1 in Appendix A are Hamiltonian vector fields relative to ω with Hamiltonian functions

$$h_1 = y, \quad h_2 = -x, \quad h_3 = xy, \quad h_4 = \tfrac{1}{2}y^2, \quad h_5 = -\tfrac{1}{2}x^2, \tag{4.132}$$

correspondingly. These functions together with $h_0 = 1$ satisfy the relations

$$\begin{aligned}
&\{h_1, h_2\}_\omega = h_0, &&\{h_1, h_3\}_\omega = -h_1, &&\{h_1, h_4\}_\omega = 0, \\
&\{h_1, h_5\}_\omega = -h_2, &&\{h_2, h_3\}_\omega = h_2, &&\{h_2, h_4\}_\omega = -h_1, \\
&\{h_2, h_5\}_\omega = 0, &&\{h_3, h_4\}_\omega = 2h_4, &&\{h_3, h_5\}_\omega = -2h_5, \\
&\{h_4, h_5\}_\omega = h_3, &&\{h_0, \cdot\}_\omega = 0.
\end{aligned} \tag{4.133}$$

Hence, $\langle h_1, \ldots, h_5, h_0 \rangle$ span a Lie algebra $\overline{\mathfrak{sl}(2, \mathbb{R}) \ltimes \mathbb{R}^2}$ which is a *non-trivial* central extension of P_5, i.e., it is not isomorphic to $P_5 \oplus \mathbb{R}$. In fact, it is isomorphic to the so-called two-photon Lie algebra $\mathfrak{h}_6$ (see [22] and references therein); this can be proven to be $\mathfrak{h}_6 \simeq \mathfrak{sl}(2, \mathbb{R}) \oplus_s \mathfrak{h}_3$, where $\mathfrak{sl}(2, \mathbb{R}) \simeq \langle h_3, h_4, h_5 \rangle$, $\mathfrak{h}_3 \simeq \langle h_1, h_2, h_0 \rangle$ is the *Heisenberg–Weyl Lie algebra*, and $\oplus_s$ stands for a semidirect sum. Furthermore, $\mathfrak{h}_6$ is also isomorphic to the (1+1)-dimensional *centrally extended Schrödinger Lie algebra* [30].

In view of Corollary 4.47, Proposition 4.48 and following the same line of reasoning than in previous cases, a Lie system admitting a minimal Lie algebra P_5 only possesses Lie–Hamilton algebras isomorphic to $\mathfrak{h}_6$.

4.4.3.2. *Imprimitive Lie algebras*

Lie algebra I_1: $\mathbb{R}$

Note that $X_1 = \partial_x$ is a modular generating system of I_1. By solving the PDE $\mathcal{L}_{X_1}\omega = 0$ with ω written in the form (4.12), we obtain that $\omega = f(y)\mathrm{d}x \wedge \mathrm{d}y$ with $f(y)$ being any non-vanishing function of y. In view of Theorem 4.43, the Lie algebra I_1 becomes a Lie algebra of Hamiltonian vector fields with respect to ω. Observe that X_1, a basis of I_1, has a Hamiltonian function, $h_1 = \int^y f(y')\mathrm{d}y'$. As h_1 spans a Lie algebra isomorphic to $\mathbb{R}$, it is obvious that a system X with $V^X \simeq I_1$ admits a Lie–Hamilton algebra isomorphic to I_1. Proposition 4.46 yields that X possesses a Lie–Hamilton algebra isomorphic to $\mathbb{R}^2$. In view of Proposition 4.45, these are the only algebraic structures for the Lie–Hamilton algebras for X_1.

Lie algebra I_4: $\mathfrak{sl}(2, \mathbb{R})$ of type II

This Lie algebra admits a modular generating system $(\mathbb{R}^2_{x \neq y}, X_1 = \partial_x + \partial_y, X_2 = x\partial_x + y\partial_y)$. Let us search for a symplectic form ω (4.12) turning X_1 and X_2 into local Hamiltonian vector fields with respect to it. In rigor of Theorem 4.43, if X_1, X_2 become locally Hamiltonian, then we can ensure that every element of I_4 is Hamiltonian with respect to ω. By imposing $\mathcal{L}_{X_i}\omega = 0$ $(i = 1, 2)$, we find that

$$\frac{\partial f}{\partial x} + \frac{\partial f}{\partial y} = 0, \quad x\frac{\partial f}{\partial x} + y\frac{\partial f}{\partial y} + 2f = 0. \tag{4.134}$$

Applying the method of characteristics to the first equation, we have that $dx = dy$. Then $f = f(x - y)$. Using this in the second equation, we obtain a particular solution $f = (x - y)^{-2}$ which gives rise to a closed and non-degenerate two-form, namely

$$\omega = \frac{dx \wedge dy}{(x - y)^2}. \tag{4.135}$$

Hence,

$$h_1 = \frac{1}{x - y}, \quad h_2 = \frac{x + y}{2(x - y)}, \quad h_3 = \frac{xy}{x - y}, \tag{4.136}$$

are Hamiltonian functions of the vector fields X_1, X_2, X_3 of the basis for I_4 given in Table A.1 in Appendix A, respectively. Using the Poisson bracket $\{\cdot, \cdot\}_\omega$ induced by (4.29), we obtain that h_1, h_2 and h_3 satisfy

$$\{h_1, h_2\}_\omega = -h_1, \quad \{h_1, h_3\}_\omega = -2h_2, \quad \{h_2, h_3\}_\omega = -h_3. \tag{4.137}$$

Then, $(\langle h_1, h_2, h_3 \rangle, \{\cdot, \cdot\}_\omega) \simeq \mathfrak{sl}(2, \mathbb{R})$. Consequently, if X is a Lie–Hamilton system admitting a minimal Lie algebra I_4, it admits a Lie–Hamilton algebra that is isomorphic to $\mathfrak{sl}(2, \mathbb{R})$ or, from Proposition 4.46, to $\mathfrak{sl}(2, \mathbb{R}) \oplus \mathbb{R}$. From Proposition 4.45, these are the only algebraic structures for its Lie–Hamilton algebras.

Lie algebra I_5: $\mathfrak{sl}(2, \mathbb{R})$ of type III

Observe that $(U, X_1 = \partial_x, X_2 = 2x\partial_x + y\partial_y)$ form a modular generating system of I_5. The conditions $\mathcal{L}_{X_1}\omega = \mathcal{L}_{X_2}\omega = 0$ ensuring that

X_1 and X_2 are locally Hamiltonian with respect to ω give rise to the equations

$$\frac{\partial f}{\partial x} = 0, \quad 2x\frac{\partial f}{\partial x} + y\frac{\partial f}{\partial y} + 3f = 0, \tag{4.138}$$

so that $f(x,y) = y^{-3}$ is a particular solution of the second equation right above, and X_1, X_2 become locally Hamiltonian vector fields relative to the symplectic form

$$\omega = \frac{\mathrm{d}x \wedge \mathrm{d}y}{y^3}. \tag{4.139}$$

In view of Theorem 4.43, this implies that every element of I_5 is Hamiltonian with respect to ω. Hamiltonian functions for the elements of the basis X_1, X_2, X_3 for I_5 given in Table A.1 in Appendix A read

$$h_1 = -\frac{1}{2y^2}, \quad h_2 = -\frac{x}{y^2}, \quad h_3 = -\frac{x^2}{2y^2}. \tag{4.140}$$

They span a Lie algebra isomorphic to $\mathfrak{sl}(2,\mathbb{R})$

$$\{h_1, h_2\}_\omega = -2h_1, \quad \{h_1, h_3\}_\omega = -h_2, \quad \{h_2, h_3\}_\omega = -2h_3. \tag{4.141}$$

Therefore, a Lie system possessing a minimal Lie algebra I_5 possesses a Lie–Hamilton algebra isomorphic to $\mathfrak{sl}(2,\mathbb{R})$ and, in view of Proposition 4.46, to $\mathfrak{sl}(2,\mathbb{R}) \oplus \mathbb{R}$. In view of Proposition 4.45, these are the only possible algebraic structures for the Lie–Hamilton algebras for X.

Lie algebra $I_8^{(\alpha=-1)}$: $B_{-1} \simeq \mathfrak{iso}(1,1)$

In view of Proposition 4.50, this Lie algebra consists of Hamiltonian vector fields with respect to the standard symplectic form $\omega = \mathrm{d}x \wedge \mathrm{d}y$. The elements of the basis for B_{-1} detailed in Table A.1

in Appendix A satisfy the commutation relations

$$[X_1, X_2] = 0, \quad [X_1, X_3] = X_1, \quad [X_2, X_3] = -X_2. \qquad (4.142)$$

Hence, these vector fields span a Lie algebra isomorphic to the $(1+1)$-dimensional Poincaré algebra $\mathfrak{iso}(1,1)$. Their corresponding Hamiltonian functions turn out to be

$$h_1 = y, \quad h_2 = -x, \quad h_3 = xy, \qquad (4.143)$$

which together with a central generator $h_0 = 1$ fulfill the commutation relations

$$\{h_1, h_2\}_\omega = h_0, \quad \{h_1, h_3\}_\omega = -h_1, \quad \{h_2, h_3\}_\omega = h_2, \quad \{h_0, \cdot\}_\omega = 0. \qquad (4.144)$$

Thus, a Lie system X admitting a minimal Lie algebra B_{-1} possesses a Lie–Hamilton algebra isomorphic to the centrally extended Poincaré algebra $\overline{\mathfrak{iso}}(1,1)$ which, in turn, is also isomorphic to the harmonic oscillator algebra $\mathfrak{h}_4$. As is well known [21], this Lie algebra is not of the form $\mathfrak{iso}(1,1) \oplus \mathbb{R}$, then Proposition 4.45 ensures that X does not admit any Lie–Hamilton algebra isomorphic to $\mathfrak{iso}(1,1)$. Moreover, Proposition 4.48 states that all Lie–Hamilton algebras of X must be isomorphic to $\overline{\mathfrak{iso}}(1,1)$.

Lie algebra I_{12}: $\mathbb{R}^{r+1}$

The vector field $X_1 = \partial_y$ is a modular generating system for I_{12} and all the elements of this Lie algebra have zero divergence. By solving the PDE $\mathcal{L}_{X_1}\omega = 0$, where we recall that ω has the form (4.12), we see that $f = f(x)$ and X_1 becomes Hamiltonian for any non-vanishing function $f(x)$. In view of Theorem 4.43, the remaining elements of I_{12} become automatically Hamiltonian with respect to ω. Then, we obtain that $X_1, \ldots, X_{r+1}$ are Hamiltonian vector fields relative to the symplectic form $\omega = f(x)\mathrm{d}x \wedge \mathrm{d}y$ with Hamiltonian functions

$$h_1 = -\int^x f(x')\mathrm{d}x', \quad h_{j+1} = -\int^x f(x')\xi_j(x')\mathrm{d}x', \qquad (4.145)$$

with $j = 1, \ldots, r$, and $r \geq 1$, which span the Abelian Lie algebra $\mathbb{R}^{r+1}$. In consequence, a Lie–Hamilton system X related to a minimal Lie algebra I_{12} possesses a Lie–Hamilton algebra isomorphic to $\mathbb{R}^{r+1}$. From Propositions 4.45 and 4.46, it only admits an additional Lie–Hamilton algebra isomorphic to $\mathbb{R}^{r+2}$.

Lie algebra I_{14}: $\mathbb{R} \ltimes \mathbb{R}^r$

The functions $\eta_1(x), \ldots, \eta_r(x)$ form a fundamental set of solutions of a system of r differential equations with constant coefficients [5, 183]. Hence, none of them vanishes in an open interval of $\mathbb{R}$ and I_{14} is such that (U_1, X_1, X_2), where X_1 and X_2 are given in Table A.1 in Appendix A, form a modular generating system. Since all the elements of I_{14} have divergence zero and using Theorem 4.43, we infer that I_{14} consists of Hamiltonian vector fields relative to a symplectic form if and only if X_1 and X_2 are locally Hamiltonian vector fields with respect to a symplectic structure. By requiring $\mathcal{L}_{X_i}\omega = 0$, with $i = 1, 2$ and ω of the form (4.12), we obtain that

$$\frac{\partial f}{\partial x} = 0, \quad \eta_j(x)\frac{\partial f}{\partial y} = 0, \quad j = 1, \ldots, r. \qquad (4.146)$$

So, I_{14} is only compatible, up to non-zero multiplicative factor, with $\omega = \mathrm{d}x \wedge \mathrm{d}y$. The Hamiltonian functions corresponding to $X_1, \ldots, X_{r+1}$ turn out to be

$$h_1 = y, \quad h_{j+1} = -\int^x \eta_j(x')\mathrm{d}x', \qquad (4.147)$$

with $j = 1, \ldots, r$, and $r \geq 1$. We remark that different Lie–Hamilton algebras, corresponding to the Lie algebras hereafter called I_{14A} and I_{14B}, are spanned by the above Hamiltonian functions:

- Case I_{14A}: If $1 \notin \langle \eta_1, \ldots, \eta_r \rangle$, then the functions (4.41) span a Lie algebra $\mathbb{R} \ltimes \mathbb{R}^r$ and, by considering Propositions 4.45 and 4.46, this case only admits an additional Lie–Hamilton algebra isomorphic to $(\mathbb{R} \ltimes \mathbb{R}^r) \oplus \mathbb{R}$.

- Case I_{14B}: If $1 \in \langle \eta_1, \dots, \eta_r \rangle$, we can choose a basis of I_{14} in such a way that there exists a function, let us say η_1, equal to 1. Then the Hamiltonian functions (4.41) turn out to be

$$h_1 = y, \quad h_2 = -x, \quad h_{j+1} = -\int^x \eta_j(x')\mathrm{d}x', \qquad (4.148)$$

with $j = 2, \dots, r$, and $r \geq 1$ which require a central generator $h_0 = 1$ in order to close a centrally extended Lie algebra $(\mathcal{H}_\Lambda, \{\cdot, \cdot\}_\omega) \simeq \overline{(\mathbb{R} \ltimes \mathbb{R}^r)}$.

In view of the above, a Lie system X with a minimal Lie algebra I_{14} is a Lie–Hamilton system. Its Lie–Hamilton algebras can be isomorphic to I_{14} or $I_{14} \oplus \mathbb{R}$ when $1 \notin \langle \eta_1, \dots, \eta_r \rangle$ (class I_{14A}). If $1 \in \langle \eta_1, \dots, \eta_r \rangle$ (class I_{14B}), a Lie–Hamilton algebra is isomorphic to $\overline{\mathbb{R} \ltimes \mathbb{R}^r}$ and since $1 \in \{\mathcal{H}_\Lambda, \mathcal{H}_\Lambda\}_\omega$, we obtain from Corollary 4.47 and Proposition 4.48 that every Lie–Hamilton algebra for X is isomorphic to it.

Lie algebra $I_{16}^{(\alpha=-1)}$: $C_{-1}^r \simeq \mathfrak{h}_2 \ltimes \mathbb{R}^{r+1}$

In view of Proposition 4.50, this Lie algebra consists of Hamiltonian vector fields with respect to the standard symplectic structure. The resulting Hamiltonian functions for $X_1, \dots, X_{r+3}$ are given by

$$h_1 = y, \quad h_2 = -x, \quad h_3 = xy, \quad \dots, \quad h_{j+3} = -\frac{x^{j+1}}{j+1}, \qquad (4.149)$$

with $j = 1, \dots, r$, and $r \geq 1$, which again require an additional central generator $h_0 = 1$ to close on a finite-dimensional Lie algebra. The commutation relations for this Lie algebra are given by

$$
\begin{aligned}
\{h_1, h_2\}_\omega &= h_0, & \{h_1, h_3\}_\omega &= -h_1, \\
\{h_2, h_3\}_\omega &= h_2, & \{h_1, h_4\}_\omega &= -h_2, \\
\{h_1, h_{k+4}\}_\omega &= -(k+1)h_{k+3}, & \{h_2, h_{j+3}\}_\omega &= 0, \qquad (4.150) \\
\{h_3, h_{j+3}\}_\omega &= -(j+1)h_{j+3}, & \{h_{j+3}, h_{k+4}\}_\omega &= 0, \\
\{h_0, \cdot\}_\omega &= 0,
\end{aligned}
$$

with $j = 1, \ldots, r$ and $k = 1, \ldots, r - 1$, which define the centrally extended Lie algebra $\overline{\mathfrak{h}_2 \ltimes \mathbb{R}^{r+1}}$. This Lie algebra is not a trivial extension of $\mathfrak{h}_2 \ltimes \mathbb{R}^{r+1}$.

Then, given a Lie system X with a minimal Lie algebra C^r_{-1}, the system is a Lie–Hamilton one which admits a Lie–Hamilton algebra isomorphic to $\overline{\mathfrak{h}_2 \ltimes \mathbb{R}^{r+1}}$. As $1 \in \left\{ \overline{\mathfrak{h}_2 \ltimes \mathbb{R}^{r+1}}, \overline{\mathfrak{h}_2 \ltimes \mathbb{R}^{r+1}} \right\}_\omega$, Corollary 4.47 and Proposition 4.48 ensure that every Lie–Hamilton algebra for X is isomorphic to $\overline{\mathfrak{h}_2 \ltimes \mathbb{R}^{r+1}}$.

4.5. Applications of Lie–Hamilton Systems on the Plane

Here, we show Lie–Hamilton systems of relevance in Science. Given the recurrent appearance of the $\mathfrak{sl}(2, \mathbb{R})$-Lie algebra, we show a few instances of Lie–Hamilton systems with a $\mathfrak{sl}(2, \mathbb{R})$ VG Lie algebra. These are: the Milne–Pinney and second-order Kummer–Schwarz equations, the complex Riccati equations with t-dependent real coefficients, the coupled Riccati equations, planar diffusion Riccati equations, etc. We will establish certain equivalences among some of the mentioned systems.

The relevance of Lie–Hamilton systems in biological applications is indubitable. Some examples will be displayed: the generalized Buchdahl equations, certain Lotka–Volterra systems, quadratic polynomial systems and models for viral infections. Another two Lie algebras worthy of mention in the realm of Lie–Hamilton systems are the two-photon and $\mathfrak{h}_2$-Lie algebras. Their respective applications are: the dissipative harmonic oscillator and the second-order Riccati equation for the two-photon Lie algebra, and the complex Bernoulli and generalized Buchdahl equations and certain Lotka–Volterra systems for the $\mathfrak{h}_2$-Lie algebra. To conclude, we will add other interesting Lie–Hamilton systems as the Cayley–Klein Riccati equation and double-Clifford or split complex Riccati equations. The forthcoming subsections show the abovementioned systems in detail.

4.5.1. $\mathfrak{sl}(2, \mathbb{R})$-*Lie–Hamilton systems*

Let us employ our techniques to study $\mathfrak{sl}(2, \mathbb{R})$-Lie–Hamilton systems [280, 311]. More specifically, we analyze Lie systems used to describe Milne–Pinney equations [102], Kummer–Schwarz equations [87] and complex Riccati equations with t-dependent real coefficients [165]. As a byproduct, our results also cover the t-dependent frequency harmonic oscillator.

Example 4.14. The *Milne–Pinney equation* is well known for its multiple properties and applications in Physics (see [254] and references therein). For example, it is useful to modelize the propagation of laser beams in nonlinear media, plasma dynamics, Bose–Einstein condensates through the so-called Gross-Pitaevskii equation, etc.

It takes the form

$$\frac{\mathrm{d}^2 x}{\mathrm{d}t^2} = -\omega^2(t)x + \frac{c}{x^3}, \tag{4.151}$$

where $\omega(t)$ is any t-dependent function and c is a real constant. It was first introduced by Ermakov as a way to find first integrals of its corresponding t-dependent harmonic oscillator when $c = 0$. By adding a new variable $y := \mathrm{d}x/\mathrm{d}t$, we can study these equations through the first-order system

$$\begin{cases} \dfrac{\mathrm{d}x}{\mathrm{d}t} = y, \\ \dfrac{\mathrm{d}y}{\mathrm{d}t} = -\omega^2(t)x + \dfrac{c}{x^3}, \end{cases} \tag{4.152}$$

which is a Lie system [102, 376]. We recall that (4.2) can be regarded as the equations of motion of the one-dimensional Smorodinsky–Winternitz oscillator [27, 173]; moreover, when the parameter c vanishes, this reduces to the harmonic oscillator (both with a t-dependent frequency). Explicitly, (4.2) is the associated system with the t-dependent vector field $X_t = X_3 + \omega^2(t)X_1$, where

$$X_1 = -x\frac{\partial}{\partial y}, \quad X_2 = \frac{1}{2}\left(y\frac{\partial}{\partial y} - x\frac{\partial}{\partial x}\right), \quad X_3 = y\frac{\partial}{\partial x} + \frac{c}{x^3}\frac{\partial}{\partial y}, \tag{4.153}$$

span a finite-dimensional real Lie algebra V of vector fields isomorphic to $\mathfrak{sl}(2, \mathbb{R})$ with commutation relations given by

$$[X_1, X_2] = X_1, \quad [X_1, X_3] = 2X_2, \quad [X_2, X_3] = X_3. \quad (4.154)$$

There are *four* classes of finite-dimensional Lie algebras of vector fields isomorphic to $\mathfrak{sl}(2, \mathbb{R})$ in the GKO classification: P_2 and I_3–I_5. To determine which one is locally diffeomorphic to V, we first find out whether V is imprimitive or not. In this respect, recall that V is *imprimitive* if there exists a one-dimensional distribution $\mathcal{D}$ invariant under the action (by Lie brackets) of the elements of V. Hence, $\mathcal{D}$ is spanned by a non-vanishing vector field

$$Y = g_x(x, y)\frac{\partial}{\partial x} + g_y(x, y)\frac{\partial}{\partial y}, \quad (4.155)$$

which must be invariant under the action of X_1, X_2 and X_3. As g_x and g_y cannot vanish simultaneously, Y can be taken either of the following local forms:

$$Y = \frac{\partial}{\partial x} + g_y\frac{\partial}{\partial y}, \quad Y = g_x\frac{\partial}{\partial x} + \frac{\partial}{\partial y}. \quad (4.156)$$

Let us assume that $\mathcal{D}$ is spanned by the first one and search for Y. Now, if $\mathcal{D}$ is invariant under the Lie brackets of the elements of V, we have that

$$\mathcal{L}_{X_1}Y = \left(1 - x\frac{\partial g_y}{\partial y}\right)\frac{\partial}{\partial y} = \gamma_1 Y, \quad (4.157a)$$

$$\mathcal{L}_{X_2}Y = \frac{1}{2}\left[\frac{\partial}{\partial x} + \left(y\frac{\partial g_y}{\partial y} - x\frac{\partial g_y}{\partial x} - g_y\right)\frac{\partial}{\partial y}\right] = \gamma_2 Y, \quad (4.157b)$$

$$\mathcal{L}_{X_3}Y = -g_y\frac{\partial}{\partial x} + \left(\frac{3c}{x^4} + y\frac{\partial g_y}{\partial x} + \frac{c}{x^3}\frac{\partial g_y}{\partial y}\right)\frac{\partial}{\partial y} = \gamma_3 Y, \quad (4.157c)$$

for certain functions $\gamma_1, \gamma_2, \gamma_3$ locally defined on $\mathbb{R}^2$. The left-hand side of (4.157a) has no term ∂_x but the right-hand one has it provided $\gamma_1 \neq 0$. Therefore, $\gamma_1 = 0$ and $g_y = y/x + G$ for a certain $G = G(x)$. Next, by introducing this result in (4.157b), we find that $\gamma_2 = 1/2$ and $2G + xG' = 0$, so that $G(x) = \mu/x^2$ for $\mu \in \mathbb{R}$. Substituting this into (4.157c), we obtain that $\gamma_3 = -(\mu + xy)/x^2$ and $\mu^2 = -4c$.

Consequently, when $c > 0$ it does not exist any non-zero Y spanning locally $\mathcal{D}$ satisfying (4.157a)–(4.157c) and V is therefore primitive, while if $c \leq 0$, there exists a vector field

$$Y = \frac{\partial}{\partial x} + \left(\frac{y}{x} + \frac{\mu}{x^2}\right)\frac{\partial}{\partial y}, \quad \mu^2 = -4c, \qquad (4.158)$$

which spans $\mathcal{D}$, so that V is imprimitive. The case of $\mathcal{D}$ being spanned by the second form of Y (4.6) can be analyzed analogously and drives to the same conclusion.

Therefore, the system (4.2) belongs to different classes within the GKO classification according to the value of the parameter c. The final result is established in the following statement.

Proposition 4.51. *The VG Lie algebra for system (4.2), corresponding to the Milne–Pinney equations, is locally diffeomorphic to P_2 for $c > 0$, I_4 for $c < 0$ and I_5 for $c = 0$.*

Proof. Since V is primitive when $c > 0$ and this is isomorphic to $\mathfrak{sl}(2, \mathbb{R})$, the GKO classification given in Table A.1 in Appendix A implies that V is locally diffeomorphic to the primitive class P_2.

Let us now consider that $c < 0$ and prove that the system is diffeomorphic to the class I_4. We do this by showing that there exists a local diffeomorphism $\phi : (x, y) \in U \subset \mathbb{R}^2_{x \neq y} \mapsto \bar{U} \subset (u, v) \in \mathbb{R}^2_{u \neq 0}$, satisfying that ϕ_* maps the basis for I_4 listed in Table A.1 in Appendix A into (4.3). Due to the Lie bracket $[X_1, X_3] = 2X_2$, verified in both bases, it is only necessary to search the map for the generators X_1 and X_3 (so for X_2 this will be automatically fulfilled). By writing in coordinates

$$\phi_*(\partial_x + \partial_y) = -x\partial_y, \quad \phi_*(x^2\partial_x + y^2\partial_y) = y\partial_x + c/x^3\partial_y, \quad (4.159)$$

we obtain

$$\begin{pmatrix} \frac{\partial u}{\partial x} & \frac{\partial u}{\partial y} \\ \frac{\partial v}{\partial x} & \frac{\partial v}{\partial y} \end{pmatrix}\begin{pmatrix} 1 \\ 1 \end{pmatrix} = \begin{pmatrix} 0 \\ -u \end{pmatrix}, \quad \begin{pmatrix} \frac{\partial u}{\partial x} & \frac{\partial u}{\partial y} \\ \frac{\partial v}{\partial x} & \frac{\partial v}{\partial y} \end{pmatrix}\begin{pmatrix} x^2 \\ y^2 \end{pmatrix} = \begin{pmatrix} v \\ c/u^3 \end{pmatrix}.$$

Hence, $\partial_x u + \partial_y u = 0 \Rightarrow u = g(x - y)$ for a certain $g : z \in \mathbb{R} \mapsto g(z) \in \mathbb{R}$. Since $x^2\partial_x u + y^2\partial_y u = v$, then $v = (x^2 - y^2)g'$, where g' is the derivative of $g(z)$ in terms of z. Using now that $\partial_x v + \partial_y v = -u$ we get $2(x-y)g' = -g$ so that $g = \lambda/|x-y|^{1/2}$ where $\lambda \in \mathbb{R}\backslash\{0\}$. Substituting this into the remaining equation $x^2\partial_x v + y^2\partial_y v = c/u^3$, we find that $\lambda^4 = -4c$. Since $c < 0$, we consistently find that

$$u = \frac{\lambda}{|x-y|^{1/2}}, \quad v = -\frac{\lambda(x+y)}{2|x-y|^{1/2}}, \quad \lambda^4 = -4c. \tag{4.160}$$

Finally, let us set $c = 0$ and search for a local diffeomorphism $\phi : (x,y) \in U \subset \mathbb{R}^2_{y\neq 0} \mapsto \bar{U} \subset (u,v) \in \mathbb{R}^2$ such that ϕ_* maps the basis corresponding to I_5 into (4.3); namely

$$\phi_*(\partial_x) = -x\partial_y, \quad \phi_*(x^2\partial_x + xy\partial_y) = y\partial_x, \tag{4.161}$$

yielding

$$\begin{pmatrix} \frac{\partial u}{\partial x} & \frac{\partial u}{\partial y} \\ \frac{\partial v}{\partial x} & \frac{\partial v}{\partial y} \end{pmatrix} \begin{pmatrix} 1 \\ 0 \end{pmatrix} = \begin{pmatrix} 0 \\ -u \end{pmatrix}, \quad \begin{pmatrix} \frac{\partial u}{\partial x} & \frac{\partial u}{\partial y} \\ \frac{\partial v}{\partial x} & \frac{\partial v}{\partial y} \end{pmatrix} \begin{pmatrix} x^2 \\ xy \end{pmatrix} = \begin{pmatrix} v \\ 0 \end{pmatrix}.$$

Hence, $\partial_x u = 0 \Rightarrow u = g_1(y)$ for a certain $g_1 : \mathbb{R} \to \mathbb{R}$. Since $\partial_x v = -u$, then $v = -g_1(y)x + g_2(y)$ for another $g_2 : \mathbb{R} \to \mathbb{R}$. Using now the PDEs of the second matrix, we see that $xy\partial_y u = xyg_1' = v = -g_1 x + g_2$, so that $g_2 = 0$ and $g_1 = \lambda/y$, where $\lambda \in \mathbb{R}\backslash\{0\}$ and $\lambda \neq 0$ to avoid ϕ not to be a diffeomorphism. It can be checked that the remaining equation is so fulfilled. Therefore, $u = \lambda/y$ and $v = -\lambda x/y$. $\qquad\square$

We remark that, since the three classes P_2, I_4 and I_5 appear in Table A.2 in Appendix A, system (4.2) can always be associated with a symplectic form turning their vector fields Hamiltonian. In this respect, recall that it was recently proven that the system (4.2) is a Lie–Hamilton one for any value of c [27]. However, we shall show that identifying it to one of the classes of the GKO classification will be useful to study the relation of this system to other ones.

Example 4.15. Let us turn now to reconsider the *second-order Kummer–Schwarz equation* (3.68) written as a first-order system

(3.70)

$$\begin{cases} \dfrac{dx}{dt} = v, \\[2mm] \dfrac{dv}{dt} = \dfrac{3}{2}\dfrac{v^2}{x} - 2c_0 x^3 + 2b_1(t)x, \end{cases} \tag{4.162}$$

on $T\mathbb{R}_0$, with $\mathbb{R}_0 = \mathbb{R} - \{0\}$, obtained by adding the new variable $v := dx/dt$ to the KS-2 equation (3.68).

It is well known that (4.162) is a Lie system [87, 280]. In fact, it describes the integral curves of the t-dependent vector field $X_t = X_3 + b_1(t)X_1$ where the vector fields X_1, X_2, X_3 correspond to those in (3.72). Since their corresponding commutation relations are the same as in (3.72). spanning a VG Lie algebra V isomorphic to $\mathfrak{sl}(2, \mathbb{R})$. Thus, V can be isomorphic to one of the four $\mathfrak{sl}(2, \mathbb{R})$-Lie algebras of vector fields in the GKO classification.

As in the previous subsection, we analyze if there exists a distribution $\mathcal{D}$ stable under V and locally generated by a vector field Y of the first form given in (4.6) (the same results can be obtained by assuming the second one). Notice that in this section we change the notation in the variable v in (4.162), as $v \leftrightarrow y$. So, imposing $\mathcal{D}$ to be stable under V yields

$$\mathcal{L}_{X_1} Y = 2\left(x\frac{\partial g_y}{\partial y} - 1\right)\frac{\partial}{\partial y} = \gamma_1 Y, \tag{4.163a}$$

$$\mathcal{L}_{X_2} Y = -\frac{\partial}{\partial x} + \left(x\frac{\partial g_y}{\partial x} + 2y\frac{\partial g_y}{\partial y} - 2g_y\right)\frac{\partial}{\partial y} = \gamma_2 Y, \tag{4.163b}$$

$$\mathcal{L}_{X_3} Y = -g_y\frac{\partial}{\partial x} + \left[X_3 g_y + \frac{3y^2}{2x^2} + 6c\,x^2 - \frac{3y}{x}g_y\right]\frac{\partial}{\partial y} = \gamma_3 Y, \tag{4.163c}$$

for certain functions $\gamma_1, \gamma_2, \gamma_3$ locally defined on $\mathbb{R}^2$. The left-hand side of (4.163a) has no term ∂_x and the right-hand one does not have it provided $\gamma_1 = 0$. Hence, $\gamma_1 = 0$ and $g_y = y/x + F$ for a $F = F(x)$. In view of (4.163b), we then obtain $\gamma_2 = -1$ and $F - xF' = 0$, that is, $F(x) = \mu x$ for $\mu \in \mathbb{R}$. Substituting g_y in (4.163c), we obtain that $\gamma_3 = -\mu x - y/x$ and $\mu^2 = -4c$. Hence, as in the Milne–Pinney equations, we find that if $c > 0$, there does not exist any Y spanning locally $\mathcal{D}$ satisfying (4.163a)–(4.163c) and V is primitive, meanwhile if $c \leq 0$, then $\mathcal{D}$ is spanned by the vector field

$$Y = \frac{\partial}{\partial x} + \left(\frac{y}{x} + \mu x\right)\frac{\partial}{\partial y}, \quad \mu^2 = -4c, \quad (4.164)$$

and V is imprimitive.

The precise classes of the GKO classification corresponding to the system (3.70) are summarized in the following proposition.

Proposition 4.52. *The VG Lie algebra for system (4.162), associated with the second-order Kummer–Schwarz equations, is locally diffeomorphic to P_2 for $c > 0$, I_4 for $c < 0$ and I_5 for $c = 0$.*

Proof. The case with $c > 0$ provides the primitive class P_2 since $Y = 0$. If $c < 0$, we look for a local diffeomorphism $\phi : (x, y) \in U \subset \mathbb{R}^2_{x \neq y} \mapsto \bar{U} \subset (u, v) \in \mathbb{R}^2_{u \neq 0}$, such that ϕ_* maps the basis of I_4 into (3.72), that is,

$$\phi_*(\partial_x + \partial_y) = 2x\partial_y, \quad \phi_*(x^2\partial_x + y^2\partial_y) = y\partial_x + (\tfrac{3}{2}y^2/x - 2c\,x^3)\partial_y. \tag{4.165}$$

Then

$$\begin{pmatrix} \frac{\partial u}{\partial x} & \frac{\partial u}{\partial y} \\ \frac{\partial v}{\partial x} & \frac{\partial v}{\partial y} \end{pmatrix}\begin{pmatrix} 1 \\ 1 \end{pmatrix} = \begin{pmatrix} 0 \\ 2u \end{pmatrix}, \quad \begin{pmatrix} \frac{\partial u}{\partial x} & \frac{\partial u}{\partial y} \\ \frac{\partial v}{\partial x} & \frac{\partial v}{\partial y} \end{pmatrix}\begin{pmatrix} x^2 \\ y^2 \end{pmatrix} = \begin{pmatrix} v \\ \tfrac{3}{2}v^2/u - 2c\,u^3 \end{pmatrix}.$$

Proceeding as in the proof of Proposition 4.51, we find that $u = g(x - y)$ and $v = (x^2 - y^2)g'$ for $g : \mathbb{R} \to \mathbb{R}$. Since now $\partial_x v + \partial_y v = 2u$, we obtain $2(x - y)g' = 2g$, so that $g = \lambda(x - y)$ with $\lambda \in \mathbb{R}\backslash\{0\}$ and assume $\lambda \neq 0$ to avoid ϕ not being a diffeomorphism. The remaining equation $x^2\partial_x v + y^2\partial_y v = \tfrac{3}{2}v^2/u - 2c\,u^3$ implies that $4\lambda^2 = -1/c$, which is consistent with the value $c < 0$. Then

$$u = \lambda(x - y), \quad v = \lambda(x^2 - y^2), \quad 4\lambda^2 = -1/c. \tag{4.166}$$

In the third possibility with $c = 0$, we require that ϕ_* maps the basis of I_5 into (3.72) fulfilling

$$\phi_*(\partial_x) = 2x\partial_y, \quad \phi_*(x^2\partial_x + xy\partial_y) = y\partial_x + \tfrac{3}{2}y^2/x\partial_y, \tag{4.167}$$

that is,

$$\begin{pmatrix} \frac{\partial u}{\partial x} & \frac{\partial u}{\partial y} \\ \frac{\partial v}{\partial x} & \frac{\partial v}{\partial y} \end{pmatrix} \begin{pmatrix} 1 \\ 0 \end{pmatrix} = \begin{pmatrix} 0 \\ 2u \end{pmatrix}, \qquad \begin{pmatrix} \frac{\partial u}{\partial x} & \frac{\partial u}{\partial y} \\ \frac{\partial v}{\partial x} & \frac{\partial v}{\partial y} \end{pmatrix} \begin{pmatrix} x^2 \\ xy \end{pmatrix} = \begin{pmatrix} v \\ \frac{3}{2}v^2/u \end{pmatrix}.$$

By taking into account the proof of Proposition 4.51, it is straightforward to check that the four PDEs are satisfied for $u = \lambda y^2$ and $v = 2\lambda xy^2$ with $\lambda \in \mathbb{R}\backslash\{0\}$. $\qquad\square$

Example 4.16. Let us return to *complex Riccati equations with t-dependent real coefficients* in the form (4.206). We have already shown that this system has a VG Lie algebra $P_2 \simeq \mathfrak{sl}(2, \mathbb{R})$. Therefore, it is locally diffeomorphic to the VG Lie algebra appearing in the above Milne–Pinney and Kummer–Schwarz equations whenever the parameter $c > 0$. In view of the GKO classification, there exist local diffeomorphisms relating to the three first-order systems associated with these equations. For instance, we can search for a local diffeomorphism $\phi : (x, y) \in U \subset \mathbb{R}^2_{y\neq 0} \mapsto \bar{U} \subset (u, v) \in \mathbb{R}^2_{u\neq 0}$ mapping every system (4.206) into one of the form (4.2), e.g., satisfying that ϕ_* maps the basis of P_2 in Table A.1 in Appendix A, related to the planar Riccati equation, into the basis (4.3) associated with the Milne–Pinney one. By writing in coordinates

$$\phi_*(\partial_x) = -x\partial_y, \quad \phi_*[(x^2 - y^2)\partial_x + 2xy\partial_y] = y\partial_x + c/x^3\partial_y,$$

$$(4.168)$$

we obtain

$$\begin{pmatrix} \frac{\partial u}{\partial x} & \frac{\partial u}{\partial y} \\ \frac{\partial v}{\partial x} & \frac{\partial v}{\partial y} \end{pmatrix} \begin{pmatrix} 1 \\ 0 \end{pmatrix} = \begin{pmatrix} 0 \\ -u \end{pmatrix}, \qquad \begin{pmatrix} \frac{\partial u}{\partial x} & \frac{\partial u}{\partial y} \\ \frac{\partial v}{\partial x} & \frac{\partial v}{\partial y} \end{pmatrix} \begin{pmatrix} x^2 - y^2 \\ 2xy \end{pmatrix} = \begin{pmatrix} v \\ c/u^3 \end{pmatrix}.$$

Similar computations to those performed in the proof of Proposition 4.51 for the three PDEs $\partial_x u = 0$, $\partial_x v = -u$ and $(x^2 - y^2)\partial_x u + 2xy\partial_y u = v$ give $u = \lambda/|y|^{1/2}$ and $v = -\lambda x/|y|^{1/2}$ with $\lambda \in \mathbb{R}\backslash\{0\}$. Substituting these results into the remaining equation, we find that $\lambda^4 = c$ which is consistent with the positive value of c. Consequently, this maps the system (4.206) into (4.2) and the solution of the first one is locally equivalent to the solutions of the second.

Summing up, we have explicitly proven that the three $\mathfrak{sl}(2, \mathbb{R})$ Lie algebras of the classes P_2, I_4 and I_5 given in Table A.2 in Appendix A cover the following $\mathfrak{sl}(2, \mathbb{R})$-Lie systems:

- P_2: Milne–Pinney and Kummer–Schwarz equations for $c > 0$ as well as complex Riccati equations with t-dependent coefficients.
- I_4: Milne–Pinney and Kummer–Schwarz equations for $c < 0$.
- I_5: Milne–Pinney and Kummer–Schwarz equations for $c = 0$ and the harmonic oscillator with t-dependent frequency.

This means that, only within each class, they are locally diffeomorphic and, therefore, there exists a local change of variables mapping one into another. Thus, for instance, there does not exist any diffeomorphism mapping the Milne–Pinney and Kummer–Schwarz equations with $c \neq 0$ to the harmonic oscillator. These results also explain from an algebraic point of view the existence of the known diffeomorphism mapping Kummer–Schwarz equations to Milne–Pinney equations [254] provided that the sign of c is the same in both systems.

4.5.2. *Lie–Hamilton biological models*

In this section, we focus on new applications of the Lie–Hamilton approach to Lotka–Volterra-type systems and to a viral infection model. We also consider here the analysis of Buchdahl equations which can be studied through a Lie–Hamilton system diffeomorphic to a precise t-dependent Lotka–Volterra system.

Example 4.17. We call *generalized Buchdahl equations* [65, 133, 141] the second-order differential equations

$$\frac{\mathrm{d}^2 x}{\mathrm{d}t^2} = a(x) \left(\frac{\mathrm{d}x}{\mathrm{d}t} \right)^2 + b(t) \frac{\mathrm{d}x}{\mathrm{d}t}, \qquad (4.169)$$

where $a(x)$ and $b(t)$ are arbitrary functions of their respective arguments. In order to analyze whether these equations can be studied through a Lie system, we add the variable $y := \mathrm{d}x/\mathrm{d}t$ and consider the first-order system of differential equations

$$\begin{cases} \dfrac{dx}{dt} = y, \\ \dfrac{dy}{dt} = a(x)y^2 + b(t)y. \end{cases} \qquad (4.170)$$

Note that if $(x(t), y(t))$ is a particular solution of this system with $y(t_0) = 0$ for a particular $t_0 \in \mathbb{R}$, then $y(t) = 0$ for every $t \in \mathbb{R}$ and $x(t) = \lambda \in \mathbb{R}$. Moreover, if $a(x) = 0$, then the solution of the above system is also trivial. As a consequence, we can restrict ourselves to studying particular solutions on $\mathbb{R}^2_{y \neq 0}$ with $a(x) \neq 0$.

Next, let us prove that (4.170) is a Lie system. Explicitly, (4.170) is associated with the t-dependent vector field $X_t = X_1 + b(t)X_2$, where

$$X_1 = y\frac{\partial}{\partial x} + a(x)y^2\frac{\partial}{\partial y}, \quad X_2 = y\frac{\partial}{\partial y}. \qquad (4.171)$$

Since

$$[X_1, X_2] = -X_1, \qquad (4.172)$$

these vector fields span a Lie algebra V isomorphic to $\mathfrak{h}_2$ leaving invariant the distribution $\mathcal{D}$ spanned by $Y := X_1$. Since the rank of $\mathcal{D}^V$ is two, V is locally diffeomorphic to the imprimitive class I_{14A} with $r = 1$ and $\eta_1(x) = e^x$ given in Table A.2 in Appendix A. This proves for the first time that generalized Buchdahl equations written as the system (4.170) are, in fact, not only a Lie system [87] but a Lie–Hamilton one.

Next, by determining a symplectic form obeying $\mathcal{L}_{X_i}\omega = 0$, with $i = 1, 2$ for the vector fields (4.171) and the generic ω (4.12), it can be shown that this reads

$$\omega = \frac{\exp\left(-\int a(x)dx\right)}{y}\, dx \wedge dy, \qquad (4.173)$$

which turns X_1 and X_2 into Hamiltonian vector fields with Hamiltonian functions

$$h_1 = y \exp\left(-\int^x a(x')\mathrm{d}x'\right), \quad h_2 = -\int^x \exp\left(-\int^{x'} a(\bar{x})\mathrm{d}\bar{x}\right)\mathrm{d}x',$$
$$(4.174)$$

respectively. Note that all the these structures are properly defined on $\mathbb{R}^2_{y\neq0}$ and hold $\{h_1, h_2\} = h_1$. Consequently, the system (4.170) has a t-dependent Hamiltonian given by $h_t = h_1 + b(t)h_2$.

Example 4.18. Consider the specific t-dependent Lotka–Volterra system [236, 355] of the form

$$\begin{cases} \dfrac{\mathrm{d}x}{\mathrm{d}t} = ax - g(t)(x - ay)x, \\ \dfrac{\mathrm{d}y}{\mathrm{d}t} = ay - g(t)(bx - y)y, \end{cases} \quad (4.175)$$

where $g(t)$ is a t-dependent function representing the variation of the seasons and a, b are certain real parameters describing the interactions among the species. We hereafter focus on the case $a \neq 0$, as otherwise the above equation becomes, up to a t-reparametrization, an autonomous differential equation that can easily be integrated. We also assume $g(t)$ to be a non-constant function and we restrict ourselves to particular solutions on $\mathbb{R}_{x,y\neq0} := \{(x,y)|x \neq 0, y \neq 0\}$ (the remaining ones can be trivially obtained).

Let us prove that (4.175) is a Lie system and that for some values of the real parameters $a \neq 0$ and b this is a Lie–Hamilton system as well. This system describes the integral curves of the t-dependent vector field $X_t = X_1 + g(t)X_2$, where

$$X_1 = ax\frac{\partial}{\partial x} + ay\frac{\partial}{\partial y}, \quad X_2 = -(x - ay)x\frac{\partial}{\partial x} - (bx - y)y\frac{\partial}{\partial y},$$
$$(4.176)$$

satisfy

$$[X_1, X_2] = aX_2, \quad a \neq 0. \quad (4.177)$$

202   *A Guide to Lie Systems with Compatible Geometric Structures*

Hence, X_1 and X_2 are the generators of a Lie algebra V of vector fields isomorphic to $\mathfrak{h}_2$ leaving invariant the distribution $\mathcal{D}$ on $\mathbb{R}_{x,y\neq0}$ spanned by $Y := X_2$. According to the values of the parameters $a \neq 0$ and b, we find the following:

- When $a = b = 1$, the rank of $\mathcal{D}^V$ on the domain of V is one. In view of Table A.1 in Appendix A, the Lie algebra V is thus isomorphic to I_2 and, by taking into account Table A.2 in Appendix A, we conclude that X is a Lie system, but not a Lie–Hamilton one.
- Otherwise, the rank of $\mathcal{D}^V$ is two, so that this Lie algebra is locally diffeomorphic to I_{14A} with $r = 1$ and $\eta_1 = e^{ax}$ given in Table A.A.2 in Appendix A and, consequently, X is a Lie–Hamilton system. As a straightforward consequence, when $a = 1$ and $b \neq 1$, the system (4.175) is locally diffeomorphic to the generalized Buchdahl equations (4.170).

Let us now derive a symplectic form (4.12) turning the elements of V into Hamiltonian vector fields by solving the system of PDEs $\mathcal{L}_{X_1}\omega = \mathcal{L}_{X_2}\omega = 0$. The first condition reads in local coordinates

$$\mathcal{L}_{X_1}\omega = (X_1 f + 2af)\mathrm{d}x \wedge \mathrm{d}y = 0. \tag{4.178}$$

So, we obtain that $f = F(x/y)/y^2$ for any function $F : \mathbb{R} \to \mathbb{R}$. By imposing that $\mathcal{L}_{X_2}\omega = 0$, we find

$$\mathcal{L}_{X_2}\omega = \left[(b-1)x^2 + (a-1)yx\right]\frac{\partial f}{\partial x} + f\left[(b-2)x + ay\right] = 0. \tag{4.179}$$

Note that, as expected, f vanishes when $a = b = 1$. We study separately the remaining cases: (i) $a \neq 1$ and $b \neq 1$; (ii) $a = 1$ and $b \neq 1$; and (iii) $a \neq 1$ and $b = 1$.

When both $a, b \neq 1$, we write $f = F(x/y)/y^2$, thus obtaining that ω reads, up to a non-zero multiplicative constant, as

$$\omega = \frac{1}{y^2}\left(\frac{x}{y}\right)^{\frac{a}{1-a}}\left[1 - a + (1-b)\frac{x}{y}\right]^{\frac{1}{a-1}+\frac{1}{b-1}}\mathrm{d}x \wedge \mathrm{d}y, \quad a, b \neq 1. \tag{4.180}$$

From this, we obtain the following Hamiltonian functions for X_1 and X_2:

$$h_1 = a(1-b)^{1+\frac{1}{a-1}+\frac{1}{b-1}}\left(\frac{x}{y}\right)^{\frac{1}{b-1}}$$
$$\times {}_2F_1\left(\frac{1}{1-b},\frac{1}{1-a}+\frac{1}{1-b};\frac{b-2}{b-1};\frac{y(1-a)}{x(b-1)}\right), \qquad (4.181)$$
$$h_2 = -y\left(\frac{x}{y}\right)^{\frac{1}{1-a}}\left[(1-a)+(1-b)\frac{x}{y}\right]^{\frac{1}{a-1}+\frac{1}{b-1}+1},$$

where ${}_2F_1(\alpha,\beta,\gamma,\zeta)$ stands for the hypergeometric function

$$ {}_2F_1(\alpha,\beta,\gamma,z) = \sum_{n=0}^{\infty}[(\alpha)_n(\beta)_n/(\gamma)_n]z^n/n!, \qquad (4.182)$$

with $(\delta)_n = \Gamma(\delta+n)/\Gamma(\delta)$ being the rising *Pochhammer symbol*. As expected, $\{h_1,h_2\}_\omega = -ah_2$.

When $a = 1$ and $b \neq 1$, the symplectic form for X becomes

$$\omega = \frac{1}{y^2}\exp\left(\frac{y-(b-2)x\ln|x/y|}{(b-1)x}\right)\,\mathrm{d}x \wedge \mathrm{d}y, \quad b \neq 1, \qquad (4.183)$$

and the Hamiltonian functions for X_1 and X_2 read

$$h_1 = -\left(\frac{1}{1-b}\right)^{\frac{1}{b-1}}\Gamma\left(\frac{1}{1-b},\frac{y}{x(1-b)}\right),$$
$$h_2 = (b-1)x\left(\frac{x}{y}\right)^{\frac{1}{b-1}}\exp\left(\frac{y}{(b-1)x}\right), \qquad (4.184)$$

with $\Gamma(u,v)$ being the *incomplete Gamma function*, which satisfy $\{h_1,h_2\}_\omega = -h_2$.

Finally, when $b = 1$ and $a \neq 1$, we have

$$\omega = \frac{1}{y^2}\left(\frac{x}{y}\right)^{\frac{a}{1-a}}\exp\left(\frac{x}{y(a-1)}\right)\,\mathrm{d}x \wedge \mathrm{d}y, \quad a \neq 1. \qquad (4.185)$$

Then, the Hamiltonian functions for X_1 and X_2 are, in this order,

$$h_1 = a(1-a)^{\frac{1}{1-a}} \Gamma\left(\frac{1}{1-a}, \frac{x}{y(1-a)}\right),$$

$$h_2 = (a-1)y \exp\left(\frac{x}{y(a-1)}\right)\left(\frac{x}{y}\right)^{\frac{1}{1-a}}. \tag{4.186}$$

Indeed, $\{h_1, h_2\}_\omega = -ah_2$.

Example 4.19. The system of differential equations [278] is known as *a class of quadratic polynomial Lie systems*

$$\begin{cases} \dfrac{dx}{dt} = b(t)x + c(t)y + d(t)x^2 + e(t)xy + f(t)y^2, \\ \dfrac{dy}{dt} = y, \end{cases} \tag{4.187}$$

where $b(t), c(t), d(t), e(t)$, and $f(t)$ are arbitrary t-dependent functions. The system (4.187) is an interacting species model of Lotka–Volterra type that belongs to the class of quadratic-linear polynomial systems with a unique singular point at the origin [278].

In general, this system is not a Lie system. For instance, consider the particular system associated with the t-dependent vector field $X_t = d(t)X_1 + e(t)X_2 + X_3$,

$$X_1 = x^2\frac{\partial}{\partial x}, \quad X_2 = xy\frac{\partial}{\partial x}, \quad X_3 = y\frac{\partial}{\partial y}, \tag{4.188}$$

where $d(t)$ and $e(t)$ are non-constant and non-proportional functions. Note that V^X contains X_1, X_2 and their successive Lie brackets, i.e., the vector fields

$$\overbrace{[X_2, \ldots [X_2, X_1] \ldots]}^{n\text{-times}} = x^2 y^n \frac{\partial}{\partial x} := Y_n. \tag{4.189}$$

Hence, $[X_2, Y_n] = Y_{n+1}$ and the family of vector fields X_1, X_2, $X_3, Y_1, Y_2, \ldots$ span an infinite-dimensional family of linearly independent vector fields over $\mathbb{R}$. Then, X is not a Lie system.

Hereafter, we analyze the cases of (4.187) with $d(t) = e(t) = 0$ which provide quadratic polynomial systems that are Lie systems.

We call them *quadratic polynomial Lie systems*; these are related to the system of differential equations. [278]

$$\begin{cases} \dfrac{\mathrm{d}x}{\mathrm{d}t} = b(t)x + c(t)y + f(t)y^2, \\ \dfrac{\mathrm{d}y}{\mathrm{d}t} = y. \end{cases} \tag{4.190}$$

Note that if a solution $(x(t), y(t))$ of the above system is such that $y(t_0) = 0$ for a certain t_0, then $y(t) = 0$ for all $t \in \mathbb{R}$ and the corresponding $x(t)$ can be then easily obtained. In view of this, we focus on those particular solutions within $\mathbb{R}^2_{y \neq 0}$. The system (4.190) is associated with the t-dependent vector field on $\mathbb{R}^2_{y \neq 0}$ of the form $X_t = X_1 + b(t)X_2 + c(t)X_3 + f(t)X_4$, where

$$X_1 = y\frac{\partial}{\partial y}, \quad X_2 = x\frac{\partial}{\partial x}, \quad X_3 = y\frac{\partial}{\partial x}, \quad X_4 = y^2\frac{\partial}{\partial x}, \tag{4.191}$$

satisfy the commutation rules

$$\begin{array}{llll} [X_1, X_2] = 0, & [X_1, X_3] = X_3, & [X_1, X_4] = 2X_4, \\ [X_2, X_3] = -X_3, & [X_2, X_4] = -X_4, & [X_3, X_4] = 0. \end{array} \tag{4.192}$$

Note that $V \simeq V_1 \ltimes V_2$ where $V_1 = \langle X_1, X_2 \rangle \simeq \mathbb{R}^2$ and $V_2 = \langle X_3, X_4 \rangle \simeq \mathbb{R}^2$. In addition, the distribution $\mathcal{D}$ spanned by $Y := \partial_x$ is invariant under the action of the above vector fields. So, V is imprimitive. In view of Table A.1 in Appendix A, we find that (4.190) is a Lie system corresponding to the imprimitive class I_{15} with $V \simeq \mathbb{R}^2 \ltimes \mathbb{R}^2$. By taking into account our classification given in Table A.2 in Appendix A, we know that this is not a Lie algebra of vector fields with respect to any symplectic structure.

Example 4.20. We now consider a subcase of (4.190) that provides a Lie–Hamilton system. We will refer to such subcase as *quadratic polynomial Lie–Hamilton systems*. In view of Table A.2 in Appendix A, the Lie subalgebra $\mathbb{R} \ltimes \mathbb{R}^2$ of V is a Lie algebra of Hamiltonian vector fields with respect to a symplectic structure, that is, $I_{14} \subset I_{15}$. So, it is natural to consider the restriction of (4.190) to

$$\begin{cases} \dfrac{\mathrm{d}x}{\mathrm{d}t} = b\,x + c(t)y + f(t)y^2, \\ \dfrac{\mathrm{d}y}{\mathrm{d}t} = y, \end{cases} \qquad (4.193)$$

where $b \in \mathbb{R}\backslash\{1,2\}$ and $c(t)$, $f(t)$ are still t-dependent functions. The system (4.193) is associated to the t-dependent vector field $X_t = X_1 + c(t)X_2 + f(t)X_3$ on $\mathbb{R}^2_{y\neq 0} := \{(x,y) \in \mathbb{R} \mid y \neq 0\}$, where

$$X_1 = b\,x\frac{\partial}{\partial x} + y\frac{\partial}{\partial y}, \quad X_2 = y\frac{\partial}{\partial x}, \quad X_3 = y^2\frac{\partial}{\partial x}, \qquad (4.194)$$

satisfy the commutation relations

$$[X_1, X_2] = (1 - b)X_2, \quad [X_1, X_3] = (2 - b)X_3, \quad [X_2, X_3] = 0. \qquad (4.195)$$

Therefore, the vector fields (4.194) generate a Lie algebra $V \simeq V_1 \ltimes V_2$, where $V_1 = \langle X_1 \rangle \simeq \mathbb{R}$ and $V_2 = \langle X_2, X_3 \rangle \simeq \mathbb{R}^2$. The domain of V is $\mathbb{R}^2_{y\neq 0}$ and the rank of $\mathcal{D}^V$ is two. Moreover, the distribution $\mathcal{D}$ spanned by the vector field $Y := \partial_x$ is stable under the action of the elements of V, which turns V into an imprimitive Lie algebra. So, V must be locally diffeomorphic to the imprimitive Lie algebra I_{14} displayed in Table A.1 in Appendix A for $r = 2$. We already know that the class I_{14} is a Lie algebra of Hamiltonian vector fields with respect to a symplectic structure.

By imposing $\mathcal{L}_{X_i}\omega = 0$ for the vector fields (4.194) and the generic symplectic form (4.12), it can be shown that ω reads

$$\omega = \frac{\mathrm{d}x \wedge \mathrm{d}y}{y^{b+1}}, \qquad (4.196)$$

which turns (4.194) into Hamiltonian vector fields with Hamiltonian functions

$$h_1 = -\frac{x}{y^b}, \quad h_2 = \frac{y^{1-b}}{1-b}, \quad h_3 = \frac{y^{2-b}}{2-b}, \quad b \in \mathbb{R}\backslash\{1,2\}. \qquad (4.197)$$

Note that all the above structures are properly defined on $\mathbb{R}^2_{y\neq 0}$. The above Hamiltonian functions span a three-dimensional Lie algebra with commutation relations

$$\{h_1, h_2\}_\omega = (b-1)h_2, \quad \{h_1, h_3\}_\omega = (b-2)h_3, \quad \{h_2, h_3\}_\omega = 0.$$
$$(4.198)$$

Consequently, V is locally diffeomorphic to the imprimitive Lie algebra I_{14A} of Table A.2 such that the Lie–Hamilton algebra is $\mathbb{R} \ltimes \mathbb{R}^2$ (also $(\mathbb{R} \ltimes \mathbb{R}^2) \oplus \mathbb{R}$). The system (4.193) has a t-dependent Hamiltonian

$$h_t = b\,h_1 + c(t)h_2 + d(t)h_3 = -b\frac{x}{y^b} + c(t)\frac{y^{1-b}}{1-b} + d(t)\frac{y^{2-b}}{2-b}.$$
$$(4.199)$$

We point out that the cases of (4.193) with either $b=1$ or $b=2$ also lead to Lie–Hamilton systems, but now belonging, both of them, to the class I_{14B} of Table A.2 as a central generator is required. For instance if $b=1$, the commutation relations (4.195) reduce to

$$[X_1, X_2] = 0, \quad [X_1, X_3] = X_3, \quad [X_2, X_3] = 0, \qquad (4.200)$$

while the symplectic form and the Hamiltonian functions are found to be

$$\omega = \frac{\mathrm{d}x \wedge \mathrm{d}y}{y^2}, \quad h_1 = -\frac{x}{y}, \quad h_2 = \ln y, \quad h_3 = y, \qquad (4.201)$$

which together with $h_0 = 1$ close the (centrally extended) Lie–Hamilton algebra $\overline{\mathbb{R} \ltimes \mathbb{R}^2}$, that is,

$$\{h_1, h_2\}_\omega = -h_0, \quad \{h_1, h_3\}_\omega = -h_3, \quad \{h_2, h_3\}_\omega = 0, \quad \{h_0, \cdot\}_\omega = 0.$$
$$(4.202)$$

A similar result can be found for $b=2$.

Example 4.21. Finally, let us consider a **simple viral infection model** given by [163]

$$\begin{cases} \dfrac{\mathrm{d}x}{\mathrm{d}t} = (\alpha(t) - g(y))x, \\ \dfrac{\mathrm{d}y}{\mathrm{d}t} = \beta(t)xy - \gamma(t)y, \end{cases} \tag{4.203}$$

where $g(y)$ is a real positive function taking into account the power of the infection. Note that if a particular solution satisfies $x(t_0) = 0$ or $y(t_0) = 0$ for a $t_0 \in \mathbb{R}$, then $x(t) = 0$ or $y(t) = 0$, respectively, for all $t \in \mathbb{R}$. As these cases are trivial, we restrict ourselves to studying particular solutions within $\mathbb{R}^2_{x,y\neq0} := \{(x,y) \in \mathbb{R}^2 \mid x \neq 0, y \neq 0\}$.

The simplest possibility consists of setting $g(y) = \delta$, where δ is a constant. Then, (4.203) describes the integral curves of the t-dependent vector field $X_t = (\alpha(t) - \delta)X_1 + \gamma(t)X_2 + \beta(t)X_3$, on $\mathbb{R}^2_{x,y\neq0}$, where the vector fields

$$X_1 = x\frac{\partial}{\partial x}, \quad X_2 = -y\frac{\partial}{\partial y}, \quad X_3 = xy\frac{\partial}{\partial y}, \tag{4.204}$$

satisfy the relations (4.200). So, X is a Lie system related to a VG Lie algebra $V \simeq \mathbb{R} \ltimes \mathbb{R}^2$ where $\langle X_1 \rangle \simeq \mathbb{R}$ and $\langle X_2, X_3 \rangle \simeq \mathbb{R}^2$. The distribution $\mathcal{D}^V$ has rank two on $\mathbb{R}^2_{x,y\neq0}$. Moreover, V is imprimitive, as the distribution $\mathcal{D}$ spanned by $Y := \partial_y$ is invariant under the action of vector fields of V. Thus, V is locally diffeomorphic to the imprimitive Lie algebra I_{14B} for $r = 2$ and, in view of Table A.2, the system X is a Lie–Hamilton one.

Next, we obtain that V is a Lie algebra of Hamiltonian vector fields with respect to the symplectic form

$$\omega = \frac{\mathrm{d}x \wedge \mathrm{d}y}{xy}. \tag{4.205}$$

Then, the vector fields X_1, X_2, and X_3 have Hamiltonian functions: $h_1 = \ln|y|, h_2 = \ln|x|, h_3 = -x$, which along with $h_0 = 1$ close the relations (4.202). If we assume $V^X = V$, the t-dependent

Hamiltonian $h_t = (\alpha(t) - \delta)h_1 + \gamma(t)h_2 + \beta(t)h_3$ gives rise to a Lie–Hamiltonian structure $(\mathbb{R}^2_{x,y\neq 0}, \omega, h)$ for X defining the Lie–Hamilton algebra $\overline{(\mathbb{R} \ltimes \mathbb{R}^2)}$.

4.5.3. *Other Lie–Hamilton systems on the plane*

Example 4.22. Let us reconsider the *Cayley–Klein Riccati equation* reviewed in (4.206),

$$\frac{\mathrm{d}x}{\mathrm{d}t} = a_0(t) + a_1(t)x + a_2(t)(x^2 - y^2), \quad \frac{\mathrm{d}y}{\mathrm{d}t} = a_1(t)y + a_2(t)2xy,$$

$$\tag{4.206}$$

where $a_0(t), a_1(t), a_2(t)$ are arbitrary t-dependent real functions with t-dependent vector field in (4.207),

$$X_t = a_0(t)X_1 + a_1(t)X_2 + a_2(t)X_3, \tag{4.207}$$

where X_1, X_2, X_3, which take the form

$$X_1 = \frac{\partial}{\partial u}, \quad X_2 = u\frac{\partial}{\partial u} + v\frac{\partial}{\partial v}, \quad X_3 = (u^2 + \iota^2 v^2)\frac{\partial}{\partial u} + 2uv\frac{\partial}{\partial v},$$

$$\tag{4.208}$$

have commutation relations (4.208),

$$[X_1, X_2] = X_1, \quad [X_1, X_3] = 2X_2, \quad [X_2, X_3] = X_3. \tag{4.209}$$

According to Table A.1 in Appendix A, system (4.206) is a LH system possessing a VG Lie algebra diffeomorphic to the primitive Lie algebra P_2 and the vector fields X_1, X_2, X_3 are Hamiltonian vector fields with respect to the symplectic form

$$\omega = \frac{\mathrm{d}x \wedge \mathrm{d}y}{y^2}. \tag{4.210}$$

Then, some Hamiltonian functions h_i of X_i (4.208) can be chosen to be

$$h_1 = -\frac{1}{y}, \quad h_2 = -\frac{x}{y}, \quad h_3 = -\frac{x^2 + y^2}{y}, \tag{4.211}$$

which satisfy the commutation relations

$$\{h_1, h_2\}_\omega = -h_1, \quad \{h_1, h_3\}_\omega = -2h_2, \quad \{h_2, h_3\}_\omega = -h_3.$$

Hence, $(\langle h_1, h_2, h_3 \rangle, \{\cdot, \cdot\}_\omega)$ is a Lie–Hamilton algebra for X isomorphic to $\mathfrak{sl}(2, \mathbb{R})$ and

$$h = a_0(t)h_1 + a_1(t)h_2 + a_2(t)h_3, \tag{4.212}$$

is a t-dependent Hamiltonian function.

The above result can be generalized by making use of analytic continuation and contractions which can also be understood as a Cayley–Klein approach [29, 193, 331, 383], which underlies the structure of the referred to as two-dimensional Cayley–Klein geometries.

Consider the real plane with coordinates $\{x, y\}$ and an "additional" unit j such that

$$j^2 \in \{-1, +1, 0\}. \tag{4.213}$$

Next, we define

$$z := x + jy, \quad \bar{z} := x - jy, \quad (x, y) \in \mathbb{R}^2. \tag{4.214}$$

Assuming that j commutes with real numbers, we can write

$$|z|^2 := z\bar{z} = x^2 - j^2 y^2, \quad z^2 = x^2 + j^2 y^2 + 2jxy. \tag{4.215}$$

In this way, we find that the number z in (4.214) comprises three possibilities

- $j^2 = -1$. In this case, we are dealing with the usual *complex numbers* $j := i$ and $z \in \mathbb{C}$. Hence,

$$|z|^2 = z\bar{z} = x^2 + y^2, \quad z^2 = x^2 - y^2 + 2ixy, \quad z \in \mathbb{C}. \tag{4.216}$$

- $j^2 = +1$. Now, we are dealing with the so-called *split-complex numbers* $z \in \mathbb{C}'$. The additional unit is usually known as the *double* or *Clifford* unit $j := e$ [383]. Thus,

$$|z|^2 = z\bar{z} = x^2 - y^2, \quad z^2 = x^2 + y^2 + 2exy, \quad z \in \mathbb{C}'. \tag{4.217}$$

- $j^2 = 0$. In this last possibility, z is known as a *dual* or *Study number* [383], which can be regarded as a *contracted* case since

$$|z|^2 = z\bar{z} = x^2, \quad z^2 = x^2 + 2\varepsilon xy, \quad z \in \mathbb{D}. \tag{4.218}$$

With these ingredients we shall call the *Cayley–Klein Riccati equation* [166] the generalization of the planar Riccati equation (4.206), which can be written as a complex Riccati equation with t-dependent real coefficients [50], to a family of Riccati equation over $z \in \{\mathbb{C}, \mathbb{C}', \mathbb{D}\}$, that is,

$$\frac{\mathrm{d}z}{\mathrm{d}t} = a_0(t) + a_1(t)z + a_2(t)z^2, \quad z := x + jy, \tag{4.219}$$

which, for real t-dependent coefficients $a_0(t), a_1(t), a_2(t)$, gives rise to a system of two differential equations

$$\frac{\mathrm{d}x}{\mathrm{d}t} = a_0(t) + a_1(t)x + a_2(t)(x^2 + j^2 y^2), \quad \frac{\mathrm{d}y}{\mathrm{d}t} = a_1(t)y + a_2(t)2xy, \tag{4.220}$$

that generalize the planar Riccati equation (4.206).

We now prove separately that the two remaining cases with $j^2 = +1$ and $j^2 = 0$ are also Lie–Hamilton systems.

Example 4.23. If we set $j^2 := +1$, the system (4.220) is known as a *Double-Clifford or split-complex Riccati equation*

$$\frac{\mathrm{d}x}{\mathrm{d}t} = a_0(t) + a_1(t)x + a_2(t)(x^2 + y^2), \quad \frac{\mathrm{d}y}{\mathrm{d}t} = a_1(t)y + a_2(t)2xy. \tag{4.221}$$

The point is to analyze whether this is also a Lie system and, in affirmative case, whether it is furthermore a Lie–Hamilton one with a VG Lie algebra diffeomorphic to a class given in Table A.1 in Appendix A.

Indeed, system (4.221) is related to the t-dependent vector field of the form (4.207) with

$$X_1 = \frac{\partial}{\partial x}, \quad X_2 = x\frac{\partial}{\partial x} + y\frac{\partial}{\partial y}, \quad X_3 = (x^2 + y^2)\frac{\partial}{\partial x} + 2xy\frac{\partial}{\partial y}. \tag{4.222}$$

These again span a VG real Lie algebra $V \simeq \mathfrak{sl}(2, \mathbb{R})$ with the same commutation rules given by (4.209). Hence, (4.221) is a Lie system.

212 *A Guide to Lie Systems with Compatible Geometric Structures*

Vector fields (4.222) do not arise exactly in the Lie algebras isomorphic to $\mathfrak{sl}(2,\mathbb{R})$ in Table A.1 in Appendix A, namely P_2, I_4 and I_5. Nevertheless, if we introduce in (4.222) the new variables $\{u,v\}$ defined by

$$u = x + y, \quad v = x - y, \qquad x = \tfrac{1}{2}(u+v), \quad y = \tfrac{1}{2}(u-v), \quad (4.223)$$

we find that

$$X_1 = \frac{\partial}{\partial u} + \frac{\partial}{\partial v}, \quad X_2 = u\frac{\partial}{\partial u} + v\frac{\partial}{\partial v}, \quad X_3 = u^2\frac{\partial}{\partial u} + v^2\frac{\partial}{\partial v}, \quad (4.224)$$

which are, exactly, the vector fields appearing in the imprimitive Lie algebra $\mathrm{I}_4 \simeq \mathfrak{sl}(2,\mathbb{R})$ of Tables A.1 and A.2 of Appendix A. Thus, these are endowed with a closed and non-degenerate 2-form; the latter and some associated Hamiltonian functions are given by

$$\omega = \frac{du \wedge dv}{(u-v)^2}, \quad h_1 = \frac{1}{u-v}, \quad h_2 = \frac{u+v}{2(u-v)}, \quad h_3 = \frac{uv}{u-v},$$
$$(4.225)$$

which satisfy the same commutation relations (4.211).

In terms of the initial variables $\{x,y\}$, the expressions (4.225) turn out to be

$$\omega = -\frac{dx \wedge dy}{2y^2}, \quad h_1 = \frac{1}{2y}, \quad h_2 = \frac{x}{2y}, \quad h_3 = \frac{x^2 - y^2}{2y}, \quad (4.226)$$

to be compared with (4.210) and (4.211).

Example 4.24. When $j^2 = 0$, the system (4.220) reduces to a *Dual-Study Riccati equation*

$$\frac{dx}{dt} = a_0(t) + a_1(t)x + a_2(t)x^2, \quad \frac{dy}{dt} = a_1(t)y + a_2(t)2xy, \quad (4.227)$$

which can be regarded as a "contracted" system from either (4.206) or (4.221). We stress that this system appears as a part of the Riccati system employed in the resolution of diffusion-type equations [346, 347]. Its relevance is due to the fact that its general solution allows us to solve the whole Riccati system and to map diffusion-type equations into an easily integrable PDE.

Let us prove that (4.227) is a Lie–Hamilton system. It is clear that this is a Lie system as it is related to the t-dependent vector field $X_t = a_0(t)X_1 + a_1(t)X_2 + a_2(t)X_3$, where

$$X_1 = \frac{\partial}{\partial x}, \quad X_2 = x\frac{\partial}{\partial x} + y\frac{\partial}{\partial y}, \quad X_3 = x^2\frac{\partial}{\partial x} + 2xy\frac{\partial}{\partial y}, \quad (4.228)$$

span a VG real Lie algebra $V \simeq \mathfrak{sl}(2,\mathbb{R})$ with commutation rules given by (4.209). These vector fields do not appear again in Table A.1 in Appendix A. Nevertheless, we can map them into a basis given in Table A.1 by choosing an appropriate change of variables.

We assume $y > 0$, while the case $y < 0$ can be studied analogously giving a similar result. Consider the new variables $\{u, v\}$ defined by

$$u = x, \quad v = \sqrt{y}, \quad x = u, \quad y = v^2. \quad (4.229)$$

By writing vector fields (4.228) in the new variables, we obtain

$$X_1 = \frac{\partial}{\partial u}, \quad X_2 = u\frac{\partial}{\partial u} + \frac{1}{2}v\frac{\partial}{\partial v}, \quad X_3 = u^2\frac{\partial}{\partial u} + uv\frac{\partial}{\partial v}, \quad (4.230)$$

which are those appearing in the imprimitive Lie algebra $I_5 \simeq \mathfrak{sl}(2,\mathbb{R})$ of Tables A.1 and A.2 and so we get

$$\omega = \frac{du \wedge dv}{v^3}, \quad h_1 = -\frac{1}{2v^2}, \quad h_2 = -\frac{u}{2v^2}, \quad h_3 = -\frac{u^2}{2v^2}. \quad (4.231)$$

These Hamiltonian functions satisfy the commutation relations (4.211). Consequently, h_1, h_2, h_3 span a Lie algebra isomorphic to $\mathfrak{sl}(2,\mathbb{R})$.

Finally we write the expressions (4.231) in terms of the initial variables $\{x, y\}$

$$\omega = \frac{dx \wedge dy}{2y^2}, \quad h_1 = -\frac{1}{2y}, \quad h_2 = -\frac{x}{2y}, \quad h_3 = -\frac{x^2}{2y}, \quad (4.232)$$

which fulfill the commutation relations (4.13). Therefore, the system (4.227) is, once more, an $\mathfrak{sl}(2,\mathbb{R})$-Lie-Hamilton system possessing a VG Lie algebra diffeomorphic to I_5 and a LH algebra $(\langle h_1, h_2, h_3 \rangle, \{\cdot, \cdot\}_\omega) \simeq \mathfrak{sl}(2,\mathbb{R})$.

214   *A Guide to Lie Systems with Compatible Geometric Structures*

We conclude that the Cayley–Klein Riccati equation (4.219) comprises in a unified way the *three* nondiffeomorphic VG Lie algebras of Hamiltonian vector fields isomorphic to $\mathfrak{sl}(2,\mathbb{R})$: P$_2$, I$_4$ and I$_5$.

Other $\mathfrak{sl}(2,\mathbb{R})$-Lie–Hamilton systems and "equivalence"

In this section, we present some $\mathfrak{sl}(2,\mathbb{R})$-Lie–Hamilton systems of mathematical and physical interest. To keep the notation simple, we say that a second-order differential equation is a Lie system (respectively, LH system) when the first-order system obtained from it by adding a new variable $y := \mathrm{d}x/\mathrm{d}t$ is a Lie system (respectively, Lie–Hamilton system).

We here study the coupled Riccati equation, Milne–Pinney and second-order Kummer–Schwarz equations, certain Lie systems appearing in the study of diffusion equations, the Smorodinsky–Winternitz oscillator and the harmonic oscillator, both with a t-dependent frequency. Next, we establish, according to Table A.1 in Appendix A, the equivalence among them and also among the three cases covered by the Cayley–Klein Riccati equation. More precisely, we establish which of all of the above systems are locally diffeomorphic.

Example 4.25. Consider the system of *coupled Riccati equations* given in (4.9) for $n = 2$ [286], with $x_1 = x$ and $x_2 = y$.

$$\frac{\mathrm{d}x}{\mathrm{d}t} = a_0(t) + a_1(t)x + a_2(t)x^2, \quad \frac{\mathrm{d}y}{\mathrm{d}t} = a_0(t) + a_1(t)y + a_2(t)y^2.$$

$$(4.233)$$

This system can be expressed as a t-dependent vector field (4.207) where

$$X_1 = \frac{\partial}{\partial x} + \frac{\partial}{\partial y}, \quad X_2 = x\frac{\partial}{\partial x} + y\frac{\partial}{\partial y}, \quad X_3 = x^2\frac{\partial}{\partial x} + y^2\frac{\partial}{\partial y},$$

$$(4.234)$$

so that these vector fields exactly reproduce those given in Table A.1 for the class I$_4 \simeq \mathfrak{sl}(2,\mathbb{R})$ which, in turn, means that this system is locally diffeomorphic to the split-complex Riccati equation (4.221).

Example 4.26. The *Milne–Pinney equation* [254] was shown in (4.2) as a first-order system

$$\begin{cases} \dfrac{\mathrm{d}x}{\mathrm{d}t} = y, \\ \dfrac{\mathrm{d}y}{\mathrm{d}t} = -\omega^2(t)x + \dfrac{c}{x^3}, \end{cases} \tag{4.235}$$

It has an associated t-dependent vector field $X = X_3 + \omega^2(t)X_1$ with vector fields

$$X_1 = -x\frac{\partial}{\partial y}, \quad X_2 = \frac{1}{2}\left(y\frac{\partial}{\partial y} - x\frac{\partial}{\partial x}\right), \quad X_3 = y\frac{\partial}{\partial x} + \frac{c}{x^3}\frac{\partial}{\partial y}. \tag{4.236}$$

The vector fields X_1, X_2, X_3 span a Lie algebra isomorphic to $\mathfrak{sl}(2,\mathbb{R})$ with Lie brackets given by (4.4).

Proposition 4.53. *The system* (4.2) *is a Lie–Hamilton system of class* P_2 *for* $c > 0$, I_4 *for* $c < 0$ *and* I_5 *for* $c = 0$ *given in Table* A.1 *in Appendix* A.

Therefore, as the Cayley–Klein Riccati equation (4.220), the Milne–Pinney equations include the three possibilities of VG Lie algebras isomorphic to $\mathfrak{sl}(2,\mathbb{R})$ of Hamiltonian vector fields.

Example 4.27. Reconsider the *second-order Kummer–Schwarz equation* studied in [87] as shown earlier in (3.68), which was rewritten in terms of a first-order system (3.70),

$$\begin{cases} \dfrac{\mathrm{d}x}{\mathrm{d}t} = v, \\ \dfrac{\mathrm{d}v}{\mathrm{d}t} = \dfrac{3}{2}\dfrac{v^2}{x} - 2c_0 x^3 + 2b_1(t)x, \end{cases} \tag{4.237}$$

on $T\mathbb{R}_0$, with $\mathbb{R}_0 = \mathbb{R} - \{0\}$, obtained by adding the new variable $v := \mathrm{d}x/\mathrm{d}t$ to the KS-2 equation (3.68).

216   *A Guide to Lie Systems with Compatible Geometric Structures*

This system has an associated t-dependent vector field $M = M_3 + \eta(t)M_1$, where the vector fields correspond with

$$M_1 = 2x\frac{\partial}{\partial v}, \quad M_2 = x\frac{\partial}{\partial x} + 2v\frac{\partial}{\partial v},$$

$$M_3 = v\frac{\partial}{\partial x} + \left(\frac{3}{2}\frac{v^2}{x} - 2c_0x^3\right)\frac{\partial}{\partial v}, \tag{4.238}$$

satisfying the commutation relations

$$[M_1, M_3] = 2M_2, \quad [M_1, M_2] = M_1, \quad [M_2, M_3] = M_3, \tag{4.239}$$

that span a Lie algebra isomorphic to $\mathfrak{sl}(2, \mathbb{R})$. It can be proven that (4.237) comprises, once more, the three VG Lie algebras of Hamiltonian vector fields isomorphic to $\mathfrak{sl}(2, \mathbb{R})$ given in Table A.1 in Appendix A according to the value of the parameter c [23].

Proposition 4.54. *The system* (4.237) *is a LH system of class* P_2 *for* $c > 0$, I_4 *for* $c < 0$ *and* I_5 *for* $c = 0$ *given in Table* A.1 *in Appendix* A.

Example 4.28. A diffusion equation can be transformed into a simpler PDE by solving a system of seven first-order ordinary differential equations (see [346] and [347, p. 104] for details). This system can be easily solved by integrating its projection to $\mathbb{R}^2$, known as the *planar diffusion Riccati system* given by

$$\frac{\mathrm{d}x}{\mathrm{d}t} = -b(t) + 2c(t)x + 4a(t)x^2 + a(t)c_0y^4, \quad \frac{\mathrm{d}y}{\mathrm{d}t} = \big(c(t) + 4a(t)x\big)y, \tag{4.240}$$

where $a(t), b(t)$ and $c(t)$ are arbitrary t-dependent functions and $c_0 \in \{0, 1\}$. We call this system planar diffusion Riccati system. This system is related to the t-dependent vector field

$$X = a(t)X_3 - b(t)X_1 + c(t)X_2, \tag{4.241}$$

where

$$X_1 = \frac{\partial}{\partial x}, \quad X_2 = 2x\frac{\partial}{\partial x} + y\frac{\partial}{\partial y}, \quad X_3 = (4x^2 + c_0y^4)\frac{\partial}{\partial x} + 4xy\frac{\partial}{\partial y}, \tag{4.242}$$

satisfy the commutation relations

$$[X_1, X_2] = 2X_1, \quad [X_1, X_3] = 4X_2, \quad [X_2, X_3] = 2X_3. \quad (4.243)$$

Consequently, they span a Lie algebra isomorphic to $\mathfrak{sl}(2, \mathbb{R})$. For $c_0 = 1$, the change of variables

$$u = 2x + y^2, \quad v = 2x - y^2, \quad x = \tfrac{1}{4}(u + v), \quad y = \sqrt{(u - v)/2}, \quad (4.244)$$

maps these vector fields to a basis of I_4 (see Table A.1). Writing the symplectic structure and the Hamiltonian functions in the coordinate system $\{x, y\}$, we obtain

$$\omega = \frac{\mathrm{d}x \wedge \mathrm{d}y}{y^3}, \quad h_1 = -\frac{1}{2y^2}, \quad h_2 = -\frac{x}{y^2}, \quad h_3 = \frac{-2x^2}{y^2} + \frac{1}{2}y^2, \quad (4.245)$$

that satisfy

$$\{h_1, h_2\}_\omega = -2h_1, \quad \{h_1, h_3\}_\omega = -4h_2, \quad \{h_2, h_3\}_\omega = -2h_3. \quad (4.246)$$

For the case $c_0 = 0$, we have that the vector fields (4.242) form a basis of I_5 (see (4.230)). Hence, their associated symplectic form and some corresponding Hamiltonian functions can easily be obtained from Table A.1 in Appendix A.

Equivalence among $\mathfrak{sl}(2, \mathbb{R})$-Lie–Hamilton systems

Consequently, by taking into account all the above results, we are led to the following statement.

Theorem 4.55. *The $\mathfrak{sl}(2, \mathbb{R})$-Lie–Hamilton systems of the form (3.70), (4.9), (4.2), (4.220), and (4.240) are equivalent through local diffeomorphisms whenever their VG Lie algebras belong to the same class in Table* A.1, *that is,*

- P_2: *Milne–Pinney and Kummer–Schwarz equations for $c > 0$ as well as complex Riccati equations.*

218   *A Guide to Lie Systems with Compatible Geometric Structures*

- I_4: *Milne–Pinney and Kummer–Schwarz equations for $c < 0$, coupled Riccati equations, split-complex Riccati equations and the planar diffusion Riccati system with $c_0 = 1$.*
- I_5: *Milne–Pinney and Kummer–Schwarz equations for $c = 0$ as well as dual-Study Riccati equations, planar diffusion Riccati system with $c_0 = 0$ and the harmonic oscillator with t-dependent frequency.*

All of the above systems are considered to have t-dependent coefficients.

Only within each class, these systems are locally diffeomorphic and, therefore, there exists a local t-independent change of variables mapping one into another. For instance, there does not exist any diffeomorphism on $\mathbb{R}^2$ mapping the Milne–Pinney and Kummer–Schwarz equations with $c \neq 0$ to the harmonic oscillator with a t-dependent frequency as the latter corresponds to set $c = 0$ and belongs to class I_5. Moreover, these results also explain the existence of the known diffeomorphism mapping Kummer–Schwarz equations to Milne–Pinney equations, which from our approach, should be understood in a "unified map" [254], that is, the value of the parameter c should be considered and everything works whenever the same sign of c is preserved.

4.5.4. Two-photon Lie–Hamilton systems

In this section, we study two different Lie–Hamilton systems that belong to the same class P_5 in Table A.1 in Appendix A; these include a dissipative harmonic oscillator and a second-order Riccati equation in Hamiltonian form. As a consequence, there exists a diffeomorphism mapping one into the other when they are written as first-order systems.

The five generators $X_1, \ldots, X_5$ written in Table A.1 in Appendix A satisfy the Lie brackets

$$[X_1, X_2] = 0, \qquad [X_1, X_3] = X_1, \qquad [X_1, X_4] = 0, \quad [X_1, X_5] = X_2,$$
$$[X_2, X_3] = -X_2, \quad [X_2, X_4] = X_1, \qquad [X_2, X_5] = 0, \quad [X_3, X_4] = -2X_4,$$
$$[X_3, X_5] = 2X_5, \quad [X_4, X_5] = -X_3,$$

$$(4.247)$$

in such a manner that they span a Lie algebra isomorphic to $\mathfrak{sl}(2, \mathbb{R}) \ltimes \mathbb{R}^2$ where $\mathbb{R}^2 = \langle X_1, X_2 \rangle$ and $\mathfrak{sl}(2, \mathbb{R}) = \langle X_3, X_4, X_5 \rangle$. Such vector fields are Hamiltonian with respect to the canonical symplectic form $\omega = \mathrm{d}x \wedge \mathrm{d}y$. Nevertheless, the corresponding Hamiltonian functions must be enlarged with a central generator $h_0 = 1$ giving rise to the centrally extended Lie algebra $\overline{\mathfrak{sl}(2, \mathbb{R}) \ltimes \mathbb{R}^2}$ which is, in fact, isomorphic to the two-photon Lie algebra $\mathfrak{h}_6 = \langle h_1, h_2, h_3, h_4, h_5, h_0 \rangle$ [22, 385]. That is why we shall call these systems *two-photon Lie–Hamilton systems*. The Lie brackets between these functions read

$$\{h_1, h_2\}_\omega = h_0, \qquad \{h_1, h_3\}_\omega = -h_1, \quad \{h_1, h_4\}_\omega = 0, \quad \{h_1, h_5\}_\omega = -h_2,$$
$$\{h_2, h_3\}_\omega = h_2, \qquad \{h_2, h_4\}_\omega = -h_1, \quad \{h_2, h_5\}_\omega = 0, \quad \{h_3, h_4\}_\omega = 2h_4,$$
$$\{h_3, h_5\}_\omega = -2h_5, \quad \{h_4, h_5\}_\omega = h_3, \qquad \{h_0, \cdot\}_\omega = 0.$$

$$(4.248)$$

Note that $\mathfrak{h}_6 \simeq \mathfrak{sl}(2, \mathbb{R}) \ltimes \mathfrak{h}_3$, where $\mathfrak{h}_3 \simeq \langle h_0, h_1, h_2 \rangle$ is the Heisenberg–Weyl–Lie algebra and $\mathfrak{sl}(2, \mathbb{R}) \simeq \langle h_3, h_4, h_5 \rangle$. Since $\mathfrak{h}_4 \simeq \langle h_0, h_1, h_2, h_3 \rangle$ is the harmonic oscillator Lie algebra (isomorphic to $\overline{\mathfrak{iso}}(1, 1)$ in the class I_8), we have the embeddings $\mathfrak{h}_3 \subset \mathfrak{h}_4 \subset \mathfrak{h}_6$.

Example 4.29. The t-dependent Hamiltonian for the *dissipative harmonic oscillator* studied in [126] is given by

$$h(t, q, p) = \alpha(t) \frac{p^2}{2} + \beta(t) \frac{pq}{2} + \gamma(t) \frac{q^2}{2} + \delta(t)p + \epsilon(t)q + \phi(t), \quad (4.249)$$

where $\alpha(t), \beta(t), \gamma(t), \delta(t), \epsilon(t), \phi(t)$ are real t-dependent functions. The corresponding Hamilton equations read

$$\frac{dq}{dt} = \frac{\partial h}{\partial p} = \alpha(t)\, p + \frac{\beta(t)}{2}\, q + \delta(t),$$

$$\frac{dp}{dt} = -\frac{\partial h}{\partial q} = -\left(\beta(t)\frac{p}{2} + \gamma(t)q + \epsilon(t)\right). \qquad (4.250)$$

This system has an associated t-dependent vector field

$$X = \delta(t)X_1 - \epsilon(t)X_2 + \frac{\beta(t)}{2}X_3 + \alpha(t)X_4 - \gamma(t)X_5, \qquad (4.251)$$

where

$$X_1 = \frac{\partial}{\partial q}, \quad X_2 = \frac{\partial}{\partial p}, \quad X_3 = q\frac{\partial}{\partial q} - p\frac{\partial}{\partial p},$$

$$X_4 = p\frac{\partial}{\partial q}, \quad X_5 = q\frac{\partial}{\partial p}, \qquad (4.252)$$

are, up to a trivial change of variables $x = q$ and $y = p$, the vector fields of the basis of P_5 given in Table A.1 in Appendix A. Hence, their Hamiltonian functions with respect to the symplectic structure $\omega = dq \wedge dp$ are indicated in Table A.1 in Appendix A.

Example 4.30. We reconsider the family of *second-order Riccati equations* (3.53) which arose by reducing third-order linear differential equations through a dilation symmetry and a t-reparametrization [131]. The crucial point is that a quite general family of second-order Riccati equations (3.53) admits a t-dependent Hamiltonian (see [131, 280] for details) given by

$$h(t, x, p) = -2\sqrt{-p} - p\left(a_0(t) + a_1(t)x + a_2(t)x^2\right), \quad p < 0, \qquad (4.253)$$

where $a_0(t), a_1(t), a_2(t)$ are certain functions related to the t-dependent coefficients of (3.53). The corresponding Hamilton equations are (3.58) and the associated t-dependent vector field has the expression

$$X = X_1 - a_0(t)X_2 - a_1(t)X_3 - a_2(t)X_4, \tag{4.254}$$

such that these vector fields correspond with (3.59) and close the commutation relations (3.60).

If we set

$$Y_1 = -\frac{1}{\sqrt{2}}X_5, \quad Y_2 = \frac{1}{\sqrt{2}}X_1, \quad Y_3 = -2X_3,$$
$$Y_4 = -X_4, \quad Y_5 = X_2, \tag{4.255}$$

we find that the commutation relations for $Y_1, \ldots, Y_5$ coincide with (4.247) and therefore span a Lie algebra isomorphic to $\mathfrak{sl}(2, \mathbb{R}) \ltimes \mathbb{R}^2$. Indeed, it is a Lie–Hamilton system endowed with a canonical symplectic form $\omega = \mathrm{d}x \wedge \mathrm{d}p$. The Hamiltonian functions corresponding to the vector fields (3.59) turn out to be

$$\tilde{h}_1 = -2\sqrt{-p}, \quad \tilde{h}_2 = p, \quad \tilde{h}_3 = xp,$$
$$\tilde{h}_4 = x^2p, \quad \tilde{h}_5 = -2x\sqrt{-p}, \tag{4.256}$$

which span along with $\tilde{h}_0 = 1$ a Lie algebra of functions isomorphic to the two-photon Lie algebra $\mathfrak{h}_6$ [27, 121] with non-vanishing Poisson brackets given by

$$\{\tilde{h}_1, \tilde{h}_3\}_\omega = -\frac{1}{2}\tilde{h}_1, \quad \{\tilde{h}_1, \tilde{h}_4\}_\omega = -\tilde{h}_5, \quad \{\tilde{h}_1, \tilde{h}_5\}_\omega = 2\tilde{h}_0,$$
$$\{\tilde{h}_2, \tilde{h}_3\}_\omega = -\tilde{h}_2, \quad \{\tilde{h}_2, \tilde{h}_4\}_\omega = -2\tilde{h}_3, \quad \{\tilde{h}_2, \tilde{h}_5\}_\omega = -\tilde{h}_1, \tag{4.257}$$
$$\{\tilde{h}_3, \tilde{h}_4\}_\omega = -\tilde{h}_4, \quad \{\tilde{h}_3, \tilde{h}_5\}_\omega = -\frac{1}{2}\tilde{h}_5.$$

Under the identification

$$h_1 = -\frac{1}{\sqrt{2}}\tilde{h}_5, \quad h_2 = \frac{1}{\sqrt{2}}\tilde{h}_1, \quad h_3 = -2\tilde{h}_3, \quad h_4 = -\tilde{h}_4,$$
$$h_5 = \tilde{h}_2, \quad h_0 = \tilde{h}_0, \tag{4.258}$$

we recover the Poisson brackets (4.248). Furthermore, the expressions for the vector fields and their Hamiltonian functions written in Table A.1 in Appendix A for the class P_5, in variables $\{u, v\}$, can be obtained from $X_1, \ldots, X_5$ by applying the change of variables defined by

$$u = \sqrt{-2p}, \quad v = x\sqrt{-2p}, \quad x = \frac{v}{u}, \quad p = -\frac{1}{2}u^2, \quad (4.259)$$

such that $\omega = \mathrm{d}x \wedge \mathrm{d}p := \mathrm{d}u \wedge \mathrm{d}v$. Note that $p/\sqrt{-p} = -\sqrt{-p}$ since $p < 0$.

The main results of this section are then summarized as follows.

Proposition 4.56. *The dissipative harmonic oscillator* (4.249) *and the Hamilton's equations* (3.58) *corresponding to the second-order Riccati equation* (3.53) *are* P_5-*Lie–Hamilton systems with Lie–Hamilton algebra isomorphic to the two-photon one* $\mathfrak{h}_6$. *Consequently, all are locally diffeomorphic.*

4.5.5. $\mathfrak{h}_2$-*Lie–Hamilton systems*

Let us study class $I_{14A} \simeq \mathbb{R} \ltimes \mathbb{R}^r$ with $r = 1$ of Table A.1 in Appendix A. It admits a basis of vector fields $X_1 = \partial_x$ and $X_2 = \eta_1(x)\partial_y$. If we require that these two vector fields close a non-abelian Lie algebra, we choose $[X_1, X_2] = X_2$, with no loss of generality, then $\eta_1(x) = \mathrm{e}^x$, that is,

$$X_1 = \frac{\partial}{\partial x}, \quad X_2 = \mathrm{e}^x \frac{\partial}{\partial y}, \quad [X_1, X_2] = X_2, \quad (4.260)$$

which we denote by $\mathfrak{h}_2 \simeq \mathbb{R} \ltimes \mathbb{R}$. This is a VG Lie algebra of Hamiltonian vector fields when endowed with the canonical symplectic form $\omega = \mathrm{d}x \wedge \mathrm{d}y$. Hence, we can choose

$$h_1 = y, \quad h_2 = -\mathrm{e}^x, \quad \{h_1, h_2\}_\omega = -h_2. \quad (4.261)$$

In the following, we prove that the complex Bernoulli equation with t-dependent real coefficients has an underlying $\mathfrak{h}_2$-Lie algebra. Furthermore, we relate this fact with the generalized Buchdahl equations and t-dependent Lotka–Volterra systems. It is remarkable that all particular cases of Cayley–Klein Riccati equations (4.220) with $a_2(t) = 0$ are $\mathfrak{h}_2$-Lie–Hamilton systems. Finally, we prove that all

Lie–Hamilton Systems 223

these systems are locally diffeomorphic and belong to the same class $\mathrm{I}_{14A}^{r=1} \simeq \mathfrak{h}_2$.

Example 4.31. The *complex Bernoulli equation* [285] with t-dependent real coefficients takes the form

$$\frac{dz}{dt} = a_1(t)z + a_2(t)z^n, \quad n \notin \{0, 1\}, \tag{4.262}$$

where $z \in \mathbb{C}$ and $a_1(t), a_2(t)$ are arbitrary t-dependent real functions. This is a generalization to the complex numbers of the usual Bernoulli equation. By writing $z := x + iy$, Equation (4.262) reads

$$\frac{dx}{dt} = a_1(t)x + a_2(t) \sum_{\substack{k=0 \\ k\text{ even}}}^{n} \binom{n}{k} x^{n-k} i^k y^k, \tag{4.263}$$

$$\frac{dy}{dt} = a_1(t)y + a_2(t) \sum_{\substack{k=1 \\ k\text{ odd}}}^{n} \binom{n}{k} x^{n-k} i^{k-1} y^k. \tag{4.264}$$

This can be studied in terms of the t-dependent vector field $X = a_1(t)X_1 + a_2(t)X_2$, where

$$X_1 = x\frac{\partial}{\partial x} + y\frac{\partial}{\partial y},$$

$$X_2 = \sum_{\substack{k=0 \\ k\text{ even}}}^{n} \binom{n}{k} x^{n-k} i^k y^k \frac{\partial}{\partial x} + \sum_{\substack{k=1 \\ k\text{ odd}}}^{n} \binom{n}{k} x^{n-k} i^{k-1} y^k \frac{\partial}{\partial y}, \tag{4.265}$$

and their commutation relation is

$$[X_1, X_2] = (n - 1)X_2, \tag{4.266}$$

which is isomorphic to $\mathfrak{h}_2$. In the GKO classification, [23, 183], there is just one Lie algebra isomorphic to $\mathfrak{h}_2$ whose vector fields are not proportional at each point: I_{14A} with $r = 1$. So, $\langle X_1, X_2 \rangle$ is a Lie algebra of Hamiltonian vector fields in view of the results of Table A.2 in Appendix A.

Let us inspect the case $n = 2$. So, the vector fields in (4.265) read $X_1 = x\partial_x + y\partial_y$ and $X_2 = (x^2 - y^2)\partial_x + 2xy\partial_y$, and become Hamiltonian with respect to the symplectic form and with Hamiltonian functions

$$\omega = \frac{dx \wedge dy}{y^2}, \quad h_1 = -\frac{x}{y}, \quad h_2 = -\frac{x^2 + y^2}{y}. \tag{4.267}$$

Additionally, $\{h_1, h_2\}_\omega = -h_2$. Note that, in this very particular case, the Hamiltonian functions become in the form given in Table A.2 for the *positive Borel subalgebra* of $P_2 \simeq \mathfrak{sl}(2, \mathbb{R})$. In fact, for $n = 2$, equation (4.262) is a particular case of the complex Riccati equation (4.206) with $a_0 = 0$.

Let us now take a closer look at $n = 3$. So, the vector field X_2 read

$$X_2 = \left(x^3 - 3xy^2\right)\frac{\partial}{\partial x} + \left(3x^2 y - y^3\right)\frac{\partial}{\partial y}, \tag{4.268}$$

and it can be shown that X_1 and X_2 become Hamiltonian vector fields with respect to the symplectic form

$$\omega = \frac{x^2 + y^2}{x^2 y^2}\, dx \wedge dy. \tag{4.269}$$

Moreover, X_1 and X_2 admit Hamiltonian functions

$$h_1 = \frac{y^2 - x^2}{xy}, \quad h_2 = -\frac{(x^2 + y^2)^2}{xy}, \tag{4.270}$$

satisfying $\{h_1, h_2\}_\omega = -2h_2$. Other new results come out by setting $n > 3$.

Example 4.32. The *generalized Buchdahl equations* [65, 133, 141] are the second-order differential equations given by (4.19) which can be rewritten in terms of a first-order system (4.20),

$$\begin{cases} \dfrac{dx}{dt} = y, \\[2mm] \dfrac{dy}{dt} = a(x)y^2 + b(t)y, \end{cases} \tag{4.271}$$

such that $X = X_1 + b(t)X_2$, where the vector fields corresponded with (4.21).

These vector fields span a Lie algebra diffeomorphic to $I_{14A}^{r=1} \simeq \mathfrak{h}_2$ so that (4.271) is a Lie–Hamilton system. The corresponding symplectic form and Hamiltonian functions can be found in [23].

Example 4.33. Finally, consider the *particular Lotka–Volterra systems* [236, 355] given by

$$\frac{dx}{dt} = ax - g(t)(x - ay)x, \quad \frac{dy}{dt} = ay - g(t)(bx - y)y, \quad a \neq 0,$$
$$(4.272)$$

where $g(t)$ determines the variation of the seasons, while a and b are constants describing the interactions among the species. System (4.272) is associated with the t-dependent vector field $X = X_1 + g(t)X_2$, where

$$X_1 = ax\frac{\partial}{\partial x} + ay\frac{\partial}{\partial y}, \quad X_2 = -(x - ay)x\frac{\partial}{\partial x} - (bx - y)y\frac{\partial}{\partial y},$$
$$(4.273)$$

satisfy

$$[X_1, X_2] = aX_2, \quad a \neq 0. \tag{4.274}$$

Hence, (4.272) is a Lie system. Moreover, it has been proven in [23] that, except for the case with $a = b = 1$, this is also a LH system belonging to the family $I_{14A}^{r=1} \simeq \mathfrak{h}_2$.

Hence, we conclude this section with the following statement.

Proposition 4.57. *The complex Bernoulli equation (4.264), the generalized Buchdahl equations (4.271) and the t-dependent Lotka–Volterra systems (4.272) (with the exception of $a = b = 1$) are LH systems with a VG Lie algebra diffeomorphic to $I_{14A}^{r=1} \simeq \mathfrak{h}_2$ in Table A.1 in Appendix A. Thus, all these systems are locally diffeomorphic.*

Chapter 5

Dirac–Lie Systems

5.1. Introduction

A step further on the study of *structure-Lie systems* is the consideration of Dirac–Lie systems. The introduction of this type of system presents some advantages over different approaches to certain physical systems. In the following paragraph, the importance of Dirac–Lie systems will be clearly stated.

5.2. Motivation for Dirac–Lie Systems

Dirac–Lie systems arise due to the existence of certain Lie systems which do not have a VG Lie algebra of Hamiltonian vector fields with respect to a Poisson structure, but they do with respect to a presymplectic one. A possible structure incorporating the presymplectic and Poisson structure is the Dirac structure. We will illustrate this fact with the following example.

Example 5.1. Let us reconsider the *third-order Kummer–Schwarz equation* [42, 87],

$$
\begin{cases}
\dfrac{dx}{dt} = v, \\[2mm]
\dfrac{dv}{dt} = a, \\[2mm]
\dfrac{da}{dt} = \dfrac{3}{2}\dfrac{a^2}{v} - 2c_0(x)v^3 + 2b_1(t)v,
\end{cases}
\tag{5.1}
$$

on the open submanifold $\mathcal{O}_2 := \{(x, v, a) \in \mathrm{T}^2\mathbb{R} \mid v \neq 0\}$ of $\mathrm{T}^2\mathbb{R} \simeq \mathbb{R}^3$. Its associated t-dependent vector field is $X^{3\mathrm{KS}} = N_3 + b_1(t)N_1$.

If $b_1(t)$ is not a constant, the VG Lie algebra is isomorphic to $\mathfrak{sl}(2, \mathbb{R})$ on $\mathcal{O}_2$ made up with vector fields (3.95). Hence, $X^{3\mathrm{KS}}$ is a Lie system.

However, $X^{3\mathrm{KS}}$ is not a Lie–Hamilton system when $b_1(t)$ is not a constant. Indeed, in this case, $\mathcal{D}^{X^{3\mathrm{KS}}}$ coincides with $T\mathcal{O}_2$ on $\mathcal{O}_2$. If $X^{3\mathrm{KS}}$ were also a Lie–Hamilton system with respect to (N, Λ), then $V^{X^{3\mathrm{KS}}}$ would consist of Hamiltonian vector fields and the characteristic distribution associated with Λ would have odd-dimensional rank on $\mathcal{O}_2$. This is impossible, as the local Hamiltonian vector fields of a Poisson manifold span a generalized distribution of even rank at each point. Our previous argument can easily be generalized to formulate the following *No-go* theorem.

Theorem 5.1 (Lie–Hamilton No-Go Theorem). *If X is a Lie system on an odd-dimensional manifold N satisfying that $\mathcal{D}^X_{x_0} = T_{x_0}N$ for a point x_0 in N, then X is not a Lie–Hamilton system on N.*

Note that from the properties of r^X, it follows that, if $\mathcal{D}^X_{x_0} = T_{x_0}N$ for a point x_0, then $\mathcal{D}^X_x = T_xN$ for x in an open neighborhood $U_{x_0} \ni x_0$. Hence, we can merely consider whether X is a Lie–Hamilton system on $N\backslash U_{x_0}$.

Despite the previous negative results, system (5.1) admits another interesting property: we can endow the manifold $\mathcal{O}_2$ with a presymplectic form $\omega_{3\mathrm{KS}}$ in such a way that $V^{X^{3\mathrm{KS}}}$ consists of Hamiltonian vector fields with respect to it. Indeed, by considering the equations

$\mathcal{L}_{N_1}\omega_{3KS} = \mathcal{L}_{N_2}\omega_{3KS} = \mathcal{L}_{N_3}\omega_{3KS} = 0$ and $d\omega_{3KS} = 0$, we can readily find the presymplectic form

$$\omega_{3KS} = \frac{dv \wedge da}{v^3}, \tag{5.2}$$

on $\mathcal{O}_2$. Additionally, we see that

$$\iota_{N_1}\omega_{3KS} = d\left(\frac{2}{v}\right), \quad \iota_{N_2}\omega_{3KS} = d\left(\frac{a}{v^2}\right),$$

$$\iota_{N_3}\omega_{3KS} = d\left(\frac{a^2}{2v^3} + 2c_0 v\right). \tag{5.3}$$

So, the system X^{3KS} becomes a Lie system with a VG Lie algebra of Hamiltonian vector fields with respect to ω_{3KS}. As seen later on, systems of this type can be studied through appropriate generalizations of the methods employed to investigate Lie–Hamilton systems.

Example 5.2. Another example of a Lie system which is not a Lie–Hamilton system but admits a VG Lie algebra of Hamiltonian vector fields with respect to a presymplectic form is the *Riccati system*,

$$\begin{cases} \dfrac{ds}{dt} = -4a(t)us - 2d(t)s, & \dfrac{dx}{dt} = \dfrac{(c(t) + 4a(t)u)x + f(t)}{-2ug(t)}, \\[2ex] \dfrac{du}{dt} = \dfrac{-b(t) + 2c(t)u}{+4a(t)u^2}, & \dfrac{dy}{dt} = (2a(t)x - g(t))v, \\[2ex] \dfrac{dv}{dt} = (c(t) + 4a(t)u)v, & \dfrac{dz}{dt} = a(t)x^2 - g(t)x, \\[2ex] \dfrac{dw}{dt} = a(t)v^2, \end{cases} \tag{5.4}$$

where $a(t)$, $b(t)$, $c(t)$, $d(t)$, $f(t)$ and $g(t)$ are arbitrary t-dependent functions. The interest of this system is due to its use in solving diffusion-type equations, Burgers' equations, and other PDEs [346].

Since every particular solution $(s(t), u(t), v(t), w(t), x(t), y(t), z(t))$ of (5.4), with $v(t_0) = 0$ ($s(t_0) = 0$) for a certain $t_0 \in \mathbb{R}$, satisfies $v(t) = 0$ ($s(t) = 0$) for every t, we can restrict ourselves to analyzing system (5.4) on the submanifold $M := \{(s, u, v, w, x, y, z) \in \mathbb{R}^7 \mid v \neq 0, s \neq 0\}$. This will simplify the application of our techniques without omitting any relevant detail.

System (5.4) describes integral curves of the t-dependent vector field

$$X_t^{RS} = a(t)X_1 - b(t)X_2 + c(t)X_3 - 2d(t)X_4 + f(t)X_5 + g(t)X_6,$$

$$(5.5)$$

where

$$X_1 = -4us\frac{\partial}{\partial s} + 4u^2\frac{\partial}{\partial u} + 4uv\frac{\partial}{\partial v} + v^2\frac{\partial}{\partial w} + 4ux\frac{\partial}{\partial x}$$
$$+ 2xv\frac{\partial}{\partial y} + x^2\frac{\partial}{\partial z},$$

$$X_2 = \frac{\partial}{\partial u}, \quad X_3 = 2u\frac{\partial}{\partial u} + v\frac{\partial}{\partial v} + x\frac{\partial}{\partial x},$$

$$X_4 = s\frac{\partial}{\partial s}, \quad X_5 = \frac{\partial}{\partial x},$$

$$X_6 = -2u\frac{\partial}{\partial x} - v\frac{\partial}{\partial y} - x\frac{\partial}{\partial z}, \quad X_7 = \frac{\partial}{\partial z}. \qquad (5.6)$$

Their commutation relations are

$$[X_1, X_2] = 4(X_4 - X_3), \quad [X_1, X_3] = -2X_1,$$
$$[X_1, X_5] = 2X_6, \quad [X_1, X_6] = 0,$$
$$[X_2, X_3] = 2X_2, \quad [X_2, X_5] = 0, \quad [X_2, X_6] = -2X_5,$$
$$[X_3, X_5] = -X_5, \quad [X_3, X_6] = X_6,$$
$$[X_5, X_6] = -X_7, \qquad (5.7)$$

and X_4 and X_7 commute with all the vector fields. Hence, system (5.4) is a Lie system associated with a VG Lie algebra V isomorphic to $(\mathfrak{sl}(2, \mathbb{R}) \ltimes \mathfrak{h}_2) \oplus \mathbb{R}$, where $\mathfrak{sl}(2, \mathbb{R}) \simeq \langle X_1, X_2, X_4 - X_3 \rangle$, $\mathfrak{h}_2 \simeq \langle X_5, X_6, X_7 \rangle$ and $\mathbb{R} \simeq \langle X_4 \rangle$. It is worth noting that this new example of Lie system is one of the few Lie systems related to remarkable PDEs until now [93].

Observe that (5.4) is not a Lie–Hamilton system when $V^{X^{RS}} = V$. In this case, $\mathcal{D}_p^{X^{RS}} = T_pM$ for any $p \in M$ and, in view of

Theorem 5.1 and the fact that $\dim T_p M = 7$, the system X^{RS} is not a Lie–Hamilton system on M.

Nevertheless, we can look for a presymplectic form turning X^{RS} into a Lie system with a VG Lie algebra of Hamiltonian vector fields. Looking for a non-trivial solution of the system of equations $\mathcal{L}_{X_k}\omega_{\mathrm{RS}} = 0$, with $k = 1, \ldots, 7$, and $d\omega_{\mathrm{RS}} = 0$, one can find the presymplectic 2-form:

$$\omega_{\mathrm{RS}} = -\frac{4wdu \wedge dw}{v^2} + \frac{dv \wedge dw}{v} + \frac{4w^2 du \wedge dv}{v^3}. \tag{5.8}$$

In addition, we can readily see that $d\omega_{\mathrm{RS}} = 0$ and $X_1, \ldots, X_r$ are Hamiltonian vector fields

$$\iota_{X_1}\omega_{\mathrm{RS}} = d\left(4uw - \frac{8u^2 w^2}{v^2} - \frac{v^2}{2}\right), \quad \iota_{X_2}\omega_{\mathrm{RS}} = -d\left(\frac{2w^2}{v^2}\right),$$

$$\iota_{X_3}\omega_{\mathrm{RS}} = d\left(w - \frac{4w^2 u}{v^2}\right), \tag{5.9}$$

and $\iota_{X_k}\omega_{\mathrm{RS}} = 0$ for $k = 4, \ldots, 7$.

Apart from the above examples, other non Lie–Hamilton systems that admit a VG Lie algebra of Hamiltonian vector fields with respect to a presymplectic form can be found in the study of certain reduced Ermakov systems [205], Wei–Norman equations for dissipative quantum oscillators [108], and $\mathfrak{sl}(2, \mathbb{R})$-Lie systems [311].

A straightforward generalization of the concept of a Lie–Hamilton system to Dirac manifolds would be a Lie system admitting a VG Lie algebra V of vector fields for which there exists a Dirac structure L such that V consists of L-Hamiltonian vector fields.

Definition 5.2. A *Dirac–Lie system* is a triple (N, L, X), where (N, L) stands for a Dirac manifold and X is a Lie system admitting a VG Lie algebra of L-Hamiltonian vector fields.

5.3. Dirac–Lie Hamiltonians

In view of Theorem 4.7 in Section 3.2, every Lie–Hamilton system admits a Lie–Hamiltonian structure. Since Dirac–Lie systems are

232 *A Guide to Lie Systems with Compatible Geometric Structures*

generalizations of these systems, it is natural to investigate whether Dirac–Lie systems admit an analogous property.

Definition 5.3. A *Dirac–Lie Hamiltonian structure* is a triple (N, L, h), where (N, L) stands for a Dirac manifold and h represents a t-parametrized family of admissible functions $h_t : N \to \mathbb{R}$ such that $\mathrm{Lie}(\{h_t\}_{t \in \mathbb{R}}, \{\cdot, \cdot\}_L)$ is a finite-dimensional real Lie algebra. A t-dependent vector field X is said to admit, to have or to possess a Dirac–Lie Hamiltonian (N, L, h) if $X_t + dh_t \in \Gamma(L)$ for all $t \in \mathbb{R}$.

Note 5.4. For simplicity, we hereafter call Dirac–Lie Hamiltonian structures Dirac–Lie Hamiltonians.

From the above definition, we see that system (5.1) related to the third-order Kummer–Schwarz equations possesses a Dirac–Lie Hamiltonian $(N, L^{\omega_{3KS}}, h^{3KS})$ and system (5.4), used to analyze diffusion equations, admits a Dirac–Lie Hamiltonian $(N, L^{\omega_{RS}}, h^{RS})$.

Consider the third-order Kummer–Schwarz equation in first-order form (5.1). Recall that N_1, N_2, and N_3 are Hamiltonian vector fields with respect to the presymplectic manifold $(\mathcal{O}_2, \omega_{3KS})$. It follows from relations (5.3) that the vector fields N_1, N_2, and N_3 have Hamiltonian functions

$$h_1 = -\frac{2}{v}, \quad h_2 = -\frac{a}{v^2}, \quad h_3 = -\frac{a^2}{2v^3} - 2c_0 v, \qquad (5.10)$$

respectively. Moreover,

$$\{h_1, h_3\} = 2h_2, \quad \{h_1, h_2\} = h_1, \quad \{h_2, h_3\} = h_3, \qquad (5.11)$$

where $\{\cdot, \cdot\}$ is the Poisson bracket on $\mathrm{Adm}(\mathcal{O}_2, \omega_{3KS})$ induced by ω_{3KS}. In consequence, h_1, h_2, and h_3 span a finite-dimensional real Lie algebra isomorphic to $\mathfrak{sl}(2, \mathbb{R})$. Thus, every X_t^{3KS} is a Hamiltonian vector field with Hamiltonian function $h_t^{3KS} = h_3 + b_1(t)h_1$ and the space $\mathrm{Lie}(\{h_t^{3KS}\}_{t \in \mathbb{R}}, \{\cdot, \cdot\})$ becomes a finite-dimensional real Lie algebra. This enables us to associate X^{3KS} to a curve in $\mathrm{Lie}(\{h_t^{3KS}\}_{t \in \mathbb{R}}, \{\cdot, \cdot\})$. The similarity of $(\mathcal{O}_2, \omega_{3KS}, h^{3KS})$ with Lie–Hamiltonians is immediate.

If we now turn to the Riccati system (5.4), we will see that we can obtain a similar result. More specifically, relations (5.9) imply that $X_1, \ldots, X_7$ have Hamiltonian functions

$$h_1 = \frac{(v^2 - 4uw)^2}{2v^2}, \quad h_2 = \frac{2\omega^2}{v^2}, \quad h_3 = \frac{4w^2 u}{v^2} - w, \qquad (5.12)$$

and $h_4 = h_5 = h_6 = h_7 = 0$. Moreover, given the Poisson bracket on admissible functions induced by ω_{3KS}, we see that

$$\{h_1, h_2\} = -4h_3, \quad \{h_1, h_3\} = -2h_1, \quad \{h_2, h_3\} = 2h_2. \qquad (5.13)$$

Hence, $h_1, \ldots, h_7$ span a real Lie algebra isomorphic to $\mathfrak{sl}(2, \mathbb{R})$ and, as in the previous case, the t-dependent vector fields X_t^{RS} possess Hamiltonian functions $h_t^{RS} = a(t)h_1 - b(t)h_2 + c(t)h_3$. Again, we can associate X^{RS} to a curve $t \mapsto h_t^{RS}$ in the finite-dimensional real Lie algebra $(\mathrm{Lie}(\{h_t^{RS}\}_{t \in \mathbb{R}}), \{\cdot, \cdot\})$.

Let us analyze the properties of Dirac–Lie structures. Observe first that there may be several systems associated with the same Dirac–Lie Hamiltonian. For instance, the systems X^{RS} and

$$X_2^{RS} = a(t)X_1 - b(t)X_2 + c(t)X_3 - 2d(t)X_4$$
$$+ f(t)z^3 X_5 + g(t)X_6 + h(t)z^2 X_7, \qquad (5.14)$$

admit the same Dirac–Lie Hamiltonian $(N, L^{\omega_{RS}}, h^{RS})$. It is remarkable that X_2^{RS} is not even a Lie system in general. Indeed, in view of

$$[z^2 X_7, z^n X_5] = nz^{n+1} X_5, \quad n = 3, 4, \ldots, \qquad (5.15)$$

we easily see that the successive Lie brackets of $z^n X_5$ and $z^2 X_7$ span an infinite set of vector fields which are linearly independent over $\mathbb{R}$. So, in those cases in which X_5 and X_7 belong to $V^{X_2^{RS}}$, this Lie algebra becomes infinite-dimensional.

In the case of a Dirac–Lie system, Proposition 2.5 shows easily the following.

Corollary 5.5. *Let (N, L, X) be a Dirac–Lie system admitting a Dirac–Lie Hamiltonian (N, L, h). Then, we have the exact sequence* *of Lie algebras*

$$0 \hookrightarrow \mathrm{Cas}(\{h_t\}_{t\in\mathbb{R}}, \{\cdot,\cdot\}_L) \hookrightarrow \mathrm{Lie}(\{h_t\}_{t\in\mathbb{R}}, \{\cdot,\cdot\}_L) \xrightarrow{B_L} \pi(V^X) \to 0,$$

(5.16)

where $\mathrm{Cas}(\{h_t\}_{t\in\mathbb{R}}, \{\cdot,\cdot\}_L) = \mathrm{Lie}(\{h_t\}_{t\in\mathbb{R}}, \{\cdot,\cdot\}_L) \cap \mathrm{Cas}(N, L)$. *In other words, we have that* $\mathrm{Lie}(\{h_t\}_{t\in\mathbb{R}}, \{\cdot,\cdot\}_L)$ *is a Lie algebra extension of the space* $\pi(V^X)$ *by* $\mathrm{Cas}(\{h_t\}_{t\in\mathbb{R}}, \{\cdot,\cdot\}_L)$.

Theorem 5.6. *Each Dirac–Lie system* (N, L, X) *admits a Dirac–Lie Hamiltonian* (N, L, h).

Proof. Since $V^X \subset \mathrm{Ham}(N, L)$ is a finite-dimensional Lie algebra, we can define a linear map $T : X_f \in V^X \mapsto f \in C^\infty(N)$ associating each L-Hamiltonian vector field in V^X with an associated L-Hamiltonian function, e.g., given a basis $X_1, \ldots, X_r$ of V^X, we define $T(X_i) = h_i$, with $i = 1, \ldots, r$, and extend T to V^X by linearity. Note that the functions $h_1, \ldots, h_r$ need not be linearly independent over $\mathbb{R}$, as a function can be Hamiltonian for two different L-Hamiltonian vector fields X_1 and X_2 when $X_1 - X_2 \in G(N, L)$. Given the system X, there exists a smooth curve $h_t = T(X_t)$ in $\mathfrak{W}_0 := \mathrm{Im}\, T$ such that $X_t + dh_t \in \Gamma(L)$. To ensure that h_t gives rise to a Dirac–Lie Hamiltonian, we need to demonstrate that $\dim \mathrm{Lie}(\{h_t\}_{t\in\mathbb{R}}, \{\cdot,\cdot\}_L) < \infty$. This will be done by constructing a finite-dimensional Lie algebra of functions containing the curve h_t.

Consider two elements $Y_1, Y_2 \in V^X$. Note that the functions $\{T(Y_1), T(Y_2)\}_L$ and $T([Y_1, Y_2])$ have the same L-Hamiltonian vector field. So, $\{T(Y_1), T(Y_2)\}_L - T([Y_1, Y_2]) \in \mathrm{Cas}(N, L)$ and, in view of Proposition 2.5, it Poisson commutes with all other admissible functions. Let us define $\Upsilon : V^X \times V^X \to C^\infty(N)$ of the form

$$\Upsilon(X_1, X_2) = \{T(X_1), T(X_2)\}_L - T[X_1, X_2].$$

(5.17)

The image of Υ is contained in a finite-dimensional real Abelian Lie subalgebra of $\mathrm{Cas}(N, L)$ of the form

$$\mathfrak{W}_{\mathcal{C}} := \langle \Upsilon(X_i, X_j) \rangle,$$

(5.18)

for $i, j = 1, \ldots, r$, and being $X_1, \ldots, X_r$ a basis for V^X. From here, it follows that

$$\{\mathfrak{W}_C, \mathfrak{W}_C\}_L = 0, \quad \{\mathfrak{W}_C, \mathfrak{W}_0\}_L = 0, \quad \{\mathfrak{W}_0, \mathfrak{W}_0\}_L \subset \mathfrak{W}_C + \mathfrak{W}_0.$$
(5.19)

Hence, $(\mathfrak{W} := \mathfrak{W}_0 + \mathfrak{W}_C, \{\cdot, \cdot\}_L)$ is a finite-dimensional real Lie algebra containing the curve h_t, and X admits a Dirac–Lie Hamiltonian $(N, L, T(X_t))$. $\square$

The following proposition is easy to verify.

Proposition 5.7. *Let (N, L, X) be a Dirac–Lie system. If (N, L, h) and $(N, L, \bar{h})$ are two Dirac–Lie Hamiltonians for (N, L, X), then*

$$h = \bar{h} + f^X, \tag{5.20}$$

where $f^X \in C^\infty(\mathbb{R} \times N)$ is a t-dependent function such that each $f_t^X : x \in N \mapsto f^X(x, t) \in \mathbb{R}$ is a Casimir function that is constant on every integral manifold $\mathcal{O}$ of $\mathcal{D}^X$.

Note that if we have a Dirac–Lie Hamiltonian (N, L, h) and we define a linear map $\widehat{T} : h \in \mathrm{Lie}(\{h_t\}_{t\in\mathbb{R}}, \{\cdot, \cdot\}) \mapsto X_h \in \mathrm{Ham}(N, L)$, the space $\widehat{T}(\mathrm{Lie}(\{h_t\}_{t\in\mathbb{R}}, \{\cdot, \cdot\})$ may span an infinite-dimensional Lie algebra of vector fields. For instance, consider again the Lie–Hamiltonian $(\mathcal{O}_2, \omega_{3\mathrm{KS}}, h_t^{3\mathrm{KS}} = h_3 + b_1(t)h_1)$ for system (5.1). The functions h_1, h_2, and h_3 are also Hamiltonian for the vector fields

$$N_1 = 2v\frac{\partial}{\partial a} + e^{v^2}\frac{\partial}{\partial x}, \quad N_2 = v\frac{\partial}{\partial v} + 2a\frac{\partial}{\partial a},$$
$$N_3 = a\frac{\partial}{\partial v} + \left(\frac{3}{2}\frac{a^2}{v} - 2c_0v^3\right)\frac{\partial}{\partial a}, \tag{5.21}$$

which satisfy

$$\overbrace{[N_2, [\ldots, [N_2, N_1]\ldots]]}^{j\text{-times}} = f_j(v)\frac{\partial}{\partial x} + 2(-1)^j v\frac{\partial}{\partial a},$$

$$f_j(v) := \overbrace{v\frac{\partial}{\partial v}\cdots v\frac{\partial}{\partial v}}^{j\text{-times}}(e^{v^2}). \tag{5.22}$$

236 *A Guide to Lie Systems with Compatible Geometric Structures*

In consequence, $\mathrm{Lie}(\widehat{T}(\mathrm{Lie}(\{h_t\}_{t\in\mathbb{R}},\{\cdot,\cdot\}))),[\cdot,\cdot])$ contains an infinite-dimensional Lie algebra of vector fields because the functions $\{f_j\}_{j\in\mathbb{R}}$ form an infinite family of linearly independent functions over $\mathbb{R}$. So, we need to impose additional conditions to ensure that the image of $\widehat{T}$ is finite-dimensional.

The following theorem yields an alternative definition of a Dirac–Lie system.

Theorem 5.8. *Given a Dirac manifold (N,L), the triple (N,L,X) is a Dirac–Lie system if and only if there exists a curve $\gamma : t \in \mathbb{R} \to \gamma_t \in \Gamma(L)$, satisfying that $\rho(\gamma_t) = X_t \in \mathrm{Ham}(N,L)$ for every $t \in \mathbb{R}$ and $\mathrm{Lie}(\{\gamma_t\}_{t\in\mathbb{R}},[[\cdot,\cdot]]_C)$ is a finite-dimensional real Lie algebra.*

Proof. Let us prove the direct part of the theorem. Assume that (N,L,X) is a Dirac–Lie system. In virtue of Theorem 5.6, it admits a Dirac–Lie Hamiltonian (N,L,h), with $h_t = T(X_t)$ and $T : V^X \to \mathrm{Adm}(N,L)$ a linear morphism associating each element of V^X with one of its L-Hamiltonian functions. We aim to prove that the curve in $\Gamma(L)$ of the form $\gamma_t = X_t + d(T(X_t))$ satisfies that $\dim \mathrm{Lie}(\{\gamma_t\}_{t\in\mathbb{R}},[[\cdot,\cdot]]_C) < \infty$.

The sections of $\Gamma(L)$ of the form

$$X_1 + dT(X_1),\ldots,\ X_r + dT(X_r),\ d\Upsilon(X_i,X_j), \tag{5.23}$$

with $i,j = 1,\ldots,r$, $X_1,\ldots,X_r$ is a basis of V^X and $\Upsilon : V^X \times V^X \to \mathrm{Cas}(N,L)$ is the map (5.17), spanning a finite-dimensional Lie algebra $(E,[[\cdot,\cdot]]_C)$. Indeed,

$$[[X_i + dT(X_i), X_j + dT(X_j)]]_C = [X_i,X_j] + d\{T(X_i),T(X_j)\}_L, \tag{5.24}$$

for $i,j = 1,\ldots,r$. Taking into account that $\{T(X_i),T(X_j)\}_L - T([X_i,X_j]) = \Upsilon(X_i,X_j)$, we see that the above is a linear combination of the generators (5.23). Additionally, we have that

$$[[X_i + dT(X_i), d\Upsilon(X_j,X_k)]]_C = d\{T(X_i),\Upsilon(X_j,X_k)\}_L = 0. \tag{5.25}$$

So, sections (5.23) span a finite-dimensional subspace E of $(\Gamma(L),[[\cdot,\cdot]]_C)$. As $\gamma_t \in E$, for all $t \in \mathbb{R}$, we conclude the direct part of the proof.

The converse is straightforward from the fact that $(L, [[\cdot, \cdot]]_C, \rho)$ is a Lie algebroid. Indeed, given the curve γ_t within a finite-dimensional real Lie algebra of sections E satisfying that $X_t = \rho(\gamma_t) \in \mathrm{Ham}(N, L)$, we have that $\{X_t\}_{t\in\mathbb{R}} \subset \rho(E)$ are L-Hamiltonian vector fields. As E is a finite-dimensional Lie algebra and ρ is a Lie algebra morphism, $\rho(E)$ is a finite-dimensional Lie algebra of vector fields and (N, L, X) becomes a Dirac–Lie system. $\qquad\square$

The above theorem shows the interest of defining a class of Lie systems related to general Lie algebroids.

5.4. Diagonal Prolongations

We analyze the properties of diagonal prolongations of Dirac–Lie systems. As a result, we discover new features that can be applied to study their superposition rules.

Let us fix m. In view of Lemma 3.9, the prolongations to N^m of the elements of a finite-dimensional real Lie algebra V of vector fields on N form a real Lie algebra $\widetilde{V}$ isomorphic to V. Similar to standard vector fields, we can define the diagonal prolongation of a t-dependent vector field X on N to N^m (see Definition 3.10 for details).

When X is a Lie–Hamilton system, its diagonal prolongations are also Lie–Hamilton systems in a natural way (see [27] and Chapters 3 and 4). Let us now focus on proving the analog of this result for Dirac–Lie systems.

Definition 5.9. Given two Dirac manifolds (N, L_N) and (M, L_M), we say that $\varphi : N \to M$ is a *forward Dirac map* between them if $(L_M)_{\varphi(x)} = \mathfrak{P}_\varphi(L_N)_x$, where

$$\mathfrak{P}_\varphi(L_N)_x = \{\varphi_{*x}X_x + \omega_{\varphi(x)} \in T_{\varphi(x)}M \oplus T^*_{\varphi(x)}M \,|\, X_x$$
$$+ (\varphi^*\omega_{\varphi(x)})_x \in (L_N)_x\}, \tag{5.26}$$

for all $x \in N$.

238   *A Guide to Lie Systems with Compatible Geometric Structures*

Proposition 5.10. *Given a Dirac structure (N, L) and the natural isomorphism*

$$(TN^m \oplus_{N^m} T^*N^m)_{(x_{(1)},\ldots,x_{(m)})}$$
$$\simeq (T_{x_{(1)}}N \oplus T^*_{x_{(1)}}N) \oplus \cdots \oplus (T_{x_{(m)}}N \oplus T^*_{x_{(m)}}N), \quad (5.27)$$

the diagonal prolongation $L^{[m]}$, viewed as a vector subbundle in $TN^m \oplus_{N^m} T^*N^m = \mathcal{P}N^{[m]}$, is a Dirac structure on N^m.

The forward image of $L^{[m]}$ through each $\pi_i : (x_{(1)},\ldots,x_{(m)}) \in N^m \to x_{(i)} \in N$, with $i = 1,\ldots,m$, equals L. Additionally, $L^{[m]}$ is invariant under the permutations $x_{(i)} \leftrightarrow x_{(j)}$, with $i,j = 1,\ldots,m$.

Proof. Being a diagonal prolongation of L, the subbundle $L^{[m]}$ is invariant under permutations $x_{(i)} \leftrightarrow x_{(j)}$ and each element of a basis $X_i + \alpha_i$ of L, with $i = 1,\ldots,n$, can naturally be considered as an element $X_i^{(j)} + \alpha_i^{(j)}$ of the jth-copy of L within $L^{[m]}$. This gives rise to a basis of $L^{[m]}$, which naturally becomes a smooth mn-dimensional subbundle of $\mathcal{P}N^m$. Considering the natural pairing $\langle \cdot, \cdot \rangle_+$ of $\mathcal{P}N^m$ and using $\langle \alpha_j^{(i)}, X_l^{(k)} \rangle = 0$ for $i \neq k$, we have

$$\left\langle \left(X_j^{(i)} + \alpha_j^{(i)}\right)(x_{(1)},\ldots,x_{(m)}), \left(X_l^{(k)} + \alpha_l^{(k)}\right)(x_{(1)},\ldots,x_{(m)}) \right\rangle_+$$
$$= \delta_k^i \left\langle (X_j + \alpha_j)(x_{(i)}), (X_l + \alpha_l)(x_{(i)}) \right\rangle_+ = 0, \quad (5.28)$$

for every $p = (x_{(1)},\ldots,x_{(m)}) \in N^m$. As the pairing is bilinear and vanishes on a basis of $L^{[m]}$, it does so on the whole $L^{[m]}$, which is therefore isotropic. Since $L^{[m]}$ has rank mn, it is maximally isotropic.

Using that

$$[X_j^{(i)}, X_l^{(k)}] = 0, \quad \iota_{X_j^{(i)}} d\alpha_l^{(k)} = 0, \quad \text{and} \quad \mathcal{L}_{X_j^{(i)}}\omega_l^{(k)} = 0 \quad (5.29)$$

for $i \neq k = 1,\ldots,m$ and $j,l = 1,\ldots,\dim N$, we obtain

$$[[X_j^{(i)} + \alpha_j^{(i)}, X_l^{(k)} + \alpha_l^{(k)}]]_C$$
$$= \delta_k^i [[X_j^{(i)} + \alpha_j^{(i)}, X_l^{(i)} + \alpha_l^{(i)}]]_C \in \Gamma(L^{[m]}). \quad (5.30)$$

So, $L^{[m]}$ is involutive. Since it is also maximally isotropic, it is a Dirac structure.

Let us prove that $\mathfrak{P}_{\pi_a}(L^{[m]}) = L$ for every π_a. Note that $(X_j^{(a)} + \alpha_j^{(a)})_p \in L_p^{[m]}$ is such that $\pi_{a*}(X_j^{(a)})_p = (X_j)_{x_{(a)}}$ and $(\alpha_j)_{x_{(a)}} \circ (\pi_{*a})_p = (\alpha_j^{(a)})_p$ for every $p \in \pi_a^{-1}(x_{(a)})$. So, $(X_j + \alpha_j)_{x_{(a)}} \in (\mathfrak{P}_{\pi_a}(L^{[m]}))_{x_{(a)}} \subset L_{x_{(a)}}$ for $j = 1, \ldots, n$ and every $x_{(a)} \in N$. Using that $X_j + \alpha_j$ is a basis for L and the previous results, we obtain $L \subset \mathfrak{P}_{\pi_a}(L^{[m]})$. Conversely, $\mathfrak{P}_{\pi_a}(L^{[m]}) \subset L$. Indeed, if $(X + \alpha)_{x_{(a)}} \in \mathfrak{P}_a(L^{[m]})$, then there exists an element $(Y + \beta)_p \in L_p^{[m]}$, with $p \in \pi^{-1}(x_{(a)})$, such that $\pi_{a*}Y_p = X_{x_{(a)}}$ and $(\alpha)_{x_{(a)}} \circ (\pi_{*a})_p = \beta_p$. Using that $(Y + \beta)_p = \sum_{ij} c_{ij}(X_j^{(i)} + \alpha_j^{(i)})_p$ for a unique set of constants c_{ij}, with $i = 1, \ldots, m$ and $j = 1, \ldots, n$, we have $\pi_{a*}(\sum_{ij} c_{ij}(X_j^{(i)})_p) = \sum_j c_{aj}(X_j)_{x_{(a)}} = X_{x_{(a)}}$. Meanwhile, $\beta_p = \alpha_{x_{(a)}} \circ (\pi_{*a})_p$ means that $\sum_j c_{aj}(\alpha_j)_{x_{(a)}} = \alpha_{x_{(a)}}$. So, $(X + \alpha)_{x_{(a)}} = \sum_j c_{aj}(X_j + \alpha_j)_{x_{(a)}} \in L_{x_{(a)}}$. $\qquad\square$

Corollary 5.11. *Given a Dirac structure (N, L), we have $\rho_m(L^{[m]}) = \rho(L)^{[m]}$, where ρ_m is the projection $\rho_m : \mathcal{P}N^m \to TN^m$. Then, if X is an L-Hamiltonian vector field with respect to L, its diagonal prolongation $\widetilde{X}^{[m]}$ to N^m is an L-Hamiltonian vector field with respect to $L^{[m]}$. Moreover, $\rho_m^*(L^{[m]}) = \rho^*(L)^{[m]}$, where ρ_m^* is the canonical projection $\rho_m^* : \mathcal{P}N^m \to T^*N^m$.*

Corollary 5.12. *If (N, L, X) is a Dirac–Lie system, then $(N^m, L^{[m]}, \widetilde{X}^{[m]})$ is also a Dirac–Lie system.*

Proof. If X admits a VG Lie algebra V of Hamiltonian vector fields with respect to (N, L), then $\widetilde{X}$ possesses a VG Lie algebra $\widetilde{V}$ given by the diagonal prolongations of the elements of V, which are $L^{[m]}$-Hamiltonian vector fields, by the construction of $L^{[m]}$ and Corollary 5.11. $\qquad\square$

Similar to the prolongations of vector fields, one can define prolongations of functions and 1-forms in an obvious way.

Proposition 5.13. *Let X be a vector field and f be a function on N. Then*

(a) If f is a L-Hamiltonian function for X, its diagonal prolongation $\widetilde{f}$ to N^m is an $L^{[m]}$-Hamiltonian function of the diagonal prolongation $\widetilde{X}$ to N^m.

(b) If $f \in \mathrm{Cas}(N, L)$, then $\widetilde{f} \in \mathrm{Cas}(N^m, L^{[m]})$.

(c) The map $\lambda : (\mathrm{Adm}(N, L), \{\cdot, \cdot\}_L) \ni f \mapsto \widetilde{f} \in (\mathrm{Adm}(N^m, L^{[m]}), \{\cdot, \cdot\}_{L^{[m]}})$ is an injective Lie algebra morphism.

Proof. Let f be an L-Hamiltonian function for X. Then, $X + df \in \Gamma(L)$ and $\widetilde{X} + d\widetilde{f} = \widetilde{X} + \widetilde{df}$ is an element of $\Gamma(L^{[m]})$. By a similar argument, if $f \in \mathrm{Cas}(N, L)$, then $\widetilde{f} \in \mathrm{Cas}(N^m, L^{[m]})$. Given $f, g \in \mathrm{Adm}(N, L)$, we have $\widetilde{\{f, g\}_L} = \widetilde{X_f g} = \widetilde{X}_{\widetilde{f}} \widetilde{g} = X_{\widetilde{f}} \widetilde{g} = \{\widetilde{f}, \widetilde{g}\}_{L^{[m]}}$, i.e., $\lambda(\{f, g\}_L) = \{\lambda(f), \lambda(g)\}_{L^{[m]}}$. Additionally, as λ is linear, it becomes a Lie algebra morphism. Moreover, it is easy to see that $\widetilde{f} = 0$ if and only if $f = 0$. Hence, λ is injective. $\qquad \square$

Note, however, that we cannot ensure that λ is a Poisson algebra morphism, as in general $\widetilde{fg} \neq \widetilde{f}\widetilde{g}$.

Using the above proposition, we can easily prove the following corollaries.

Corollary 5.14. *If $h_1, \ldots, h_r : N \to \mathbb{R}$ is a family of functions on a Dirac manifold (N, L) spanning a finite-dimensional real Lie algebra of functions with respect to the Lie bracket $\{\cdot, \cdot\}_L$, then their diagonal prolongations $\widetilde{h}_1, \ldots, \widetilde{h}_r$ to N^m close an isomorphic Lie algebra of functions with respect to the Lie bracket $\{\cdot, \cdot\}_{L^{[m]}}$ induced by the Dirac structure $(N^m, L^{[m]})$.*

Corollary 5.15. *If (N, L, X) is a Dirac–Lie system admitting a Dirac–Lie Hamiltonian (N, L, h), then $(N^m, L^{[m]}, \widetilde{X}^{[m]})$ is a Dirac–Lie system with a Dirac–Lie Hamiltonian $(N^m, L^{[m]}, h^{[m]})$, where $h_t^{[m]} = \widetilde{h}_t^{[m]}$ is the diagonal prolongation of h_t to N^m.*

5.5. Superposition Rules and t-Independent Constants of Motion

Let us give a first straightforward application of Dirac–Lie systems to obtain constants of the motion.

Proposition 5.16. *Given a Dirac–Lie system* (N, L, X), *the elements of* $\mathrm{Cas}(N, L)$ *are constants of the motion for* X. *Moreover, the set* $\mathcal{I}_L^X$ *of its admissible t-independent constants of the motion form a Poisson algebra* $(\mathcal{I}_L^X, \cdot, \{\cdot, \cdot\}_L)$.

Proof. Two admissible functions f and g are t-independent constants of the motion for X if and only if $X_t f = X_t g = 0$ for every $t \in \mathbb{R}$. Using that every X_t is a derivation of the associative algebra $C^\infty(N)$, we see that given $f, g \in \mathcal{I}_L^X$, then $f + g$, λf, and $f \cdot g$ are also constants of the motion for X for every $\lambda \in \mathbb{R}$. Since the sum and product of admissible functions are admissible functions, then $\mathcal{I}_L^X$ is closed under the sum and product of elements and real constants. So, $(\mathcal{I}_L^X, \cdot)$ is an associative subalgebra of $(C^\infty(N), \cdot)$.

As (N, L, X) is a Dirac–Lie system, the vector fields $\{X_t\}_{t \in \mathbb{R}}$ are L-Hamiltonian. Therefore,

$$X_t \{f, g\}_L = \{X_t f, g\}_L + \{f, X_t g\}_L. \tag{5.31}$$

As f and g are constants of the motion for X, then $\{f, g\}_L$ is so also. Using that $\{f, g\}_L$ is also an admissible function, we finish the proof. $\square$

The following can easily be proven.

Proposition 5.17. *Let* (N, L, X) *be a Dirac–Lie system admitting a Dirac–Lie Hamiltonian* (N, L, h). *An admissible function* $f : N \to \mathbb{R}$ *is a constant of the motion for* X *if and only if it Poisson commutes with all the elements of* $\mathrm{Lie}(\{h_t\}_{t \in \mathbb{R}}, \{\cdot, \cdot\}_L)$.

Consider a Dirac–Lie system (N, L^ω, X) with ω being a symplectic structure and X being an autonomous system. Consequently, $\mathrm{Adm}(N, L) = C^\infty(N)$ and the above proposition entails that $f \in C^\infty(N)$ is a constant of the motion for X if and only if it Poisson commutes with a Hamiltonian function h associated with X. This shows that Proposition 5.17 recovers as a particular case this well-known result [3]. Additionally, Proposition 5.17 suggests us that the role played by autonomous Hamiltonians for autonomous Hamiltonian

242   *A Guide to Lie Systems with Compatible Geometric Structures*

systems is performed by finite-dimensional Lie algebras of admissible functions associated with a Dirac–Lie Hamiltonian for Dirac–Lie systems. This fact can be employed, for instance, to study t-independent first integrals of Dirac–Lie systems, e.g., the maximal number of such first integrals in involution, which would lead to the interesting analysis of integrability/superintegrability and action/angle variables for Dirac–Lie systems [387].

Another reason to study t-independent constants of motion for Lie systems is their use in deriving superposition rules [91]. More explicitly, a superposition rule for a Lie system can be obtained through the t-independent constants of the motion of one of its diagonal prolongations [93]. The following proposition provides some ways of obtaining such constants.

Proposition 5.18. *If X is a system possessing a t-independent constant of the motion f, then*

1. *the diagonal prolongation $\widetilde{f}^{[m]}$ is a t-independent constant of the motion for $\widetilde{X}^{[m]}$;*
2. *if Y is a t-independent Lie symmetry of X, then $\widetilde{Y}^{[m]}$ is a t-independent Lie symmetry of $\widetilde{X}^{[m]}$;*
3. *if h is a t-independent constant of the motion for $\widetilde{X}^{[m]}$, then $\widetilde{Y}^{[m]}h$ is another t-independent constant of the motion for $\widetilde{X}^{[m]}$.*

Proof. This result is a straightforward application of Proposition 3.8 and the properties of the diagonal prolongations of t-dependent vector fields.   □

Proposition 5.19. *Given a Dirac–Lie system (N, L, X) that admits a Dirac–Lie Hamiltonian (N, L, h) such that $\{h_t\}_{t\in\mathbb{R}}$ is contained in a finite-dimensional Lie algebra of admissible functions $(\mathfrak{W}, \{\cdot,\cdot\}_L)$. Given the momentum map $J : N^m \to \mathfrak{W}^*$ associated with the Lie algebra morphism $\phi : f \in \mathfrak{W} \mapsto \widetilde{f} \in \mathrm{Adm}(N^m, L^{[m]})$, the pull-back $J^*(C)$ of any Casimir function C on $\mathfrak{W}^*$ is a constant of the motion for the diagonal prolongation $\widetilde{X}^{[m]}$. If $\mathfrak{W} \simeq \mathrm{Lie}(\{\widetilde{h}_t\}_{t\in\mathbb{R}}, \{\cdot,\cdot\}_{L^{[m]}})$, the function $J^*(C)$ Poisson commutes with all $L^{[m]}$-admissible constants of the motion of $\widetilde{X}^{[m]}$.*

Example 5.3. Let us use the above results to devise a superposition rule for the *third-order Kummer–Schwarz equation* in first-order form (5.1) with $c_0 = 0$, the so-called *Schwarzian equations* [339, 353]. To simplify the presentation, we will always assume $c_0 = 0$ in this section. It is known (cf. [87]) that the derivation of a superposition rule for this system can be reduced to obtaining certain three t-independent constants of the motion for the diagonal prolongation $\widetilde{X}^{3\mathrm{KS}}$ of $X^{3\mathrm{KS}}$ to $\mathcal{O}_2^2$. In [87] such constants were worked out through the method of characteristics which consists in solving a series of systems of ODEs. Nevertheless, we can determine such constants more easily through Dirac–Lie systems. System (5.1) is represented by the t-dependent vector field

$$X_t^{3\mathrm{KS}} = v\frac{\partial}{\partial x} + a\frac{\partial}{\partial v} + \left(\frac{3}{2}\frac{a^2}{v} + 2b_1(t)v\right)\frac{\partial}{\partial a}$$
$$= Y_3 + b_1(t)Y_1, \tag{5.32}$$

where the vector fields on $\mathcal{O}_2 = \{(x,v,a) \in T^2\mathbb{R} \mid v \neq 0\}$ given by (3.95). If so, the t-dependent vector field $\widetilde{X}^{3\mathrm{KS}}$ is spanned by a linear combination of the diagonal prolongations of N_1, N_2, and N_3 to $\mathcal{O}_2^2$. From (3.95), we have

$$\widetilde{N}_1 = \sum_{i=1}^{2} v_i\frac{\partial}{\partial a_i}, \quad \widetilde{N}_2 = \sum_{i=1}^{2}\left(v_i\frac{\partial}{\partial v_i} + 2a_i\frac{\partial}{\partial a_i}\right),$$
$$\widetilde{N}_3 = \sum_{i=1}^{2}\left(v_i\frac{\partial}{\partial x_i} + a_i\frac{\partial}{\partial v_i} + \frac{3}{2}\frac{a_i^2}{v_i}\frac{\partial}{\partial a_i}\right). \tag{5.33}$$

From Proposition 5.13 and functions (5.10), the vector fields $\widetilde{N}_1, \widetilde{N}_2, \widetilde{N}_3$ are $L^{[2]}$-Hamiltonian with $L^{[2]}$-Hamiltonian functions

$$\widetilde{h}_1 = -\frac{2}{v_1} - \frac{2}{v_2}, \quad \widetilde{h}_2 = -\frac{a_1}{v_1^2} - \frac{a_2}{v_2^2}, \quad \widetilde{h}_3 = -\frac{a_1^2}{2v_1^3} - \frac{a_2^2}{2v_2^3}. \tag{5.34}$$

Indeed, these are the diagonal prolongations to $\mathcal{O}_2^2$ of the L-Hamiltonian functions of N_1, N_2, and N_3. Moreover, they span a real Lie algebra of functions isomorphic to that one spanned by h_1, h_2, h_3 and to $\mathfrak{sl}(2,\mathbb{R})$. We can then define a Lie algebra morphism

$\phi : \mathfrak{sl}(2,\mathbb{R}) \to C^\infty(N^2)$ of the form $\phi(e_1) = \tilde{h}_1$, $\phi(e_2) = \tilde{h}_2$ and $\phi(e_3) = \tilde{h}_3$, where $\{e_1, e_2, e_3\}$ is the standard basis of $\mathfrak{sl}(2,\mathbb{R})$. Using that $\mathfrak{sl}(2,\mathbb{R})$ is a simple Lie algebra, we can compute the Casimir invariant on $\mathfrak{sl}(2,\mathbb{R})^*$ as $e_1 e_3 - e_2^2$ (where e_1, e_2, e_3 can be considered as functions on $\mathfrak{sl}(2,\mathbb{R})$). Proposition 5.19 ensures then that $\tilde{h}_1 \tilde{h}_3 - \tilde{h}_2^2$ Poisson commutes with $\tilde{h}_1, \tilde{h}_2$ and $\tilde{h}_3$. In this way, we obtain a constant of the motion for $\widetilde{X}^{3\mathrm{KS}}$ given by

$$I = \tilde{h}_1 \tilde{h}_3 - \tilde{h}_2^2 = \frac{(a_2 v_1 - a_1 v_2)^2}{v_1^3 v_2^3}. \tag{5.35}$$

Schwarzian equations admit a Lie symmetry $Z = x^2 \partial/\partial x$ [305]. Its prolongation to $\mathrm{T}^2\mathbb{R}$, i.e.,

$$Z_P = x^2 \frac{\partial}{\partial x} + 2vx \frac{\partial}{\partial v} + 2(ax + v^2) \frac{\partial}{\partial a} \tag{5.36}$$

is a Lie symmetry of $X^{3\mathrm{KS}}$. From Proposition 5.13, we get that $\widetilde{Z}_P$ is a Lie symmetry of $\widetilde{X}^{3\mathrm{KS}}$. So, we can construct constants of the motion for $\widetilde{X}^{3\mathrm{KS}}$ by applying $\widetilde{Z}_P$ to any of its t-independent constants of the motion. In particular,

$$F_2 := -\widetilde{Z}_P \log|I| = x_1 + x_2 + \frac{2v_1 v_2 (v_1 - v_2)}{a_2 v_1 - a_1 v_2} \tag{5.37}$$

is constant on particular solutions $(x_{(1)}(t), v_{(1)}(t), a_{(1)}(t), x_{(2)}(t),$ $v_{(2)}(t), a_{(2)}(t))$ of $\widetilde{X}^{3\mathrm{KS}}$. If $(x_{(2)}(t), v_{(2)}(t), a_{(2)}(t))$ is a particular solution for $X^{3\mathrm{KS}}$, its opposite is also a particular solution. So, the function

$$F_3 := x_1 - x_2 + \frac{2v_1 v_2 (v_1 + v_2)}{a_2 v_1 - a_1 v_2} \tag{5.38}$$

is also constant along solutions of $\widetilde{X}^{3\mathrm{KS}}$, i.e., it is a new constant of the motion. In consequence, we get three t-independent constants of the motion: $\Upsilon_1 = I$ and

$$\Upsilon_2 = \frac{F_2 + F_3}{2} = x_1 + \frac{2v_1^2 v_2}{a_2 v_1 - a_1 v_2},$$

$$\Upsilon_3 = \frac{F_2 - F_3}{2} = x_2 - \frac{2v_1 v_2^2}{a_2 v_1 - a_1 v_2}. \tag{5.39}$$

This gives rise to three t-independent constants of the motion for $\widetilde{X}^{3KS}$. Taking into account that $\partial(\Upsilon_1, \Upsilon_2, \Upsilon_3)/\partial(x_1, v_1, a_1) \neq 0$, the expressions $\Upsilon_1 = \lambda_1$, $\Upsilon_2 = \lambda_2$, and $\Upsilon_3 = \lambda_3$ allow us to obtain the expressions of x_1, v_1, a_1 in terms of the remaining variables and $\lambda_1, \lambda_2, \lambda_3$. More specifically,

$$x_1 = \frac{4}{\lambda_1(\lambda_3 - x_2)} + \lambda_2, \quad v_1 = \frac{4v_2}{\lambda_1(\lambda_3 - x_2)^2},$$

$$a_1 = \frac{8v_2^2 + 4a_2(\lambda_3 - x_2)}{\lambda_1(\lambda_3 - x_2)^3}. \tag{5.40}$$

According to the theory of Lie systems [93], the map Φ : $(x_2, v_2, a_2; \lambda_1, \lambda_2, \lambda_3) \in \mathcal{O}_2^2 \times \mathbb{R}^3 \mapsto (x_1, v_1, a_1) \in \mathcal{O}_2^2$ enables us to write the general solution of (5.1) into the form

$$(x(t), v(t), a(t)) = \Phi(x_2(t), v_2(t), a_2(t); \lambda_1, \lambda_2, \lambda_3). \tag{5.41}$$

This is the known superposition rule for Schwarzian equations (in first-order form) derived in [280] by solving a system of PDEs. Meanwhile, our present techniques enable us to obtain the same result without any integration. Note that $x(t)$, the general solution of Schwarzian equations, can be written as $x(t) = \tau \circ \Phi(x_2(t), \lambda_1, \lambda_2, \lambda_3)$, with τ the projection $\tau : (x_2, v_2, a_2) \in T^2\mathbb{R} \mapsto x_2 \in \mathbb{R}$, from a unique particular solution of (5.1), recovering a known feature of these equations [305].

5.6. Bi-Dirac–Lie Systems

It can happen that a Lie system X on a manifold N possesses VG Lie algebras of vector fields with respect to two different Dirac structures. This results in defining two Dirac–Lie systems. For instance, the system of coupled Riccati equations (4.9) admits two associated Dirac–Lie systems [23]: the one previously given, $(\mathcal{O}, L^\omega, X)$, where ω is given by (4.11), and a second one, $(\mathcal{O}, L^{\bar{\omega}}, X)$, with

$$\bar{\omega} = \sum_{i<j=1}^{4} \frac{dx_i \wedge x_j}{(x_i - x_j)^2}. \tag{5.42}$$

Definition 5.20. A *bi-Dirac–Lie system* is a four-tuple (N, L_1, L_2, X), where (N, L_1) and (N, L_2) are two different Dirac manifolds and X is a Lie system on N such that $V^X \subset \mathrm{Ham}(N, L_1) \cap \mathrm{Ham}(N, L_2)$.

Given a Bi-Dirac–Lie system (N, L_1, L_2, X), we can apply indistinctly the methods of the previous sections to (N, L_1, X) and (N, L_2, X) to obtain superposition rules, constants of the motion, and other properties of X. This motivates studies on constructions of this type of structures.

Let us depict a new procedure to build up Bi-Dirac–Lie systems from (N, L^ω, X) whose X possesses a t-independent Lie symmetry Z. This method is a generalization to non-autonomous systems, associated with presymplectic manifolds, of the method for autonomous Hamiltonian systems devised in [123].

Consider a Dirac–Lie system (N, L^ω, X), where ω is a presymplectic structure, and a t-independent Lie symmetry Z of X, i.e., $[Z, X_t] = 0$ for all $t \in \mathbb{R}$. Under the above assumptions, $\omega_Z = \mathcal{L}_Z \omega$ satisfies $d\omega_Z = d\mathcal{L}_Z \omega = \mathcal{L}_Z d\omega = 0$, so (N, ω_Z) is a presymplectic manifold. The vector fields of V^X are still Hamiltonian with respect to (N, ω_Z). Indeed, we can see that Theorem 5.6 ensures that X admits a Dirac–Lie Hamiltonian (N, L^ω, h) and

$$[Z, X_t] = 0 \implies \iota_{X_t} \circ \mathcal{L}_Z = \mathcal{L}_Z \circ \iota_{X_t},$$

$$\Downarrow$$

$$\iota_{X_t} \omega_Z = \iota_{X_t} \mathcal{L}_Z \omega = \mathcal{L}_Z \iota_{X_t} \omega = -\mathcal{L}_Z dh_t = -d(Zh_t), \quad (5.43)$$

$\forall t \in \mathbb{R}$. So, the vector fields $\{X_t\}_{t \in \mathbb{R}}$ are L^{ω_Z}-Hamiltonian. Since the successive Lie brackets and linear combinations of L-Hamiltonian vector fields and elements of V^X are L-Hamiltonian vector fields, the whole Lie algebra V^X is Hamiltonian with respect to ω_Z. Consequently, (N, L^{ω_Z}, X) is also a Dirac–Lie system. Also, (N, L^{ω_Z}, Zh_t) is also a Lie–Hamiltonian for X. Moreover,

$$\{\bar{h}_t, \bar{h}_{t'}\}_{L^{\omega_Z}} = X_t(\bar{h}_{t'}) = X_t(Zh_{t'}) = Z(X_t h_{t'})$$
$$= Z\{h_t, h_{t'}\}_{L^\omega}, \forall t \in \mathbb{R}. \quad (5.44)$$

Summarizing, we have the following proposition.

Proposition 5.21. *If (N, L^ω, X) is a Dirac–Lie system for which X admits a t-independent Lie symmetry Z, then $(N, L^\omega, L^{\mathcal{L}_Z\omega}, X)$ is a bi-Dirac–Lie system. If (N, L^ω, h) is a Dirac–Lie Hamiltonian for X, then $(N, L^{\mathcal{L}_Z\omega}, Zh)$ is a Dirac–Lie Hamiltonian for X and there exists an exact sequence of Lie algebras*

$$(\{h_t\}_{t\in\mathbb{R}}, \{\cdot,\cdot\}_{L^\omega}) \xrightarrow{Z} (\{Zh_t\}_{t\in\mathbb{R}}, \{\cdot,\cdot\}_{L^{\omega_Z}}) \to 0. \qquad (5.45)$$

Note that, given a Lie–Hamilton system (N, L^ω, X), the triple (N, L^{ω_Z}, X) need not be a Lie–Hamilton system: ω_Z may fail to be a symplectic 2-form (cf. [123]). This causes that the theory of Lie–Hamilton systems cannot be applied to study (N, L^{ω_Z}, X), while the methods of our work do.

Let us illustrate the above theory with an example. Recall that Schwarzian equations admit a Lie symmetry $Z = x^2\partial/\partial x$. As a consequence, system (3.94), with $c_0 = 0$, possesses a t-independent Lie symmetry Z_P given by (5.36) and

$$\omega_{Z_P} := \mathcal{L}_{Z_P}\omega_{3KS} = -\frac{2}{v^3}(x dv \wedge da + v da \wedge dx + a dx \wedge dv).$$
$$(5.46)$$

Moreover,

$$\iota_{Y_1}\omega_{Z_P} = -d(Z_P h_1) = -d\left(\frac{4x}{v}\right),$$

$$\iota_{Y_2}\omega_{Z_P} = -d(Z_P h_2) = d\left(2 - \frac{2ax}{v^2}\right),$$

$$\iota_{Y_3}\omega_{Z_P} = -d(Z_P h_3) = d\left(\frac{2a}{v} - \frac{a^2 x}{v^3}\right). \qquad (5.47)$$

So, Y_1, Y_2, and Y_3 are Hamiltonian vector fields with respect to ω_{Z_P}. Moreover, since

$$\{Z_P h_1, Z_P h_2\}_{L^{\omega_{Z_P}}} = Z_P h_1,$$

$$\{Z_P h_2, Z_P h_3\}_{L^{\omega_{Z_P}}} = Z_P h_3,$$

$$\{Z_P h_1, Z_P h_3\}_{L^{\omega_{Z_P}}} = 2Z_P h_2, \qquad (5.48)$$

we see that $Z_P h_1$, $Z_P h_2$, and $Z_P h_3$ span a new finite-dimensional real Lie algebra. So, if $(\mathcal{O}_2, \omega, h)$ is a Lie–Hamiltonian structure for X, then $(\mathcal{O}_2, L^{\omega_{Z_P}}, Z_P h)$ is a Dirac–Lie Hamiltonian for X.

Let us devise a more general method to construct bi-Dirac–Lie systems. Given a Dirac manifold (N, L) and a closed 2-form ω on N, the sections on $TN \oplus_N T^*N$ of the form

$$X + \alpha - \iota_X \omega, \tag{5.49}$$

where $X + \alpha \in \Gamma(L)$, span a new Dirac structure (N, L^ω) [67]. When two Dirac structures are connected by a transformation of this type, it is said that they are *gauge equivalent*. Using this, we can prove the following propositions.

Proposition 5.22. *Let Z be a vector field on N. Then, the Dirac structures L^ω and L^{ω_Z}, with $\omega_Z = \mathcal{L}_Z \omega$, are gauge equivalent.*

Proof. The Dirac structure L^ω is spanned by sections of the form $X - \iota_X \omega$, with $X \in \Gamma(N)$, and the Dirac structure L^{ω_Z} is spanned by sections of the form $X - \iota_X \omega_Z$. Recall that $d\omega = d\omega_Z = 0$. So, L^{ω_Z} is of the form

$$X - \iota_X \omega - \iota_X (\omega_Z - \omega), \quad X - \iota_X \omega \in \Gamma(L^\omega). \tag{5.50}$$

As $d(\omega_Z - \omega) = 0$, then L^ω and L^{ω_Z} are connected by a gauge transformation. $\square$

5.7. Dirac–Lie Systems and Schwarzian–KdV Equations

Let us give some final relevant applications of our methods. In particular, we devise a procedure to construct traveling wave solutions for some relevant nonlinear PDEs by means of Dirac–Lie systems. For simplicity, we hereafter denote the partial derivatives of a function $f : (x_1, \ldots, x_n) \in \mathbb{R}^n \mapsto f(x_1, \ldots, x_n) \in \mathbb{R}$ by $\partial_{x_i} f$.

Example 5.4. Consider the so-called *Schwarzian Korteweg–de Vries equation* (SKdV equation) [18]

$$\{\Phi, x\}\partial_x \Phi = \partial_t \Phi, \tag{5.51}$$

where $\Phi : (t, x) \in \mathbb{R}^2 \to \Phi(t, x) \in \mathbb{R}$ and

$$\{\Phi, x\} := \frac{\partial_x^3 \Phi}{\partial_x \Phi} - \frac{3}{2}\left(\frac{\partial_x^2 \Phi}{\partial_x \Phi}\right)^2, \tag{5.52}$$

which is equivalent to that introduced in (3.109). This PDE has been attracting some attention due to its many interesting properties [18, 190, 313]. For instance, Dorfman established a bi-symplectic structure for this equation [159], and many others have been studying its solutions and generalizations [18, 313]. As a relevant result, we can mention that, given a solution Φ of the SKdV equation, the function $\{\Phi, x\}$ is a particular solution of the Korteweg de Vries equation (KdV equation) [37]

$$\partial_t u = \partial_x^3 u + 3u\partial_x u. \tag{5.53}$$

We now look for traveling wave solutions of (5.51) of the type $\Phi(t, x) = g(x - f(t))$ for a certain fixed t-dependent function f with $df/dt = v_0 \in \mathbb{R}$. Substituting $\Phi = g(x - f(t))$ within (5.51), we obtain that g is a particular solution of the Schwarzian equation

$$\frac{d^3 g}{dz^3} = \frac{3}{2}\frac{(d^2 g/dz^2)^2}{dg/dz} - v_0\frac{dg}{dz}, \tag{5.54}$$

where $z := x - f(t)$. We already know that the Schwarzian equations can be studied through the superposition rule (5.40), which can better be obtained by using that Schwarzian equations can be studied through a Dirac–Lie system, as seen in this section. More specifically, we can generate all their solutions from a known one as

$$g_2(z) = \frac{\alpha g_1(z) + \beta}{\gamma g_1(z) + \delta}, \quad \alpha\delta - \beta\gamma \neq 0, \ \alpha, \beta, \gamma, \delta \in \mathbb{R}. \tag{5.55}$$

In addition, (5.54) is a HODE Lie system, i.e., when written as a first-order system by adding the variables $v = dx/dz$ and $a = dv/dz$, it becomes a Lie system X, namely one of the form (5.32). It can be proven that (5.54) can be integrated for any $v_0 = df/dt$. For instance, particular solutions of this equation read

$$\bar{g}_1(z) = \text{th}\left[\sqrt{v_0/2}z\right] \quad (v_0 > 0), \quad g_1(z) = \frac{1}{z+1} \quad (v_0 = 0). \quad (5.56)$$

Note that $g_1(z)$ has the shape of a solitary stationary solution, i.e., $\lim_{x \to \pm\infty} g_1(x - \lambda_0) = 0$ for every $\lambda_0 \in \mathbb{R}$. Meanwhile, $\bar{g}_1$ is a traveling wave solution. Moreover, the general solution of (5.54) in both cases can be obtained from (5.55).

Chapter 6

Jacobi–Lie Systems

6.1. Introduction

Following the research line of previous chapters, we here study Lie systems with VG Lie algebras of Hamiltonian vector fields with respect to Jacobi manifolds. Roughly speaking, a *Jacobi manifold* is a manifold N endowed with a local Lie algebra $(C^\infty(N), \{\cdot, \cdot\})$ [241, 264, 333]. Understanding that Poisson manifolds are a particular case of Jacobi manifolds, we can consider Jacobi–Lie systems as a generalization of Lie–Hamilton systems. Although each Jacobi manifold gives rise to an associated Dirac manifold, not all Hamiltonian vector fields with respect to the Jacobi manifold become Hamiltonian with respect to its associated Dirac manifold (cf. [150]). Hence, not every Jacobi–Lie system can be straightforwardly understood as a Dirac–Lie system. Even in this case, the Jacobi manifold allows us to construct a Dirac manifold to study the system. The main difference between Jacobi–Lie systems and Lie–Hamilton systems is that Jacobi manifolds do not naturally give rise to Poisson brackets on a space of smooth functions on the manifold, which makes it difficult to prove analogs and/or extensions of the results for Lie–Hamilton systems.

In this section, we extend to Jacobi–Lie systems some of the main structures found for Lie–Hamilton systems, e.g., Lie–Hamiltonians

[121], and we classify all Jacobi–Lie systems on $\mathbb{R}$ and $\mathbb{R}^2$ by determining all VG Lie algebras of Hamiltonian vector fields with respect to Jacobi manifolds on $\mathbb{R}$ and $\mathbb{R}^2$. This is achieved by using the local classification of Lie algebras of vector fields on $\mathbb{R}$ and $\mathbb{R}^2$ derived by Lie [267] and improved by González–López, Kamran, and Olver (GKO) [183] (see also [23]). As a result, we obtain that every Lie system on the real line is a Jacobi–Lie system and we show that every Jacobi–Lie system on the plane admits a VG Lie algebra diffeomorphic to one of the 14 classes indicated in Table A.4 in Appendix A.

6.2.   Jacobi–Lie Systems

We now introduce Jacobi–Lie systems as Lie systems admitting a VG Lie algebra of Hamiltonian vector fields relative to a Jacobi manifold.

Definition 6.1. A *Jacobi–Lie system* (N, Λ, R, X) consists of the Jacobi manifold (N, Λ, R) and a Lie system X admitting a VG Lie algebra of Hamiltonian vector fields with respect to (N, Λ, R).

Example 6.1. We reconsider the *continuous Heisenberg group* [370] given in Chapter 2, which describes the space of matrices

$$
\mathbb{H} = \left\{ \begin{pmatrix} 1 & x & z \\ 0 & 1 & y \\ 0 & 0 & 1 \end{pmatrix} \middle| \, x, y, z \in \mathbb{R} \right\}, \tag{6.1}
$$

endowed with the standard matrix multiplication, where $\{x, y, z\}$ is the natural coordinate system on $\mathbb{H}$ induced by (6.1).

A straightforward calculation shows that the Lie algebra $\mathfrak{h}$ of left-invariant vector fields on $\mathbb{H}$ is spanned by

$$
X_1^L = \frac{\partial}{\partial x}, \quad X_2^L = \frac{\partial}{\partial y} + x \frac{\partial}{\partial z}, \quad X_3^L = \frac{\partial}{\partial z}. \tag{6.2}
$$

Consider now the system on $\mathbb{H}$ given by

$$
\frac{\mathrm{d}h}{\mathrm{d}t} = \sum_{\alpha=1}^{3} b_\alpha(t) X_\alpha^L(h), \quad h \in \mathbb{H}, \tag{6.3}
$$

for arbitrary t-dependent functions $b_1(t), b_2(t)$ and $b_3(t)$. Since the associated t-dependent vector field $X_t^{\mathbb{H}} = b_1(t)X_1^L + b_2(t)X_2^L + b_3(t)X_3^L$ takes values in a finite-dimensional Lie algebra of vector fields, the system $X^{\mathbb{H}}$ is a Lie system. The interest of $X^{\mathbb{H}}$ is due to its appearance in the solution of the so-called *quantum Lie systems* as well as Lie systems admitting a VG Lie algebra isomorphic to $\mathfrak{h}$ [108].

Let us show that system (6.3) gives rise to a Jacobi–Lie system. Consider the bivector field $\Lambda_{\mathbb{H}}$ given by

$$\Lambda_{\mathbb{H}} := -y\frac{\partial}{\partial y} \wedge \frac{\partial}{\partial z} + \frac{\partial}{\partial x} \wedge \frac{\partial}{\partial y}, \tag{6.4}$$

and the vector field $R_{\mathbb{H}} := \partial/\partial z$. Then,

$$X_1^L = [\Lambda_{\mathbb{H}}, -y]_{SN} - yR_{\mathbb{H}}, \quad X_2^L = [\Lambda_{\mathbb{H}}, x]_{SN} + xR_{\mathbb{H}},$$
$$X_3^L = [\Lambda_{\mathbb{H}}, 1]_{SN} + R_{\mathbb{H}}, \tag{6.5}$$

that is, X_1^L, X_2^L, and X_3^L are Hamiltonian vector fields with Hamiltonian functions $h_1 = -y$, $h_2 = x$ and $h_3 = 1$, respectively. Hence, we obtain that $(\mathbb{H}, \Lambda_{\mathbb{H}}, R_{\mathbb{H}}, X^{\mathbb{H}})$ is a Jacobi–Lie system. It is remarkable that each Hamiltonian function h_i is a first integral of X_i^L and $R_{\mathbb{H}}$ for $i = 1, 2, 3$, respectively.

Lemma 6.2. *The space of good Hamiltonian functions $G(N, \Lambda, R)$ with respect to a Jacobi structure (N, Λ, R) is a Poisson algebra with respect to the Jacobi bracket $\{\cdot, \cdot\}_{\Lambda, R}$. Additionally, $\star_g : f \in C^\infty(N) \mapsto \{g, f\}_{\Lambda, R} \in C^\infty(N)$, for any $g \in G(N, \Lambda, R)$, is a derivation on $(C^\infty(N), \cdot)$.*

Proof. First, we prove that the Jacobi bracket of two good Hamiltonian functions is a good Hamiltonian function. For general functions $u_1, u_2 \in C^\infty(N)$, we have

$$R\{u_1, u_2\}_{\Lambda, R} = R(\Lambda(du_1, du_2) + u_1 Ru_2 - u_2 Ru_1). \tag{6.6}$$

If $u_1, u_2 \in G(N, \Lambda, R)$, then $Ru_1 = Ru_2 = 0$. Using this and $[\Lambda, R]_{SN} = 0$, we obtain

$$R\{u_1, u_2\}_{\Lambda,R} = R(\Lambda(\mathrm{d}u_1, \mathrm{d}u_2)) = [R, [[\Lambda, u_1]_{SN}, u_2]_{SN}]_{SN} = 0. \tag{6.7}$$

Hence, $\{u_1, u_2\}_{\Lambda,R} \in G(N, \Lambda, R)$, which becomes a Lie algebra relative to $\{\cdot, \cdot\}_{\Lambda,R}$. Note also that $R(u_1 \cdot u_2) = 0$ and $u_1 \cdot u_2 \in G(N, \Lambda, R)$.

Given an arbitrary $u_1 \in G(N, \Lambda, R)$ and any $u_2, u_3 \in C^\infty(N)$, we have that

$$\begin{aligned}
\star_{u_1}(u_2 \cdot u_3) &= \Lambda(\mathrm{d}u_1, \mathrm{d}(u_2u_3)) + u_1 R(u_2u_3) - u_2u_3 Ru_1 \\
&= X_{u_1}(u_2 \cdot u_3) = u_3 X_{u_1} u_2 + u_2 X_{u_1} u_3 \\
&= u_3 \star_{u_1} u_2 + u_2 \star_{u_1} u_3.
\end{aligned} \tag{6.8}$$

So, $\star_{u_1}$ is a derivation on $(C^\infty(N), \cdot)$ and also on $(G(N, \Lambda, R), \{\cdot, \cdot\}_{\Lambda,R})$. From this, it trivially follows that $(G(N, \Lambda, R), \cdot, \{\cdot, \cdot\}_{\Lambda,R})$ is a Poisson algebra. $\square$

6.3. Jacobi–Lie Hamiltonians

Definition 6.3. We call *Jacobi–Lie Hamiltonian* a quadruple (N, Λ, R, h), where (N, Λ, R) is a Jacobi manifold and $h : (t, x) \in \mathbb{R} \times N \mapsto h_t(x) \in N$ is a t-dependent function such that $(\{h_t\}_{t\in\mathbb{R}}, \{\cdot, \cdot\}_{\Lambda,R})$ is a finite-dimensional real Lie algebra. Given a system X on N, we say that X admits a *Jacobi–Lie Hamiltonian* (N, Λ, R, h) if X_t is a Hamiltonian vector field with Hamiltonian function h_t (with respect to (N, Λ, R)) for each $t \in \mathbb{R}$.

Example 6.2. So, we find that $h_t = b_1(t)h_1 + b_2(t)h_2 + b_3(t)h_3 = -b_1(t)y + b_2(t)x + b_3(t)$ is a Hamiltonian function for each $X_t^{\mathbb{H}}$ in (6.3), with $t \in \mathbb{R}$. In addition,

$$\{h_1, h_2\}_{\Lambda_{\mathbb{H}}, R_{\mathbb{H}}} = h_3, \quad \{h_1, h_3\}_{\Lambda_{\mathbb{H}}, R_{\mathbb{H}}} = 0, \quad \{h_2, h_3\}_{\Lambda_{\mathbb{H}}, R_{\mathbb{H}}} = 0. \tag{6.9}$$

In other words, the functions $\{h_t\}_{t\in\mathbb{R}}$ span a finite-dimensional real Lie algebra of functions with respect to the Poisson bracket (2.3) induced by the Jacobi manifold. Thus, $X^\mathbb{H}$ admits a Jacobi–Lie Hamiltonian $(N, \Lambda_\mathbb{H}, R_\mathbb{H}, h)$.

Theorem 6.4. *Given a Jacobi–Lie Hamiltonian (N, Λ, R, h), the system X of the form $X_t = X_{h_t}$, $\forall t \in \mathbb{R}$, is a Jacobi–Lie system. If X is a Lie system whose $\{X_t\}_{t\in\mathbb{R}}$ are good Hamiltonian vector fields, then it admits a Jacobi–Lie Hamiltonian.*

Proof. Let us prove the direct part. By assumption, the Hamiltonian functions $\{h_t\}_{t\in\mathbb{R}}$ are contained in a finite-dimensional Lie algebra $\mathrm{Lie}(\{h_t\}_{t\in\mathbb{R}}, \{\cdot,\cdot\}_{\Lambda,R})$. The Lie algebra morphism $\phi_{\Lambda,R}$: $f \in C^\infty(N) \mapsto X_f \in \mathrm{Ham}(N, \Lambda, R)$ maps the curve h_t into a curve X_t within $\phi_{\Lambda,R}(\mathrm{Lie}(\{h_t\}_{t\in\mathbb{R}}, \{\cdot,\cdot\}_{\Lambda,R}))$.

Since $\mathrm{Lie}(\{h_t\}_t, \{\cdot,\cdot\}_{\Lambda,R})$ is finite-dimensional and $\phi_{\Lambda,R}$ is a Lie algebra morphism,

$$\phi_{\Lambda,R}(\mathrm{Lie}(\{h_t\}_t, \{\cdot,\cdot\}_{\Lambda,R}))$$

is a finite-dimensional Lie algebra. Hence, X takes values in a finite-dimensional Lie algebra of Hamiltonian vector fields and it is a Jacobi–Lie system.

Let us prove the converse. Since the elements of $\{X_t\}_{t\in\mathbb{R}}$ are good Hamiltonian vector fields by assumption and $\mathrm{Lie}(\{X_t\}_{t\in\mathbb{R}}) = V^X$, every element of V^X is a good Hamiltonian vector field and we can choose a basis $X_1, \ldots, X_r$ of V^X with good Hamiltonian functions $h_1, \ldots, h_r$. The Jacobi bracket $\{h_i, h_j\}_{\Lambda,R}$ is a good Hamiltonian function for $[X_i, X_j]$.

Since $[X_i, X_j] = \sum_{k=1}^r c_{ijk} X_k$ for certain constants c_{ijk}, we obtain that each function,

$$s_{ij} = \{h_i, h_j\}_{\Lambda,R} - \sum_{k=1}^r c_{ijk} h_k, \quad i < j, \tag{6.10}$$

is the difference of two good Hamiltonian functions with the same Hamiltonian vector field. Hence, $\{s_{ij}, h\}_{\Lambda,R} = 0$ for all $h \in C^\infty(N)$. Using this, we obtain that the linear space generated by

$h_1, \ldots, h_r, s_{ij}$, with $1 \leq i < j \leq r$, is a finite-dimensional Lie algebra relative to $\{\cdot, \cdot\}_{\Lambda,R}$. If $X = \sum_{\alpha=1}^{r} b_\alpha(t) X_\alpha$, then $(N, \Lambda, R, h = \sum_{\alpha=1}^{r} b_\alpha(t) h_\alpha)$ is a Jacobi–Lie Hamiltonian for X. $\qquad\square$

One of the relevant properties of the Jacobi–Lie Hamiltonians is given by the following proposition, whose proof is straightforward.

Proposition 6.5. *Let* (N, Λ, R, X) *be a Jacobi–Lie system admitting a Jacobi–Lie Hamiltonian* (N, Λ, R, h) *of good Hamiltonian functions* $\{h_t\}_{t\in\mathbb{R}}$. *A function* $f : N \to \mathbb{R}$ *is a t-independent constant of motion for* X *if and only if* f *commutes with all the elements of* $\mathrm{Lie}(\{h_t\}_{t\in\mathbb{R}}, \{\cdot, \cdot\}_{\Lambda,R})$ *relative to* $\{\cdot, \cdot\}_{\Lambda,R}$.

6.4. Jacobi–Lie Systems on Low-Dimensional Manifolds

Let us prove that every Lie system on the real line gives rise to a Jacobi–Lie system. In the case of two-dimensional manifolds, we display, with the aid of the GKO classification [23, 183] of Lie algebras of vector fields on the plane given in Table A.1 in Appendix A, all the possible VG Lie algebras related to Jacobi–Lie systems on the plane given in Table A.4 in Appendix A.

Example 6.3. Let us show that the *coupled Riccati equation* (4.9) can be associated with a Jacobi–Lie system for $n = 1$, which proves that every Lie system on the real line can be considered as a Jacobi–Lie system. Recall that (4.9) is a Lie system with a VG Lie algebra V spanned by (4.10). Observe that V consists of Hamiltonian vector fields with respect to a Jacobi manifold on $\mathbb{R}$ given by $\Lambda = 0$ and $R = \frac{\partial}{\partial x_1}$. Indeed, the vector fields $X_1, X_2, X_3 \in V$ admit Hamiltonian functions

$$ h_1 = 1, \quad h_2 = x_1, \quad h_3 = x_1^2. \tag{6.11} $$

We now classify Jacobi–Lie systems $(\mathbb{R}^2, \Lambda, R, X)$, where we may assume Λ and R to be locally equal or different from zero. There exists just one Jacobi–Lie system with $\Lambda = 0$ and $R = 0$: $(\mathbb{R}^2, \Lambda = 0,$

$R = 0, X = 0$). Jacobi–Lie systems of the form $(\mathbb{R}^2, \Lambda \neq 0, R = 0)$ are Lie–Hamilton systems, whose VG Guldberg Lie algebras were obtained in [23]. In Table A.4 in Appendix A, we indicate these cases by writing *Poisson*. A Jacobi–Lie system $(\mathbb{R}^2, \Lambda = 0, R \neq 0, X)$ is such that if $Y \in V^X$, then $Y = fR$ for certain $f \in C^\infty(\mathbb{R}^2)$. All cases of this type can be easily obtained out of the bases given in Table A.1 in Appendix A. We describe them by writing $(0, R)$ in the last column.

Propositions 6.9 and 6.10 show that the VG Lie algebras of Table A.4 in Appendix A that do not fall into the mentioned categories are not VG Lie algebras of Hamiltonian vector fields with respect to Jacobi manifolds $(\mathbb{R}^2, \Lambda \neq 0, R \neq 0)$. This means that every $(\mathbb{R}^2, \Lambda, R, X)$ admits a VG Lie algebra belonging to one of the previously mentioned classes.[a]

Lemma 6.6. *Every Jacobi manifold on the plane with $R \neq 0$ and $\Lambda \neq 0$ admits a local coordinate system $\{s, t\}$ where $R = \partial_s$ and $\Lambda = \partial_s \wedge \partial_t$.*

Proof. Since it is assumed $R \neq 0$, there exist local coordinates $\{s, t_0\}$ on which $R = \partial_s$. Meanwhile, we have that $\Lambda = \Lambda(s, t_0)\partial_s \wedge \partial_{t_0}$. Since $[R, \Lambda]_{SN} = 0$, we get that $\partial_s \Lambda = 0$ and $\Lambda = \Lambda(t_0)$. Hence, $\Lambda = \Lambda(t_0)\partial_s \wedge \partial_{t_0}$. As we consider $\Lambda \neq 0$, we can define a new variable $t = t(t_0)$ such that $dt/dt_0 := \Lambda^{-1}(t_0)$. Finally, $\Lambda = \partial_s \wedge \partial_t$. $\square$

Definition 6.7. We call the local coordinate variables $\{s, t\}$ of the above lemma *local rectifying coordinates* of the Jacobi manifold on the plane.

Lemma 6.8. *Let $(\mathbb{R}^2, \Lambda, R)$ be a Jacobi manifold with $R_\xi \neq 0$ and $\Lambda_\xi \neq 0$ at every $\xi \in \mathbb{R}^2$. The Lie algebra morphism $\phi : C^\infty(\mathbb{R}^2) \to \mathrm{Ham}(\mathbb{R}^2, \Lambda, R)$ has non-trivial kernel. On local rectifying coordinates, we have $\ker \phi = \langle e^t \rangle$.*

---

[a]To exclude P_1 with $\alpha \neq 0$, we need a trivial modification of Proposition 6.10 using exactly the same line of thought.

Proof. If $f \in C^\infty(\mathbb{R}^2)$ belongs to $\ker \phi$, then $\widehat{\Lambda}(df) + fR = 0$. In local rectifying coordinates, we get

$$\partial_s f \partial_t - \partial_t f \partial_s + f \partial_s = 0 \Rightarrow \begin{cases} \partial_s f = 0 \\ \partial_t f = f \end{cases} \Rightarrow f = \lambda e^t, \quad \lambda \in \mathbb{R}.$$

$\square$

Proposition 6.9. *If a Lie system on the plane is related to a VG Lie algebra V of vector fields containing two non-zero vector fields X_1, X_2 satisfying $[X_1, X_2] = X_1$ and $X_1 \wedge X_2 = 0$, then V does not consist of Hamiltonian vector fields relative to any Jacobi manifold with $R \neq 0$ and $\Lambda \neq 0$.*

Proof. Let us take a rectifying coordinate system for $(\mathbb{R}^2, \Lambda, R)$. Since ϕ is a morphism of Lie algebras, we get that X_1, X_2 amount to the existence of non-zero functions h_1 and h_2 such that $\{h_1, h_2\}_{\Lambda,R} = h_1 + g$, where $g \in \ker \phi$. So,

$$\{h_1, h_2\}_{\Lambda,R} = \Lambda(dh_1, dh_2) + h_1 R h_2 - h_2 R h_1 = h_1 + g. \quad (6.12)$$

Meanwhile, $X_1 \wedge X_2 = 0$ implies that

$$\widehat{\Lambda}(dh_1) \wedge \widehat{\Lambda}(dh_2) + R \wedge [h_1 \widehat{\Lambda}(dh_2) - h_2 \widehat{\Lambda}(dh_1)] = 0. \quad (6.13)$$

Using local rectifying coordinates, we see that $\widehat{\Lambda}(dh_i) = (Rh_i)\partial_t - \partial_t h_i R$ and $R \wedge \widehat{\Lambda}(dh_i) = (Rh_i)\Lambda$ for $i = 1, 2$. Hence,

$$[(Rh_1)\partial_t - \partial_t h_1 R] \wedge [(Rh_2)\partial_t - \partial_t h_2 R] + [h_1(Rh_2) - h_2(Rh_1)]\Lambda = 0, \quad (6.14)$$

and

$$0 = (Rh_1 \partial_t h_2 - Rh_2 \partial_t h_1)\Lambda + (h_1 Rh_2 - h_2 Rh_1)\Lambda,$$

$$\Updownarrow$$

$$\Lambda(dh_1, dh_2) + h_1 Rh_2 - h_2 Rh_1 = 0. \quad (6.15)$$

This amounts to $\{h_1, h_2\}_{\Lambda,R} = 0$, which implies that $0 = h_1 + g$ and $X_1 = 0$. This is impossible by assumption and X_1 and X_2 cannot be Hamiltonian.

$\square$

$$\text{Jacobi–Lie Systems} \qquad 259$$

Proposition 6.10. *There exists no Jacobi manifold on the plane with $\Lambda \neq 0$ and $R \neq 0$ turning the elements of a Lie algebra diffeomorphic to $\langle \partial_x, \partial_y, x\partial_x + \alpha y\partial_y \rangle$, with $\alpha \notin \{0, -1\}$, into Hamiltonian vector fields.*

Proof. Let us proceed by reduction to absurd. Assume (N, Λ, R) to be a Jacobi manifold turning the above-mentioned vector fields into Hamiltonian ones. Taking local rectifying coordinates for the Jacobi manifold, it turns out that every Lie algebra diffeomorphic to the previous one can be written in terms of three vector fields X_1, X_2, X_3 diffeomorphic to $\partial_x, \partial_y, x\partial_x + \alpha y\partial_y$. Then, they satisfy $[X_1, X_2] = 0, [X_1, X_3] = X_1, [X_2, X_3] = \alpha X_2, X_1 \wedge X_2 \neq 0$ and

$$X_3 = \mu_2 X_1 + \alpha\mu_1 X_2, \qquad (6.16)$$

for certain first integrals μ_1 and μ_2 for X_1 and X_2, respectively. From (6.16) and using that X_1, X_2, X_3 are Hamiltonian for certain Hamiltonian functions h_1, h_2, h_3, correspondingly, we obtain

$$\widehat{\Lambda}(dh_3) + h_3 R = \mu_2 \widehat{\Lambda}(dh_1) + \mu_2 h_1 R + \alpha\mu_1 \widehat{\Lambda}(dh_2) + \alpha\mu_1 h_2 R, \qquad (6.17)$$

and, by means of the rectified expression for Λ and R, we get

$$\frac{\partial h_3}{\partial s} = \mu_2 \frac{\partial h_1}{\partial s} + \alpha\mu_1 \frac{\partial h_2}{\partial s},$$

$$\frac{\partial h_3}{\partial t} = \mu_2 \frac{\partial h_1}{\partial t} + \alpha\mu_1 \frac{\partial h_2}{\partial t} - \mu_2 h_1 - \alpha\mu_1 h_2 + h_3. \qquad (6.18)$$

Since $[X_1, X_3] = X_1$, then $\{h_1, h_3\}_{\Lambda,R} = h_1 + \lambda_1 e^t$, where e^t is a function with zero Hamiltonian vector field and $\lambda_1 \in \mathbb{R}$. Hence,

$$h_1 + \lambda_1 e^t = \{h_1, h_3\}_{\Lambda,R} = \Lambda(dh_1, dh_3) + h_1(Rh_3) - h_3(Rh_1). \qquad (6.19)$$

Simplifying and using previous expressions (6.18), we find that

$$h_1 + \lambda_1 e^t = \mu_1 \left(\frac{\partial h_2}{\partial t} \frac{\partial h_1}{\partial s} - \frac{\partial h_1}{\partial t} \frac{\partial h_1}{\partial s} + h_1 \frac{\partial h_2}{\partial s} - h_2 \frac{\partial h_1}{\partial s} \right)$$

$$= \alpha\mu_1 \{h_1, h_2\}_{\Lambda,R}. \qquad (6.20)$$

As $[X_1, X_2] = 0$, then $\{h_1, h_2\}_{\Lambda,R} = \lambda e^t$ for a certain constant $\lambda \in \mathbb{R}$. Hence, $h_1 = (\alpha \mu_1 \lambda - \lambda_1) e^t$ and $\lambda \neq 0$. Proceeding analogously for $\{h_2, h_3\}_{\Lambda,R} = \alpha(h_2 + \lambda_2 e^t)$, we obtain

$$h_2 + \lambda_2 e^t = \mu_2 \{h_2, h_1\}_{\Lambda,R} \Rightarrow h_2 = (-\mu_2 \lambda/\alpha - \lambda_2) e^t. \qquad (6.21)$$

Writing the compatibility condition for system (6.18), we reach

$$(\alpha + 1)\lambda \left(\frac{\partial \mu_1}{\partial s} \frac{\partial \mu_2}{\partial t} - \frac{\partial \mu_2}{\partial s} \frac{\partial \mu_1}{\partial t} \right) = 0. \qquad (6.22)$$

This implies that $\mathrm{d}\mu_1 \wedge \mathrm{d}\mu_2 = 0$. Since μ_1 is a first integral for X_1 and μ_2 is a first integral for X_2, this means that $X_1 \wedge X_2 = 0$, which is impossible by assumption. This finishes the proof. $\qquad \square$

Chapter 7

k-Symplectic Lie Systems

7.1. Introduction

After studying Lie systems with compatible Poisson, Dirac, and Jacobi structures, we will devote this chapter to studying Lie systems with compatible k-symplectic structures [20, 203, 261]. To motivate their usefulness, recall that while inspecting Dirac–Lie systems in Chapter 6, we described Lie systems admitting VG Lie algebras of Hamiltonian vector fields with respect to several presymplectic forms. Here, we will unveil a new characteristic of many of these Lie systems: the kernels of their compatible presymplectic structures have trivial intersection, i.e., they form a k-symplectic structure [203]. Using this new compatible structure, we will describe Schwarzian equations [90] and coupled Riccati equations [27, 93].

Motivated by previous examples, we will define here a new type of Lie systems, the *k-symplectic Lie systems* [281], possessing VG Lie algebras of Hamiltonian vector fields relative to the presymplectic forms of a k-symplectic structure. Since k-symplectic structures generalize symplectic structures, it will make sense to define, analogous to the case of symplectic structures [78], k-symplectic Hamiltonian functions and k-symplectic vector fields.

Some attention has lately been paid to Lie systems admitting a VG Lie algebra of Hamiltonian vector fields with respect to several geometric structures [27, 50, 90, 121]. Surprisingly, studying these particular types of Lie systems led to the investigation of much more Lie systems and applications than before.

Not until the work [281] had k-symplectic structures been naturally related to Poisson algebras of functions. This has helped in the better understanding of Lie systems. In fact, [281] is one of the first works linking k-symplectic structures with certain Poisson structures on functions, the derived Poisson algebras, which were key to derive superposition rules for k-symplectic Lie systems. This method did not involve integration of PDEs or ODEs as standard methods [108, 376].

So far, k-symplectic geometry has been mainly applied to studying first-order classical field theories [259] and, more particularly, to the Euler–Lagrange and Hamilton–De Donder–Weyl field equations and the study of their constraints, symmetries, conservation laws, reductions, etc. [259]. Instead, this chapter uses k-symplectic structures to investigate ODEs, which is a very different field of study.

In this chapter, we will show that k-symplectic Lie systems can be considered as Dirac–Lie systems in several non-equivalent ways. Despite that, the techniques devised for k-symplectic Lie systems are more powerful, since they allow for the use of non-equivalent Dirac–Lie systems for the same Lie system simultaneously.

To illustrate the versatility of k-symplectic Lie systems, we show that Schwarzian equations [45, 305] can be either analyzed through them or by using Dirac–Lie systems. The k-symplectic Lie system provides a superposition rule for Schwarz equations directly, whereas we showed in the previous chapter that one needs to relate Schwarzian equations with different Dirac–Lie systems and use Lie symmetries to derive a superposition rule with the other approach.

Let us outline the contents of this chapter. Section 7.2 analyzes several remarkable Lie systems admitting a VG Lie algebra of Hamiltonian vector fields relative to a k-symplectic structure. This induces a natural definition of k-symplectic Lie systems. Subsequently, we propose in Section 7.3 a very useful no-go theorem to determine which Lie systems can be considered as k-symplectic Lie systems. Sections 7.4 and 7.5 present some useful geometric structures to study k-symplectic Lie systems. In particular, Section 7.4 introduces Ω-Hamiltonian functions as a generalization of Hamiltonian functions in symplectic geometry [78]. Section 7.5 relates k-symplectic structures with various Poisson algebras: their derived Poisson algebras. Next, we introduce k-symplectic Lie–Hamiltonian structures in Section 7.6, and then, we analyze general properties of k-symplectic Lie systems in Section 7.7. Finally, Section 7.8 illustrates how to derive superposition rules for k-symplectic Lie systems algebraically. It is worth noting that this chapter uses the sign convention of Geometric Mechanics.

7.2. The Need of k-Symplectic Lie Systems

This section describes several Lie systems admitting VG Lie algebras of Hamiltonian vector fields with respect to presymplectic forms of a certain k-symplectic structure. All following examples are resumed, along with other results to be proved in the following sections, in Table 7.1. These examples lead the definition of k-symplectic Lie systems in Section 7.2.7. In the following sections, we study their fundamental properties, as well as their physical and mathematical applications.

7.2.1. *Schwarzian equation*

Consider again the *Schwarzian equation* [45, 305]

$$\{x, t\} = \frac{\mathrm{d}^3 x}{\mathrm{d}t^3} \left(\frac{\mathrm{d}x}{\mathrm{d}t} \right)^{-1} - \frac{3}{2} \left(\frac{\mathrm{d}^2 x}{\mathrm{d}t^2} \right) \left(\frac{\mathrm{d}x}{\mathrm{d}t} \right)^{-2} = 2b_1(t), \qquad (7.1)$$

where $\{x, t\}$ is the so-called *Schwarzian derivative* of the function $x(t)$ relative to the variable t and $b_1(t)$ is any t-dependent function. Schwarz equations can be considered as particular cases of the so-called third-order Kummer–Schwarz equations [87] that appear in the study of iterative differential equations [292] and the analysis

Table 7.1. Lie systems admitting a Lie algebra of Hamiltonian vector fields relative to a k-symplectic form (for further details, see [283]). For simplicity, we define $\omega_{ij} := dx_i \wedge dx_j$, $\partial_{x_i} = \partial_i$.

Application	Basis of vector fields X_i	Ω-Hamiltonian functions h_i	k-symplectic structure ω_i
Superposition rules for Riccati equations $\sum_{\alpha=1}^{3} a_\alpha(t)X_\alpha$	$\sum_{i=1}^{4}\partial_i$	$\left(\frac{1}{x_1-x_2}+\frac{1}{x_3-x_4}\right)\otimes e_1 + \left(\sum_{i<j=1}^{4}\frac{1}{x_i-x_j}\right)\otimes e_2$	$\frac{\omega_{12}}{(x_1-x_2)^2}+\frac{\omega_{34}}{(x_3-x_4)^2}$
	$\sum_{i=1}^{4}x_i\partial_i$	$\frac{1}{2}\left(\frac{x_1+x_2}{x_1-x_2}+\frac{x_3+x_4}{x_3-x_4}\right)\otimes e_1 + \frac{1}{2}\left(\sum_{i<j=1}^{4}\frac{x_i+x_j}{x_i-x_j}\right)\otimes e_2$	$\sum_{i<j=1}^{4}\frac{\omega_{ij}}{(x_i-x_j)^2}$
	$\sum_{i=1}^{4}x_i^2\partial_i$	$\left(\frac{x_1x_2}{x_1-x_2}+\frac{x_3x_4}{x_3-x_4}\right)\otimes e_1 + \left(\sum_{i<j=1}^{4}\frac{x_ix_j}{x_i-x_j}\right)\otimes e_2$	
Control system $\sum_{\alpha=1}^{2} a_\alpha(t)X_\alpha$	∂_1	$x_2\otimes e_1 + x_3\otimes e_2 + x_4\otimes e_3 + \frac{1}{3}x_2^3\otimes e_4$	ω_{12}
	$\partial_2 + x_1(\partial_3 + x_1\partial_4 + 2x_2\partial_5)$	$-x_1\otimes e_1 - \frac{1}{2}x_1^2\otimes e_2 - \frac{1}{3}x_1^3\otimes e_3 + (x_5 - x_1x_2^2)\otimes e_4$	ω_{13}
	$\partial_3 + 2x_1\partial_4 + 2x_2\partial_5$	$x_1\otimes e_2 - x_1^2\otimes e_3 - x_2^2\otimes e_4$	ω_{14}
	∂_4	$-x_1\otimes e_3$	$\omega_{25} + x_2^2\omega_{12}$
	∂_5	$-x_2\otimes e_4$	
Control system $\sum_{\alpha=1}^{2} a_\alpha(t)X_\alpha$	$\partial_1 - x_2\partial_3 + x_2^2\partial_5$	$x_2\otimes e_1 - \frac{1}{3}x_2^3\otimes e_2 + x_4\otimes e_3 + (x_1x_2 + x_3)\otimes e_4$	ω_{12}
	∂_{x_4}	$x_1\otimes e_1 + x_5\otimes e_2 - \frac{1}{3}x_1^3\otimes e_3 - x_1^2\otimes e_4$	ω_{25}
	∂_5	$\frac{1}{2}x_2^2\otimes e_2 - \frac{1}{2}x_1^2\otimes e_3 - x_1\otimes e_4$	ω_{13}
	$\partial_2 + x_1\partial_3 + x_1^2\partial_{x_4}$	$-x_1\otimes e_3$	$\omega_{13} + x_1\omega_{12}$
	$\partial_3 + x_1\partial_{x_4} - x_2\partial_5$	$-x_2\otimes e_2$	
Diffusion equations $\sum_{\alpha=1}^{3} a_\alpha(t)X_\alpha$	$4x_1^2\partial_1 + 4x_1x_2\partial_2 + x_2^2\partial_3$	$\left(4x_1x_3 - 8\frac{x_1^2x_3^2}{x_2^2} - \frac{x_2^2}{2}\right)\otimes e_1 + \left(x_1 - 4\frac{x_1^2x_3}{x_2^2}\right)\otimes e_2$	$\frac{\omega_{23}}{x_2} + \frac{4x_3^2\omega_{12}}{x_2^3} - \frac{4x_3\omega_{13}}{x_2^2}$
	$2x_1\partial_1 + x_2\partial_2$	$-2\frac{x_3^2}{x_2^2}\otimes e_1 - 4\frac{x_3}{x_2^2}\otimes e_2$	$-\frac{4\omega_{13}}{x_2^2} + \frac{8x_3\omega_{12}}{x_2^3}$
	∂_1	$\left(x_3 - 4\frac{x_1x_3^2}{x_2^2}\right)\otimes e_1 - 8\frac{x_1x_3}{x_2^2}\otimes e_2$	
Lotka-Volterra system $a(t)X_1 + b(t)X_2$	$\sum_{i=1}^{5}x_i\partial_i$	$\left(\frac{x_1+x_2}{x_1-x_2}+\frac{x_3+x_4}{x_3-x_4}\right)\otimes e_1 + \left(\frac{x_1+x_2}{x_1-x_2}+\frac{x_3+x_5}{x_3-x_5}\right)\otimes e_2$	$\frac{\omega_{12}}{(x_1-x_2)^2}+\frac{\omega_{34}}{(x_3-x_4)^2}$
		$\left(\frac{x_1+x_2}{x_1-x_2}+\frac{x_4+x_5}{x_4-x_5}\right)\otimes e_3 + \left(\frac{x_1+x_3}{x_1-x_3}+\frac{x_4+x_5}{x_4-x_5}\right)\otimes e_4$	$\frac{\omega_{12}}{(x_1-x_2)^2}+\frac{\omega_{35}}{(x_3-x_5)^2}$
	$\sum_{i=1}^{5}x_i^2\partial_i$	$\left(\frac{x_1x_2}{x_1-x_2}+\frac{x_3x_4}{x_3-x_4}\right)\otimes e_1 + \left(\frac{x_1x_2}{x_1-x_2}+\frac{x_3x_5}{x_3-x_5}\right)\otimes e_2$	$\frac{\omega_{12}}{(x_1-x_2)^2}+\frac{\omega_{45}}{(x_4-x_5)^2}$
		$\left(\frac{x_1x_2}{x_1-x_2}+\frac{x_4x_5}{x_4-x_5}\right)\otimes e_3 + \left(\frac{x_1x_3}{x_1-x_3}+\frac{x_4x_5}{x_4-x_5}\right)\otimes e_4$	$\frac{\omega_{13}}{(x_1-x_3)^2}+\frac{\omega_{45}}{(x_4-x_5)^2}$

of Riccati and second-order Kummer–Schwarz equations [280]. For simplicity, we hereafter assume $b_1(t)$ to be non-constant.

Let us recall the Lie systems approach to Schwarzian equations given in Chapter 6 (see also [87]). The Schwarz equation can be considered as a first-order system of differential equations by adding the variables $v := \mathrm{d}x/\mathrm{d}t$ and $a := \mathrm{d}^2x/\mathrm{d}t^2$ to (7.1), i.e.,

$$\begin{cases} \dfrac{\mathrm{d}x}{\mathrm{d}t} = v, \\[2mm] \dfrac{\mathrm{d}v}{\mathrm{d}t} = a, \\[2mm] \dfrac{\mathrm{d}a}{\mathrm{d}t} = \dfrac{3}{2}\dfrac{a^2}{v} + 2b_1(t)v. \end{cases} \tag{7.2}$$

Let us show that (7.1) is indeed a Lie system on the manifold $\mathcal{O}_2 = \{(x, v, a) \in \mathrm{T}^2\mathbb{R} \,|\, v \neq 0\}$, with $\mathrm{T}^2\mathbb{R}$ being the second tangent bundle to $\mathbb{R}$ [264]. The t-dependent vector field associated with (7.2) reads

$$X_t^{\mathrm{Sc}} = v\frac{\partial}{\partial x} + a\frac{\partial}{\partial v} + \left(\frac{3}{2}\frac{a^2}{v} + 2b_1(t)v\right)\frac{\partial}{\partial a} = X_3 + b_1(t)X_1,$$

where X_1, X_2, X_3 are vector fields on $\mathcal{O}$ given by

$$X_1 = 2v\frac{\partial}{\partial a}, \quad X_2 = v\frac{\partial}{\partial v} + 2a\frac{\partial}{\partial a}, \quad X_3 = v\frac{\partial}{\partial x} + a\frac{\partial}{\partial v} + \frac{3}{2}\frac{a^2}{v}\frac{\partial}{\partial a}. \tag{7.3}$$

Since

$$[X_1, X_2] = X_1, \quad [X_1, X_3] = 2X_2, \quad [X_2, X_3] = X_3, \tag{7.4}$$

the vector fields X_1, X_2, and X_3 span a Lie algebra of vector fields, V^{Sc}, isomorphic to $\mathfrak{sl}(2, \mathbb{R})$. As X^{Sc} takes values in V^{Sc}, one has that X^{Sc} is a Lie system admitting a VG Lie algebra V^{Sc}.

Since the vector fields X_1, X_2, X_3 satisfy that $X_1 \wedge X_2 \wedge X_3 \neq 0$ and $\mathcal{O}_2$ is odd-dimensional, the no-go theorem for Lie-Hamilton systems explains that X^{Sc} cannot be studied through Lie–Hamilton systems when V^{Sc} is indeed a minimal Lie algebra for X^{Sc}. It is immediate to show that this happens if and only if $b_1(t)$ is not constant. In fact, if $b_1(t) = c$ is a constant, then $V^{X^{\mathrm{Sc}}} = \langle X_3 + cX_1 \rangle \neq V^{\mathrm{Sc}}$.

Otherwise, there exist $t_1, t_2 \in \mathbb{R}$ such that $X_{t_1}^{\mathrm{Sc}} \neq X_{t_2}^{\mathrm{Sc}}$ and therefore $V^{X^{\mathrm{Sc}}} \supset \langle X_{t_1}^{\mathrm{Sc}} - X_{t_2}^{\mathrm{Sc}}, X_{t_2}^{\mathrm{Sc}} \rangle$. This implies that $X_1, X_3 \in V^{\mathrm{Sc}}$ and V^{Sc} also contains X_2.

Let us prove that V^{Sc} is a finite-dimensional Lie algebra of Hamiltonian vector fields with respect to the presymplectic forms of a two-symplectic manifold $(\mathcal{O}_2, \omega_1, \omega_2)$. To do so, we look for presymplectic forms ω satisfying that X_1, X_2, and X_3, are Hamiltonian vector fields with respect to it, i.e., $\mathcal{L}_{X_\alpha}\omega = 0$ for $\alpha = 1, 2, 3$ and $d\omega = 0$ (see [189] for an algebraic method to obtain such presymplectic forms without solving systems of PDEs). By solving the latter system of PDEs for ω, we find the presymplectic forms

$$\omega_1 := \frac{dv \wedge da}{v^3}, \quad \omega_2 := -\frac{2}{v^3}(x\,dv \wedge da + v\,da \wedge dx + a\,dx \wedge dv).$$

$$(7.5)$$

As any point $(x, v, a) \in \mathcal{O}_2$ satisfies that $v \neq 0$, one has that ω_1 and ω_2 have constant rank equal to two. Since $\mathcal{O}_2$ is additionally a three-dimensional manifold, $\ker\omega_1$ and $\ker\omega_2$ are distributions of rank one. Moreover, the distribution $\ker\omega_1 \cap \ker\omega_2$ has rank only at points where ω_1 and ω_2 are proportional. It stems from expressions (7.5) that ω_1 and ω_2 are proportional at no point of $\mathcal{O}_2$ and $\ker\omega_1 \cap \ker\omega_2 = \{0\}$. Alternatively, this fact can also be proved by the direct calculation of the kernels of ω_1, ω_2, which read

$$\ker\omega_1 = \left\langle \frac{\partial}{\partial x} \right\rangle, \quad \ker\omega_2 = \left\langle x\frac{\partial}{\partial x} + v\frac{\partial}{\partial v} + a\frac{\partial}{\partial a} \right\rangle. \quad (7.6)$$

Consequently, (ω_1, ω_2) forms a two-symplectic structure.

In addition, X_1, X_2, and X_3 are Hamiltonian vector fields with respect to the presymplectic forms ω_1, ω_2:

$$\iota_{X_1}\omega_1 = d\left(\frac{2}{v}\right), \quad \iota_{X_2}\omega_1 = d\left(\frac{a}{v^2}\right), \quad \iota_{X_3}\omega_1 = d\left(\frac{a^2}{2v^3}\right), \quad (7.7)$$

and

$$\iota_{X_1}\omega_2 = -\mathrm{d}\left(\frac{4x}{v}\right), \quad \iota_{X_2}\omega_2 = \mathrm{d}\left(2 - \frac{2ax}{v^2}\right),$$

$$\iota_{X_3}\omega_2 = \mathrm{d}\left(\frac{2a}{v} - \frac{a^2 x}{v^3}\right). \tag{7.8}$$

The use of the previous two-symplectic structure will allow us to study Schwarz equation through similar techniques to those developed for Lie–Hamilton systems [90]. Although the Schwarzian equations could additionally be studied through the Dirac structure induced by ω_1 or ω_2, it will be clear in the following sections that the use of a two-symplectic structure gives more possibilities of analysis than the use of each one of its presymplectic forms separately.

7.2.2. Riccati systems

Let us now show that the system of Riccati equations

$$\frac{\mathrm{d}x_i}{\mathrm{d}t} = a(t) + b(t)x_i + c(t)x_i^2, \quad i = 1, 2, 3, 4, \tag{7.9}$$

for arbitrary t-dependent functions $a(t), b(t), c(t)$ also admits a VG Lie algebra of Hamiltonian vector fields relative to a two-symplectic structure. Recall that system (7.9) is relevant due to the fact that its t-independent constants of motion can be used to derive a superposition rule for Riccati equations (see Chapter 3).

Let us prove that (7.9) is a Lie system. This can be achieved by using the Lie–Scheffers theorem or by applying the theory of diagonal prolongations explained in Chapter 2.

One the one hand, system (7.9) can be understood as the diagonal prolongation of the t-dependent vector field related to a Riccati equation. As a Riccati equation is a Lie system (see Chapter 3), one has that the prolongation of its associated t-dependent vector field to $\mathbb{R}^4$ is also associated with a Lie system.

Alternatively, one can show that system (7.9) is a Lie system by noting that its associated t-dependent vector field

$$X_t^{\mathrm{Ric}} = a(t)X_1 + b(t)X_2 + c(t)X_3,$$

is such that the vector fields on $\mathbb{R}^4$ of the form

$$X_1 = \sum_{i=1}^{4} \frac{\partial}{\partial x_i}, \quad X_2 = \sum_{i=1}^{4} x_i \frac{\partial}{\partial x_i}, \quad X_3 = \sum_{i=1}^{4} x_i^2 \frac{\partial}{\partial x_i}, \qquad (7.10)$$

have commutation relations

$$[X_1, X_2] = X_1, \quad [X_1, X_3] = 2X_2, \quad [X_2, X_3] = X_3, \qquad (7.11)$$

and span a Lie algebra isomorphic to $\mathfrak{sl}(2, \mathbb{R})$. Hence, X^{Ric} takes values in a VG Lie algebra $V^{\mathrm{Ric}} = \langle X_1, X_2, X_3 \rangle$ and (7.9) becomes a Lie system.

For the sake of simplicity, we will hereafter consider system (7.9) to be restricted to a submanifold $\mathcal{O} := \{(x_1, x_2, x_3, x_4) : \prod_{i<j}(x_i - x_j) \neq 0\}$. Since the restriction of a Lie system on a manifold N to an open subset $U \subset N$ is a Lie system, the restriction of the Lie system (7.9) on $\mathbb{R}^4$ to the open subset $\mathcal{O} \subset \mathbb{R}^4$ is also a Lie system. This restriction will allow the simplification of the restriction through k-symplectic structures.

Let us show that (7.9) admits a VG Lie algebra of Hamiltonian vector fields relative to a k-symplectic structure on $\mathcal{O}$. Let us define the symplectic forms

$$\omega_1 = \frac{\mathrm{d}x_1 \wedge \mathrm{d}x_2}{(x_1 - x_2)^2} + \frac{\mathrm{d}x_3 \wedge \mathrm{d}x_4}{(x_3 - x_4)^2}, \quad \omega_2 = \sum_{\substack{i,j=1 \\ i<j}}^{4} \frac{\mathrm{d}x_i \wedge \mathrm{d}x_j}{(x_i - x_j)^2}. \qquad (7.12)$$

Therefore, $\ker \omega_1 \cap \ker \omega_2 = \{0\}$ and (ω_1, ω_2) becomes a two-symplectic structure on $\mathcal{O}$. It is worth noting that this two-symplectic structure is not well defined away from $\mathcal{O}$ because the denominators in (7.12) are not properly defined.

The vector fields X_1, X_2, and X_3 are Hamiltonian relative to ω_1 and ω_2. In fact,

$$\iota_{X_1}\omega_1 = \mathrm{d}\left(\frac{1}{x_1 - x_2} + \frac{1}{x_3 - x_4}\right), \quad \iota_{X_2}\omega_1 = \frac{1}{2}\mathrm{d}\left(\frac{x_1 + x_2}{x_1 - x_2} + \frac{x_3 + x_4}{x_3 - x_4}\right),$$
$$\iota_{X_3}\omega_1 = \mathrm{d}\left(\frac{x_1 x_2}{x_1 - x_2} + \frac{x_3 x_4}{x_3 - x_4}\right), \tag{7.13}$$

and

$$\iota_{X_1}\omega_2 = \mathrm{d}\left(\sum_{\substack{i,j=1 \\ i<j}}^{4} \frac{1}{x_i - x_j}\right), \quad \iota_{X_2}\omega_2 = \frac{1}{2}\mathrm{d}\left(\sum_{\substack{i,j=1 \\ i<j}}^{4} \frac{x_i + x_j}{x_i - x_j}\right),$$
$$\iota_{X_3}\omega_2 = \mathrm{d}\left(\sum_{\substack{i,j=1 \\ i<j}}^{4} \frac{x_i x_j}{x_i - x_j}\right). \tag{7.14}$$

Hence, V^{Ric} becomes a Lie algebra of Hamiltonian vector fields relative to the k-symplectic structure (ω_1, ω_2).

7.2.3. Control problems

Let us now describe another example given by a control system [293, 316] given by the system of differential equations on $\mathbb{R}^5$ of the form

$$\frac{\mathrm{d}x_1}{\mathrm{d}t} = b_1(t), \quad \frac{\mathrm{d}x_2}{\mathrm{d}t} = b_2(t), \quad \frac{\mathrm{d}x_3}{\mathrm{d}t} = b_2(t)x_1,$$
$$\frac{\mathrm{d}x_4}{\mathrm{d}t} = b_2(t)x_1^2, \quad \frac{\mathrm{d}x_5}{\mathrm{d}t} = 2b_2(t)x_1 x_2, \tag{7.15}$$

where $b_1(t)$ and $b_2(t)$ are arbitrary t-dependent functions. This control system appeared in [293], under a slightly different form, in order to illustrate techniques for the description of the explicit calculation of controls in the types of $\hat{\mathrm{C}}$aplying polynomial systems.

System (7.15) is associated with the t-dependent vector field $X = b_1(t)X_1 + b_2(t)X_2$, where

$$X_1 = \frac{\partial}{\partial x_1}, \quad X_2 = \frac{\partial}{\partial x_2} + x_1\frac{\partial}{\partial x_3} + x_1^2\frac{\partial}{\partial x_4} + 2x_1x_2\frac{\partial}{\partial x_5},$$

$$X_3 = \frac{\partial}{\partial x_3} + 2x_1\frac{\partial}{\partial x_4} + 2x_2\frac{\partial}{\partial x_5}, \quad X_4 = \frac{\partial}{\partial x_4}, \quad X_5 = \frac{\partial}{\partial x_5}, \quad (7.16)$$

span a Lie algebra of vector fields whose non-vanishing commutation relations read

$$[X_1, X_2] = X_3, \quad [X_1, X_3] = 2X_4, \quad [X_2, X_3] = 2X_5. \quad (7.17)$$

Consequently, X is a Lie system as indicated in [316] and it admits a VG Lie algebra V^{cl}. Additionally, V^{cl} consists of Hamiltonian vector fields relative to the presymplectic forms

$$\omega_1 = \mathrm{d}x_1 \wedge \mathrm{d}x_2, \quad \omega_2 = \mathrm{d}x_1 \wedge \mathrm{d}x_3,$$

$$\omega_3 = \mathrm{d}x_1 \wedge \mathrm{d}x_4, \quad \omega_4 = \mathrm{d}x_2 \wedge \mathrm{d}x_5 + x_2^2\mathrm{d}x_1 \wedge \mathrm{d}x_2. \quad (7.18)$$

The kernels of the above presymplectic forms are

$$\ker \omega_1 = \left\langle \frac{\partial}{\partial x_3}, \frac{\partial}{\partial x_4}, \frac{\partial}{\partial x_5} \right\rangle, \quad \ker \omega_2 = \left\langle \frac{\partial}{\partial x_2}, \frac{\partial}{\partial x_4}, \frac{\partial}{\partial x_5} \right\rangle,$$

$$\ker \omega_3 = \left\langle \frac{\partial}{\partial x_2}, \frac{\partial}{\partial x_3}, \frac{\partial}{\partial x_5} \right\rangle, \quad \ker \omega_4 = \left\langle \frac{\partial}{\partial x_3}, \frac{\partial}{\partial x_4}, \frac{\partial}{\partial x_1} + x_2^2\frac{\partial}{\partial x_5} \right\rangle.$$

$$(7.19)$$

Then, $\cap_{i=1}^4 \ker \omega_i = \{0\}$ and $(\omega_1, \ldots, \omega_4)$ becomes a four-symplectic structure on $\mathbb{R}^5$. In addition, $X_1, X_2, X_3, X_4,$ and X_5 are Hamiltonian vector fields with respect to $\omega_1, \ldots, \omega_5$. In fact,

$$\iota_{X_1}\omega_1 = \mathrm{d}x_2, \quad \iota_{X_1}\omega_2 = \mathrm{d}x_3, \quad \iota_{X_1}\omega_3 = \mathrm{d}x_4, \quad \iota_{X_1}\omega_4 = \tfrac{1}{3}\mathrm{d}x_2^3,$$

$$\iota_{X_2}\omega_1 = -\mathrm{d}x_1, \quad \iota_{X_2}\omega_2 = -\tfrac{1}{2}\mathrm{d}x_1^2, \quad \iota_{X_2}\omega_3 = -\tfrac{1}{3}\mathrm{d}x_1^3, \quad \iota_{X_2}\omega_4 = \mathrm{d}(x_5 - x_1x_2^2),$$

$$\iota_{X_3}\omega_1 = 0, \quad \iota_{X_3}\omega_2 = \mathrm{d}x_1, \quad \iota_{X_3}\omega_3 = -\mathrm{d}x_1^2, \quad \iota_{X_3}\omega_4 = -\mathrm{d}x_2^2,$$

$$\iota_{X_4}\omega_1 = 0, \quad \iota_{X_4}\omega_2 = 0, \quad \iota_{X_4}\omega_3 = -\mathrm{d}x_1, \quad \iota_{X_4}\omega_4 = 0,$$

$$\iota_{X_5}\omega_1 = 0, \quad \iota_{X_5}\omega_2 = 0, \quad \iota_{X_5}\omega_3 = 0, \quad \iota_{X_5}\omega_4 = -\mathrm{d}x_2.$$

$$(7.20)$$

Hence, V^{c1} is a VG Lie algebra for (7.15) consisting of Hamiltonian vector fields relative to a four-symplectic structure.

Let us now consider a second control system in $\mathbb{R}^5$ [316] given by

$$\frac{\mathrm{d}x_1}{\mathrm{d}t} = b_1(t), \qquad\qquad \frac{\mathrm{d}x_2}{\mathrm{d}t} = b_2(t),$$

$$\frac{\mathrm{d}x_3}{\mathrm{d}t} = b_2(t)x_1 - b_1(t)x_2, \qquad \frac{\mathrm{d}x_4}{\mathrm{d}t} = b_2(t)x_1^2,$$

$$\frac{\mathrm{d}x_5}{\mathrm{d}t} = b_1(t)x_2^2. \qquad\qquad\qquad\qquad (7.21)$$

The interest of this system can be motivated by its study in [64] so as to solve the problem of determining the controls, namely the functions $b_1(t)$ and $b_2(t)$, allowing the dynamics of (7.21) to connect two points in $\mathbb{R}^5$ while optimizing a cost functional depending on the controls. System (7.21) is, with respect to this optimization problem, equivalent to a plate-ball system (see [64] for details).

System (7.21) is associated with the t-dependent vector field $X_t = b_1(t)X_1 + b_2(t)X_2$, where

$$X_1 = \frac{\partial}{\partial x_1} - x_2\frac{\partial}{\partial x_3} + x_2^2\frac{\partial}{\partial x_5}, \quad X_2 = \frac{\partial}{\partial x_2} + x_1\frac{\partial}{\partial x_3} + x_1^2\frac{\partial}{\partial x_4},$$

$$X_3 = \frac{\partial}{\partial x_3} + x_1\frac{\partial}{\partial x_4} - x_2\frac{\partial}{\partial x_5}, \quad X_4 = \frac{\partial}{\partial x_4}, \quad X_5 = \frac{\partial}{\partial x_5}, \quad (7.22)$$

span a finite-dimensional Lie algebra of vector fields. In fact, the only non-vanishing commutation relations between the previous vector fields read

$$[X_1, X_2] = 2X_3, \quad [X_1, X_3] = X_4, \quad [X_2, X_3] = -X_5. \quad (7.23)$$

Consequently, X is a Lie system with a VG Lie algebra V^{c2}.

Additionally, the elements of V^{c2} are Hamiltonian relative to the presymplectic forms on $\mathbb{R}^5$

$$\omega_1 = \mathrm{d}x_1 \wedge \mathrm{d}x_2, \quad \omega_2 = \mathrm{d}x_2 \wedge \mathrm{d}x_5,$$

$$\omega_3 = \mathrm{d}x_1 \wedge \mathrm{d}x_4, \quad \omega_4 = \mathrm{d}x_1 \wedge \mathrm{d}x_3 + x_1\mathrm{d}x_1 \wedge \mathrm{d}x_2. \quad (7.24)$$

272   *A Guide to Lie Systems with Compatible Geometric Structures*

Since

$$\ker \omega_1 = \left\langle \frac{\partial}{\partial x_3}, \frac{\partial}{\partial x_4}, \frac{\partial}{\partial x_5} \right\rangle, \quad \ker \omega_2 = \left\langle \frac{\partial}{\partial x_1}, \frac{\partial}{\partial x_3}, \frac{\partial}{\partial x_4} \right\rangle,$$

$$\ker \omega_3 = \left\langle \frac{\partial}{\partial x_2}, \frac{\partial}{\partial x_3}, \frac{\partial}{\partial x_5} \right\rangle, \quad \ker \omega_4 = \left\langle \frac{\partial}{\partial x_4}, \frac{\partial}{\partial x_5}, \frac{\partial}{\partial x_2} - x_1 \frac{\partial}{\partial x_3} \right\rangle,$$

$$(7.25)$$

one gets that $\cap_{i=1}^{4} \ker \omega_i = \{0\}$ and $(\omega_1, \ldots, \omega_4)$ becomes a four-symplectic structure on $\mathbb{R}^5$.

Note now that the vector fields X_1, X_2, X_3, X_4, and X_5 are Hamiltonian with respect to $\omega_1, \ldots, \omega_4$. More specifically,

$$
\begin{aligned}
&\iota_{X_1}\omega_1 = \mathrm{d}x_2, \quad \iota_{X_1}\omega_2 = -\tfrac{1}{3}\mathrm{d}x_2^3, \ \iota_{X_1}\omega_3 = \mathrm{d}x_4, \quad \iota_{X_1}\omega_4 = \mathrm{d}(x_1 x_2 + x_3), \\
&\iota_{X_2}\omega_1 = -\mathrm{d}x_1, \ \iota_{X_2}\omega_2 = \mathrm{d}x_5, \quad \iota_{X_2}\omega_3 = -\tfrac{1}{3}\mathrm{d}x_1^3, \ \iota_{X_2}\omega_4 = -\mathrm{d}x_1^2, \\
&\iota_{X_3}\omega_1 = 0, \quad\quad \iota_{X_3}\omega_2 = \tfrac{1}{2}\mathrm{d}x_2^2, \ \iota_{X_3}\omega_3 = -\tfrac{1}{2}\mathrm{d}x_1^2, \ \iota_{X_3}\omega_4 = -\mathrm{d}x_1, \quad (7.26) \\
&\iota_{X_4}\omega_1 = 0, \quad\quad \iota_{X_4}\omega_2 = 0, \quad\quad \iota_{X_4}\omega_3 = -\mathrm{d}x_1, \quad \iota_{X_4}\omega_4 = 0, \\
&\iota_{X_5}\omega_1 = 0, \quad\quad \iota_{X_5}\omega_2 = -\mathrm{d}x_2, \ \iota_{X_5}\omega_3 = 0, \quad\quad \iota_{X_5}\omega_4 = 0.
\end{aligned}
$$

Therefore, V^{c2} becomes a VG Lie algebra for (7.21) consisting of Hamiltonian vector fields relative to a four-symplectic structure.

It is worth noting that the previous examples are only a few cases of all Lie systems admitting a VG Lie algebra of Hamiltonian vector fields relative to a k-symplectic structure that one can find in the works [64, 293].

7.2.4. *Riccati diffusion system*

It was recently proved that diffusion equations and other PDEs can be approached by integrating the following system of differential equations on $\mathbb{R}^3$ (cf. [250, 283, 346]):

$$
\begin{cases}
\dfrac{\mathrm{d}u}{\mathrm{d}t} = -b(t) + 2c(t)u + 4a(t)u^2, \\[2mm]
\dfrac{\mathrm{d}v}{\mathrm{d}t} = (c(t) + 4a(t)u)v, \\[2mm]
\dfrac{\mathrm{d}w}{\mathrm{d}t} = a(t)v^2.
\end{cases}
\qquad (7.27)
$$

In particular, we hereafter restrict ourselves to studying its particular solutions within $\mathcal{O}_2 = \{(u, v, w) \in \mathbb{R}^3 : v \neq 0\}$. Note that the particular solutions of (7.27) with $v(t_0) = 0$ for a certain value $t_0 \in \mathbb{R}$ satisfy that $v(t) = 0$ for every value of t and makes $v(t)$ to be t-independent. Hence, system (7.27) becomes essentially a Riccati equation, which can be studied separately. In consequence, it makes sense to study (7.27) on $\mathcal{O}$. Moreover, this condition will allow us to simplify the further description of (7.27) through k-symplectic structures.

System (7.27) is a Lie system (cf. [90]). In fact, system (7.27) describes the integral curves of the t-dependent vector field on $\mathcal{O}$ given by $X^{\mathrm{RS}} = a(t)X_1 - b(t)X_2 + c(t)X_3$, where

$$X_1 = 4u^2 \frac{\partial}{\partial u} + 4uv \frac{\partial}{\partial v} + v^2 \frac{\partial}{\partial w}, \quad X_2 = \frac{\partial}{\partial u}, \quad X_3 = 2u \frac{\partial}{\partial u} + v \frac{\partial}{\partial v},$$

$$\tag{7.28}$$

satisfy the commutation relations

$$[X_1, X_2] = -4X_3, \quad [X_1, X_3] = -2X_1, \quad [X_2, X_3] = 2X_2. \tag{7.29}$$

Then, system (7.27) admits a VG Lie algebra $V^{\mathrm{Ricc}} = \langle X_1, X_2, X_3 \rangle$.

Let us prove that V^{Ricc} consists of Hamiltonian vector fields relative to a two-symplectic structure on $\mathcal{O}_2$. Consider the presymplectic forms

$$\omega_{\mathrm{RS}-1} = -\frac{4wdu \wedge dw}{v^2} + \frac{dv \wedge dw}{v} + \frac{4w^2 du \wedge dv}{v^3},$$

$$\omega_{\mathrm{RS}-2} = -\frac{4du \wedge dw}{v^2} + \frac{8wdu \wedge dv}{v^3}. \tag{7.30}$$

These presymplectic forms are well defined because $v \neq 0$ on $\mathcal{O}_2$. Their kernels read

$$\ker \omega_{\mathrm{RS}-1} = \left\langle v^2 \frac{\partial}{\partial u} + 4wv \frac{\partial}{\partial v} + 4w^2 \frac{\partial}{\partial w} \right\rangle,$$

$$\ker \omega_{\mathrm{RS}-2} = \left\langle v \frac{\partial}{\partial v} + 2w \frac{\partial}{\partial w} \right\rangle, \tag{7.31}$$

and $\ker \omega_{\mathrm{RS}-1} \cap \ker \omega_{\mathrm{RS}-2} = \{0\}$. Alternatively, one could also prove this fact as in Section 7.2.1. The presymplectic forms $\omega_{\mathrm{RS}-i}$, with

$i = 1, 2$, have constant rank two and $\mathcal{O}_2$ is a three-dimensional manifold. Therefore, $\ker \omega_{\mathrm{RS}-1} \cap \ker \omega_{\mathrm{RS}-2}$ will be a distribution of rank one only in points where $\omega_{\mathrm{RS}-i}$ are proportional. Otherwise, $\ker \omega_{\mathrm{RS}-1} \cap \ker \omega_{\mathrm{RS}-2} = \{0\}$. Since $\omega_{\mathrm{RS}-1}$ and $\omega_{\mathrm{RS}-2}$ in (7.30) are not proportional at any point, one has that $\ker \omega_{\mathrm{RS}-1} \cap \ker \omega_{\mathrm{RS}-2} = \{0\}$.

Using any of the above proofs, we get finally that $(\omega_{\mathrm{RS}-1}, \omega_{\mathrm{RS}-2})$ is a two-symplectic structure on $\mathbb{R}^3$.

Moreover, X_1, X_2, and X_3 are Hamiltonian vector fields with respect to $\omega_{\mathrm{RS}-1}$, $\omega_{\mathrm{RS}-2}$:

$$\iota_{X_1}\omega_{\mathrm{RS}-1} = \mathrm{d}\left(4uw - 8\frac{u^2w^2}{v^2} - \frac{v^2}{2}\right), \ \iota_{X_2}\omega_{\mathrm{RS}-1}$$

$$= -2\mathrm{d}\left(\frac{w^2}{v^2}\right), \iota_{X_3}\omega_{\mathrm{RS}-1} = \mathrm{d}\left(w - 4\frac{uw^2}{v^2}\right), \quad (7.32)$$

$$\iota_{X_1}\omega_{\mathrm{RS}-2} = 4\,\mathrm{d}\left(u - 4\frac{u^2w}{v^2}\right), \ \iota_{X_2}\omega_{\mathrm{RS}-2} = -4\,\mathrm{d}\left(\frac{w}{v^2}\right),$$

$$\iota_{X_3}\omega_{\mathrm{RS}-2} = -8\,\mathrm{d}\left(\frac{uw}{v^2}\right). \quad (7.33)$$

Then, V^{Ric} is a VG Lie algebra consisting of Hamiltonian vector fields relative to the presymplectic forms of $(\omega_{\mathrm{RS}-1}, \omega_{\mathrm{RS}-2})$.

7.2.5. Lotka–Volterra systems

Let us consider a family of Lotka–Volterra systems that can be studied as Lie systems, the *Lie–Lotka–Volterra systems* [283]. More specifically, consider the system on $\mathcal{F} = \{(x_1, x_2, x_3, x_4, x_5) \in \mathbb{R}^5 : \prod_{i<j=1}^{5}(x_i - x_j) \neq 0\} \subset \mathbb{R}^5$ of the form

$$\frac{\mathrm{d}x_1}{\mathrm{d}t} = a(t)x_1 + b(t)x_1^2, \quad \frac{\mathrm{d}x_2}{\mathrm{d}t} = a(t)x_2 + b(t)x_2^2,$$

$$\frac{\mathrm{d}x_3}{\mathrm{d}t} = a(t)x_3 + b(t)x_3^2, \quad \frac{\mathrm{d}x_4}{\mathrm{d}t} = a(t)x_4 + b(t)x_4^2,$$

$$\frac{\mathrm{d}x_5}{\mathrm{d}t} = a(t)x_5 + b(t)x_5^2. \quad (7.34)$$

This system is associated with the t-dependent vector field given by $X = a(t)X_1 + b(t)X_2$, where

$$X_1 = \sum_{i=1}^{5} x_i \frac{\partial}{\partial x_i}, \quad X_2 = \sum_{i=1}^{5} x_i^2 \frac{\partial}{\partial x_i}, \tag{7.35}$$

and $[X_1, X_2] = X_2$. Therefore, X admits a VG Lie algebra V^{Con} isomorphic to the Lie algebra of affine transformations on the real line. Let us show that this system is a four-symplectic Lie system. Consider the presymplectic forms on $\mathcal{F}$ given by

$$\omega_1 = \frac{\mathrm{d}x_1 \wedge \mathrm{d}x_2}{(x_1 - x_2)^2} + \frac{\mathrm{d}x_3 \wedge \mathrm{d}x_4}{(x_3 - x_4)^2}, \quad \omega_2 = \frac{\mathrm{d}x_1 \wedge \mathrm{d}x_2}{(x_1 - x_2)^2} + \frac{\mathrm{d}x_3 \wedge \mathrm{d}x_5}{(x_3 - x_5)^2},$$

$$\omega_3 = \frac{\mathrm{d}x_1 \wedge \mathrm{d}x_2}{(x_1 - x_2)^2} + \frac{\mathrm{d}x_4 \wedge \mathrm{d}x_5}{(x_4 - x_5)^2}, \quad \omega_4 = \frac{\mathrm{d}x_1 \wedge \mathrm{d}x_3}{(x_1 - x_3)^2} + \frac{\mathrm{d}x_4 \wedge \mathrm{d}x_5}{(x_4 - x_5)^2}.$$

$$\tag{7.36}$$

Observe that the definition of $\mathcal{F}$ allows indeed for the proper definition of all previous presymplectic forms.

Since

$$\ker \omega_1 = \left\langle \frac{\partial}{\partial x_5} \right\rangle, \quad \ker \omega_2 = \left\langle \frac{\partial}{\partial x_4} \right\rangle, \quad \ker \omega_3 = \left\langle \frac{\partial}{\partial x_3} \right\rangle,$$

$$\ker \omega_4 = \left\langle \frac{\partial}{\partial x_2} \right\rangle, \tag{7.37}$$

one gets that $\cap_{i=1}^{4} \ker \omega_i = \{0\}$ and $(\omega_1, \omega_2, \omega_3, \omega_4)$ is a four-symplectic structure on $\mathcal{F}$.

Additionally, X_1 and X_2 are Hamiltonian vector fields with respect to ω_i with $i = 1, \ldots, 4$:

$$\iota_{X_1}\omega_1 = \frac{1}{2}\mathrm{d}\left(\frac{x_1 + x_2}{x_1 - x_2} + \frac{x_3 + x_4}{x_3 - x_4} \right), \quad \iota_{X_2}\omega_1 = \mathrm{d}\left(\frac{x_1 x_2}{x_1 - x_2} + \frac{x_3 x_4}{x_3 - x_4} \right),$$

$$\iota_{X_1}\omega_2 = \frac{1}{2}\mathrm{d}\left(\frac{x_1 + x_2}{x_1 - x_2} + \frac{x_3 + x_5}{x_3 - x_5} \right), \quad \iota_{X_2}\omega_2 = \mathrm{d}\left(\frac{x_1 x_2}{x_1 - x_2} + \frac{x_3 x_5}{x_3 - x_5} \right),$$

$$\iota_{X_1}\omega_3 = \frac{1}{2}\mathrm{d}\left(\frac{x_1+x_2}{x_1-x_2}+\frac{x_4+x_5}{x_4-x_5}\right), \quad \iota_{X_2}\omega_3 = \mathrm{d}\left(\frac{x_1x_2}{x_1-x_2}+\frac{x_4x_5}{x_4-x_5}\right),$$

$$\iota_{X_1}\omega_4 = \frac{1}{2}\mathrm{d}\left(\frac{x_1+x_3}{x_1-x_3}+\frac{x_4+x_5}{x_4-x_5}\right), \quad \iota_{X_2}\omega_4 = \mathrm{d}\left(\frac{x_1x_3}{x_1-x_3}+\frac{x_4x_5}{x_4-x_5}\right).$$

$$(7.38)$$

Hence, V^{Con} is a VG Lie algebra of four-Hamiltonian vector fields relative to $(\omega_1,\ldots,\omega_4)$.

7.2.6. A k-symplectic Lie system on $SL(2,\mathbb{R})$

This section concerns a type of Lie system on a Lie group that admits a VG Lie algebra of Hamiltonian vector fields relative to a k-symplectic structure that additionally can retrieve, as particular cases, some of the previous examples.

Consider the Lie group $SL(2,\mathbb{R})$ and a basis of right-invariant vector fields X_1^R, X_2^R, X_3^R. Then, we define the Lie system X^R on $SL(2,\mathbb{R})$ of the form

$$\frac{\mathrm{d}g}{\mathrm{d}t} = \sum_{\alpha=1}^{3} b_\alpha(t)X_\alpha^R(g), \quad g \in G, \tag{7.39}$$

where the functions $b_1(t), b_2(t), b_3(t)$ are arbitrary. The relevance of (7.39) is due to the fact that the integration of general Lie systems admitting a VG Lie algebra isomorphic to $\mathfrak{sl}(2,\mathbb{R})$ can be reduced to obtain a particular solution of (7.39) (see Chapter 2 or [91, 108]). Let us assume, with no loose of generality, that the elements of the base satisfy

$$[X_1^R, X_2^R] = X_1^R, \quad [X_1^R, X_3^R] = 2X_2^R, \quad [X_2^R, X_3^R] = X_3^R. \tag{7.40}$$

To construct a compatible k-symplectic structure, one can recall that $SL(2,\mathbb{R})$ has a basis of left-invariant vector fields, X_1^L, X_2^L, X_3^L. Assume that $X_i^L(e) = X_i^R(e)$ for $i = 1,2,3$, hence

$$[X_1^L, X_2^L] = -X_1^L, \quad [X_1^L, X_3^L] = -2X_2^L, \quad [X_2^L, X_3^L] = -X_3^L. \tag{7.41}$$

The dual differential 1-forms, $\eta_1^L, \eta_2^L, \eta_3^L$ are left-invariant. Since $\mathrm{d}\eta_i^L(X_j^L, X_k^L) = X_j^L \eta_i^L(X_k^L) - X_k^L \eta_i^L(X_j^L) - \eta_i^L([X_j^L, X_j^L])$ by the standard formula for the exterior differential, one gets

$$\mathrm{d}\eta_1^L = -\eta_1^L \wedge \eta_2^L, \quad \mathrm{d}\eta_2^L = -2\eta_1^L \wedge \eta_3^L, \quad \mathrm{d}\eta_3^L = -\eta_2^L \wedge \eta_3^L. \quad (7.42)$$

Therefore, $\omega_{G1} = \eta_1^L \wedge \eta_2^L$, $\omega_{G2} = \eta_1^L \wedge \eta_3^L$ are presymplectic forms of rank two and their kernels, $\ker \omega_{G1} = \langle X_3^L \rangle$, $\ker \omega_{G2} = \langle X_2^L \rangle$, satisfy that $\ker \omega_{G1} \cap \ker \omega_{G2} = \{0\}$. Hence, $(\omega_{G1}, \omega_{G2})$ is a two-symplectic structure on $SL(2, \mathbb{R})$. Similarly, $(\omega_{G1}, \omega_{G3})$ and $(\omega_{G2}, \omega_{G3})$ are also two-symplectic structures on $SL(2, \mathbb{R})$.

Additionally, one has that $\mathrm{d}\iota_{X_\alpha^R} \omega_{G1} = \mathcal{L}_{X_\alpha^R} \omega_{G1} = 0$, for $\alpha = 1, 2, 3$. Therefore, $\iota_{X_\alpha^R} \omega_{G1}$ is closed and, locally around the neutral element e of $SL(2, \mathbb{R})$, the vector fields X_α^R are locally Hamiltonian.

Let us use the above to prove the following proposition, which explains why many Lie systems admit a VG Lie algebra of Hamiltonian vector fields relative to the presymplectic forms of a k-symplectic form.

Proposition 7.1. *Every Lie system X on a three-dimensional manifold N admitting a VG Lie algebra V isomorphic to $\mathfrak{sl}(2, \mathbb{R})$ and such that $\mathcal{D}^V = TN$ is such that V is a Lie algebra of k-symplectic vector fields locally around every $x \in N$.*

Proof. Since V is isomorphic to $\mathfrak{sl}(2, \mathbb{R})$, one has that V admits a basis Y_1, Y_2, Y_3, satisfying that

$$[Y_1, Y_2] = Y_1, \quad [Y_1, Y_3] = 2Y_2, \quad [Y_2, Y_3] = Y_3. \quad (7.43)$$

Since the elements of V span TN, one has that $Y_1 \wedge Y_2 \wedge Y_3 \neq 0$. Therefore, one can integrate these vector fields to give rise to a local Lie group action $\Psi : SL(2, \mathbb{R}) \times \mathcal{O}_2 \to \mathcal{O}_2$ whose fundamental vector fields are spanned by V^{Ricc}. Since the vector fields of V^{Ricc} span $T\mathcal{O}_2$, the orbits of Ψ are indeed open subsets of $\mathcal{O}_2$. Hence, the map $\Psi_\xi : g \in SL(2, \mathbb{R}) \mapsto \Psi(g, \xi) \in \mathcal{O}_2$ is a local diffeomorphism onto the orbits of Φ which maps a basis of right-invariant vector fields on G

to the vector fields Y_1, Y_2, Y_3:

$$(\Phi_{x*g}X_i^R)_g = \frac{d}{dt}\Big|_{t=0} \Phi_x(\exp(tv_i)g)$$

$$= \frac{d}{dt}\Big|_{t=0} \Phi(\exp(tv_i)g, x) = \frac{d}{dt}\Big|_{t=0} \Phi(\exp(tv_i), \Phi(g, x))$$

$$= (Y_i)_{\Phi_x(g)}.$$

In consequence, $\Phi_{x*}^{-1}X$ is a t-dependent vector field of the form (7.39). Since the latter is a k-symplectic Lie system relative to some two-symplectic structure (ω_1, ω_2) on an open U close to the neutral element $e \in SL(2, \mathbb{R})$, the diffeomorphism Φ_x ensures that X is also a k-symplectic Lie system relative to $(\Phi_x^{-1*}\omega_1, \Phi^{-1*}\omega_2)$ relative to an open subset containing x. $\qquad\square$

Note that the above proposition can be applied to the case of the Riccati diffusion system (7.27) and (7.2).

7.2.7. Definition of k-symplectic Lie systems

Lie systems described in Sections 7.2.1–7.2.6 possess a VG Lie algebra of Hamiltonian vector fields relative to the presymplectic forms of a k-symplectic structure. This suggests us the following definition.

Definition 7.2. Given a k-symplectic structure $(\omega_1, \ldots, \omega_k)$ on an $n = p(k+1)$-dimensional manifold N, we say that a vector field Y on N is k-*Hamiltonian* if it is a Hamiltonian vector field with respect to the presymplectic forms $\omega_1, \ldots, \omega_k$.

Then, X is a k-Hamiltonian vector field if and only if it is Hamiltonian for all the presymplectic forms of the space $\langle \omega_1, \ldots, \omega_k \rangle$. In view of Theorem 2.6, two k-symplectic structures $(\omega_1, \ldots, \omega_k)$ and $(\omega'_1, \ldots, \omega'_k)$ are equivalent if and only if $\langle \omega_1, \ldots, \omega_k \rangle = \langle \omega'_1, \ldots, \omega'_k \rangle$. So, if X is k-Hamiltonian for a k-symplectic manifold, then it is also k-Hamiltonian for all equivalent k-symplectic manifolds. It also makes sense to say that X is Ω-Hamiltonian for a polysymplectic form Ω if X is k-Hamiltonian for a k-symplectic manifold possessing Ω as associated polysymplectic form. From now on, we will talk about *k-Hamiltonian* and/or *Ω-Hamiltonian vector fields* indistinctly. We write $\mathrm{Ham}(\Omega)$, where Ω is a polysymplectic form induced by $(\omega_1, \ldots, \omega_k)$ for the space of k-Hamiltonian vector fields.

Now, it makes sense to define the following notion of k-symplectic Lie system.

Definition 7.3. A *k-symplectic Lie system* is a Lie system whose V^X consists of k-Hamiltonian vector fields with respect to a k-symplectic structure $(\omega_1, \ldots, \omega_k)$. We call $(\omega_1, \ldots, \omega_k)$ a *compatible* k-symplectic structure of X.

Since every k-symplectic structure amounts to a polysymplectic one and vice versa, it also makes sense to talk about polysymplectic Lie systems as those Lie systems admitting a minimal Lie algebra consisting of Ω-Hamiltonian vector fields relative to a polysymplectic form Ω.

Definition 7.3 is equivalent to saying that a k-symplectic Lie system X on a manifold N is a Lie system admitting a VG Lie algebra of k-Hamiltonian vector fields with respect to a k-symplectic structure on N. Consequently, Lie–Hamilton systems [121] can be regarded as k-symplectic Lie systems. On the other hand, as illustrated by Schwarz equations (7.2), not every k-symplectic Lie system is a Lie–Hamilton system (for more details, see [90]).

Every k-symplectic Lie system can be considered as a Dirac–Lie system [90]. Let us explain this fact. If X is a k-symplectic Lie system relative to the k-symplectic structure $(\omega_1, \ldots, \omega_k)$, then V^X is a family of Hamiltonian vector fields with respect to the presymplectic forms $\omega_1, \ldots, \omega_k$. Consequently, V^X is a Lie algebra of Hamiltonian vector fields relative to each Dirac structure L^{ω_i} given by the graph of the $\omega_i^\flat$ (see Chapter 5 and [90, 151] for details). Then, the triple (N, L^{ω_r}, X) is a Dirac–Lie system.

On the other hand, not every Dirac–Lie system can be considered as a k-symplectic Lie system. For instance, a Lie system given by an autonomous vector field $X \neq 0$ on the real line can be considered as a Dirac–Lie system relative to $(\mathbb{R}, \mathfrak{X}(N), X)$, while there is no k-symplectic structure on the real line and, therefore, X cannot be considered as a k-symplectic Lie system.

It is worth noting that the main advantage of k-symplectic Lie systems is that they can be considered as Dirac–Lie systems in different ways. This suggests us to find a natural approach to the study of these systems, which is given by k-symplectic structures.

7.3. A No-Go Theorem for k-Symplectic Lie Systems

Determining whether a Lie system X is a k-symplectic Lie system generally requires solving a system of PDEs to find a k-symplectic structure compatible with the smallest VG Lie algebra, V^X, related to the Lie system. More specifically, if we want to ensure that X is a k-symplectic Lie system, we have to find k solutions, $\omega_1, \ldots, \omega_k$ of

$$\mathcal{L}_Y \omega = 0, \quad \forall Y \in V^X, \tag{7.44}$$

such that $\cap_{i=1}^k \ker \omega_i = \{0\}$. In many cases, it can be difficult to establish whether the above system of PDEs has enough solutions to generate a k-symplectic structure. This is why it is important to find simple necessary and/or sufficient conditions to ensure or to discard that a Lie system is a k-symplectic Lie system.

This section provides a no-go theorem detailing conditions ensuring that a Lie system is not a k-symplectic Lie system. The main idea is that the smallest Lie algebra of the Lie system under study must leave stable, in the sense given next, the kernels of the presymplectic forms of any k-symplectic structure compatible with the Lie system. This condition is easier to verify than finding a compatible k-symplectic structure. Although we here provide only one main result, it is easy to develop further no-go theorems from our ideas.

Let us start giving a standard notion in the literature [183].

Definition 7.4. A distribution $\mathcal{D}$ is *stable* under the action of a Lie algebra V of vector fields when $[X, Y] \in \mathcal{D}$ for every $Y \in \mathcal{D}$ and $X \in V$.

Meanwhile, the following definition is a generalization of the notion of a primitive Lie algebra of vector fields on the plane given in [183] adapted so as to prove our no-go theorem.

Definition 7.5. A VG Lie algebra V on N is *s-primitive* when there exists no distribution $\mathcal{D}$ of rank s stable under the action of V. We call V *odd–primitive* when V is s-primitive for every odd value of $s < \dim N$.

Note 7.6. The definition of s-primitive VG Lie algebra is a generalization introduced in [281] of the notion of a primitive Lie algebra of vector fields on the plane given in [183].

Let us now provide a sufficient condition to ensure that a Lie system is not a k-symplectic one.

Theorem 7.7 (No-go k-symplectic Lie systems theorem). *If X is a Lie system on an odd-dimensional manifold N and V^X is odd-primitive, then X is not a k-symplectic Lie system.*

Proof. Let us suppose that X admits a compatible k-symplectic structure $(\omega_1, \ldots, \omega_k)$. On an odd-dimensional manifold, every 2-form of the k-symplectic structure has non-trivial odd-dimensional kernel. Let $Z \neq 0$ be a vector field $Z \in \ker \omega_i$. As the elements of V^X are Hamiltonian with respect to $\omega_1, \ldots, \omega_k$, we have that $\mathcal{L}_Y \omega_i = 0$ for every $Y \in V^X$ and

$$0 = \mathcal{L}_Y \iota_Z \omega_i = \iota_Z \mathcal{L}_Y \omega_i + \iota_{[Y,Z]} \omega_i = \iota_{[Y,Z]} \omega_i. \qquad (7.45)$$

Therefore, $[Y, Z] \in \ker \omega_i$ and the kernel of ω_i is a generalized distribution, which is a regular one around neighborhoods of generic points, that is stable under the action of the elements of V^X. As V^X is odd-primitive and $\ker \omega_i$ is odd-dimensional, this is a contradiction. Then, X has no compatible k-symplectic structure. $\qquad \square$

282   *A Guide to Lie Systems with Compatible Geometric Structures*

Example 7.1 (Lie systems on Lie groups). A specially relevant type of Lie systems related to the integrability of Lie systems (see [108]) are of the form

$$\frac{\mathrm{d}g}{\mathrm{d}t} = \sum_{\alpha=1}^{r} \left(b_\alpha^R(t) X_\alpha^R(g) + b_\alpha^L(t) X_\alpha^L(g) \right), \quad g \in G, \tag{7.46}$$

where G is a Lie group, $X_1^R, \ldots, X_r^R$ and $X_1^L, \ldots, X_r^L$ are bases of right- and left-invariant vector fields on G, respectively, and $b_1^L(t), \ldots, b_r^L(t), b_1^R(t), \ldots, b_r^R(t)$ are arbitrary t-dependent functions. Systems of the form (7.46) are relevant as they appear in the study of the integrability of Lie systems (see [108]).

An interesting question is to determine if systems (7.46) can be endowed with a compatible k-symplectic structure. As proved next, the answer is negative for a large subclass.

Assume that G is odd-dimensional and its Lie algebra, $\mathfrak{g}$, has no proper odd-dimensional ideals, e.g., $\mathfrak{g}$ is simple. Suppose also that the minimal Lie algebra V^X of the system X given by (7.46) is isomorphic (as a Lie algebra) to $\mathfrak{g}$. Let us prove that the V^X is odd-primitive by reduction to absurd.

Consider an odd-dimensional distribution $\mathcal{D}$ on G invariant under the action of V^X. Since G is connected, the invariance of $\mathcal{D}$ implies that the vector fields taking values in $\mathcal{D}$ are invariant under the diffeomorphisms given by the right-hand multiplications on the group, namely the mappings $R_g : g' \in G \mapsto g' \cdot g \in G$ with $g \in G$. Hence, given $\mathcal{D}_e \subset T_e G$, i.e., the subspace of the distribution $\mathcal{D}$ at the neutral element e of G, we obtain that $R_{g*e} Y_e \in \mathcal{D}_g$ for every $Y_e \in \mathcal{D}_e$. Indeed, since R_g is a diffeomorphism, then $\mathcal{D}_e \simeq \mathcal{D}_g$.

If Y^R is a right-invariant vector field on G with $Y_e^R = Y$, we have that $Y_g^R = R_{g*e} Y_e^R \in \mathcal{D}_g$, for every $g \in G$. Consequently, Y^R takes values in $\mathcal{D}$. As each R_g is a diffeomorphism, the vector fields $Y_1^R, \ldots, Y_s^R$ whose values $(Y_1^R)_e, \ldots, (Y_s^R)_e$ form a basis for $\mathcal{D}_e$, then $(Y_1^R)_g, \ldots, (Y_s^R)_g$ form a basis of $\mathcal{D}_g$ for every $g \in G$. Consequently, $\mathcal{D}$ admits a global basis of right-invariant vector fields $Y_1^R, \ldots, Y_s^R$. As $\mathcal{D}$ is invariant under V^X, then $\mathcal{L}_{X_\alpha^R} Y_j^R$, with $\alpha = 1, \ldots, \dim G$

and $j = 1, \ldots, s$, is a right-invariant vector field taking values in $\mathcal{D}$. Hence,

$$[X_\alpha^R, Y_j^R] \in \langle Y_1^R \ldots, Y_s^R \rangle,$$

and $\langle Y_1^R, \ldots, Y_s^R \rangle$ is an odd-dimensional ideal of V^X. By assumption, V^X has no odd-dimensional ideals. This is a contradiction and we have that $\mathcal{D}$ is not invariant under V^X. Theorem 7.7 then ensures that (7.46) is not a k-symplectic Lie system.

7.4. Ω-Hamiltonian Functions

Each k-Hamiltonian vector field X relative to a k-symplectic structure $(\omega_1, \ldots, \omega_k)$ gives rise to a family $h_1, \ldots, h_k$ of Hamiltonian functions satisfying $\iota_X \omega = \mathrm{d}h_i$ for $i = 1, \ldots, k$. The study of k-symplectic Lie systems makes it convenient to introduce a generalization of the Hamiltonian function notion for presymplectic forms so as to analyze simultaneously $h_1, \ldots, h_k$. This section presents and analyze such a generalization. The theory presented represents an extension to k-symplectic structures of results devised by Awane in [20] for a more particular type of k-symplectic structures.

Definition 7.8. Let $\Omega = \sum_{i=1}^k \omega_i \otimes e^i$ be a polysymplectic structure on N. An Ω-*Hamiltonian function* is an $\mathbb{R}^k$-valued function on N of the form $h = \sum_{i=1}^k h_i \otimes e^i$, where $\iota_{X_h} \omega_i = \mathrm{d}h_i$ for $i = 1, \ldots, k$ and a vector field X_h on N. We call h an Ω-*Hamiltonian function* for X_h and we write $C^\infty(\Omega)$ for the space of Ω-Hamiltonian functions.

Several comments are now in order. Note that if Ω and $\tilde{\Omega}$ are two gauge equivalent polysymplectic forms, then $C^\infty(\Omega)$ and $C^\infty(\tilde{\Omega})$ are the same up to a change of variables on $\mathbb{R}^k$. If we are given the associated k-symplectic structure $(\omega_1, \ldots, \omega_k)$ instead of Ω, the function h in Definition 7.8 is called a k-*Hamiltonian function* for X_h. Analogously, the work [289] calls the vector field X_h the k-*Hamiltonian system* associated with the $\mathbb{R}^k$-valued Hamiltonian h. Meanwhile, Awane [20] calls h a *Hamiltonian map* of X in case that X is additionally an infinitesimal automorphism of a certain distribution on

284   *A Guide to Lie Systems with Compatible Geometric Structures*

which it is assumed that the presymplectic forms of the k-symplectic distribution vanish.

Example 7.2. In view of the relations (7.32) and (7.33), the vector fields $X_1 = 4u^2\partial/\partial u + 4uv\partial/\partial v + v^2\partial/\partial w$, $X_2 = \partial/\partial u$ and $X_3 = 2u\partial/\partial u + v\partial/\partial v$ have Ω-Hamiltonian functions

$$f = \left(4uw - 8\frac{u^2w^2}{v^2} - \frac{v^2}{2}\right) \otimes e^1 + \left(4u - 16\frac{u^2w}{v^2}\right) \otimes e^2,$$

$$g = -2\frac{w^2}{v^2} \otimes e^1 - 4\frac{w}{v^2} \otimes e^2, \quad h = \left(w - 4\frac{uw^2}{v^2}\right) \otimes e^1 - 8\frac{uw}{v^2} \otimes e^2,$$

$$\tag{7.47}$$

relative to the polysymplectic structure $\Omega = \omega_{\text{RS}-1} \otimes e^1 + \omega_{\text{RS}-2} \otimes e^2$ obtained from the two-symplectic structure $(\omega_{\text{RS}-1}, \omega_{\text{RS}-2})$ constructed from the presymplectic forms (7.30).

Proposition 7.9. *Let $\Omega = \sum_{i=1}^{k} \omega_i \otimes e^i$ be a polysymplectic structure. Every Ω-Hamiltonian vector field admits an Ω-Hamiltonian function. Conversely, every Ω-Hamiltonian function induces a unique Ω-Hamiltonian vector field.*

Proof. The direct part is obvious. Let us prove the converse. By definition, each Ω-Hamiltonian function $h = \sum_{i=1}^{k} h_i \otimes e^i$ is associated with, at least, one vector field X_h. Suppose that there exist two Ω-Hamiltonian vector fields X_1 and X_2 associated with h. Then,

$$\iota_{X_1}\omega_i = \iota_{X_2}\omega_i = \mathrm{d}h_i, \quad \Longrightarrow \quad \iota_{X_1 - X_2}\omega_i = 0, \quad i = 1, \ldots, k.$$

Since $\bigcap_{i=1}^{k} \ker \omega_i = \{0\}$, it turns out that $X_1 = X_2$. $\square$

Since every Ω-Hamiltonian function, h, induces a unique Ω-Hamiltonian vector field, it makes sense to represent it by X_h.

Proposition 7.10. *The space $C^\infty(\Omega)$ is a linear space over $\mathbb{R}$ with the natural sum $h + g := \sum_{i=1}^{k}(h_i + g_i) \otimes e^i$ and the multiplication by scalars $\lambda \cdot h := \sum_{i=1}^{k} \lambda h_i \otimes e^i$, where $h = \sum_{i=1}^{k} h_i \otimes e^i$, $g = \sum_{i=1}^{k} g_i \otimes e^i \in C^\infty(\Omega)$ and $\lambda \in \mathbb{R}$. Moreover, $C^\infty(\Omega)$ becomes a Lie* *algebra relative to the Lie bracket $\{\cdot,\cdot\}_\Omega : C^\infty(\Omega) \times C^\infty(\Omega) \to C^\infty(\Omega)$ of the form*

$$\left\{ \sum_{i=1}^{k} h_i \otimes e^i, \sum_{i=1}^{k} g_i \otimes e^i \right\}_\Omega = \sum_{i=1}^{k} \{h_i, g_i\}_{\omega_i} \otimes e^i, \qquad (7.48)$$

where $\{\cdot,\cdot\}_{\omega_i}$ is the Poisson bracket associated with the presymplectic form ω_i, with $i = 1,\ldots,k$.

Proof. Let X_h and X_g be the Ω-Hamiltonian vector fields associated to h and g, respectively. Any linear combination $\lambda h + \mu g$, with $\lambda, \mu \in \mathbb{R}$, is an Ω-Hamiltonian function associated with $\lambda X_h + \mu X_g$. Indeed,

$$\iota_{\lambda X_h + \mu X_g} \omega_i = \mathrm{d}(\lambda h_i + \mu g_i), \quad i = 1,\ldots,k. \qquad (7.49)$$

Then, $C^\infty(\Omega)$ is closed with respect to the defined addition of elements and multiplication by scalars. These operations also give rise to a vector space structure on $C^\infty(\Omega)$.

Let us now prove that $C^\infty(\Omega)$ is a Lie algebra relative to the (7.48). Since X_h and X_g are Hamiltonian vector fields relative to the presymplectic form ω_i with Hamiltonian functions h_i and g_i, respectively, one has that

$$\iota_{[X_h, X_g]} \omega_i = \mathrm{d}\{g_i, h_i\}_{\omega_i}, \quad i = 1,\ldots,k. \qquad (7.50)$$

Hence, $\{g, h\}_\Omega$ is an Ω-Hamiltonian function with Hamiltonian vector field $[X_h, X_g]$. Hence, $C^\infty(\Omega)$ is closed with respect to (7.48), which is trivially antisymmetric. Let us prove that it also satisfies the Jacobi identity. If $\sum_{i=1}^{k} l_i \otimes e^i$ is another Ω-Hamiltonian function, then one has that

$$\left\{ \left\{ \sum_{i=1}^{k} g_i \otimes e^i, \sum_{i=1}^{k} h_i \otimes e^i \right\}_\Omega, \sum_{i=1}^{k} l_i \otimes e^i \right\}_\Omega = \sum_{i=1}^{k} \{\{g_i, h_i\}_{\omega_i}, l_i\}_{\omega_i} \otimes e^i.$$
$$(7.51)$$

Summing the three cyclic permutations of the arguments of the left-hand side, we obtain

$$\sum_{i=1}^{k} (\{\{g_i, h_i\}_{\omega_i}\}, l_i\}_{\omega_i} + c.p.) \otimes e^i, \qquad (7.52)$$

and, using the Jacobi identity for the $\{\cdot,\cdot\}_{\omega_i}$, we obtain that $\{\cdot,\cdot\}_\Omega$ satisfies the Jacobi identity.

Summing all previous results, $(C^\infty(\Omega),\{\cdot,\cdot\}_\Omega)$ becomes a Lie algebra. $\qquad\square$

Every polysymplectic structure taking values in $\mathbb{R}$ is just a symplectic forms. In this case, $C^\infty(\Omega) = C^\infty(N)$ and $C^\infty(\Omega)$ is a Poisson algebra relative to $\{\cdot,\cdot\}_\Omega$ and the natural multiplication of functions on N. Meanwhile, if Ω takes values in $\mathbb{R}^k$ and $k \neq 1$, then it is not in general possible to ensure that $C^\infty(\Omega)$ is a Poisson algebra relative to $\{\cdot,\cdot\}_\Omega$ and some naturally defined product of Ω-functions. Indeed, if $h = \sum_{i=1}^k h_i \otimes e^i$ and $g = \sum_{i=1}^k g_i \otimes e^i \in C^\infty(\Omega)$, then it is reasonable to define the product of the Ω-functions h and g by

$$h \cdot g = \sum_{i=1}^k (h_i g_i) \otimes e^i. \qquad (7.53)$$

In fact, this definition recovers the natural product of functions for $k = 1$. Nevertheless, (7.53) is not in general a $C^\infty(\Omega)$-Hamiltonian function. Indeed,

$$\mathrm{d}\,(h_i g_i) = g_i \mathrm{d}\, h_i + h_i \mathrm{d}\, g_i = \iota_{g_i X_h}\omega_i + \iota_{h_i X_g}\omega_i = \iota_{(g_i X_h + h_i X_g)}\omega_i,$$
$$i = 1, \ldots, k. \qquad (7.54)$$

and, generally, $g_i X_h + h_i X_g$ is different for each i, which shows that $h \cdot g$ need not be an Ω-Hamiltonian function when $k > 1$. Let us provide a very particular example of this. Let us reconsider Example 7.2. The function

$$h \cdot g = -2\frac{w^2}{v^2}\left(w - 4\frac{uw^2}{v^2}\right) \otimes e^1 + 32\frac{uw^2}{v^4} \otimes e^2, \qquad (7.55)$$

is not an Ω-Hamiltonian function for $\Omega = \sum_{i=1}^2 \omega_{\mathrm{RS}-i} \otimes e^i$ because $-2\frac{w^2}{v^2}\left(w - 4\frac{uw^2}{v^2}\right)$ and $32uw^2/v^4$ are related to the vector fields

$$g_1 X_h + h_1 X_g = -\frac{2w^2}{v^2}\left(2u\frac{\partial}{\partial u} + v\frac{\partial}{\partial v}\right) + \left(w - 4\frac{uw^2}{v^2}\right)\frac{\partial}{\partial u},$$

$$g_2 X_h + h_2 X_g = -\frac{4w}{v^2}\left(2u\frac{\partial}{\partial u} + v\frac{\partial}{\partial v}\right) - \frac{8uw}{v^2}\frac{\partial}{\partial u}, \qquad (7.56)$$

which are different.

Since $(C^\infty(\Omega), \cdot, \{\cdot,\cdot\}_\Omega)$ is not in general a Poisson algebra, it is not possible to say that $\{h, \cdot\}_\Omega : g \in C^\infty(\Omega) \mapsto \{g, h\}_\Omega \in C^\infty(\Omega)$, with $h \in C^\infty(\Omega)$, is a derivation with respect to the product (7.53) of Ω-Hamiltonian functions. This illustrates that k-symplectic geometry is quite different from Poisson and presymplectic geometry, where one can naturally attach these structures to Poisson algebras of functions. Nevertheless, $\{h, g\}_\Omega = 0$ for every locally constant Ω-function g and, moreover, there are other interesting properties of these Lie algebras. Let us consider, for instance, the following feature.

Proposition 7.11. *Let (N, Ω) be a polysymplectic manifold. Every Ω-Hamiltonian vector field X_h acts as a derivation on the Lie algebra $(C^\infty(\Omega), \{\cdot,\cdot\}_\Omega)$ in the form*

$$X_h f = \{f, h\}_\Omega, \quad \forall f \in C^\infty(\Omega). \qquad (7.57)$$

Proof. Let us show that (7.57) does not depend on the Ω-Hamiltonian function for X_h. In fact, if $X_h = X_{h'}$, then

$$\{f, h - h'\}_\Omega = \sum_{i=1}^k \{f_i, h_i - h_i'\}_{\omega_i} \otimes e^i = \sum_{i=1}^k (X_h - X_{h'})f_i \otimes e^i = 0$$

$$\Rightarrow \quad X_h f = X_{h'} f, \qquad (7.58)$$

and $X_h f$ becomes well defined. Let us show that X_h is a derivation relative to $\{\cdot,\cdot\}_\Omega$. Using the definition of X_h and the Jacobi identity for $\{\cdot,\cdot\}_\Omega$, we obtain

$$X_h\{f, g\}_\Omega = \{\{f, g\}_\Omega, h\}_\Omega = \{\{f, h\}_\Omega, g\}_\Omega + \{f, \{g, h\}_\Omega\}_\Omega$$

$$= \{X_h f, g\} + \{f, X_h g\}_\Omega, \quad \forall f, g \in C^\infty(\Omega). \qquad (7.59)$$

288   *A Guide to Lie Systems with Compatible Geometric Structures*

Consequently, X_h satisfies the Leibniz rule relative to $\{\cdot,\cdot\}_\Omega$. Since X_h acts linearly on $C^\infty(\Omega)$, the X_h acts as a derivation on $(C^\infty(\Omega), \{\cdot,\cdot\}_\Omega)$ and the proposition follows. $\qquad\square$

Theorem 7.12. *Let* $\Omega = \sum_{i=1}^k \omega_i \otimes e^i$ *be a polysymplectic form on* N. *There exists an exact sequence of Lie algebras:*

$$0 \hookrightarrow \mathrm{H}^0_{\mathrm{dH}}(N) \oplus \overset{k}{\cdots} \oplus \mathrm{H}^0_{\mathrm{dH}}(N) \hookrightarrow C^\infty(\Omega) \xrightarrow{B_\Omega} \mathrm{Ham}(\Omega) \to 0, \quad (7.60)$$

where $B_\Omega(f) = -X_f$ *and* $\mathrm{H}^0_{\mathrm{dH}}(N)$ *is the zero De Rham cohomology group of* N.

Proof. Let us prove that $\mathrm{Ham}(\Omega)$ is a Lie algebra. Every two Ω-Hamiltonian vector fields X_h and X_g satisfy $\iota_{X_h}\omega_i = \mathrm{d}h_i, \iota_{X_g}\omega_i = \mathrm{d}g_i$ for each $i = 1, \ldots, k$. Hence,

$$\begin{aligned} \iota_{\lambda X_h + \mu X_g}\omega_i &= \mathrm{d}(\lambda h_i + \mu g_i), \\ \iota_{[X_h, X_g]}\omega_i &= (\mathcal{L}_{X_h}\iota_{X_g} - \iota_{X_g}\mathcal{L}_{X_h})\omega_i \qquad i = 1, \ldots, k, \quad \forall \lambda, \mu \in \mathbb{R}. \\ &= \{g_i, h_i\}_{\omega_i} \end{aligned} \qquad (7.61)$$

Consequently, the linear combinations and Lie brackets of Ω-Hamiltonian vector fields are Ω-Hamiltonian vector fields, which makes $\mathrm{Ham}(\Omega)$ into a Lie algebra. In particular, the second expression in (7.61) shows that $[X_h, X_g]$ is an Ω-Hamiltonian vector field with an Ω-Hamiltonian function given by $-\{h, g\}_\Omega$.

Proposition 7.10 states that $C^\infty(\Omega)$ is a Lie algebra relative to $\{\cdot,\cdot\}_\Omega$. The elements of the space $\mathrm{H}^0_{\mathrm{dH}}(N) \oplus \overset{k}{\cdots} \oplus \mathrm{H}^0_{\mathrm{dH}}(N)$, via its injection in $C^\infty(\Omega)$, can be considered as constant Ω-Hamiltonian functions. Hence, they span an Abelian Lie algebra. Therefore, all the spaces of the sequence (7.60) are Lie algebras.

Let us now turn to showing that the sequence (7.60) is exact. The inclusions of 0 in $\mathrm{H}^0_{\mathrm{dH}}(N) \oplus \overset{k}{\cdots} \oplus \mathrm{H}^0_{\mathrm{dH}}(N)$ and of $\mathrm{H}^0_{\mathrm{dH}}(N) \oplus \overset{k}{\cdots} \oplus \mathrm{H}^0_{\mathrm{dH}}(N)$ in $C^\infty(\Omega)$ as well as the projection of $\mathrm{Ham}(\Omega)$ onto 0 are Lie algebra morphisms. Recall that $\{h, g\}_\Omega$ is an Ω-Hamiltonian function of the Ω-Hamiltonian vector field $-[X_f, X_g]$. Hence, $B_\Omega(\{f, g\}_\Omega) = [X_f, X_g]$ and B_Ω is a Lie algebra morphism. Summing up, all morphisms in (7.60) are Lie algebra morphisms.

Finally, let us show that the kernel of B_Ω is given by those Ω-Hamiltonian functions h related to a zero vector field. In fact, if $X_h = 0$, this implies that $dh_i = 0$ for $i = 1, \ldots, k$. So, every h_i is a constant on each connected component, O_j, with $j = 1, \ldots, p$, of N. Since $\dim \mathrm{H}^0_{\mathrm{dH}}$ is the number of connected components of N, the elements of $\ker B_\Omega$ can be determined by a series of $k \times \dim \mathrm{H}^0_{\mathrm{dH}}$ numbers and vice versa. This gives the isomorphism

$$(h_1(O_1), \ldots, h_1(O_p), \ldots, h_k(O_1), \ldots, h_k(O_p))$$

$$\in \mathrm{H}^0_{\mathrm{dH}}(N) \oplus \overset{k}{\ldots} \oplus \mathrm{H}^0_{\mathrm{dH}}(N) \longmapsto h \in \ker B_\Omega, \qquad (7.62)$$

which finishes our proof. $\qquad\qquad\square$

7.5. Derived Poisson Algebras

We already showed that a polysymplectic manifold (N, Ω) do not allow for a natural structure of Poisson algebra in $C^\infty(\Omega)$. Instead, we will use the polysymplectic form to build Poisson algebras on certain subspaces of $C^\infty(N)$, hereafter called derived Poisson algebras. In the following sections, such Poisson algebras will be employed to study the constants of motion and superposition rules of k-symplectic Lie systems.

Consider the polysymplectic form $\Omega = \sum_{i=1}^{k} \omega_i \otimes e^i$ and an element $\theta \in (\mathbb{R}^k)^*$. Since the ω_i are closed forms, the contraction

$$\Omega_\theta := \langle \Omega, \theta \rangle = \sum_{i=1}^{k} \theta(e^i)\omega_i$$

is a presymplectic form on N. Let $\mathrm{Adm}(\Omega_\theta)$ be the set of admissible functions relative to the presymplectic manifold (N, Ω_θ). We hereafter denote by X_f, with $f \in C^\infty(N)$, a Hamiltonian vector field of f relative to a presymplectic form. Recall that when h is a k-Hamiltonian function, X_h denotes the k-Hamiltonian vector field associated to the k-Hamiltonian function h. This notation does not lead to confusion as soon as we know what kind of function h is. Moreover, it also makes notation more natural and compact.

Note that a vector field X is k-Hamiltonian if and only if it is Hamiltonian for all the presymplectic forms of the space $\langle \omega_1, \ldots, \omega_k \rangle$. In particular, X is Hamiltonian for any presymplectic form Ω_θ with $\theta \in (\mathbb{R}^k)^*$. This gives rise to the following proposition.

Proposition 7.13. *Let $\Omega = \sum_{i=1}^k \omega_i \otimes e^i$ be a polysymplectic form and $\theta \in (\mathbb{R}^k)^*$. Every Ω-Hamiltonian function h gives rise to an admissible function $\langle h, \theta \rangle := \sum_{i=1}^k \theta(e^i) h_i$ relative to (N, Ω_θ).*

Proof. If $h = \sum_{i=1}^n h_i \otimes e^i$ is an Ω-Hamiltonian function, then there exists an Ω-Hamiltonian vector field X_h such that $\iota_{X_h} \omega_i = \mathrm{d}h_i$ for $i = 1, \ldots, k$. Thus, one has

$$\iota_{X_h} \Omega_\theta = \sum_{i=1}^k \theta(e^i) \iota_{X_h} \omega_i = \sum_{i=1}^k \theta(e^i) \mathrm{d}h_i = \mathrm{d}h_\theta,$$

where $h_\theta = \langle h, \theta \rangle$, and therefore $h_\theta \in \mathrm{Adm}(\Omega_\theta)$. $\qquad\square$

Since the space of admissible functions for a presymplectic form is a Poisson algebra, one gets the following proposition (for details on presymplectic forms, see [358] or Chapter 2).

Proposition 7.14. *Every polysymplectic form $\Omega = \sum_{i=1}^k \omega_i \otimes e^i$ gives rise to a family of Poisson algebras $(\mathrm{Adm}(\Omega_\theta), \cdot, \{\cdot, \cdot\}_\theta)$, where $\{\cdot, \cdot\}_\theta$ is the Poisson bracket induced by the presymplectic form Ω_θ on its space of admissible functions.*

Proposition 7.15. *If $\Omega = \sum_{i=1}^k \omega_i \otimes e^i$ is a polysymplectic form, then every Ω-Hamiltonian vector field X_h is a derivation on all the Lie algebras $(\mathrm{Adm}(\Omega_\theta), \{\cdot, \cdot\}_\theta)$, with $\theta \in (\mathbb{R}^k)^*$, in the form*

$$X_h f = \{f, h_\theta\}_\theta, \quad \forall f \in \mathrm{Adm}(\Omega_\theta). \tag{7.63}$$

Proof. Let us first prove the form of (7.63). By definition of Ω-Hamiltonian vector fields, one has that X_h is a Hamiltonian vector field relative to each ω_i, i.e., $\iota_{X_h}\omega_i = \mathrm{d}\theta_i$ for $i = 1, \ldots, k$. Hence,

$$\iota_{X_h}\Omega_\theta = \iota_{X_h}\left[\sum_{i=1}^k \theta(e^i)\omega_i\right] = \sum_{i=1}^k \theta(e^i)\mathrm{d}h_i = \mathrm{d}\langle h, \theta\rangle,$$

and X_h is a Hamiltonian vector field with Hamiltonian function h_θ relative to ω_θ. Then, expression (7.63) follows from the standard expression for the Poisson bracket for the presymplectic form ω_θ.

To see that X_h is a derivation relative to $\{\cdot, \cdot\}_\theta$, one has to recall that X_h acts linearly and make use of the Jacobi identity for $\{\cdot, \cdot\}_\theta$, namely

$$X_h\{f, g\}_\theta = \{\{f, g\}_\theta, h_\theta\}_\theta = \{\{f, h_\theta\}_\theta, g\}_\theta + \{f, \{g, h_\theta\}_\theta\}_\theta$$
$$= \{X_h f, g\}_\theta + \{f, X_h g\}_\theta,$$

which proves the Leibniz identity relative to $\{\cdot, \cdot\}_\theta$. $\qquad\square$

Proposition 7.16. *Let $\Omega = \sum_{i=1}^k \omega_i \otimes e^i$ be a polysymplectic form. Then,*

$$\phi_\theta : (C^\infty(\Omega), \{\cdot, \cdot\}_\Omega) \to (\mathrm{Adm}(\Omega_\theta), \{\cdot, \cdot\}_\theta)$$
$$h \mapsto h_\theta = \langle h, \theta\rangle$$

is a Lie algebra morphism. Moreover, every finite-dimensional Lie algebra $(\mathcal{W} \subset C^\infty(\Omega), \{\cdot, \cdot\}_\Omega)$ is a Lie algebra extension of the Lie algebra $(\phi_\theta(\mathcal{W}), \{\cdot, \cdot\}_\theta)$.

Proof. The map ϕ_θ is clearly linear. To prove that it is also a Lie algebra morphism, we show that, if g, h are two Ω-Hamiltonian functions, then $\phi_\theta(\{g, h\}_\Omega) = \{\phi_\theta(g), \phi_\theta(h)\}_\theta$. From (7.48) and (7.63), we obtain

$$\phi_\theta(\{h, g\}_\Omega) = \sum_{i=1}^k \theta(e^i)\{h_i, g_i\}_{\omega_i} = \sum_{i=1}^k \theta(e^i)X_g h_i = X_g h_\theta = \{h_\theta, g_\theta\}_\theta,$$

and ϕ_θ becomes a Lie algebra morphism.

The morphisms of (7.60) can be restricted to give rise to a new sequence of Lie algebra morphisms

$$0 \hookrightarrow \left(\mathrm{H}^0_{\mathrm{dR}}(N) \oplus \overset{k}{\ldots} \oplus \mathrm{H}^0_{\mathrm{dR}}(N)\right) \cap \mathcal{W} \hookrightarrow \mathcal{W} \xrightarrow{\phi_\theta|_\mathcal{W}} \phi_\theta(\mathcal{W}) \to 0.$$

Therefore, $(\mathcal{W}, \{\cdot, \cdot\}_\Omega)$ is a Lie algebra extension of $(\phi_\theta(\mathcal{W}), \{\cdot, \cdot\}_\theta)$. $\square$

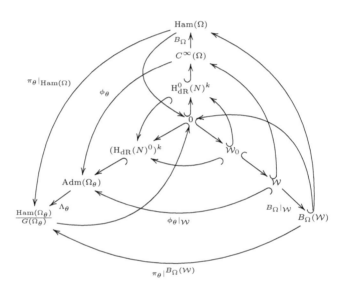

Proposition 7.17. *Every polysymplectic manifold* (N, Ω) *gives rise to the commutative diagram above, where* $\mathcal{W}_0 = \mathrm{H}^0_{\mathrm{dR}}(N)^k \cap \mathcal{W}$, *the space of gauge vector fields of* Ω_θ *is represented by* $\mathrm{G}(\Omega_\theta)$, *we call* $\pi_\theta : X \in \mathrm{Ham}(\Omega_\theta) \mapsto [X] \in \mathrm{Ham}(\Omega_\theta)/\mathrm{G}(\Omega_\theta)$ *the quotient map onto* $\mathrm{Ham}(\Omega_\theta)/\mathrm{G}(\Omega_\theta)$, *the mapping* $B_\Omega : f \in C^\infty(\Omega) \mapsto -X_f \in \mathrm{Ham}(\Omega)$ *maps every* Ω-*Hamiltonian function onto its* Ω-*Hamiltonian vector field, and* $\Lambda_\theta : \mathrm{Adm}(\Omega_\theta) \to \mathrm{Ham}(\Omega_\theta)/\mathrm{G}(\Omega_\theta)$ *is the Lie algebra morphism mapping each* $f \in \mathrm{Adm}(\Omega_\theta)$ *to the class* $[-X_f]$.

Proof. Most parts of the above diagrams are consequences of previous results given in this chapter. The only non-trivial result is to prove that the diagram

$$\begin{array}{ccc} C^\infty(\Omega) & \xrightarrow{\;B_\Omega\;} & \mathrm{Ham}(\Omega) \\ \downarrow{\scriptstyle \phi_\theta} & & \downarrow{\scriptstyle \pi_\theta|\mathrm{Ham}(\Omega)} \\ \mathrm{Adm}(\Omega_\theta) & \xrightarrow{\;\Lambda_\theta\;} & \dfrac{\mathrm{Ham}(\Omega_\theta)}{G(\Omega_\theta)} \end{array}$$

is commutative. Since $\mathrm{Ham}(\Omega) \subset \mathrm{Ham}(\Omega_\theta)$, one has that

$$\mathrm{d}h_\theta = \mathrm{d}\phi_\theta(h) = \langle \mathrm{d}h, \theta \rangle = \langle \iota_{X_h}\Omega, \theta \rangle = \iota_{X_h}\Omega_\theta \Rightarrow [X_{h_\theta}] = [X_h],$$

for an arbitrary $h \in C^\infty(\Omega)$. Consequently,

$$\begin{aligned} \pi_\theta \circ B_\Omega(h) &= \pi_\theta(-X_h) = [-X_h] = [-X_{h_\theta}] = \Lambda_\theta(h_\theta) \\ &= \Lambda_\theta \circ \phi_\theta(h), \quad \forall h \in C^\infty(\Omega) \Rightarrow \pi_\theta \circ B_\Omega = \Lambda_\theta \circ \phi_\theta. \end{aligned}$$

$\square$

7.6. k-Symplectic Lie–Hamiltonian Structures

In Chapter 5, we proved that every Dirac–Lie system was endowed with a Dirac–Lie structure that led to the study of many of their properties. Here, we show that k-symplectic Lie systems admit an analog of such a property too.

For this, let us reconsider the Schwarzian equations in first-order form (7.2). Recall that Y_1, Y_2, and Y_3 are Hamiltonian vector fields relative to the presymplectic forms ω_1 and ω_2 given by (7.5). Indeed, relations (7.7) and (7.8) showed that X_1, X_2, and X_3 have Hamiltonian functions

$$h_1^1 = \frac{2}{v}, \quad h_1^2 = \frac{a}{v^2}, \quad h_1^3 = \frac{a^2}{2v^3}, \tag{7.64}$$

and

$$h_2^1 = -\frac{4x}{v}, \quad h_2^2 = 2 - \frac{2ax}{v^2}, \quad h_2^3 = \frac{2a}{v} - \frac{a^2 x}{v^3}, \tag{7.65}$$

relative to the presymplectic forms ω_1 and ω_2, respectively. Moreover,

$$\left\{h_i^1, h_i^2\right\}_{\omega_i} = -h_i^1, \quad \left\{h_i^1, h_i^3\right\}_{\omega_i} = -2h_i^2,$$

$$\left\{h_i^2, h_i^3\right\}_{\omega_i} = -h_i^3, \quad i = 1, 2,$$

where $\{\cdot, \cdot\}_{\omega_i}$ stands for the Poisson bracket related to ω_i. Then, the functions h_i^α, with $\alpha = 1, 2, 3$ and a fixed i, span a finite-dimensional real Lie algebra of functions isomorphic to $\mathfrak{sl}(2, \mathbb{R})$. The same applies to $h_1^\alpha + h_2^\alpha$, with $\alpha = 1, 2, 3$, and in general for any linear combination $\mu_1 h_1^\alpha + \mu_2 h_2^\alpha$, with fixed $(\mu_1, \mu_2) \in \mathbb{R}^2 \backslash \{(0, 0)\}$. Hence, a k-symplectic structure gives rise to many different Lie algebras of functions, which will play a role in the analysis of the properties of the system.

Let us analyze the space $C^\infty(\Omega)$ of Ω-Hamiltonian functions related to the two-symplectic structure (ω_1, ω_2). Relations (7.7) and (7.8) suggest that the functions $h^\alpha = h_1^\alpha \otimes e^1 + h_2^\alpha \otimes e^2$, with $\alpha = 1, 2, 3$, span a finite-dimensional Lie algebra relative to the Lie bracket (7.48).

In view of the above, every X_t^{Sc} is an Ω-Hamiltonian vector field with Ω-Hamiltonian function

$$h_t^{\mathrm{Sc}} = (h_1^3 + b_1(t)h_1^1) \otimes e^1 + (h_2^3 + b_1(t)h_2^1) \otimes e^2.$$

Since $b_1(t)$ is non-constant because of our initial assumption, the space $\mathrm{Lie}(\{h_t^{\mathrm{Sc}}\}_{t \in \mathbb{R}}, \{\cdot, \cdot\}_\Omega)$ becomes a real Lie algebra isomorphic to $\mathfrak{sl}(2, \mathbb{R})$.

Let us now show a second example with a similar structure appearing for Schwarzian equations. Consider the system of Riccati equations (7.9) associated with a t-dependent vector field X^{Ric} and admitting a VG Lie algebra spanned by the vector fields X_1, X_2, and X_3. The relations (7.13) and (7.14) imply that X_1, X_2, and X_3 have Hamiltonian functions

$$h_1^1 = \frac{1}{x_1 - x_2} + \frac{1}{x_3 - x_4}, \quad h_1^2 = \frac{1}{2}\left(\frac{x_1 + x_2}{x_1 - x_2} + \frac{x_3 + x_4}{x_3 - x_4}\right),$$

$$h_1^3 = \frac{x_1 x_2}{x_1 - x_2} + \frac{x_3 x_4}{x_3 - x_4}, \tag{7.66}$$

and

$$h_2^1 = \sum_{\substack{i,j=1 \\ i\leq j}}^{4} \frac{1}{x_i - x_j}, \quad h_2^2 = \frac{1}{2}\left(\sum_{\substack{i,j=1 \\ i\leq j}}^{4} \frac{x_i + x_j}{x_i - x_j}\right), \quad h_2^3 = \sum_{\substack{i,j=1 \\ i\leq j}}^{4} \frac{x_i x_j}{x_i - x_j},$$

(7.67)

relative to the presymplectic forms ω_1, ω_2 given by (7.12), respectively. Moreover, we have that

$$\{h_i^1, h_i^2\}_{\omega_i} = -h_i^1, \quad \{h_i^1, h_i^3\}_{\omega_i} = -2h_i^2,$$
$$\{h_i^2, h_i^3\}_{\omega_i} = -h_i^3, \quad i = 1, 2.$$

Consequently, the functions h_i^α, with $\alpha = 1, 2, 3$ and a fixed i span a finite-dimensional real Lie algebra of functions.

Now, we consider the space $C^\infty(\Omega)$ of Ω-Hamiltonian functions given by the two-symplectic structure (ω_1, ω_2). From the relations (7.13) and (7.14), the functions $h^\alpha = h_1^\alpha \otimes e^1 + h_2^\alpha \otimes e^2$, with $\alpha = 1, 2, 3$ span a finite-dimensional Lie algebra relative to the Lie bracket (7.48).

Moreover, every X_t^{Ric} is an Ω-Hamiltonian vector field admitting an Ω-Hamiltonian function

$$h_t^{\mathrm{Ric}} = a(t)h^1 + b(t)h^2 + c(t)h^3 .$$

Again, X^{Ric} gives rise to a curve $t \to h_t^{\mathrm{Ric}}$ in a finite-dimensional real Lie algebra of the form $\mathrm{Lie}(\{h_t^{\mathrm{Ric}}\}_{t\in\mathbb{R}}, \{\cdot, \cdot\}_\Omega)$.

Previous examples motivate the following notions.

Definition 7.18. A *polysymplectic Lie–Hamiltonian structure* is a triple (N, Ω, h) where (N, Ω) is a polysymplectic manifold and h represents a t-parametrized family of Ω-Hamiltonian functions $h_t \colon N \to \mathbb{R}^k$ such that $\mathrm{Lie}(\{h_t\}_{t\in\mathbb{R}}, \{\cdot, \cdot\}_\Omega)$ is a finite-dimensional real Lie algebra.

Since a polysymplectic structure Ω amounts to a k-symplectic one, we can also talk indistinctly about k-symplectic Lie–Hamiltonian structures.

296 *A Guide to Lie Systems with Compatible Geometric Structures*

Definition 7.19. We say that a t-dependent vector field X admits a polysymplectic Lie–Hamiltonian structure (N, Ω, h) if $B_\Omega(h_t) = -X_t$ for all $t \in \mathbb{R}$.

Example 7.3. Let us turn to the *continuous Heisenberg group* [370], i.e., the matrix Lie group given in (2.2). A basis of right-invariant vector fields read $X_1^R = \partial/\partial x + y\partial/\partial z, X_2^R = \partial/\partial y, X_3^R = \partial/\partial z$. Let us analyze the Lie system of the form $X^R = \sum_{\alpha=1}^r b_\alpha(t) X_\alpha^R$ for arbitrary functions $b_1(t), b_2(t), b_3(t)$. Systems of this type appear, for instance, in the solution of certain quantum systems [108].

Consider the presymplectic differential 2-forms

$$\omega_1 = \mathrm{d}x \wedge \mathrm{d}y, \quad \omega_1 = \mathrm{d}y \wedge \mathrm{d}z.$$

We have that $\ker \omega_1 \cap \ker \omega_2 = \{0\}$ and (ω_1, ω_2) becomes a two-symplectic structure on $\mathbb{H}$.

Then, the vector fields X_1^R, X_2^R, X_3^R are Hamiltonian relative to ω_1 and ω_2 with Hamiltonian functions $h_1^1 = y, h_2^1 = -x, h_3^1 = 0$ and $h_1^2 = -y^2/2, h_2^2 = z, h_3^2 = -y$, respectively. In consequence, each X_t^R admits a t-dependent Ω-Hamiltonian function

$$h(t) = b_1(t)(y \otimes e^1 - y^2/2 \otimes e^2) + b_2(t)(-x \otimes e^1 + z \otimes e^2) - b_3(t) y \otimes e^2.$$

Additionally, if $h^1 = y \otimes e^1 - y^2 \otimes e^2/2, h^2 = -x \otimes e^1 + z \otimes e^2, h^3 = -y \otimes e^2$, then

$$\{h^1, h^2\}_\Omega = e^1 + h^3, \quad \{h^1, h^3\}_\Omega = 0, \quad \{h^2, h^3\}_\Omega = e^2.$$

Hence, $h(t)$ takes values in the finite-dimensional Lie algebra of Ω-Hamiltonian functions $\langle h^1, h^2, h^3, e^1, e^2 \rangle$. Summing up, X^R admits a polysymplectic Lie–Hamiltonian structure $(\mathbb{H}, \Omega, h(t))$.

The following theorem shows that every k-symplectic Lie system admits a k-symplectic Lie–Hamiltonian structure and vice versa.

Theorem 7.20. *A system X admits a k-symplectic Lie–Hamiltonian structure if and only if it is a k-symplectic Lie system.*

Proof. Let (N, Ω, h) be a k-symplectic Lie–Hamiltonian structure for X. Then, $\mathrm{Lie}(\{h_t\}_{t \in \mathbb{R}}, \{\cdot, \cdot\}_\Omega)$ is a finite-dimensional real

Lie algebra. Since (7.60) is a sequence of Lie algebra morphisms, B_Ω is a Lie algebra morphism and $V = B_\Omega(\mathrm{Lie}(\{h_t\}_{t\in\mathbb{R}}))$ is a finite-dimensional real Lie algebra. As each vector field $-X_t$ is Ω-Hamiltonian with an Ω-Hamiltonian function within $\{h_t\}_{t\in\mathbb{R}}$, then $\{X_t\}_{t\in\mathbb{R}} \subset V$. Therefore, $V^X = \mathrm{Lie}(\{X_t\}_{t\in\mathbb{R}}) \subset V$ and X is a k-symplectic Lie system.

Conversely, if X is a k-symplectic Lie system, then the vector fields $\{X_t\}_{t\in\mathbb{R}}$ belong to a finite-dimensional Lie algebra V^X of Ω-Hamiltonian vector fields. Consequently, $X_t = \sum_{\alpha=1}^r b_\alpha(t)X_\alpha$ for a basis $\{X_1, \ldots, X_r\}$ for V^X of Ω-Hamiltonian vector fields and certain t-dependent functions $b_1(t), \ldots, b_r(t)$. The sequence (7.60) shows that $B_\Omega^{-1}(V^X)$ is a finite-dimensional real Lie algebra of Ω-Hamiltonian functions. If $h_1, \ldots, h_r$ is a set of elements of $C^\infty(\Omega)$ with associated Ω-Hamiltonian vector fields $X_1, \ldots, X_r$, then each $h_t = \sum_{\alpha=1}^r b_\alpha(t)h_\alpha$ is an Ω-Hamiltonian function and $-B_\Omega(h_t) = X_t$ for every $t \in \mathbb{R}$. Hence, $h_1, \ldots, h_r$ are contained within the finite-dimensional Lie algebra $B_\Omega^{-1}(V^X)$ and (N, Ω, h) becomes a k-symplectic Lie–Hamiltonian structure for X. $\qquad\square$

Despite the result given by the previous theorem, it is worth noting k-symplectic Lie systems are more important than k-symplectic Lie–Hamilton structures. In fact, k-symplectic Lie–Hamiltonian structures are used next only as a tool to study the dynamics of k-symplectic Lie systems. Moreover, a k-symplectic Lie system may have different k-symplectic Lie–Hamilton structures related to different Lie algebras of k-Hamiltonian functions and not all of them, as shown in the following sections, are equally appropriate to study the properties of k-symplectic Lie systems.

7.7. General Properties of k-Symplectic Lie Systems

Let us describe the analog for k-symplectic Lie systems of the general features of Lie systems. Remarkably, we will show that the derived Poisson algebras allow for the derivation of t-independent constants of motion and superposition rules for k-symplectic Lie systems in an algebraic manner.

As for every Lie system, the general solution of a k-symplectic Lie system X on N, let us say $x(t)$ can be written as $x(t) = \varphi(g(t), x_0)$, where $x_0 \in N$ and $\varphi \colon G \times N \to N$ is a Lie group action. The Lie group action φ can be employed to devise a Bäcklund transformation-like theory for Lie systems. In particular, if G is connected, every curve $\bar{g}(t)$ in G gives rise to a t-dependent change of variables, $x(t) = \varphi(g(t), x_0)$, mapping a Lie system X taking values in a Lie algebra V^X into another Lie system Y, with general solution $y(t) = \varphi(\bar{g}(t), x(t))$, taking values in the same Lie algebra V^X [85, 94, 99]. If X is additionally a k-symplectic Lie system, then V^X is a Lie algebra of k-Hamiltonian vector fields relative to some k-symplectic structure. As $\{Y_t\}_{t\in\mathbb{R}}$ belong to V^X, they are k-Hamiltonian vector fields and Y also becomes a k-symplectic Lie system.

Using again that $x(t) = \varphi(g(t), x_0)$, we see that the each particular solution of a Lie system X is contained within an orbit S of φ. Indeed, it is easy to see that the vector fields $\{X_t\}_{t\in\mathbb{R}}$ are tangent to such orbits and it makes sense to define the restriction $X|_S$ of X to each orbit S. Therefore, the integration of a Lie system X reduces to integrating its restrictions to each orbit of φ, which are also Lie systems. Thus, it is interesting to know whether $X|_S$ is again a k-symplectic Lie system. More generally, we want to know whether the restriction $X|_S$ of a k-symplectic Lie system X to a submanifold $S \subset N$, where it has a sense to define $X|_S$, is again a k-symplectic Lie system. This requires studying the notion of l-symplectic submanifold ($l \leq k$) of a k-symplectic manifold $(N, \omega_1, \ldots, \omega_k)$.

Proposition 7.21. *Let X be a k-symplectic Lie system relative to the k-symplectic structure $(\omega_1, \ldots, \omega_k)$ on N. If S is an l-symplectic submanifold such that $\mathcal{D}^X \subset TS$, then the restriction of X to S is an l-symplectic Lie system.*

Proof. As X is a k-symplectic Lie system with respect to the k-symplectic structure $(\omega_1, \ldots, \omega_k)$ on N, one has that V^X is a finite-dimensional real Lie algebra of k-Hamiltonian vector fields. In consequence, if $Y \in V^X$, then Y is a Hamiltonian vector field with respect to $\omega_1, \ldots, \omega_k$. Since $\mathcal{D}^X \subset TS$ by assumption, the t-dependent vector field X is tangent to S and its restriction to S gives rise to a new t-dependent vector field $X|_S$ on S satisfying that $V^{X|_S} = \mathrm{Lie}(\{X_t|_S\})$ is a finite-dimensional real Lie algebra of Hamiltonian vector fields with respect to $\iota^*\omega_1, \ldots, \iota^*\omega_l$, where ι is the embedding $\iota : S \hookrightarrow N$. Therefore, $X|_S$ to S is an l-symplectic Lie system. $\qquad\square$

Example 7.4. Consider the Lie subgroup of the Heisenberg group (2.2). This also represents a k-symplectic submanifold since the pullback of ω_1, ω_2 relative to the embedding $\iota : \mathbb{B} \hookrightarrow \mathbb{H}$ gives rise to a symplectic form $\iota^*\omega_2$ on $\mathbb{B}$. If we consider the k-symplectic Lie system on $\mathbb{H}$ given by $X = b_2(t)X_2^R + b_3(t)X_3^R$, where $X_2^R = \partial/\partial y, X_3^R = \partial/\partial z$, one has that X takes values in $T\mathbb{B}$. Consequently, the application of Proposition 7.21 shows that X gives rise to a restriction $X|_\mathbb{B}$ which is also a k-symplectic Lie system relative to $\iota^*\omega_2$, indeed a symplectic one.

Let us now turn to describing several properties of constants of motion for Lie systems.

Proposition 7.22. *Let X be a k-symplectic Lie system on N admitting a k-symplectic Lie–Hamiltonian structure (N, Ω, h) and let $\theta \in (\mathbb{R}^k)^*$. Then, the space $\mathcal{I}_\theta^X$ of t-independent constants of motion of X that are admissible relative to the presymplectic form Ω_θ is a Poisson algebra with respect to its associated Poisson bracket $\{\cdot, \cdot\}_\theta$.*

Proof. If $f_1, f_2 \in \mathcal{I}_\theta^X$, then $X_t f_i = 0$ for $i = 1, 2$ and every $t \in \mathbb{R}$. As X is a k-symplectic Lie system, the elements of V^X are Hamiltonian vector fields with respect to each Ω_θ with $\theta \in (\mathbb{R}^k)^*$. In view of Proposition 7.15, one has that $X_t\{f, g\}_\theta = \{X_t f, g\}_\theta + \{f, X_t g\}_\theta$ for every $f, g \in \mathrm{Adm}(\Omega_\theta)$ and $t \in \mathbb{R}$. In particular, $X_t\{f_1, f_2\}_\theta = \{X_t f_1, f_2\}_\theta + \{f_1, X_t f_2\}_\theta = 0$ for every $t \in \mathbb{R}$. Consequently, the Poisson bracket of t-independent admissible constants of motion relative to Ω_θ is a new one. Since $\lambda f_1 + \mu f_2$ and $f_1 \cdot f_2$ are also t-independent

300   *A Guide to Lie Systems with Compatible Geometric Structures*

constants of motion for every $\lambda, \mu \in \mathbb{R}$ and they are admissible functions relative to Ω_θ, one has that $\mathcal{I}_\theta^X$ is a Poisson algebra relative to the Poisson bracket $\{\cdot, \cdot\}_\theta$ associated with the presymplectic form Ω_θ and the natural product of functions. $\qquad \square$

Example 7.5. Consider again the Heisenberg group $\mathbb{H}$ and the k-symplectic Lie system given by $X^R = f(t)X_1^R = f(t)(\partial/\partial x + y \partial/\partial x)$ for any $f(t)$. Let $\theta = e_1$. Since $\Omega = \mathrm{d}x \wedge \mathrm{d}y \otimes e^1 + \mathrm{d}y \wedge \mathrm{d}z \otimes e^2$, one has that $\langle \Omega, \theta \rangle = \mathrm{d}x \wedge \mathrm{d}y$ and the space $C^\infty(\Omega_\theta)$ is given by the functions depending only on x and y. The constants of motion for X^R are functions only depending on y. Hence, $\{f_1(y), f_2(y)\}_\theta = 0$ for every $f_1(y), f_2(y) \in C^\infty(\Omega_\theta)$ and they span a Poisson algebra.

Let us prove some final interesting results about the t–independent constants of motion for k-symplectic Lie systems.

Proposition 7.23. *If X is a k-symplectic Lie system on N with a k-symplectic Lie–Hamiltonian structure (N, Ω, h) and $\theta \in (\mathbb{R}^k)^*$, then function $f \in C^\infty(N) \in \mathcal{I}_\theta^X$ if and only if f Poisson commutes with all elements of $\phi_\theta(\mathrm{Lie}(\{h_t\}_{t\in\mathbb{R}}, \{\cdot, \cdot\}_\Omega))$.*

Proof. Assume that $f \in \mathrm{Adm}(\Omega_\theta)$. The function f is a t-independent constant of motion for X if and only if

$$0 = X_t f = \{f, \langle h_t, \theta \rangle\}_\theta, \quad \forall t \in \mathbb{R}. \tag{7.68}$$

Using the above, we obtain that

$$\{f, \{\langle h_t, \theta \rangle, \langle h_{t'}, \theta \rangle\}_\theta\}_\theta = \{\{f, \langle h_t, \theta \rangle\}_\theta, \langle h_{t'}, \theta \rangle\}_\theta + \{\langle h_t, \theta \rangle,$$
$$\{f, \langle h_{t'}, \theta \rangle\}_\theta\}_\theta = 0, \quad \forall t, t' \in \mathbb{R},$$

and, recurrently, f Poisson commutes with all successive Poisson brackets of elements of $\{\langle h_t, \theta \rangle\}_{t\in\mathbb{R}}$. The same happens with respect to all linear combinations of the elements $\langle h_t, \theta \rangle$ and their successive Lie brackets. As these elements span $\mathrm{Lie}(\phi_\theta(\{h_t\}_{t\in\mathbb{R}}), \{\cdot, \cdot\}_\theta) = \phi_\theta(\mathrm{Lie}(\{h_t\}_{t\in\mathbb{R}}, \{\cdot, \cdot\}_\Omega))$, we get that f Poisson commutes with $\phi_\theta(\mathrm{Lie}(\{h_t\}_{t\in\mathbb{R}}, \{\cdot, \cdot\}_\Omega))$.

Conversely, if f Poisson commutes with $\phi_\theta(\mathrm{Lie}(\{h_t\}_{t\in\mathbb{R}}, \{\cdot, \cdot\}_\Omega))$, it Poisson commutes, in particular, with the elements $\langle \{h_t\}_{t\in\mathbb{R}}, \theta \rangle$, and, in view of (7.68), it becomes a constant of motion for X admissible relative to Ω_θ. □

Each autonomous Hamiltonian system is a k-symplectic Lie system with respect to unique symplectic form ω. It therefore possesses a k-Hamiltonian structure (N, Ω, h), where h is a t-independent Hamiltonian. In this case, Proposition 7.23 states that the time-independent first integrals for a Hamiltonian system are the functions that Poisson commute with its Hamiltonian, retrieving this particular well-known result.

7.8. Diagonal Prolongations of k-Symplectic Lie Systems and Superposition Rules

This section shows that the diagonal prolongations of k-symplectic Lie systems are also k-symplectic Lie systems. This allows us to use the techniques depicted so far in this chapter to derive t-independent constants of motion and superposition rules for k-symplectic Lie systems [93]. To do so, let us define the diagonal prolongation of a section of a vector bundle (see [90] for details).

Let us recall certain notions concerning the diagonal prolongations of sections of vector bundles, say $\tau : E \to N$. The *diagonal prolongation* of $E \xrightarrow{\tau} N$ to N^m is the Cartesian product bundle $E^{[m]} = E \times \cdots \times E$ of m copies of E, viewed as a vector bundle over N^m in a natural way:

$$E^{[m]}_{(x_{(1)},\ldots,x_{(m)})} \simeq E_{x_{(1)}} \oplus \cdots \oplus E_{x_{(m)}} .$$

Every section $e : N \to E$ of the bundle E over N has a natural *diagonal prolongation* to a section $e^{[m]}$ of $E^{[m]}$ of the form

$$e^{[m]}(x_{(1)},\ldots,x_{(m)}) = e(x_{(1)}) + \cdots + e(x_{(m)}) .$$

Given a function $f : N \to \mathbb{R}$, the *diagonal prolongation* of f to N^m the function

$$\widetilde{f}^{[m]}(x_{(1)},\ldots,x_{(m)}) = f(x_{(1)}) + \cdots + f(x_{(m)}).$$

The above construction can also be used to define the diagonal prolongation of a t-dependent vector field X on N, let us say

$$X_t = \sum_{l=1}^{n} X^l(t, x) \frac{\partial}{\partial x^l}.$$

Its diagonal prolongation to N^m is the unique t-dependent vector field $\widetilde{X}^{[m]}$ on N^m such that $\widetilde{X}_t^{[m]} = X_t^{[m]}$ for each $t \in \mathbb{R}$, namely

$$X_t^{[m]} = \sum_{a=1}^{m} \sum_{l=1}^{n} X^l(t, x_{(a)}) \frac{\partial}{\partial x_{(a)}^l},$$

where $\{x_{(a)}^l \mid a = 1, \ldots, m, l = 1, \ldots, n = \dim N\}$ is the coordinate system on N^m given by defining $x_{(a)}^l(x_{(1)}, \ldots, x_{(m)}) = x^l(x_{(a)})$ for points $x_{(1)}, \ldots, x_{(m)} \in N$.

Let us use the above notions to show the following relevant result.

Proposition 7.24. *If X is a k-symplectic Lie system relative to $(\omega_1, \ldots, \omega_k)$, then $\widetilde{X}^{[m]}$ is a k-symplectic Lie system relative to $(\omega_1^{[m]}, \ldots, \omega_k^{[m]})$.*

Proof. The differential of the diagonal prolongation of a differential form is the prolongation of the differential form (cf. [90]). Then, since $\omega_1, \ldots, \omega_k$ are closed, so are the diagonal prolongations $\omega_1^{[m]}, \ldots, \omega_k^{[m]}$ to N^m. Hence, $\omega_1^{[m]}, \ldots, \omega_k^{[m]}$ are closed. Let us prove that $\omega_1^{[m]}, \ldots, \omega_k^{[m]}$ give rise to a k-symplectic structure. This amounts to proving that $\mathcal{D} := \bigcap_{i=1}^{k} \ker \omega_i^{[m]} = \{0\}$. Let us show it.

Let $\pi_j : N^m \to N$ be the projection of N^m onto the j-th component of N^m. If X takes values in $\mathcal{D}$, then

$$0 = (\omega_i^{[m]})_p \left(X_p, \frac{\partial}{\partial x_{(j)}^i} \bigg|_p \right) = (\pi_j^* \omega_i)_p \left(X_p, \frac{\partial}{\partial x_{(j)}^i} \bigg|_p \right),$$

$$i = 1, \ldots, k, \quad \forall p \in N^m.$$

Hence, $(\pi_j)_{*p}X_p \in \cap_{i=1}^k \ker(\omega_i)_{\pi_j(p)} = 0$. So, $(\pi_j)_{*p}X_p = 0$. Repeating the same for each j, we obtain $X = 0$. Therefore, $(\omega_1^{[m]}, \ldots, \omega_k^{[m]})$ is a k-symplectic structure.

If X was a k-symplectic Lie system relative to $(\omega_1, \ldots, \omega_k)$, then the minimal Lie algebra V^X consists of k-Hamiltonian vector fields relative to $(\omega_1, \ldots, \omega_k)$. In consequence, $X^{[m]}$ takes values in $(V^X)^{[m]}$, whose elements are the diagonal prolongations to N^m of the elements of V^X. If Y is a k-Hamiltonian vector field relative to $(\omega_1, \ldots, \omega_k)$, then its diagonal prolongation, $X^{[m]}$, to N^m is a k-Hamiltonian vector field relative to $(\omega_1^{[m]}, \ldots, \omega_k^{[m]})$. In consequence, the elements of $(V^X)^{[m]}$ are k-Hamiltonian relative to $(\omega_1^{[m]}, \ldots, \omega_k^{[m]})$ and $X^{[m]}$ becomes a k-symplectic Lie system. $\qquad\square$

Previous proposition gives the first justification of the utility of the following definition.

Definition 7.25. Given a polysymplectic form $\Omega = \sum_{i=1}^k \omega_i \otimes e^i$ on N, its *diagonal prolongation* to N^m is the polysymplectic form $\Omega^{[m]} = \sum_{i=1}^k \omega_i^{[m]} \otimes e^i$.

Let us illustrate previous results and their relevance by deriving through the distributional method (see Chapter 2) a superposition rule for Schwarzian equations written as a first-order system, X^{Sc}, given by (7.2).

Recall that Schwarzian equations are k-symplectic Lie systems admitting a VG Lie algebra of k-Hamiltonian vector fields $\langle X_1, X_2, X_3 \rangle$, given by (7.7), relative to a two-symplectic structure (ω_1, ω_2) on $\mathcal{O}_2$ described in (7.5).

Proposition 7.24 shows that the diagonal prolongation to $\mathcal{O}_2^{[2]}$ of the k-symplectic structure (ω_1, ω_2) gives rise to a new two-symplectic structure on this latter manifold. More particularly, the prolongations of ω_1 and ω_2 to $\mathcal{O}_2^{[2]}$ are closed forms given by

$$\omega_1^{[2]} = \sum_{i=1}^2 \frac{\mathrm{d}v_{(i)} \wedge \mathrm{d}a_{(i)}}{v_{(i)}}, \quad \omega_2^{[2]} = -\sum_{i=1}^2 \frac{2}{v_{(i)}^3}(x_{(i)}\mathrm{d}v_{(i)} \wedge \mathrm{d}a_{(i)}$$
$$+ v_{(i)}\mathrm{d}a_{(i)} \wedge \mathrm{d}x_{(i)} + a_{(i)}\mathrm{d}x_{(i)} \wedge \mathrm{d}v_{(i)}),$$

while their kernels are of the form

$$\ker \omega_1^{[2]} = \left\langle \frac{\partial}{\partial x_{(1)}}, \frac{\partial}{\partial x_{(2)}} \right\rangle,$$

$$\ker \omega_2^{[2]} = \bigoplus_{i=1}^{2} \left\langle x_{(i)} \frac{\partial}{\partial x_{(i)}} + v_{(i)} \frac{\partial}{\partial v_{(i)}} + a_{(i)} \frac{\partial}{\partial a_{(i)}} \right\rangle.$$

Proposition 7.24 ensures that both kernels have zero intersection. This can also be proved by a short direct calculation. Hence, $\left(\omega_1^{[2]}, \omega_2^{[2]} \right)$ becomes a k-symplectic structure on $\mathcal{O}_2^{[2]}$.

Proposition 7.24 also states that the diagonal prolongations of X_1, X_2, X_3 to $\mathcal{O}^2$ are k-Hamiltonian vector fields relative to $(\omega_1^{[2]}, \omega_2^{[2]})$. In fact, it is simply to see that the k-Hamiltonian functions for the diagonal prolongations $X_1^{[2]}, X_2^{[2]}, X^{[2]}$ read

$$h^{1,[2]} = \sum_{i=1}^{2} \left(\frac{2}{v_{(i)}} \otimes e^1 - \frac{4x_{(i)}}{v_{(i)}} \otimes e^2 \right),$$

$$h^{2,[2]} = \sum_{i=1}^{2} \left[\frac{a_{(i)}}{v_{(i)}^2} \otimes e^1 + \left(2 - \frac{2a_{(i)} x_{(i)}}{v_{(i)}^2} \right) \otimes e^2 \right],$$

and

$$h^{3,[2]} = \sum_{i=1}^{2} \left[\frac{a_{(i)}^2}{2v_{(i)}^3} \otimes e^1 + \left(\frac{2a_{(i)}}{v_{(i)}} - \frac{a_{(i)}^2 x_{(i)}}{v_{(i)}^3} \right) \otimes e^2 \right].$$

Moreover,

$$\left\{ h^{1,[2]}, h^{2,[2]} \right\}_{\Omega^{[2]}} = h^{1,[2]}, \quad \left\{ h^{1,[2]}, h^{3,[2]} \right\}_{\Omega^{[2]}} = 2h^{2,[2]},$$

$$\left\{ h^{2,[2]}, h^{3,[2]} \right\}_{\Omega^{[2]}} = h^{3,[2]}.$$

Therefore, the k-Hamiltonian functions $h^{1,[2]}, h^{2,[2]}, h^{3,[2]}$ span a Lie algebra isomorphic to $\mathfrak{sl}(2, \mathbb{R})$.

Let us now use the derived Poisson algebras of $\Omega^{[2]}$ to calculate t-independent constants of motion for $(X^{\mathrm{Sc}})^{[2]}$. The polysymplectic form $\Omega^{[2]}$ gives rise to several presymplectic form $\Omega_\xi^{[2]}$ by contracting $\Omega^{[2]}$ with any element of $\xi \in (\mathbb{R}^2)^*$, i.e., $\Omega_\xi^{[2]} = \langle \Omega^{[2]}, \xi \rangle$.

For instance, let $\{\theta_1, \theta_2\}$ be the dual basis to $\{e^1, e^2\}$. Let us define the presymplectic forms

$$\Omega_{\xi_1} := \langle \Omega^{[2]}, \theta_1 \rangle = \omega_1^{[2]}, \quad \Omega_{\xi_2} := \langle \Omega^{[2]}, \theta_2 \rangle = \omega_2^{[2]}.$$

Then, Proposition 7.16 shows that the Hamiltonian functions $(h^{1,[2]})_\xi, (h^{2,[2]})_\xi, (h^{3,[2]})_\xi$, for every $\xi \in (\mathbb{R}^2)^*$, span a real Lie algebra $\mathfrak{W}$ so that $\mathfrak{sl}(2, \mathbb{R})$ is a Lie algebra extension. Since $\mathfrak{sl}(2, \mathbb{R})$ is simple, $\mathfrak{W}$ is isomorphic to $\mathfrak{sl}(2, \mathbb{R})$ or zero.

Let $\{\cdot, \cdot\}_\xi$ be the Poisson bracket on the space of admissible functions of $\Omega_\xi^{[2]}$. If $\mathfrak{W}$ is isomorphic to $\mathfrak{sl}(2, \mathbb{R})$, it was proved in [90] that $\{C_\xi, (h^{i,[2]})_\xi\}_\xi = 0$ for $i = 1, 2, 3$ and

$$C_\xi = (h^{1,[2]})_\xi (h^{3,[2]})_\xi - (h^{2,[2]})_\xi^2.$$

Note that C_ξ can be obtained from a Casimir element of a Lie algebra isomorphic to $\mathfrak{sl}(2, \mathbb{R})$, constructing the basis by $(h^{1,[2]})_\xi$, $(h^{2,[2]})_\xi, (h^{3,[2]})_\xi$. In view of the above, C_ξ is a t-independent constant of motion for the diagonal prolongation $(X^{\text{Sc}})^{[2]}$.

If we write $\xi = \lambda_1 \theta_1 + \lambda_2 \theta_2$, with $\lambda_1, \lambda_2 \in \mathbb{R}$, then

$$C_\xi = [\lambda_1 (h^{1,[2]})_{\xi_1} + \lambda_2 (h^{1,[2]})_{\xi_2}][\lambda_1 (h^{3,[2]})_{\xi_1} + \lambda_2 (h^{3,[2]})_{\xi_2}]$$
$$- [\lambda_1 (h^{2,[2]})_{\xi_1} + \lambda_2 (h^{2,[2]})_{\xi_2}]^2,$$

and

$$C_\xi = \lambda_1^2 [(h^{1,[2]})_{\xi_1} (h^{3,[2]})_{\xi_1} - (h^{2,[2]})_{\xi_1}^2] + \lambda_2^2 [(h^{1,[2]})_{\xi_2} (h^{3,[2]})_{\xi_2}$$
$$- (h^{2,[2]})_{\xi_2}^2] + \lambda_1 \lambda_2 [(h^{1,[2]})_{\xi_1} (h^{3,[2]})_{\xi_2} + (h^{3,[2]})_{\xi_1} (h^{1,[2]})_{\xi_2}$$
$$- 2(h^{2,[2]})_{\xi_2} (h^{2,[2]})_{\xi_1}].$$

Hence, $C_\xi = \lambda_1^2 C_{\xi_1} + \lambda_2^2 C_{\xi_2} + \lambda_1 \lambda_2 F_{\xi_1 \xi_2}$ where

$$C_{\xi_1} = (h^{1,[2]})_{\xi_1} (h^{3,[2]})_{\xi_1} - (h^{2,[2]})_{\xi_1}^2 = \frac{(a_2 v_1 - a_1 v_2)^2}{v_1^3 v_2^3},$$

$$C_{\xi_2} = (h^{1,[2]})_{\xi_2} (h^{3,[2]})_{\xi_2} - (h^{2,[2]})_{\xi_2}^2$$
$$= -4 \left(-x_1 x_2 + \frac{2 v_1 v_2 (v_1 x_2 - v_2 x_1)}{a_1 v_2 - v_1 a_2} \right) \frac{(a_2 v_1 - a_1 v_2)^2}{v_1^3 v_2^3} - 4^2,$$

$$F_{\xi_1\xi_2} = (h^{1,[2]})_{\xi_1}(h^{3,[2]})_{\xi_2} + (h^{3,[2]})_{\xi_1}(h^{1,[2]})_{\xi_2} - 2(h^{2,[2]})_{\xi_2}(h^{2,[2]})_{\xi_1}$$

$$= -\frac{2(a_2v_1 - v_2a_1)^2}{v_1^3 v_2^3}\left(x_1 + x_2 - \frac{2v_1v_2(v_1 - v_2)}{a_1v_2 - v_1a_2}\right),$$

are the three constants of motion for $(X^{Sc})^{[2]}$.

Since C_{ξ_1} is a t-independent constant of motion, one can construct from $C_{\xi_2}, F_{\xi_1\xi_2}$ three new simpler t-independent constants of motion F_1, F_3, F_4:

$$F_1 = x_1 x_2 - \frac{2v_1v_2(v_1x_2 - v_2x_1)}{a_1v_2 - v_1a_2}, \quad F_3 = x_1 + x_2 - \frac{2v_1v_2(v_1 - v_2)}{a_1v_2 - v_1a_2},$$

$$F_4 = \sqrt{F_3^2 - 4F_1 + \frac{16}{C_{\xi_1}}} = x_1 - x_2 - \frac{2v_1v_2(v_1 + v_2)}{a_1v_2 - v_1a_2},$$

which were employed in [90, 280] to obtain the superposition rule for Schwarzian equation in first-order form.

To conclude the chapter, we would like to comment on the relevance of all the spread-out results. As we have pointed out, the complicated part in the derivation of a superposition rule through the distributional method resides in the derivation of three t-independent constants of motion for the diagonal prolongation $(X^{Sc})^{[2]}$, satisfying condition (3.43). These constants of motion C_{ξ_2}, F_3, F_4 were first obtained by the method of characteristics in [280], which was, as pictured, very tedious, or secondly, by using different Dirac–Lie systems and Lie symmetries of Schwarzian equations. The second approach was simpler because the use of Dirac–Lie systems allowed for the derivation of the superposition rule algebraically without integrating systems of PDEs as standard methods in [108, 376]. Instead, in this chapter, we have derived these constants of motion by only using the prolongation of the k-symplectic structure related to Schwarzian equations and algebraic techniques. This shows that k-symplectic structures provide the most straightforward method of derivation of such first integrals among previous methods.

For more applications of k-symplectic structures in the derivations of superposition rules, we refer the reader to [281].

Chapter 8

Conformal and Killing Lie Algebras on the Plane

8.1. Introduction

The local classification of the finite-dimensional real Lie algebras of vector fields on the plane was accomplished by Lie [5, 267]. González-López, Kamran, and Olver retrieved his classification via modern differential geometric techniques while solving unclear points in Lie's work that had been misunderstood in the previous literature [183]. We hereupon call their classification the *GKO classification*.

Our chapter focuses upon finite-dimensional Lie algebras of conformal and Killing vector fields on $\mathbb{R}^2$ relative to a pseudo-Riemannian metric. A *conformal vector field* relative to a pseudo-Riemannian metric g on a manifold M is a vector field X satisfying that $\mathcal{L}_X g = fg$ for a function f on M. If $f = 0$, then X is called a *Killing vector field* relative to g. Lie algebras of conformal and Killing vector fields are relevant due to their applications to Einstein equations [373], covariant quantizations [160, 256], and differential equations [73, 74, 215, 254].

The problem of classifying Lie algebras of conformal and/or Killing vector fields on types of manifolds has drawn certain attention [240]. For instance, the local form of Lie algebras of conformal vector fields relative to flat pseudo-Riemannian metrics on a manifold M is known. The case $\dim M = 2$ is the most puzzling one, as it leads to an infinite-dimensional Lie algebra of conformal vector fields [57, 351]. To this respect, Boniver and Lecomte proved that the Lie algebras of conformal polynomial vector fields on $\mathbb{R}^2$ relative to $\eta_\pm := \mathrm{d}x \otimes \mathrm{d}x \pm \mathrm{d}y \otimes \mathrm{d}y$ are maximal in the Lie algebra of polynomial vector fields in the variables x, y [57]. It is also interesting to study which finite-dimensional Lie algebras of conformal or Killing vector fields determine second-order ordinary differential equations [73, 74].

The local structure of finite-dimensional Lie algebras of conformal vector fields relative to a flat pseudo-Riemannian metric on $\mathbb{R}^2$ was studied in [197]. This result is here extended to finite-dimensional Lie algebras of conformal vector fields relative to any pseudo-Riemannian metric on $\mathbb{R}^2$ by using their conformal flatness [197, 351]. Our work also performs a local classification of Lie algebras of Killing vector fields on the plane relative to an arbitrary pseudo-Riemannian metric. Simple arguments are given so as to classify the so-called invariant distributions of finite-dimensional Lie algebras of vector fields on $\mathbb{R}^2$, which much simplifies the straightforward but long approach proposed in [197]. This result is interesting as it appears in the analysis of finite-dimensional Lie algebras of conformal vector fields on the plane [183, 197].

More specifically, we here prove that all conformal Lie algebras of vector fields relative to an arbitrary pseudo-Riemannian metric on $\mathbb{R}^2$ are, up to a local diffeomorphism, the Lie subalgebras of the Lie algebras I_7 and I_{11} of the GKO classification. Meanwhile, the Lie algebras of Killing vector fields on $\mathbb{R}^2$ are locally diffeomorphic to the Lie subalgebras of the Lie algebras I_4, $P_1^{\alpha=0}$, P_2, and P_3 of the GKO classification. The pseudo-Riemannian metrics associated with these Lie algebras are constructed via a certain type of tensor fields, the so-called *Casimir tensor fields* [23], derived by means of quadratic *Casimir elements* of the above-mentioned Lie algebras [382]. This result represents a new application of the theory of Casimir tensor fields initiated in [23]. Our classifications are detailed in Table A.2.

Finally, our findings are applied to Milne–Pinney equations and projective t-dependent Schrödinger equations, which are relevant differential equations frequently occurring in physics. These are types of *Lie–Hamilton systems* [121], namely they are differential equations describing the integral curves of a t-dependent vector field taking values in a finite-dimensional Lie algebra of Hamiltonian vector fields relative to a *Poisson bivector*. The Poisson bivectors associated with above-mentioned differential equations were obtained in previous works [23] by means of tedious calculations or *ad hoc* considerations. In this work, it is shown that they can be derived geometrically in an easy manner by our here developed application of *Casimir tensor fields*.

The structure of this chapter goes as follows. Section 8.2 addresses an introduction to finite-dimensional Lie algebras of vector fields. Section 8.3 surveys the theory of conformal and Killing Lie algebras of vector fields on $\mathbb{R}^2$. Section 8.4 is devoted to the classification of finite-dimensional Lie algebras of conformal and Killing vector fields on the plane, respectively. Section 8.5 addresses the calculation of invariant distributions for finite-dimensional Lie algebras of vector fields. Finally, Section 8.6 illustrates several applications of our results to Milne–Pinney and projective Schrödinger equations on the complex projective space $\mathbb{C}P^1$.

8.2. VG Lie Algebras and Their Distributions

This section surveys known results on the theory of Lie algebras of vector fields on the plane. Special attention is paid to finite-dimensional Lie algebras of vector fields, the so-called *VG Lie algebras* [99]. To simplify our presentation, manifolds are hereafter assumed to be connected.

Recall that *Stefan-Sussmann distribution*, also called generalized distribution, on M is a subset $\mathcal{D} \subset TM$ such that $\mathcal{D}_\xi := T_\xi M \cap \mathcal{D}$ is not empty for every $\xi \in M$. To simplify our terminology, we will refer to Stefan-Sussmann distributions as distributions. The distribution

$\mathcal{D}$ is *regular* at $\xi \in M$ if the rank of $\mathcal{D}$ is constant at points of an open $U \subset M$ containing ξ. The *domain* of $\mathcal{D}$ is the set $\mathrm{Dom}(\mathcal{D})$ of its regular points. If $\mathrm{Dom}(\mathcal{D}) = M$, then $\mathcal{D}$ is called *regular*. If a vector field X takes values in $\mathcal{D}$, it is written $X \in \mathcal{D}$. We write $\mathfrak{X}(M)$ for the space of vector fields on M.

Definition 8.1. Let V be a VG Lie algebra on M. The so-called *distribution* $\mathcal{D}^V$ *associated with* V takes the form

$$\mathcal{D}^V_\xi := \{X_\xi : X \in V\} \subset TM, \quad \forall \xi \in M.$$

A *generic point* for V is a regular point of $\mathcal{D}^V$. The *domain* of V is the set, $\mathrm{Dom}\, V$, of generic points of V.

Example 8.1. Consider the Lie algebra of vector fields on $\mathbb{R}^2$ given by

$$I_4 := \langle \partial_x + \partial_y, x\partial_x + y\partial_y, x^2\partial_x + y^2\partial_y \rangle. \tag{8.1}$$

The rank of the distribution $\mathcal{D}^{I_4}$ associated with I_4 at $(x, y) \in \mathbb{R}^2$ is given by the rank of

$$M(x, y) := \begin{pmatrix} 1 & x & x^2 \\ 1 & y & y^2 \end{pmatrix}.$$

The rank of $M(x, y)$ is two if and only if $x \neq y$. Hence, $\mathrm{Dom}(V) = \{(x, y) \in \mathbb{R}^2 : x \neq y\}$.

Definition 8.2. An *invariant distribution* of a Lie algebra V of vector fields on M is a distribution $\mathcal{D}$ on M different from $M \times \{0\}$ and TM satisfying that for every vector field $Y \in \mathcal{D}$ and $X \in V$, the vector field $[Y, X]$ takes values in $\mathcal{D}$.

It is straightforward to see that a distribution $\mathcal{D}$ is invariant relative to V if and only if the Lie bracket of any element from a fixed basis of V and any vector field of a fixed family of vector fields spanning $\mathcal{D}$ takes values in $\mathcal{D}$.

Example 8.2. The Lie algebra I_4 on $\mathbb{R}^2$ admits two-invariant distributions $\mathcal{D}^x$ and $\mathcal{D}^y$ generated by ∂_x or ∂_y, correspondingly. Indeed,

$\mathcal{D}^x$ is invariant relative to I_4 because the Lie bracket of a generating element of $\mathcal{D}^x$, e.g., ∂_x, and any element of the basis (8.1) of I_4 belongs to $\mathcal{D}^x$:

$$[\partial_x, \partial_x + \partial_y] = 0, \quad [\partial_x, x\partial_x + y\partial_y] = \partial_x, \quad [\partial_x, x^2\partial_x + y^2\partial_y] = 2x\partial_x.$$

Similarly, it can be proved that $\mathcal{D}^y$ is an invariant distribution relative to I_4.

Definition 8.3. A finite-dimensional Lie algebra of vector fields V on $\mathbb{R}^2$ is *imprimitive*, if it admits an invariant distribution. A Lie algebra V is *one-imprimitive* if it has only one invariant distribution, and it is *multi-imprimitive* if it admits more than one. If V has not invariant distributions, then V is called *primitive*.

Lie proved [5, 183] that every VG Lie algebra on $\mathbb{R}^2$ is locally diffeomorphic around a generic point to one of the Lie algebras described in Table A.2.

8.3. Conformal Geometry and Lie Algebras of Vector Fields on the Plane

Let us now survey the fundamental notions on conformal geometry and related VG Lie algebras of conformal and Killing vector fields to be used in following sections.

Definition 8.4. A *pseudo-Riemannian manifold* is a pair (M, g), where M is a manifold and g is a symmetric non-degenerate two-covariant tensor field on M: the *pseudo-Riemannian metric* of (M, g).

To simplify the notation, we will simply call g a *metric*. The manifold M will be assumed to be m-dimensional. In coordinates $\{x^i\}$, we can write $g = \sum_{i,j=1}^{m} g_{ij} \mathrm{d}x^i \otimes \mathrm{d}x^j$.

Example 8.3. Consider $\mathbb{R}^2$ with the global coordinates $\{x, y\}$. The metrics on $\mathbb{R}^2$ given by $g_E := \mathrm{d}x \otimes \mathrm{d}x + \mathrm{d}y \otimes \mathrm{d}y$ and $g_H := \mathrm{d}x \otimes \mathrm{d}y + \mathrm{d}y \otimes \mathrm{d}x$ will play a relevant role in this chapter. This is due to the fact that every flat metric on $\mathbb{R}^2$ can be mapped into one of

312 *A Guide to Lie Systems with Compatible Geometric Structures*

them, up to a non-zero multiplicative constant, by an appropriate diffeomorphism [134, 242].

Definition 8.5. Let (M, g) be a pseudo-Riemannian manifold, a vector field X on M is *conformal* relative to g if $\mathcal{L}_X g = f_X g$ for some function $f_X \in C^\infty(M)$. The function f_X is called the *potential* of X. If $f_X = 0$, it is said that X is a *Killing vector field* of g.

Example 8.4. Let us consider the *Kerr metric* [61] given by

$$g_K := -\left(1 - \frac{2GMr}{c^2\Sigma}\right)c^2 dt \otimes dt - \frac{4aGMr\sin^2\theta}{c^3\Sigma}c dt \otimes d\phi$$

$$+ \frac{\Sigma}{\Delta} dr \otimes dr + \Sigma d\theta \otimes d\theta$$

$$+ \left(r^2 + \frac{a^2}{c^2} + \frac{2a^2 Mr\sin^2\theta}{c^4\Sigma}\right)\sin^2\theta d\phi \otimes d\phi, \; G, c, M, \Sigma > 0.$$

This metric appears in the description of rotating black holes [134]. The vector fields ∂_t, ∂_φ are Killing vector fields relative to the Kerr metric, namely $\triangle\mathcal{L}_{\partial_t} g_K = \mathcal{L}_{\partial_\varphi} g_S = 0$.

Definition 8.6. Two metrics g_1 and g_2 on M are *conformally equivalent* if

$$\exists \sigma \in C^\infty(M), \quad g_1 = e^\sigma g_2.$$

Definition 8.7. A pseudo-Riemannian manifold (M, g) is *conformally flat* if g is conformally equivalent to a flat metric, i.e., there exists for each $x \in M$ an open $U^x \ni x$ and a function $f \in C^\infty(U^x)$ such that $g = e^{2f} g_f$ on U^x for a flat metric g_f on U^x.

Let us recall a relevant result concerning conformally flat metrics on the plane [242, 266, 348].

Theorem 8.8. *Every metric on the plane is conformally flat.*

Let us now discuss Lie algebras of conformal and Killing vector fields relative to a flat metric on $\mathbb{R}^2$. Definition 8.4, describing conformal and Killing vector fields, shows that Lie algebras of conformal vector fields relative to a metric g on M contain, as Lie subalgebras, the Killing vector fields relative to g.

The following proposition entails that the classification of Lie algebras of conformal vector fields on $\mathbb{R}^2$ relative to a metric amounts to classifying Lie algebras of conformal vector fields relative to the metrics g_E and g_H. This fact much simplifies the posterior determination of non-diffeomorphic Lie algebras of conformal Lie algebras relative to an arbitrary metric.

Proposition 8.9. *Lie algebras of conformal vector fields on $\mathbb{R}^2$ relative to definite (respectively, indefinite) metrics are diffeomorphic.*

Proof. Let X be a conformal vector field relative to a metric g_1. If g_2 is a metric on $\mathbb{R}^2$ conformally equivalent to g_1, then there exists around each point of $\mathbb{R}^2$ a function $f \in C^\infty(\mathbb{R}^2)$ such that $g_1 = e^f g_2$. Therefore,

$$
\begin{aligned}
f_X g_1 &= \mathcal{L}_X g_1 = \mathcal{L}_X e^f g_2 = (X e^f) g_2 + e^f \mathcal{L}_X g_2 \\
&\implies \mathcal{L}_X g_2 = (f_X - Xf) g_2,
\end{aligned}
$$

and X is a conformal vector field relative to g_2.

In view of the above, if V is a Lie algebra of conformal vector fields relative to g, then V is also a Lie algebra of conformal vector fields relative to any other conformally equivalent metric.

Since, by Theorem 8.8, all metrics on the plane are conformally flat, the Lie algebra of conformal vector fields of a general metric on $\mathbb{R}^2$ is the Lie algebra of conformal vector fields of a flat metric. Flat metrics can be mapped into g_E and g_H through a local diffeomorphism [369]. Therefore, the Lie algebra of conformal vector fields relative to a flat metric is, up to a diffeomorphism, the Lie algebra of conformal vector fields with respect to g_E or g_H depending on whether the initial metric was definite or indefinite, respectively.

In consequence, the Lie algebra of conformal vector fields relative to a metric on $\mathbb{R}^2$ is diffeomorphic to the Lie algebra of conformal vector fields relative to g_E, if the metric is definite, and to g_H, otherwise. $\qquad\square$

The above proposition allows us to slightly generalize the typical definition of conformal Lie algebras on $\mathbb{R}^2$ in terms of flat metrics as follows.

Definition 8.10. We call $\mathfrak{conf}(p, q)$ the abstract Lie algebra isomorphic to the Lie algebra of conformal vector fields relative to a metric on $\mathbb{R}^2$ with signature (p, q).

Let us now analyze the Lie algebras of conformal vector fields relative to a definite and indefinite metric on $\mathbb{R}^2$. This amounts to studying the Lie algebras of conformal vector fields relative to g_E or g_H, respectively.

Let $X = X^x \partial_x + X^y \partial_y$ be a conformal vector field relative to the flat metric g_E on $\mathbb{R}^2$. Then, $\mathcal{L}_X g_E = f_X g_E$ for some $f_X \in C^\infty(\mathbb{R}^2)$ and

$$\mathcal{L}_X g_E = 2\partial_x X^x \mathrm{d}x \otimes \mathrm{d}x + 2\partial_y X^y \mathrm{d}y \otimes \mathrm{d}y + \left(\partial_y X^x + \partial_x X^y\right)$$

$$\times (\mathrm{d}x \otimes \mathrm{d}y + \mathrm{d}y \otimes \mathrm{d}x) = f_X g_E.$$

Consequently, $\partial_x X^y + \partial_y X^x = 0, \partial_x X^x = \partial_y X^y = f_X/2$. and X is a conformal vector field relative to g_E if and only if the complex function $f : \mathbb{C} \ni z := x + \mathrm{i}y \mapsto X^x(x, y) + \mathrm{i}X^y(x, y) \in \mathbb{C}$, where $x, y \in \mathbb{R}$, is holomorphic, i.e., it satisfies the *Cauchy–Riemann conditions* [332]. Summing up,

$$\mathfrak{conf}(2, 0) = \mathfrak{conf}(0, 2) \simeq \left\{ f\partial_z \,|\, f : z \in \mathbb{C} \mapsto f(z) \in \mathbb{C} \text{ is holomorphic} \right\},$$

and $\mathfrak{conf}(2, 0)$, which is a realification of the so-called *Witt algebra* on the real line [360], becomes infinite-dimensional.

We now turn to studying the Lie algebra of conformal vector fields of the hyperbolic metric g_H. If X is a conformal vector field relative to g_H, then

$$\mathcal{L}_X g_H = \left(\partial_x X^x + \partial_y X^y\right) g_H + 2\partial_y X^x \mathrm{d}y \otimes \mathrm{d}y + 2\partial_x X^y \mathrm{d}x \otimes \mathrm{d}x = f_X g_H,$$

which occurs if and only if $\partial_y X^x = \partial_x X^y = 0, \partial_x X^x + \partial_y X^y = f_X$. Hence,

$$X = X^x(x)\partial_x + X^y(y)\partial_y \implies \mathfrak{conf}(1, 1) \simeq \mathfrak{X}(\mathbb{R}) \oplus \mathfrak{X}(\mathbb{R}).$$

Conformal and Killing Lie Algebras on the Plane 315

The study of previous Lie algebras and their use throughout this chapter entails the analysis of the so-called *pseudo-orthogonal Lie algebras* or *indefinite orthogonal Lie algebras*. These are matrix Lie algebras of the form

$$\mathfrak{so}(p,q) := \{A \in \mathfrak{gl}(p+q) : A^T \eta + \eta A = 0, \ \eta := \mathrm{diag}(\underbrace{+ \ldots +}_{p} \underbrace{- \ldots -}_{q})\},$$

where $\mathfrak{gl}(p+q)$ is the vector space of $(p+q) \times (p+q)$ matrices with real entries.

Example 8.5. It is possible to see that $P_7 \simeq \mathfrak{so}(3,1)$, where P_7 is given in Table A.2. In fact, it can be proved that the matrices of $\mathfrak{so}(3,1)$ can be parametrized through vectors $\mathbf{a}, \mathbf{b} \in \mathbb{R}^2$ and constants $\omega, \lambda \in \mathbb{R}$ in the form given in the left hand-side of the linear isomorphism morphism (see [14] for details)

$$\Psi : \begin{pmatrix} 0 & (-\mathbf{a}^T - \mathbf{c}^T)\eta & \lambda \\ \mathbf{c} - \mathbf{a} & -\omega & -\mathbf{a} - \mathbf{c} \\ \lambda & (\mathbf{a}^T - \mathbf{c}^T)\eta & 0 \end{pmatrix} \in \mathfrak{so}(3,1)$$

$$\mapsto [\mathbf{a} + \lambda\mathbf{x} + \omega\mathbf{x} + \mathbf{c}\mathbf{x}^T \eta\mathbf{x} - 2(\mathbf{c}^T \eta\mathbf{x})\mathbf{x}]\frac{\partial}{\partial\mathbf{x}} \in \mathfrak{conf}(2,0).$$

Similarly, one obtains that $I_{11} \simeq \mathfrak{so}(2,2)$ is a Lie algebra of conformal vector fields with respect to g_H. The Lie algebras P_7 and I_{11} are some of the most relevant Lie algebras to be treated hereafter.

The following proposition states that the curvature for metrics with a sufficient number of Killing vector fields must be constant. This shows that conformally equivalent metrics on the plane do not need to have the same Lie algebra of Killing vector fields, e.g., any of the many metrics with non-constant curvature on $\mathbb{R}^2$ and the conformally equivalent flat metric given by Proposition 8.8.

Proposition 8.11. *If V is a Lie algebra of Killing vector fields of a metric g on a connected manifold M and $\mathcal{D}^V = TM$, then the scalar curvature R of g is constant.*

Proof. It is known that the Killing vector fields of g are symmetries of the scalar curvature R thereof [369]. Since the vector fields in V

span the tangent bundle TM and R is a function, R must be a first integral of every vector field on M and, consequently, becomes a constant. $\square$

8.4. VG Lie Algebras of Conformal and Killing Vector Fields on $\mathbb{R}^2$

This section provides a characterization of VG Lie algebras of conformal and Killing vector fields on the plane. The results given in here represent a reinterpretation of the findings discovered in [197, 263]. Some previous findings have been written in a slightly more general form than in previous chapters while simpler proofs have been sometimes devised.

Lemma 8.12. *There exist no conformal vector fields X_1, X_2 relative to a conformally flat Riemannian metric on $\mathbb{R}^n$ such that $n > 1$ and $X_1 \wedge X_2 = 0$.*

Proof. Let us prove our claim by reduction to absurd. Assume then that there exist linearly independent (over $\mathbb{R}$) vector fields X_1, X_2 on $\mathbb{R}^n$ satisfying

$$X_1 \wedge X_2 = 0, \quad \mathcal{L}_{X_1} g = f_1 g, \quad \mathcal{L}_{X_2} g = f_2 g,$$

for some $f_1, f_2 \in C^\infty(\mathbb{R}^n)$ and a conformally flat metric g. Since X_1, X_2 are linearly independent over $\mathbb{R}$ and $X_1 \wedge X_2 = 0$, there exists around any point in $\mathbb{R}^n$ a non-constant function f such that $X_2 = f X_1$. Using this, we have

$$\begin{aligned}
f_2 g(X, Y) &= [\mathcal{L}_{X_2} g](X, Y) = [\mathcal{L}_{f X_1} g](X, Y) \\
&= f[\mathcal{L}_{X_1} g](X, Y) + g(X_1, (Xf)Y + (Yf)X),
\end{aligned}$$

for all $X, Y \in \mathfrak{X}(\mathbb{R}^n)$. Therefore,

$$(f_2 - f f_1) g(X, Y) = g(X_1, (Xf)Y + (Yf)X), \quad \forall X, Y \in \mathfrak{X}(\mathbb{R}^n). \tag{8.2}$$

Since $n > 1$, we can choose $X = Y \neq 0$ and $Xf = 0$. Substituting these values into the above equations, we obtain

$(f_2 - f f_1)g(X, X) = 0$. As g is a conformally flat Euclidean metric, it follows that $g(X, X) \neq 0$ for $X \neq 0$ and $f_2 = f f_1$. Substituting the latter in (8.2), we obtain

$$0 = g(X_1, (Xf)Y + (Yf)X), \quad \forall X, Y \in \mathfrak{X}(\mathbb{R}^n). \qquad (8.3)$$

The above can be rewritten as

$$0 = g(X_1, \iota_{df}(X \otimes Y)), \quad \forall X, Y \in \mathfrak{X}(\mathbb{R}^n), \qquad (8.4)$$

where $\iota_{df} : X \otimes Y \in TM \otimes_M TM \mapsto (Xf)Y + (Yf)X \in TM$ has kernel $Y \otimes Y$, with $Yf = 0$ for $df \neq 0$. Hence, every $Z \in TM$ can be written as $Z = \iota_{df}(X \otimes Y)$ and, according to (8.4), X_1 belongs to the kernel of g, which is a contradiction. $\qquad \square$

It is worth noting that Lemma 8.12 cannot be extended to $\mathbb{R}$. A contra-example is given by assuming the vector fields $X_1 := \partial_u$ and $X_2 := u\partial_u$, which are linearly independent and conformal relative to $du \otimes du$ on $\mathbb{R}$ whereas $X_1 \wedge X_2 = 0$.

Lemma 8.13. *Let V be a Lie algebra of conformal vector fields relative to a metric g on $\mathbb{R}^2$ and let $\mathcal{D}$ be an invariant distribution relative to V. Then,*

1. *the distribution $\mathcal{D}^\perp$ perpendicular to $\mathcal{D}$, i.e.,*

$$\mathcal{D}_\xi^\perp := \{X_\xi \in T_\xi M : g_\xi(X_\xi, \bar{X}_\xi) = 0, \forall \bar{X}_\xi \in \mathcal{D}_\xi\}, \quad \forall \xi \in \mathbb{R}^2,$$

is invariant relative to V.

2. *If V is a Lie algebra of conformal symmetries relative to a Riemannian metric on $\mathbb{R}^2$, then it is either primitive or multi-imprimitive.*

3. *The Lie algebra of conformal vector fields relative to an indefinite metric on $\mathbb{R}^2$ has, at least, two invariant distributions generated by commuting vector fields Y_1, Y_2 with $Y_1 \wedge Y_2 \neq 0$. Then, a conformal vector field can be brought into the form $Z = f_Z^1 Y_1 + f_Z^2 Y_2$, where $Y_1 f_Z^2 = Y_2 f_Z^1 = 0$ for some $f_Z^1, f_Z^2 \in C^\infty(\mathbb{R}^2)$.*

Proof. Point 1. If $X \in \mathcal{D}$ and $X^\perp \in \mathcal{D}^\perp$, then

$$0 = \mathcal{L}_Y[g(X, X^\perp)] = f_Y g(X, X^\perp) + g(\mathcal{L}_Y X, X^\perp) + g(X, \mathcal{L}_Y X^\perp)$$
$$= g(X, \mathcal{L}_Y X^\perp), \quad \forall Y \in V. \tag{8.5}$$

Therefore, $\mathcal{L}_Y X^\perp \in \mathcal{D}^\perp$ for every $Y \in V$ and $\mathcal{D}^\perp$ is invariant under V.

Point 2. Any one-dimensional distribution of rank one on $\mathbb{R}^2$ that is invariant relative to V satisfies, due to the non-degeneracy of g, that $\mathcal{D}_\xi \cap \mathcal{D}_\xi^\perp = \{0\}$ for every $\xi \in \mathbb{R}^2$. For the same reason, the distribution $\mathcal{D}^\perp$ has rank one. It stems from point 1 that $\mathcal{D}^\perp$ is invariant relative to V. In consequence, V must admit no invariant distribution of rank one, or two different invariant distributions or rank one.

Point 3. Every indefinite metric has, at least, a non-zero vector field of zero module. Let X be the non-zero vector field of module zero related to g and let $\mathcal{D}_X$ be the distribution generated on $\mathbb{R}^2$ by X. Let V be a Lie algebra of conformal vector fields relative to g. We first want to prove that $\mathcal{D}_X$ is an invariant distribution relative to V.

Since the rank of $\mathcal{D}_X^\perp$ is equal to the codimension of $\mathcal{D}_X$, namely $\dim \mathcal{D}_X^\perp = 1$, and $g(X, X) = 0$ by assumption, we get $\mathcal{D}_X^\perp = \mathcal{D}_X$. If we write $X^\perp = X$ in (8.5), then we obtain that $\mathcal{L}_Y X$ is perpendicular to X for every $Y \in V$. Consequently, $\mathcal{L}_Y X$ takes values in $\mathcal{D}_X^\perp = \mathcal{D}_X$ and $\mathcal{D}_X$ becomes invariant under the action of V.

At a fixed point $\xi \in \mathbb{R}^2$, there always exist two linearly independent tangent vectors with module zero relative to the indefinite metric g. It is simple to prove that such tangent vectors can be extended to two well-defined vector fields X_1, X_2 of module zero on a neighborhood of ξ spanning two distributions, $\mathcal{D}_{X_1}$ and $\mathcal{D}_{X_2}$, such that $\mathcal{D}_{X_1}$ and $\mathcal{D}_{X_2}$ have zero intersection on an open neighbourhood of ξ. This shows that $X_1 \wedge X_2 \neq 0$. As proved in point 1, the distributions $\mathcal{D}_{X_1}$ and $\mathcal{D}_{X_2}$ are invariant under the action of V. Hence, V is multi-imprimitive.

Let us prove that $\mathcal{D}_{X_1}, \mathcal{D}_{X_2}$ are spanned by two commuting vector fields Y_1, Y_2, respectively. Since $X_1 \wedge X_2 \neq 0$ and $X_1, X_2 \in \mathfrak{X}(\mathbb{R}^2)$, then $[X_1, X_2] = f_1 X_1 + f_2 X_2$ for functions $f_1, f_2 \in C^\infty(\mathbb{R}^2)$. For arbitrary functions h_1, h_2, we have $[h_1 X_1, h_2 X_2] = h_2 (h_1 f_1 - X_2 h_1) X_1 + h_1 (h_2 f_2 + X_1 h_2) X_2$. Let us prove that there exist local non-vanishing solutions h_1, h_2 of $X_2 h_1 = h_1 f_1$ and $X_1 h_2 = -h_2 f_2$ on an open interval of ξ. Since X_2 is a non-vanishing vector field, there exist coordinates ξ_1, ξ_2 on $\mathbb{R}^2$ such that $X_2 = \partial/\partial \xi_1$. Then, $\partial_{\xi_1} \ln |h_1| = f_1$ and we can choose $h_1 = \exp(\int f_1(\xi_1, \xi_2) \mathrm{d}\xi_1)$. The equation $X_1 h_2 = -h_2 f_2$ can be solved similarly. Hence, $Y_1 := h_1 X_1, Y_2 := h_2 X_2$ commute and span the distributions $\mathcal{D}_{X_1}$ and $\mathcal{D}_{X_2}$, respectively.

Every vector field $X \in V$ is a conformal one and then must leave the distributions $\mathcal{D}_{X_1}, \mathcal{D}_{X_2}$ invariant. Imposing that $\mathcal{L}_X Y_i$ belongs to the distribution $\mathcal{D}_{X_i}$ spanned by Y_i and recalling that $Y_1 \wedge Y_2 \neq 0$, we obtain that $X = f_X^1 Y_1 + f_X^2 Y_2$ with $Y_2 f_X^1 = Y_1 f_X^2 = 0$. $\qquad \square$

The following two theorems classify all Lie algebras of conformal vector fields on the plane up to diffeomorphisms.

Propositions 7.3 and 7.5 in [197] prove, in a quite technical way using a particular case of Lemma 8.13, analogs of Propositions 8.14 and 8.15 for Lie algebras of conformal vector fields relative to the metrics g_E and g_H. We already proved in the chapter that, since every metric on the plane is conformally flat, a Lie algebra consists of conformal vector fields relative to a (respectively, pseudo-Riemannian) Riemannian metric if and only if it is a Lie algebra of conformal vector fields relative to (respectively, g_H) g_E. In view of this, we straightforwardly get the following generalizations (see [263] for another formulation of the same result).

Proposition 8.14. *Every VG Lie algebra of conformal vector fields relative to a definite metric on the plane is locally diffeomorphic to the Lie algebras* I_1, $P_1, P_2, P_3, P_4, P_7, I_8^{\alpha=1}, I_{14}^{r=1}$ *of the GKO classification. The latter constitute, up to a diffeomorphism, the Lie subalgebras of* P_7.

Proposition 8.15. *Every VG Lie algebra of conformal vector fields relative to an indefinite metric on the plane is locally diffeomorphic* to the Lie algebras $I_1 - I_4, I_6, I_8^{\alpha=1}, I_9 - I_{11}, I_{14B}, I_{15B}$. The latter constitute, up to a diffeomorphism, the Lie subalgebras of I_{11}.

Let us now classify Lie algebras of Killing vector fields on $\mathbb{R}^2$ relative to a metric. Our findings are summarized in Table 8.1.

Theorem 8.16. *Let X_1, X_2, Y be Killing vector fields relative to a metric g on $\mathbb{R}^2$ such that $X_1 \wedge X_2 \neq 0$ and $[X_1, X_2] = 0$. Then:*

(1) *The functions $g(X_i, X_j)$, $i, j = 1, 2$, are constant.*
(2) *If $Y \wedge X_i = 0$ for a fixed $i \in \{1, 2\}$, then Y and the X_j, with $j = 1, 2$, are orthogonal or commute.*

Proof. Point 1. Since X is a Killing vector field for g and $[X_1, X_2] = 0$ by assumption, it follows that

$$X_i g(X_j, X_k) = (\mathcal{L}_{X_i} g)(X_j, X_k) + g(\mathcal{L}_{X_i} X_j, X_k)$$

$$+ g(X_j, \mathcal{L}_{X_i} X_k) = 0, \quad i, j, k = 1, 2. \quad (8.6)$$

As $X_1 \wedge X_2 \neq 0$ and $X_1, X_2 \in \mathfrak{X}(\mathbb{R}^2)$, the tangent vectors $(X_1)_\xi$, $(X_2)_\xi$ span $T_\xi \mathbb{R}^2$ at every point $\xi \in \mathbb{R}^2$. Then, expression (8.6) entails that $d[g(X_j, X_k)] = 0$ and $g(X_j, X_k)$ is a constant for $j, k = 1, 2$.

Point 2. Let us assume $Y \wedge X_1 = 0$. Since X_1 does not vanish, $Y = fX_1$ for a function $f \in C^\infty(\mathbb{R}^2)$. It stems from part 1 that $0 = \mathcal{L}_{fX_1} g(X_j, X_j)$ for $j = 1, 2$. Since Y is a Killing vector field for g and $[X_1, X_2] = 0$, it additionally follows that

$$0 = \mathcal{L}_{fX_1} g(X_j, X_j) = g(\mathcal{L}_{fX_1} X_j, X_j) + g(X_j, \mathcal{L}_{fX_1} X_j)$$

$$= -2(X_j f) g(X_1, X_j), \quad j = 1, 2.$$

Therefore, X_j and Y are orthogonal, i.e., $g(X_j, Y) = 0$, or $[X_j, Y] = (X_j f) X_1 = 0$. $\square$

The following lemma gives a simple form for a metric admitting a special type of Lie algebra of Killing vector fields frequently appearing in the GKO classification. This fact will be useful to classify VG Lie algebras consisting of Killing vector fields relative to a metric.

Lemma 8.17. *Let V be a Lie algebra of Killing vector fields relative to a metric g on $\mathbb{R}^2$ and let $X_1, X_2 \in V$ be such that $[X_1, X_2] = 0$*

Table 8.1.   GKO Classification of VG Lie algebras on $\mathbb{R}^2$.

#	Primitive	Basis X_i	Dom V	Inv. distribution	Kill.	Conf.
P_1	$A_\alpha \simeq \mathbb{R} \ltimes \mathbb{R}^2$	$\partial_x, \partial_y, \alpha(x\partial_x + y\partial_y) + y\partial_x - x\partial_y, \quad \alpha \geq 0$	$\mathbb{R}^2$	—	$+(\alpha = 0)$	g_E
P_2	$\mathfrak{sl}(2,\mathbb{R})$	$\partial_x, x\partial_x + y\partial_y, (x^2 - y^2)\partial_x + 2xy\partial_y$	$\mathbb{R}^2_{y\neq 0}$	—	$+$	g_E
P_3	$\mathfrak{so}(3)$	$y\partial_x - x\partial_y, (1+x^2-y^2)\partial_x + 2xy\partial_y, 2xy\partial_x + (1+y^2-x^2)\partial_y$	$\mathbb{R}^2$	—	$+$	g_E
P_4	$\mathbb{R}^2 \ltimes \mathbb{R}^2$	$\partial_x, \partial_y, x\partial_x + y\partial_y, y\partial_x - x\partial_y$	$\mathbb{R}^2$	—	—	g_E
P_5	$\mathfrak{sl}(2,\mathbb{R}) \ltimes \mathbb{R}^2$	$\partial_x, \partial_y, x\partial_x - y\partial_y, y\partial_x, x\partial_y$	$\mathbb{R}^2$	—	—	—
P_6	$\mathfrak{gl}(2,\mathbb{R}) \ltimes \mathbb{R}^2$	$\partial_x, \partial_y, x\partial_x, y\partial_x, x\partial_y, y\partial_y$	$\mathbb{R}^2$	—	—	—
P_7	$\mathfrak{so}(3,1)$	$\partial_x, \partial_y, x\partial_x + y\partial_y, y\partial_x - x\partial_y, (x^2 - y^2)\partial_x + 2xy\partial_y, 2xy\partial_x + (y^2 - x^2)\partial_y$	$\mathbb{R}^2$	—	—	g_E
P_8	$\mathfrak{sl}(3,\mathbb{R})$	$\partial_x, \partial_y, x\partial_x, y\partial_x, x\partial_y, y\partial_y, x^2\partial_x + xy\partial_y, xy\partial_x + y^2\partial_y$	$\mathbb{R}^2$	—	—	—

#	One-imprimitive	Basis X_i	Dom V	Inv. distribution	Kill.	Conf.
I_5	$\mathfrak{sl}(2,\mathbb{R})$	$\partial_x, 2x\partial_x + y\partial_y, x^2\partial_x + xy\partial_y$	$\mathbb{R}^2_{y\neq 0}$	∂_y	—	—
I_7	$\mathfrak{gl}(2,\mathbb{R})$	$\partial_x, y\partial_y, x\partial_x, x^2\partial_x + xy\partial_y$	$\mathbb{R}^2_{y\neq 0}$	∂_y	—	—
I_{12}	$\mathbb{R}^{r+1}$	$\partial_y, \xi_1(x)\partial_y, \ldots, \xi_r(x)\partial_y, \quad r \geq 1$	$\mathbb{R}^2$	∂_y	—	—
I_{13}	$\mathbb{R} \ltimes \mathbb{R}^{r+1}$	$\partial_y, y\partial_y, \xi_1(x)\partial_y, \ldots, \xi_r(x)\partial_y, \quad r \geq 1$	$\mathbb{R}^2$	∂_y	—	—
I_{14}	$\mathbb{R} \ltimes \mathbb{R}^r$	$\partial_x, \eta_1(x)\partial_y, \eta_2(x)\partial_y, \ldots, \eta_r(x)\partial_y, \quad (r > 1, \ r = 1, \ \eta_1'(x) \neq \eta_1(x))$	$\mathbb{R}^2$	∂_y	—	—
I_{15}	$\mathbb{R}^2 \ltimes \mathbb{R}^r$	$\partial_x, y\partial_y, \eta_1(x)\partial_y, \ldots, \eta_r(x)\partial_y, \quad (r > 1, \ r = 1, \ \eta_1'(x) \neq \eta_1(x))$	$\mathbb{R}^2$	∂_y	—	—
I_{16}	$C_\alpha^r \simeq \mathfrak{h}_2 \ltimes \mathbb{R}^{r+1}$	$\partial_x, \partial_y, x\partial_x + \alpha y\partial_y, x\partial_y, \ldots, x^r\partial_y, \quad r \geq 1, \quad \alpha \in \mathbb{R}$	$\mathbb{R}^2$	∂_y	—	—
I_{17}	$\mathbb{R} \ltimes (\mathbb{R} \ltimes \mathbb{R}^r)$	$\partial_x, \partial_y, x\partial_x + (ry + x^r)\partial_y, x\partial_y, \ldots, x^{r-1}\partial_y, \quad r \geq 1$	$\mathbb{R}^2$	∂_y	—	—
I_{18}	$(\mathfrak{h}_2 \oplus \mathbb{R}) \ltimes \mathbb{R}^{r+1}$	$\partial_x, \partial_y, x\partial_x, x\partial_y, y\partial_y, x^2\partial_y, \ldots, x^r\partial_y, \quad r \geq 1$	$\mathbb{R}^2$	∂_y	—	—
I_{19}	$\mathfrak{sl}(2,\mathbb{R}) \ltimes \mathbb{R}^{r+1}$	$\partial_x, \partial_y, x\partial_y, 2x\partial_x + ry\partial_y, x^2\partial_x + rxy\partial_y, x^2\partial_y, \ldots, x^r\partial_y, \quad r \geq 1$	$\mathbb{R}^2$	∂_y	—	—
I_{20}	$\mathfrak{gl}(2,\mathbb{R}) \ltimes \mathbb{R}^{r+1}$	$\partial_x, \partial_y, x\partial_x, x\partial_y, y\partial_y, x^2\partial_x + rxy\partial_y, x^2\partial_y, \ldots, x^r\partial_y, \quad r \geq 1$	$\mathbb{R}^2$	∂_y	—	—

(*Continued*)

Table 8.1. (*Continued*)

#	Multi-imprimitive	Basis X_i	Dom V	Inv. distribution	Kill.	Conf.
I_1	$\mathbb{R}$	∂_x	$\mathbb{R}^2$	$\partial_y, \partial_x + h(y)\partial_y$	–	g_E, g_H
I_2	$\mathfrak{h}_2$	$\partial_x, x\partial_x$	$\mathbb{R}^2$	∂_x, ∂_y	–	g_H
I_3	$\mathfrak{sl}(2,\mathbb{R})$	$\partial_x, x\partial_x, x^2\partial_x$	$\mathbb{R}^2$	∂_x, ∂_y	–	g_H
I_4	$\mathfrak{sl}(2,\mathbb{R})$	$\partial_x + \partial_y, x\partial_x + y\partial_y, x^2\partial_x + y^2\partial_y$	$\mathbb{R}^2_{x\neq y}$	∂_x, ∂_y	+	g_H
I_6	$\mathfrak{gl}(2,\mathbb{R})$	$\partial_x, \partial_y, x\partial_x, x^2\partial_x$	$\mathbb{R}^2$	∂_x, ∂_y	–	g_H
$I_8^{\alpha\neq1}$	$B_{\alpha\neq1} \simeq \mathbb{R}\ltimes\mathbb{R}^2$	$\partial_x, \partial_y, x\partial_x + \alpha y\partial_y, \quad 0<|\alpha|<1$	$\mathbb{R}^2$	∂_x, ∂_y	–	–
$I_8^{\alpha=1}$	$B_1 \simeq \mathbb{R}\ltimes\mathbb{R}^2$	$\partial_x, \partial_y, x\partial_x + y\partial_y$	$\mathbb{R}^2$	$\lambda_x \partial_x + \lambda_y \partial_y$	–	g_E, g_H
I_9	$\mathfrak{h}_2 \oplus \mathfrak{h}_2$	$\partial_x, \partial_y, y\partial_y$	$\mathbb{R}^2$	∂_x, ∂_y	–	g_H
I_{10}	$\mathfrak{sl}(2,\mathbb{R}) \oplus \mathfrak{h}_2$	$\partial_x, \partial_y, x\partial_x, y\partial_y, x^2\partial_x$	$\mathbb{R}^2$	∂_x, ∂_y	–	g_H
I_{11}	$\mathfrak{sl}(2,\mathbb{R}) \oplus \mathfrak{sl}(2,\mathbb{R})$	$\partial_x, \partial_y, x\partial_x, y\partial_y, x^2\partial_x, y^2\partial_y$	$\mathbb{R}^2$	∂_x, ∂_y	+	g_H
I_{14A}	$\mathbb{R}\ltimes\mathbb{R}$	$\partial_x, e^{cx}\partial_y, \quad c\in\mathbb{R}\setminus 0$	$\mathbb{R}^2$	$e^{cx}\partial_y, \partial_x + cy\partial_y$	+	g_E
I_{14B}	$\mathbb{R}\ltimes\mathbb{R}$	∂_x, ∂_y	$\mathbb{R}^2$	$\lambda_x \partial_x + \lambda_y \partial_y$	+	g_E, g_H
I_{15A}	$\mathbb{R}^2\ltimes\mathbb{R}$	$\partial_x, y\partial_y, e^{cx}\partial_y, \quad c\in\mathbb{R}\setminus 0$	$\mathbb{R}^2$	$e^{cx}\partial_y, \partial_x + cy\partial_y$	–	–
I_{15B}	$\mathbb{R}^2\ltimes\mathbb{R}$	$\partial_x, y\partial_y, \partial_y$	$\mathbb{R}^2$	∂_x, ∂_y	–	g_H

Notes: Functions $\xi_1(x),\ldots,\xi_r(x)$ are linearly independent, $\eta_1(x),\ldots,\eta_r(x)$ form a base of solutions to a linear system of r linear differential equations with constant coefficients. We write $\mathfrak{g} = \mathfrak{g}_1 \ltimes \mathfrak{g}_2$ to indicate that $\mathfrak{g}$ is the direct sum of $\mathfrak{g}_1$ and $\mathfrak{g}_2$, where $\mathfrak{g}_2$ is an ideal $\mathfrak{g}$. The symbol "+" in the column Kill. indicates that a Lie algebra consists of Killing vector fields relative to metric and "–" is written otherwise. The column Conf. details when a Lie algebra consists of conformal vector fields relative to a definite metric, (g_E), or a indefinite metric (g_H). The symbol "–" means that a Lie algebra does not consist of conformal vector fields relative to any metric.

Conformal and Killing Lie Algebras on the Plane   323

and $X_1 \wedge X_2 \neq 0$. Then, $g = c_{ij}\theta^i \otimes \theta^j$, where the c_{ij} are constant and θ^1, θ^2 are dual one-forms to X_1, X_2, i.e., $\theta^i(X_j) = \delta^i_j$, $i, j = 1, 2$.

Proof. The assumption $X_1 \wedge X_2 \neq 0$ implies that $\theta^1 \wedge \theta^2 \neq 0$. Then, the metric g can be brought into the form

$$g = g_{11}\theta^1 \otimes \theta^1 + g_{12}(\theta^1 \otimes \theta^2 + \theta^2 \otimes \theta^1) + g_{22}\theta^2 \otimes \theta^2,$$

for some uniquely defined functions $g_{ij} \in C^\infty(\mathbb{R}^2)$. Since $[X_i, X_j] = 0$ for $i, j = 1, 2$ by assumption,

$$(\mathcal{L}_{X_i}\theta^j)(X_k) = X_i[\theta^j(X_k)] - \theta^j([X_i, X_k]) = 0, \quad i, j = 1, 2.$$

Since $X_1 \wedge X_2 \neq 0$, one has that $\mathcal{L}_{X_i}\theta^j = 0$ for $i, j = 1, 2$. As X_i is a Killing vector field relative to g, it follows that $\mathcal{L}_{X_i}g = 0$, $i = 1, 2$. Due to this and as $\mathcal{L}_{X_i}\theta^j = 0$ for $i, j = 1, 2$, we get

$$\mathcal{L}_{X_i}g = (\mathcal{L}_{X_i}g_{11})\theta^1 \otimes \theta^1 + (\mathcal{L}_{X_i}g_{12})(\theta^1 \otimes \theta^2 + \theta^2 \otimes \theta^1)$$
$$+(\mathcal{L}_{X_i}g_{22})\theta^2 \otimes \theta^2 = 0, \quad i = 1, 2.$$

The last equality holds if and only if $\mathcal{L}_{X_i}g_{11} = \mathcal{L}_{X_i}g_{12} = \mathcal{L}_{X_i}g_{22} = 0$ for $i = 1, 2$. Since $X_1 \wedge X_2 \neq 0$, this means that the g_{ij} are constant for $i, j = 1, 2$. $\qquad\square$

Among the VG Lie algebras on the plane (see Table A.2), we aim to classify those Lie algebras V consisting of Killing vector fields relative to a metric g, namely $\mathcal{L}_X g = 0$, $\forall X \in V$. The following proposition is a consequence of Lemma 8.17.

Proposition 8.18. *A VG Lie algebra diffeomorphic to* I_{14B} *consists of Killing vector fields only relative to flat metrics on* $\mathbb{R}^2$.

Proof. Let us consider first the particular case of I_{14B} and then we will analyze the case of a VG Lie algebra V diffeomorphic to it.

If the elements of $\mathrm{I}_{14B} := \langle \partial_x, \partial_y \rangle$ are Killing vector fields relative to a metric g, then Lemma 8.17 ensures that $g = c_{xx}\mathrm{d}x \otimes \mathrm{d}x + c_{xy}(\mathrm{d}x \otimes \mathrm{d}y + \mathrm{d}y \otimes \mathrm{d}x) + c_{yy}\mathrm{d}y \otimes \mathrm{d}y$ for some constants c_{xx}, c_{xy}, c_{yy}. Then, if I_{14B} is a Lie algebra of Killing vector fields relative to a metric, then the metric must be flat. Additionally, the previous form of g ensures that $\mathcal{L}_Y g = 0$ for any $Y \in \mathrm{I}_{14B}$ and constants c_{xx}, c_{xy}, c_{yy}.

324 *A Guide to Lie Systems with Compatible Geometric Structures*

It is enough then to choose c_{xx}, c_{xy}, c_{yy} so that g is non-degenerate to verify that I_{14B} is a Lie algebra of Killing vector fields relative to it.

Let ϕ be the diffeomorphism on $\mathbb{R}^2$ mapping I_{14B} on V. Then, V will be a Lie algebra of Killing vector fields relative to a metric g if and only if I_{14B} is diffeomorphic to $\phi^* g$. Since $\phi^* g$ must be flat, so is g. $\qquad\square$

Proposition 8.19. *The Lie algebras on the plane diffeomorphic to $P_1^{\alpha \neq 0}, P_4 - P_8, I_6 - I_{11}, I_{16} - I_{20}$ do not consist of Killing vector fields relative to any metric on $\mathbb{R}^2$.*

Proof. If V is a VG Lie algebra on the plane diffeomorphic to a Lie algebra B of the GKO classification, then V consists of Killing vector fields relative a metric if and only if B does. Consequently, the enunciate of this proposition reduces to proving that the Lie algebras $P_1^{\alpha \neq 0}, P_4 - P_8, I_6 - I_{11}, I_{16} - I_{20}$ do not consist of Killing vector fields to any metric.

Propositions 8.14 and 8.15 state that P_5, P_6, P_8, I_7, $I_8^{\alpha \neq 1}$, I_{10}, and I_{16}–I_{20} are not among the Lie algebras of conformal vector fields. Hence, they cannot be Lie algebras of Killing vector fields. Let us then focus on the remaining cases, namely

$$P_1^{\alpha \neq 0}, P_4, P_7, I_6, I_8^{\alpha = 1}, I_9, I_{11}. \tag{8.7}$$

Let us proceed by reduction to the absurd, and we assume that previous Lie algebras consist of Killing vector fields relative to a metric on $\mathbb{R}^2$. Lie algebras (8.7) satisfy the conditions given in Lemma 8.17 for $X_1 = \partial_x$ and $X_2 = \partial_y$. The dual one-forms to X_1, X_2 read $\theta_1 = \mathrm{d}x$ and $\theta_2 = \mathrm{d}y$. Hence,

$$g = c_{xx}\mathrm{d}x \otimes \mathrm{d}x + c_{xy}(\mathrm{d}x \otimes \mathrm{d}y + \mathrm{d}y \otimes \mathrm{d}x) + c_{yy}\mathrm{d}y \otimes \mathrm{d}y, \tag{8.8}$$

for certain constants c_{xx}, c_{xy}, c_{yy}.

- *Lie algebra $P_1^{\alpha \neq 0}$*: Let us take $X_3 := \alpha(x\partial_x + y\partial_y) + y\partial_x - x\partial_y \in P_1$ where $\alpha > 0$. Since X_3 is a Killing vector field relative to g, then

$$\mathcal{L}_{X_3}g = 2(\alpha c_{xx} - c_{xy})\mathrm{d}x \otimes \mathrm{d}x + (c_{xx} + \alpha c_{xy} - c_{yy})(\mathrm{d}x \otimes \mathrm{d}y$$
$$+ \mathrm{d}y \otimes \mathrm{d}x) + 2(c_{xy} + \alpha c_{yy})\mathrm{d}y \otimes \mathrm{d}y = 0,$$

and therefore condition $\mathcal{L}_{X_3}g = 0$ amounts to

$$2(\alpha c_{xx} - c_{xy}) = (c_{xx} + \alpha c_{xy} - c_{yy}) = 2(c_{xy} + \alpha c_{yy}) = 0$$
$$\Rightarrow \alpha^2 c_{xx}(2 + \alpha^2) = 0.$$

Since $\alpha \neq 0$ by assumption, $c_{xx} = c_{xy} = c_{yy} = 0$ and $g = 0$. This is a contradiction and $\mathrm{P}_1^{\alpha \neq 0}$ does not consist of Killing vector fields for any g on $\mathbb{R}^2$.

- *Lie algebras* P_4, P_6, P_7, *and* I_8: If we choose $X_3 = x\partial_x + y\partial_y \in$ P_4, P_6, P_7, I_8, then

$$\mathcal{L}_{X_3}g = \mathcal{L}_{x\partial_x + y\partial_y}(c_{xx}\mathrm{d}x \otimes \mathrm{d}x + c_{xy}(\mathrm{d}x \otimes \mathrm{d}y + \mathrm{d}y \otimes \mathrm{d}x)$$
$$+ c_{yy}\mathrm{d}y \otimes \mathrm{d}y) = 2g. \tag{8.9}$$

Therefore, X_3 cannot be a Killing vector field. Hence, P_4, P_6, P_7, I_8 do not consist of Killing vector fields relative to any metric on the plane.

- *Lie algebras* I_6, I_9, I_{10}: All these Lie algebras contain the vector field $X_3 = x\partial_x$. In view of (8.8) and requiring X_3 to be a Killing vector field, one has that

$$\mathcal{L}_{X_3}g = 2c_{xx}\mathrm{d}x \otimes \mathrm{d}x + c_{xy}(\mathrm{d}x \otimes \mathrm{d}y + \mathrm{d}y \otimes \mathrm{d}x) = 0.$$

This implies that $c_{xy} = c_{xx}$ and g cannot be a metric, which is a contradiction. Hence, none of the previous Lie algebras consists of Killing vector fields relative to any g on $\mathbb{R}^2$. $\qquad\square$

Corollary 8.20. *If V is a VG Lie algebra on $\mathbb{R}^2$ containing linearly independent X_1, X_2, X_3 such that $[X_1, X_2] = [X_2, X_3] = 0$, $[X_1, X_3] \neq 0$, $X_2 \wedge X_3 \neq 0$, $X_1 \wedge X_3 = 0$, then V is not a Lie algebra of Killing vector fields related to any metric.*

Proof. Let us proceed by reduction to the absurd and assume that V is a VG Lie algebra of Killing vector fields relative to a metric g.

Since $X_1 \wedge X_3 = 0$ but $X_2 \wedge X_3 \neq 0$, there exists a non-zero function $f \in C^\infty(\mathbb{R}^2)$ such that, $X_1 = f(\xi)X_3$, $\forall \xi \in \mathbb{R}^2$. Since X_1 is a Killing vector field by assumption, one has $\mathcal{L}_{X_1}g = 0$. Also from assumption $X_3 \wedge X_2 \neq 0$. Hence, using Lemma 8.17 and $[X_3, X_2] = 0$, we find that

$$0 = \mathcal{L}_{fX_3}[g(X_3, X_3)] = -2(X_3 f)g(X_3, X_3), \quad 0 = \mathcal{L}_{fX_3}[g(X_3, X_2)]$$
$$= -(X_3 f)g(X_3, X_2).$$

Since $[X_1, X_3] \neq 0$ by assumption, if follows that $X_3 f \neq 0$ and $g(X_3, X_3) = g(X_3, X_2) = 0$. From this result and as $X_2 \wedge X_3 \neq 0$, it turns out that g is degenerate. This is a contradiction, which finishes the proof. $\square$

Proposition 8.21. *VG Lie algebras diffeomorphic to* $\mathrm{P}_1^{\alpha=0}$, P_3, *and* I_4 *are Lie algebras of Killing vector fields relative to some metrics on the plane.*

Proof. As in previous results, the proof of the statement for VG Lie algebras diffeomorphic to $\mathrm{P}_1^{\alpha=0}$, P_3, and I_4 follows from the fact that the latter Lie algebras are itself Lie algebras of Killing vector fields relative to some metric g. Let us then analyze the cases concerning $\mathrm{P}_1^{\alpha=0}$, P_3, and I_4.

In coordinates x, y on $\mathbb{R}^2$,

$$g = g_{xx}\mathrm{d}x \otimes \mathrm{d}x + g_{xy}(\mathrm{d}x \otimes \mathrm{d}y + \mathrm{d}y \otimes \mathrm{d}x) + g_{yy}\mathrm{d}y \otimes \mathrm{d}y, \quad (8.10)$$

for certain functions $g_{xx}, g_{xy}, g_{yy} \in C^\infty(\mathbb{R}^2)$. Let us analyze the possible values of g making $\mathrm{P}_1^{\alpha=0}$, P_3, and I_4 consist of Killing vector fields.

- *Lie algebra* $\mathrm{P}_1^{\alpha=0}$: Let us search functions $g_{xx}, g_{xy}, g_{yy} \in C^\infty(\mathbb{R}^2)$ turning the vector fields of

$$\mathrm{P}_1^{\alpha=0} = \langle X_1 := \partial_x, \ X_2 := \partial_y, \ X_3 := y\partial_x - x\partial_y \rangle,$$

into Killing vector fields relative to g, i.e., $\mathcal{L}_{X_k} g = 0$ for $k = 1, 2, 3$. By applying Lemma 8.17 with $X_1 = \partial_x, X_2 = \partial_y$, we obtain

$$g = c_{xx}\mathrm{d}x \otimes \mathrm{d}x + c_{xy}(\mathrm{d}x \otimes \mathrm{d}y + \mathrm{d}y \otimes \mathrm{d}x) + c_{yy}\mathrm{d}y \otimes \mathrm{d}y,$$
$$c_{xx}, c_{xy}, c_{yy} \in \mathbb{R}.$$

Meanwhile, the vector field X_3 is Killing if and only if

$$\mathcal{L}_{y\partial_x - x\partial_y} g = (c_{xx} - c_{yy})(\mathrm{d}x \otimes \mathrm{d}y + \mathrm{d}y \otimes \mathrm{d}x)$$
$$+ 2c_{xy}(\mathrm{d}y \otimes \mathrm{d}y - \mathrm{d}x \otimes \mathrm{d}x) = 0.$$

The last equality is satisfied if and only if $c_{xx} = c_{yy}, c_{xy} = 0$. Hence, the Lie algebra $\mathrm{P}_1^{a=0}$ is a Lie algebra of Killing vector fields only relative to a flat Riemannian metric

$$g = c_{xx}(\mathrm{d}x \otimes \mathrm{d}x + \mathrm{d}y \otimes \mathrm{d}y), \quad c_{xx} \in \mathbb{R}\backslash\{0\}. \tag{8.11}$$

- *Lie algebra* P_3: Let us determine the functions $g_{xx}, g_{xy}, g_{yy} \in C^\infty(\mathbb{R}^2)$ ensuring that

$$\mathrm{P}_3 = \langle X_1 := y\partial_x - x\partial_y, \ X_2 := (1 + x^2 - y^2)\partial_x + 2xy\partial_y,$$
$$X_3 := 2xy\partial_x + (1 + y^2 - x^2)\partial_y \rangle$$

consists of Killing vector fields relative to a metric g, i.e., $\mathcal{L}_{X_k} g = 0$ for $k = 1, 2, 3$. This condition for $k = 1$ takes the form

$$\begin{cases} yg_{xx,x} - xg_{xx,y} - 2g_{xy} = 0, \\ 2g_{xy} + yg_{yy,x} - xg_{yy,y} = 0, \\ g_{xx} - g_{yy} + yg_{xy,x} - xg_{xy,y} = 0. \end{cases}$$

$$\Longleftrightarrow \begin{cases} y(g_{xx} + g_{yy})_{,x} - x(g_{xx} + g_{yy})_{,y} = 0, \\ y(g_{xx} - g_{yy})_{,x} - x(g_{xx} - g_{yy})_{,y} - 4g_{xy} = 0, \\ g_{xx} - g_{yy} + yg_{xy,x} - xg_{xy,y} = 0, \end{cases}$$

where subscripts given by a coordinate after a comma determine a derivative with respect to that coordinate. Above equations can exactly be solved by using polar coordinates and defining $h(x, y) = g_{xx} + g_{yy}$ and $g(x, y) = g_{xx} - g_{yy}$. For our purposes, it is nevertheless enough to consider the case $g(x, y) = 0$, i.e.,

$g_{xx} = g_{yy} =: f(x,y)$. Then, $yf_{,x} - xf_{+,y} = 0$ and $g_{xy} = 0$. Consequently, X_1 is a Killing vector field for

$$g_f := f(x,y)[\mathrm{d}x \otimes \mathrm{d}x + \mathrm{d}y \otimes \mathrm{d}y].$$

Let us determine $f \in C^\infty(\mathbb{R}^2)$ so that $\mathcal{L}_{X_k} g_f = 0$ for $k = 2,3$. Hence,

$$\begin{cases} \mathcal{L}_{X_2} g = ((1 + x^2 - y^2)f_{,x} + 2xyf_{,y} + 4x)(\mathrm{d}x \otimes \mathrm{d}x + \mathrm{d}y \otimes \mathrm{d}y) = 0, \\ \mathcal{L}_{X_3} g = (2xyf_{,x} + (1 + y^2 - x^2)f_{,y} + 4y)(\mathrm{d}x \otimes \mathrm{d}x + \mathrm{d}y \otimes \mathrm{d}y) = 0. \end{cases}$$
$$(8.12)$$

From the system (8.12), one gets

$$\begin{cases} (1 + x^2 - y^2)f_{,x} + 2xyf_{,y} + 4x = 0 \\ 2xyf_{,x} + (1 + y^2 - x^2)f_{,y} + 4y = 0 \end{cases}$$

$$\implies \begin{pmatrix} f_{,x} \\ f_{,y} \end{pmatrix} = -\frac{4}{1 + x^2 + y^2} \begin{pmatrix} x \\ y \end{pmatrix}. \qquad (8.13)$$

As commented, to prove that P_3 is a VG Lie algebra of Killing vector fields, one only needs to provide a solution of previous equations, e.g., $f(x,y) = \ln[\lambda/(1 + x^2 + y^2)^2]$ for a constant $\lambda \in \mathbb{R}\setminus\{0\}$. Hence, P_3 consists of Killing vector fields relative to the Riemannian metric

$$g = \ln[\lambda/(1 + x^2 + y^2)^2](\mathrm{d}x \otimes \mathrm{d}x + \mathrm{d}y \otimes \mathrm{d}y). \qquad (8.14)$$

Since the Lie algebra P_3 is primitive (see Table 8.1), it does not admit any invariant distribution. Moreover, Lemma 8.13 ensures then that P_3 is not a Lie algebra of Killing vector fields relative to any indefinite metric on $\mathbb{R}^2$.

- *Lie algebra* I_4: We now study the Lie algebra I_4. In this case, $g_{xx}, g_{xy}, g_{yy} \in C^\infty(\mathbb{R}^2)$ must be chosen so that the Lie algebra

$$I_4 = \langle X_1 := \partial_x + \partial_y, \ X_2 := x\partial_x + y\partial_y, X_3 := x^2\partial_x + y^2\partial_y \rangle$$

will consist of Killing vector fields relative to g, i.e., $\mathcal{L}_{X_k} g = 0$ for $k = 1,2,3$. In particular, a straightforward calculation shows that

$$\mathcal{L}_{X_1} g = 0 \implies X_1 g_{xx} = X_1 g_{yy} = X_1 g_{xy} = 0.$$

Introducing coordinates $\xi_1 := x - y$ and $\xi_2 := x + y$, we obtain that $X_1 f = 2\partial_{\xi_2} f = 0$ and $f = f(x - y)$. Consequently, $g_{xx} = h_{xx}(x - y), h_{yy} = h_{yy}(x - y), g_{xy} = h_{xy}(x - y)$ for certain $h_{xx}, h_{yy}, h_{xy} \in C^\infty(\mathbb{R})$.

Meanwhile, the condition $\mathcal{L}_{X_2} g = 0$ amounts to the fact that

$$X_2 g_{xx} + 2g_{xx} = X_2 g_{xy} + 2g_{xy} = X_2 g_{yy} + 2g_{yy} = 0.$$

And recalling the explicit dependence of g_{xx}, g_{yy}, g_{xy} we obtain

$$(x - y)h'_{xx} + 2h_{xx} = (x - y)h'_{xy} + 2h_{xy} = (x - y)h'_{yy} + 2h_{yy} = 0.$$

Since the solution to $(x - y)f'(x - y) + 2f(x - y) = 0$ is $f(x - y) = \lambda/(x - y)^2$, $\lambda \in \mathbb{R}$, the metric g takes the form

$$g = \frac{1}{(x - y)^2} [c_{xx} \mathrm{d}x \otimes \mathrm{d}x + c_{xy}(\mathrm{d}x \otimes \mathrm{d}y + \mathrm{d}y \otimes \mathrm{d}x) + c_{yy} \mathrm{d}y \otimes \mathrm{d}y],$$

for some constants c_{xx}, c_{xy}, c_{yy}. If we finally impose on the previous expression that $\mathcal{L}_{X_3} g = 0$, we are led to

$$\mathcal{L}_{X_3} g = \frac{2}{(x - y)} [c_{xx} \mathrm{d}x \otimes \mathrm{d}x + c_{yy} \mathrm{d}y \otimes \mathrm{d}y] = 0.$$

And this implies that $c_{xx} = c_{yy} = 0$, which in turn causes I_4 to be a Lie algebra of Killing vector fields relative to an indefinite metric. $\qquad\square$

Proposition 8.21 ensures that $P_1^{\alpha=0}$ is a VG Lie algebra of Killing vector fields relative only to a metric proportional, up to a certain function, to g_E. Meanwhile, P_3 consists of Killing vector fields only with respect to Riemannian metrics. Finally, I_4 is a Lie algebra of Killing vector fields only relative to indefinite metrics on $\mathbb{R}^2$ taking, up to a non-zero proportional constant, the form $g = (\mathrm{d}x \otimes \mathrm{d}y + \mathrm{d}y \otimes \mathrm{d}x)/(x - y)^2$. Since all previous Lie algebras have associated distributions of rank two and the curvature tensor R for each metric is invariant under Killing vector fields, it follows that R is covariant invariant and the corresponding spaces are *locally Riemannian*.

330 *A Guide to Lie Systems with Compatible Geometric Structures*

In general, one sees that Proposition 8.21 along with Propositions 8.18 and 8.19 almost complete the classification of Lie algebras of Killing vector fields on the plane. It is only left to analyze P_2 and I_{14A}. To study these last cases and to understand some of the previous results more geometrically, we will use the algebraic properties of universal enveloping algebras.

It was already proved in previous chapters that Casimir elements play a rôle in the determination of invariants for VG Lie algebras. In the same spirit, Proposition 8.21 can be reinterpreted as the result of the existence of a type of quadratic Casimir element for the Lie algebras $P_1^{\alpha=0}$, I_4, and P_3. This section aims to explain in detail this relevant fact, which will also enable the description of all Lie algebras of Killing vector fields on $\mathbb{R}^2$ relative to arbitrary metrics.

Let $\phi : \mathfrak{g} \to \mathfrak{X}(M)$ be a Lie algebra morphism. We already commented in Chapter 2 that the *universal enveloping algebra*, $U(\mathfrak{g})$, of $\mathfrak{g}$ is isomorphic to the *symmetric tensor algebra*, $S(\mathfrak{g})$, of $\mathfrak{g}$ (see also [359]). This allows for the extension of ϕ to a unique morphism of associative algebras $\Upsilon : U(\mathfrak{g}) \simeq S(\mathfrak{g}) \to S(M)$, where $S(M)$ stands for the space of symmetric tensor fields on M. A Lie algebra representation $\rho_{\mathfrak{g}} : \mathfrak{g} \to \text{End}(U(\mathfrak{g}))$ can be obtained by extending the derivation $\text{ad}_v : w \in \mathfrak{g} \mapsto [v, w] \in \mathfrak{g}$, with $v \in \mathfrak{g}$, to a derivation $[v, \cdot]_{U(\mathfrak{g})}$ on $U(\mathfrak{g})$. If $V := \phi(\mathfrak{g})$, then it is possible to define a second Lie algebra representation $\rho_V : X \in V \mapsto \mathcal{L}_X \in \text{End}(S(M))$. Then, it is immediate to verify that

$$\Upsilon([v, C]_{U(\mathfrak{g})}) = \mathcal{L}_{\Upsilon(v)}\Upsilon(C), \quad \forall v \in \mathfrak{g}, \forall C \in U(\mathfrak{g}).$$

Therefore, if $C \in U(\mathfrak{g})$ is a *Casimir element* of $\mathfrak{g}$, i.e., $[v, C]_{U(\mathfrak{g})} = 0$ for all $v \in \mathfrak{g}$, then $\mathcal{L}_X\Upsilon(C) = 0$ for every $X \in \phi(\mathfrak{g})$.

Particular types of symmetric tensor fields of the form $\Upsilon(C)$, where C is a Casimir for $\mathfrak{sl}(2, \mathbb{R})$, have appeared previously in this work (see also [23, 189, 263]), where they were called *Casimir tensor fields*. According to this terminology, we will call the $\Upsilon(C)$, for C being a Casimir for a certain Lie algebra, *Casimir tensor fields*.

Theorem 8.22. *Let V be a VG Lie algebra whose isomorphic abstract Lie algebra $\mathfrak{g}$ admits a quadratic Casimir element $C \in U(\mathfrak{g})$*

Conformal and Killing Lie Algebras on the Plane 331

such that $\Upsilon(C)$ *is non-degenerate. Then, V consists of Killing vector fields relative to $\Upsilon(C)^{-1}$.*

Proof. Since C is a Casimir element of $\mathfrak{g}$, one has that $\mathcal{L}_X \Upsilon(C) = 0$ for every $X \in V$, which turns $\Upsilon(C)$ into a symmetric tensor field on M invariant relative to the vector fields of V. In local coordinates, $G := \Upsilon(C) = g^{\mu\nu} \partial_\mu \otimes \partial_\nu$ for certain functions $g^{\mu\nu} \in C^\infty(M)$. Then $\mathcal{L}_X G = 0$ for every $X \in V$ is equivalent to say, if $X = X^\alpha \partial_\alpha$ in local coordinates, that

$$(\mathcal{L}_X G)^{\mu\nu} = X^\alpha \partial_\alpha g^{\mu\nu} - (\partial_\alpha X^\mu) g^{\alpha\nu} - (\partial_\alpha X^\nu) g^{\alpha\mu} = 0. \qquad (8.15)$$

Since G is non-degenerate by assumption, the matrix $g^{\mu\nu}$ has an inverse $g_{\mu\nu}$. If we write $g := g_{\mu\nu} dx^\mu \otimes dx^\nu$ in coordinates, then

$$(\mathcal{L}_X g)_{\mu\nu} = X^\alpha \partial_\alpha g_{\mu\nu} + (\partial_\mu X^\alpha) g_{\alpha\nu} + (\partial_\nu X^\alpha) g_{\alpha\mu}. \qquad (8.16)$$

Substituting the equality $\partial_\alpha g_{\mu\nu} = -g_{\mu\pi}(\partial_\alpha g^{\pi\kappa}) g_{\kappa\nu}$ into (8.16) and using (8.15), it turns out that $\mathcal{L}_X g = 0$ and V becomes a Lie algebra of Killing vector fields relative to the metric g. $\qquad \square$

Example 8.6. Let us use Theorem 8.22 to show that $P_1^{\alpha=0}$ consists of Killing vector fields relative to a metric on $\mathbb{R}^2$. Let $\mathfrak{g}$ be an abstract Lie algebra isomorphic to $P_1^{\alpha=0}$ admitting a basis $\{v_1, v_2, v_3\}$ obeying the same commutation relations as the basis of vector fields $\{X_1, X_2, X_3\}$ for $P_1^{\alpha=0}$ given in Table A.2. This leads to a Lie algebra morphism $\phi : \mathfrak{g} \to \mathfrak{X}(\mathbb{R}^2)$ mapping each v_i into X_i. The Lie algebra $\mathfrak{g}$ admits a quadratic Casimir element $v_1 \otimes v_1 + v_2 \otimes v_2$. If $\Upsilon : U(\mathfrak{g}) \to S(\mathbb{R}^2)$ is the associative algebra morphism induced by ϕ, then $\Upsilon(v_1 \otimes v_1 + v_2 \otimes v_2) = X_1 \otimes X_1 + X_2 \otimes X_2 = \partial_x \otimes \partial_x + \partial_y \otimes \partial_y$. This tensor field is non-degenerate and its inverse reads

$$g = dx \otimes dx + dy \otimes dy.$$

In view of Theorem 8.22, this metric is invariant relative to the elements of $P_1^{\alpha=0}$. In fact, g is essentially the metric (8.11) obtained in Proposition 8.21 so that the elements of $P_1^{\alpha=0}$ become Killing vector fields relative to it.

Proposition 8.23. *The Lie algebras* P_2 *and* I_{14A} *consist of Killing vector fields relative to a metric on* $\mathbb{R}^2$.

Proof. Let us now apply Theorem 8.22 to showing that P_2 and I_{14A} consist of Killing vector fields relative to a metric on $\mathbb{R}^2$. It stems from Table A.2 that P_2 is isomorphic to $\mathfrak{sl}(2, \mathbb{R})$. Choose a basis $\{v_1, v_2, v_3\}$ thereof satisfying the same commutation relations as the basis $\{X_1, X_2, X_3\}$ for P_2 detailed in Table A.2. This gives rise to a Lie algebra morphism $\phi : \mathfrak{sl}(2, \mathbb{R}) \to \mathfrak{X}(\mathbb{R}^2)$ mapping each v_i into X_i. The Lie algebra $\mathfrak{sl}(2, \mathbb{R})$ admits a quadratic Casimir element $C := v_1 \otimes v_3 + v_3 \otimes v_1 - 2v_2 \otimes v_2$ (as shown in [27, 359]). If $\Upsilon : U(\mathfrak{sl}(2, \mathbb{R})) \to S(\mathbb{R}^2)$ is the corresponding associative algebra morphism, then $\Upsilon(C) = X_1 \otimes X_3 + X_3 \otimes X_1 - 2X_2 \otimes X_2 = -2y^2(\partial_x \otimes \partial_x + \partial_y \otimes \partial_y)$. Then, $\Upsilon(C)$ is non-degenerate and its inverse takes the form

$$g = \frac{-1}{2y^2}(\mathrm{d}x \otimes \mathrm{d}x + \mathrm{d}y \otimes \mathrm{d}y).$$

In view of Theorem 8.22, this metric is invariant relative to the elements of P_2. In fact, it is straightforward to verify that g is invariant under the elements of P_2.

Since X_1, X_2 span a Lie algebra diffeomorphic to I_{14A} (cf. [23] and Table A.2), it follows that this Lie algebra also consists of Killing vector fields relative to g. □

8.5. Invariant Distributions for VG Lie Algebras on $\mathbb{R}^2$

It was proved by Lie that the Lie algebras $\{P_i\}_{i=1,\ldots,8}$ do not admit any invariant distribution, while Lie algebras $\{I_i\}_{i=1,\ldots,20}$ do [183, 267]. The knowledge of invariant distributions for VG Lie algebras on $\mathbb{R}^2$ is relevant in several problems related to them, e.g., the characterization of VG Lie algebras of conformal vector fields on $\mathbb{R}^2$ (cf. [197, 263]). For the sake of completeness of this work, let us

Conformal and Killing Lie Algebras on the Plane 333

review the properties of these distributions obtained in [197, 263]. The first step to accomplish this aim is to prove the following lemma.

Lemma 8.24. *If a VG Lie algebra V on $\mathbb{R}^2$ contains two vector fields X_1, X_2 such that $[X_1, X_2] = 0$ and $X_1 \wedge X_2 \neq 0$, then every invariant distribution $\mathcal{D}$ for V is spanned by a linear combination $\lambda_1 X_1 + \lambda_2 X_2$, with $\lambda_1, \lambda_2 \in \mathbb{R}$.*

Proof. Since $X_1 \wedge X_2 \neq 0$, the vector fields X_1, X_2 are linearly invariant at each point of $\mathbb{R}^2$ and the invariant distribution $\mathcal{D}$ for V can be spanned via a vector field of the form X_2 or $X_1 + \mu X_2$ for a certain function $\mu \in C^\infty(\mathbb{R}^2)$. If $\mathcal{D}$ is generated by X_2, then the lemma follows immediately. Otherwise, there exist functions $f_i \in C^\infty(\mathbb{R}^2)$, with $i = 1, 2$, such that

$$[X_i, X_1 + \mu X_2] = (X_i \mu) X_2 = f_i (X_1 + \mu X_2), \quad i = 1, 2.$$

Since $X_1 \wedge X_2 \neq 0$, one has $f_1 = f_2 = 0$. Moreover, $X_i \mu = 0$ for $i = 1, 2$ and $\mu = \text{const}$. Therefore, $\mathcal{D}$ is generated by $\lambda_1 X_1 + \lambda_2 X_2$ for certains $\lambda_1, \lambda_2 \in \mathbb{R}$. $\square$

Note that Lemma 8.24 can be applied to Lie algebras I_{14B} and $I_8^{\alpha=1}$. Then, one can see that all distributions of the form $\lambda_1 X_1 + \lambda_2 X_2$ are invariant relative to I_{14B} and $I_8^{\alpha=1}$.

Theorem 8.25. *If V is a VG Lie algebra on $\mathbb{R}^2$ containing two linearly independent vector fields X_1, X_2 such that $[X_1, X_2] = 0$ and $X_1 \wedge X_2 = 0$, then each invariant distribution $\mathcal{D}$ relative to V is spanned by X_1.*

Proof. We can assume that at generic point of $\mathbb{R}^2$ the vector field X_1 is not zero. Since X_1, X_2 are linearly independent vector fields of V satisfying $X_1 \wedge X_2 = 0$ and $[X_1, X_2] = 0$, there exists $f \in C^\infty(\mathbb{R}^2)$ such that $X_2 = f X_1$ and $X_1 f = 0$. It is immediate that there exists a vector field X_3 satisfying $X_1 \wedge X_3 \neq 0$. As $\mathcal{D}$ is a distribution of rank one and $X_1 \wedge X_3 \neq 0$, it is clear that $\mathcal{D}$ is therefore generated by X_3 or $X_1 + \mu X_3$ for a certain $\mu \in C^\infty(\mathbb{R}^2)$.

If $\mathcal{D}$ is generated by X_3, then $[f X_1, X_3] = f_3 X_3$ for a certain $f_3 \in C^\infty(\mathbb{R}^2)$ and $[X_1, X_3] = g X_3$ for a certain $g \in C^\infty(\mathbb{R}^2)$ by assumption. In other words,

$$[fX_1, X_3] = -(X_3 f)X_1 + fgX_3, \quad [X_1, X_3] = gX_3.$$

Since $X_1 \wedge X_3 \neq 0$, it follows that $X_3 f = 0$ and $g = 0$. Then, f is a constant and X_2 and X_1 are linearly independent, which is a contradiction and shows that $\mathcal{D}$ cannot be spanned by X_3.

Let us assume that $\mathcal{D}$ is generated by $X_1 + \mu X_3$. If $\mu = 0$, the theorem follows. Let us assume that $\mu \neq 0$ and we will show that this leads to contradiction. Since $\mathcal{D}$ is invariant relative to X_1, X_2, there exist functions $f_1, f_2 \in C^\infty(\mathbb{R}^2)$ such that

$$[X_1, X_1 + \mu X_3] = (X_1\mu)X_3 + \mu[X_1, X_3] = f_1(X_1 + \mu X_3), \quad (8.17)$$

$$[fX_1, X_1 + \mu X_3] = f[(X_1\mu)X_3 + \mu[X_1, X_3]]$$
$$- \mu(X_3 f)X_1 = f_2(X_1 + \mu X_3). \quad (8.18)$$

Substituting (8.17) in (8.18) and recalling that $X_1 \wedge X_3 \neq 0$, we obtain that

$$f f_1(X_1 + \mu X_3) - \mu(X_3 f)X_1 = f_2(X_1 + \mu X_3) \Rightarrow (f f_1 - f_2 - \mu X_3 f)$$
$$X_1 + \mu(f f_1 - f_2)X_3 = 0.$$

Since $X_3 \wedge X_1 \neq 0$, the coefficients of X_1, X_3 in the above expression must vanish. In particular, $\mu(f f_1 - f_2) = 0$ and, since $\mu \neq 0$ by assumption, then $f f_1 = f_2$ and $\mu X_3 f = 0$. As $X_1 \wedge X_3 \neq 0$ and $X_1 f = 0$, which is a consequence of the assumption $[X_2, X_1] = 0$, we obtain that f is a constant. This goes against our assumption that X_1, X_2 are linearly independent. In consequence, $\mu = 0$ and $\mathcal{D}$ is generated by X_1. $\square$

Corollary 8.26. *The Lie algebras $I_{12}, I_{13}, I_{16} - I_{20}$ and I_{14}, I_{15} for $r > 1$ admit only one invariant distribution generated by ∂_y.*

Proof. In view of Table A.2, the above-mentioned Lie algebras contain the vector fields $Y_1 := \partial_y, Y_2 := \eta_1(x)\partial_y$. The application of Theorem 8.25 shows that their invariant distributions are spanned by X_1. $\square$

Conformal and Killing Lie Algebras on the Plane 335

Theorem 8.27. *Let V be a Lie algebra containing some vector fields X_1, X_2, X_3 on $\mathbb{R}^2$ such that $X_1 \wedge X_2 \neq 0$, $[X_1, X_2] = 0$ and $[X_3, X_2] = 0$. Let $\mathcal{D}$ be an invariant distribution on $\mathbb{R}^2$ relative to V. Hence, if $[X_1, X_3] = X_1$, then $\mathcal{D}$ is generated by X_1 or X_2.*

Proof. From the assumptions and Lemma 8.24 follow that $\mathcal{D}$ is generated by a linear combination with real coefficients of X_1, X_2. Hence, there exist $c_1, c_2 \in \mathbb{R}$ with $c_1^2 + c_2^2 \neq 0$ such that $\mathcal{D}$ is spanned by $c_1 X_2 + c_2 X_2$.

Since $\mathcal{D}$ is invariant relative to X_3, there exist $f_1 \in C^\infty(\mathbb{R}^2)$ such that

$$[X_3, c_1 X_1 + c_2 X_2] = -c_1 X_1 = f_1(c_1 X_1 + c_2 X_2)$$
$$\Rightarrow c_1(f_1 + 1)X_1 + f_1 c_2 X_2 = 0.$$

As $X_1 \wedge X_2 \neq 0$, one has that $f_1 c_2 + c_1 = f_1 c_1 = 0$. There exist again two possibilities: $f_1 = 0$ and therefore $c_1 = 0$, which implies $\mathcal{D}$ is generated by X_2; or $f_1 \neq 0$, which gives $c_2 = 0$ and $\mathcal{D}$ is generated by X_1. $\qquad \square$

Corollary 8.28. *The Lie algebras I_6, I_9, I_{10}, and I_{11} have only two invariant distributions spanned by ∂_x and ∂_y. The Lie algebra I_7 has only one invariant distribution spanned by $X = \partial_y$. The Lie algebra I_{15B} has two invariant distributions spanned by ∂_x and ∂_y, whereas I_{15A}, and its Lie subalgebra I_{14A}, have two invariant distributions spanned by $\partial_x + cy\partial_y$ and $e^{cx}\partial_y$.*

Proof. The vector fields of $I_7 = \langle X_1, X_2, X_3, X_4 \rangle$, where $X_1, \ldots, X_4$ are given in Table A.2, are such that X_1, X_2, X_3 satisfy the conditions of the subcase b of Theorem 8.27. Consequently, the invariant distributions of I_7 are generated by X_1 or X_2. A short computation shows that the only invariant distribution is $X_2 = \partial_y$.

Meanwhile, the vector fields X_1, X_2, X_3 of the Lie algebras I_6, I_9, I_{10}, and I_{11} given in Table A.2 satisfy the conditions of subcase b of Theorem 8.27. It can be then proved that their invariant distributions are generated by X_1 or X_2. A simple calculation shows that each of these vector fields generates an invariant distribution for the mentioned Lie algebras.

The Lie algebra I_{15B} admits a basis $\partial_x, y\partial_y, \partial_y$ satisfying the conditions of the part b of Theorem 8.27. Hence, every invariant distribution must be spanned by ∂_x, ∂_y. A simple calculation shows that both vector fields span invariant distributions.

Let us consider the Lie algebra I_{15A}. The basis $e^{cx}\partial_y, \partial_x + cy\partial_y, y\partial_y$ satisfies the conditions of part b of Theorem 8.27. In consequence, invariant distributions must be spanned by $e^{cx}\partial_y$ and/or $\partial_x + cy\partial_y$. A simple computation shows that both vector fields span invariant distributions. The I_{14A} is a Lie subalgebra of I_{15A}. Hence, it also admits the previous invariant distributions. A simple calculation shows that there are no more invariant distributions. $\qquad\square$

The distributions I_1, I_2, and I_3 need no detailed analysis. Indeed, they are spanned by vector fields depending on a single variable and their invariant distributions are immediate. For instance, the invariant distributions of I_2 and I_3 are spanned by ∂_x and ∂_y. Meanwhile, the distribution I_5 can be embedded in I_{19}, which admits an invariant distribution ∂_y. Hence, I_5 admits this invariant distribution. A direct calculation using this knowledge shows that I_5 admits no other invariant distribution.

8.6. Applications in Physics

Let us finally illustrate in this chapter the physical relevance of systems of differential equations whose dynamics can be determined by VG Lie algebras of conformal and Killing vector fields on $\mathbb{R}^2$ relative to a certain metric g. In particular, the findings of previous sections will be used to construct g and to prove that VG Lie algebras consisting of Killing vector fields relative to g are also Lie algebras of Hamiltonian vector fields relative to the symplectic structure induced by g. This provides an explanation of previous results obtained, by solving involved partial differential equations, in [23].

8.6.1. Milne–Pinney equations

Let us reconsider the Milne–Pinney equation, which has been several times treated in this book (see [73, 254] for some of the physical applications), takes the form

$$\frac{d^2 x}{dt^2} = -\omega^2(t)x + \frac{c}{x^3}, \tag{8.19}$$

where $\omega(t)$ is any function depending on t and $c \in \mathbb{R}$. If we add a new variable $y := dx/dt$, the above differential equation can be rewritten as

$$\begin{cases} \dfrac{dx}{dt} = y, \\ \dfrac{dy}{dt} = -\omega^2(t)x + \dfrac{c}{x^3}. \end{cases} \tag{8.20}$$

System (8.20) is associated with the t-dependent vector field (cf. [99]) $X := X_3 + \omega^2(t)X_1$, where

$$X_1 = -x\partial_y, \quad X_2 = \frac{1}{2}\left(y\partial_y - x\partial_x\right), \quad X_3 = y\partial_x + \frac{c}{x^3}\partial_y$$

form a basis of a Lie algebra V_{MP}. The matrix of its Killing form, κ, in the basis $\mathcal{B} := \{X_1, X_2, X_3\}$ of V_{MP} takes the form

$$[\kappa]_{\mathcal{B}} = \begin{pmatrix} 0 & 0 & -4 \\ 0 & 2 & 0 \\ -4 & 0 & 0 \end{pmatrix}.$$

Hence, κ is non-degenerate and indefinite. The Cartan criterion [354] ensures then that V_{MP} is semi-simple. Table A.2 shows that every three-dimensional semi-simple Lie algebra of vector fields on the plane is isomorphic to $\mathfrak{sl}(2, \mathbb{R})$ or to $\mathfrak{so}(3)$. Since V_{MP} is indefinite, V_{MP} is isomorphic to $\mathfrak{sl}(2, \mathbb{R})$.

Consider the Lie algebra $\mathfrak{sl}(2, \mathbb{R})$ and a basis $\{v_1, v_2, v_3\}$ thereof satisfying the same commutation relations as X_1, X_2, X_3. This gives rise to a Lie algebra morphism $\phi : \mathfrak{sl}(2, \mathbb{R}) \to \mathfrak{X}(\mathbb{R}^2)$ mapping each v_i onto X_i. In turn, this leads to an associative algebra morphism

$\Upsilon : U(\mathfrak{sl}(2,\mathbb{R})) \to S(\mathbb{R}^2)$. Since the Lie algebra $\mathfrak{sl}(2,\mathbb{R})$ admits a quadratic Casimir element

$$
\begin{aligned}
C : \;=\; & v_1 \otimes v_3 + v_3 \otimes v_1 - 2v_1 \otimes v_1 \\
\implies \;\; & G := \Upsilon(C) = X_1 \otimes X_3 + X_3 \otimes X_1 - 2X_2 \otimes X_2.
\end{aligned}
$$

The coordinate expression for X_1, X_2, X_3 ensures that

$$
G = -\frac{x^2}{2}\partial_x \otimes \partial_x - \left(\frac{2c}{x^2} + \frac{y^2}{2}\right)\partial_y \otimes \partial_y - \frac{1}{2}xy(\partial_x \otimes \partial_y + \partial_y \otimes \partial_x)
$$
$$
\Rightarrow \det G = c.
$$

Hence, G is non-degenerate for $c \neq 0$ and Theorem 8.22 states that V_{MP} consists of Killing vector fields relative to

$$
g := G^{-1} = -\left(\frac{2}{x^2} + \frac{y^2}{2c}\right)\mathrm{d}x \otimes \mathrm{d}x
$$
$$
+\frac{xy}{2c}(\mathrm{d}x \otimes \mathrm{d}y + \mathrm{d}y \otimes \mathrm{d}x) - \frac{x^2}{2c}\mathrm{d}y \otimes \mathrm{d}y.
$$

The associated symplectic form is given by $\omega := \star 1$, i.e.,

$$
\omega = \sqrt{|c|}\mathrm{d}x \wedge \mathrm{d}y.
$$

The vector fields of V_{MP} become Hamiltonian relative to ω. This produces a symplectic form turning the elements of V_{MP} into Hamiltonian vector fields algebraically. This is much easier than obtaining ω by solving a system of PDEs or guessing ω as in [23, 121].

8.6.2. Schrödinger equation on $\mathbb{C}^2$

Let $\mathcal{H}$ be an n-dimensional Hilbert space with a scalar product $\langle \cdot, \cdot \rangle$, let $H(t) \subset \mathrm{End}(\mathcal{H})$ be a Hermitian Hamiltonian operator on $\mathcal{H}$ for every $t \in \mathbb{R}$, and let $\{\psi_i\}_{i \in \overline{1,n}} \in \mathcal{H}$ be an orthonormal basis of $\mathcal{H}$. The space $\mathcal{H}_0 := \mathcal{H}\backslash\{0\}$ admits an equivalence relation given by

$$
\psi_1 \sim \psi_2 \Leftrightarrow \exists \lambda \in \mathbb{C}\backslash\{0\} : \psi_1 = \lambda\psi_2,
$$

whose space of equivalence classes is the so-called complex projective space $\mathcal{PH} := \mathcal{H}_0/\sim$. As $\mathcal{PH}$ is also the space of orbits of the free and proper multiplicative action of the Lie group $\mathbb{C}_0 := \mathbb{C}\backslash\{0\}$ on

$\mathbb{C}_0^n := \mathbb{C}^n\setminus\{0\}$, the space $\mathcal{PH}$ becomes a manifold and, for $n = 2$, the projection $\mathbb{C}_0^2 \ni \psi \mapsto [\psi] \in \mathbb{C}P^1 \simeq \mathbb{C}_0^2/\mathbb{C}_0$, $\psi := (z_1, z_2)$ is a submersion (see [4] for details).

A t-dependent Schrödinger equation on $\mathcal{H}$ induced by a t-dependent Hamiltonian $H(t)$ takes the form

$$\frac{\mathrm{d}\psi}{\mathrm{d}t} = -\mathrm{i}H(t)\psi \Leftrightarrow \frac{\mathrm{d}}{\mathrm{d}t}\begin{pmatrix} z_1 \\ z_2 \end{pmatrix} = -\mathrm{i}H(t)\begin{pmatrix} z_1 \\ z_2 \end{pmatrix}$$

$$= -\mathrm{i}\begin{pmatrix} \lambda_1(t) & b(t) \\ \bar{b}(t) & \lambda_2(t) \end{pmatrix}\begin{pmatrix} z_1 \\ z_2 \end{pmatrix},$$

for $b(t) := b_1(t) + \mathrm{i}b_2(t)$, $\lambda_i, b_i \in \mathbb{R}$. If $\mu := z_1 z_2^{-1}$, $z_1 \in \mathbb{C}$, $z_2 \in \mathbb{C}_0$, then

$$\frac{\mathrm{d}\mu}{\mathrm{d}t} = \mathrm{i}[\bar{b}(t)\mu^2 + (\lambda_2(t) - \lambda_1(t))\mu - b(t)].$$

Writing $\mu = x + \mathrm{i}y$, $x, y \in \mathbb{R}$ and gathering the parts real and imaginary of the previous system in the new variables, we obtain

$$\begin{cases} \dfrac{\mathrm{d}x}{\mathrm{d}t} = b_2(t)(x^2 - y^2 + 1) - (\lambda_2(t) - \lambda_1(t))y - 2b_1(t)xy, \\ \dfrac{\mathrm{d}y}{\mathrm{d}t} = b_1(t)(x^2 - y^2 - 1) + (\lambda_2(t) - \lambda_1(t))x + 2b_2(t)xy. \end{cases}$$

This system of differential equations is associated with the t-dependent vector field on $\mathbb{C}P^1$ of the form

$$X = b_1(t)X_1 + b_2(t)X_2 + (\lambda_2(t) - \lambda_1(t))X_3,$$

with

$$-X_1 := 2xy\partial_x + (1 + y^2 - x^2)\partial_y, \qquad X_2 := (x^2 - y^2 + 1)\partial_x + 2xy\partial_y,$$

$$-X_3 := y\partial_x - x\partial_y.$$

The vector fields X_0, X_1, X_2 span a Lie algebra $V_Q = \mathrm{P}_3$. The Killing form, κ, of P_3 in the basis $\mathcal{B} := \{X_1, X_2, X_3\}$ reads

$$[\kappa]_{\mathcal{B}} = \begin{pmatrix} -8 & 0 & 0 \\ 0 & -8 & 0 \\ 0 & 0 & -2 \end{pmatrix}.$$

This Killing form is non-degenerate and negative-definite. In view of Table A.2, the Lie algebra V_Q must be diffeomorphic to P$_3$.

The vector fields X_1, X_2, X_3 are exactly those ones of P$_3$ given in Table A.2. Consider a basis v_1, v_2, v_3 of $\mathfrak{so}(3)$ satisfying the same commutation relations as X_1, X_2, X_3. This leads to an associative algebra morphism $\Upsilon : U(\mathfrak{so}(3)) \to S(\mathbb{R}^2)$. Meanwhile, $\mathfrak{so}(3)$ admits a quadratic Casimir element

$$C = v_1 \otimes v_1 + v_2 \otimes v_2 + 4v_3 \otimes v_3.$$

Then,

$$G_0 : = \Upsilon(C) = X_1 \otimes X_1 + X_2 \otimes X_2 + 4X_3 \otimes X_3$$
$$= (1 + x^2 + y^2)^2 (\partial_x \otimes \partial_x + \partial_y \otimes \partial_y).$$

This G_0 is non-degenerate and Theorem 8.22 leads to the construction of a Riemannian metric g turning the elements of V_Q into Killing vector fields relative to

$$g = G_0^{-1} = \frac{\mathrm{d}x \otimes \mathrm{d}x + \mathrm{d}y \otimes \mathrm{d}y}{(1 + x^2 + y^2)^2}.$$

The symplectic form related to g takes the form

$$\omega = \frac{\mathrm{d}x \wedge \mathrm{d}y}{(1 + x^2 + y^2)^2} = \star 1.$$

Theorem 8.22 ensures that V_Q is a Lie algebra of Killing vector fields relative to ω. As in the previous section, this symplectic form is obtained algebraically, which is much simple than working out ω by solving a system PDEs as done, for instance, in [23].

Chapter 9

Lie Symmetry for Differential Equations

9.1. Introduction

During the last decades of the 20th century, attention has been focused on Lie symmetry computation methods for nonlinear evolution equations of hydrodynamic type in Plasma Physics, Cosmology and other fields [2, 54, 108]. These equations have solutions in the form of solitons. In order to study and derive solutions of this kind of equations, the classical and non-classical Lie symmetry approaches have proven their efficiency. Given their recurrent appearance in the Physics and Mathematics literature, it is important to settle an algorithmic method for the calculation of their symmetries, eventual reduction and search of their solutions. For this reason, we will devote this part of the book to calculating symmetries and reducing differential equations, ranging from ODEs to hierarchies of PDEs and Lax pairs. In this way, the plan of the chapter goes as follows.

In Section 9.2, we inspect certain types of Lie symmetries for Lie systems. We prove that every Lie system admits a Lie algebra of Lie symmetries whose geometric properties are determined by the VG Lie algebra of the Lie system. To illustrate our results, we analyze a particular type of Lie symmetries for $\mathfrak{sl}(2, \mathbb{R})$-Lie systems and we apply our findings to inspect Riccati equations,

Cayley–Klein Riccati equations, quaternionic Riccati equations with real t-dependent coefficients, generalized Darboux–Brioschi–Halphen systems and second-order Kummer–Schwarz equations. In Section 9.3, we analyze a particular type of Lie symmetries for Aff($\mathbb{R}$)-Lie systems and we apply our results to Buchdahl equations. To conclude, we generalize the search of this type of Lie symmetries to PDE Lie systems in Section 9.4 . In a similar fashion, we search for a particular type of Lie symmetries and illustrate the interest of our theory by studying partial Riccati equations.

It is remarkable that the application of our methods to Lie systems leads to obtaining Lie symmetries for all Lie systems sharing isomorphic VG Lie algebras. In this way, we are concluding the results for several different Lie systems with the same VG Lie algebra simultaneously.

9.2. Lie Symmetries for Lie Systems

Given the interest of Lie symmetries for the study of differential equations, we devote this section to the study of Lie symmetries for Lie systems. We pioneer the study and application of certain types of Lie symmetries for higher-order Lie systems and PDE Lie systems, i.e., the generalizations of Lie systems to the realms of higher-order ODEs and PDEs [87, 93, 102, 166]. As a byproduct, we develop one of the few applications of the theory of Lie systems in the investigation of systems of PDEs. Additionally, only a few particular results about Lie symmetries for Lie systems had appeared before in [121]. We accomplish here a careful and exhaustive study and calculation of such Lie symmetries [166].

9.2.1. *Certain Lie symmetries for Lie systems*

We are now concerned with the study of certain Lie symmetries for Lie systems. We prove that the features of these Lie symmetries are determined by the algebraic structure of a VG Lie algebra of the Lie system.

Lie Symmetry for Differential Equations　　　343

Consider a general Lie system X given by

$$X = \sum_{\alpha=1}^{r} b_\alpha(t) X_\alpha, \tag{9.1}$$

where $b_1, \ldots, b_r$ are some t-dependent functions. Then, (9.1) admits a VG Lie algebra V with basis $X_1, \ldots, X_r$ and structure constants $c_{\alpha\beta\gamma}$. Let us study the Lie symmetries of X of the form

$$Y = f_0(t) \frac{\partial}{\partial t} + \sum_{\alpha=1}^{r} f_\alpha(t) X_\alpha, \tag{9.2}$$

where $f_0, \ldots, f_r$ are certain t-dependent functions. We denote by $\mathcal{S}_X^V$ *the space of Lie symmetries*. Recall that $Y \in \mathcal{S}_X^V$ if and only if

$$[Y, \bar{X}] = h\bar{X}, \tag{9.3}$$

for a function $h \in C^\infty(\mathbb{R} \times N)$, with $\bar{X}$ being the autonomization of X [302]. From this, it immediately follows that $\mathcal{S}_X^V$ is a real Lie algebra of vector fields.

Using the properties of $\mathcal{S}_X^V$, we now characterize the elements of $\mathcal{S}_X^V$ as particular solutions of a family of Lie systems.

Lemma 9.1. *The vector field Y of the form (9.2) is a Lie symmetry for the Lie system (9.1) if and only if the t-dependent functions $f_0, \ldots, f_r$ satisfy the system of differential equations*

$$\frac{\mathrm{d}f_0}{\mathrm{d}t} = b_0(t), \quad \frac{\mathrm{d}f_\alpha}{\mathrm{d}t} = f_0 \frac{\mathrm{d}b_\alpha}{\mathrm{d}t}(t) + b_\alpha(t) b_0(t) + \sum_{\beta,\gamma=1}^{r} b_\beta(t) f_\gamma c_{\gamma\beta\alpha}, \tag{9.4}$$

for a certain t-dependent function b_0 and $\alpha = 1, \ldots, r$.

Proof. From (9.1–9.3), we have that

$$
\begin{aligned}
[Y, \bar{X}] &= \left[f_0 \frac{\partial}{\partial t} + \sum_{\alpha=1}^{r} f_\alpha X_\alpha, \ \frac{\partial}{\partial t} + \sum_{\beta=1}^{r} b_\beta X_\beta \right] \\
&= -\frac{\mathrm{d} f_0}{\mathrm{d} t} \frac{\partial}{\partial t} + \sum_{\alpha=1}^{r} \left(\left(f_0 \frac{\mathrm{d} b_\alpha}{\mathrm{d} t} - \frac{\mathrm{d} f_\alpha}{\mathrm{d} t} \right) X_\alpha + \sum_{\beta=1}^{r} b_\beta f_\alpha [X_\alpha, X_\beta] \right) \\
&= -\frac{\mathrm{d} f_0}{\mathrm{d} t} \frac{\partial}{\partial t} + \sum_{\alpha=1}^{r} \left(\left(f_0 \frac{\mathrm{d} b_\alpha}{\mathrm{d} t} - \frac{\mathrm{d} f_\alpha}{\mathrm{d} t} \right) X_\alpha + \sum_{\beta,\gamma=1}^{r} b_\beta f_\alpha c_{\alpha\beta\gamma} X_\gamma \right) \\
&= -\frac{\mathrm{d} f_0}{\mathrm{d} t} \frac{\partial}{\partial t} + \sum_{\alpha=1}^{r} \left(f_0 \frac{\mathrm{d} b_\alpha}{\mathrm{d} t} - \frac{\mathrm{d} f_\alpha}{\mathrm{d} t} + \sum_{\beta,\gamma=1}^{r} b_\beta f_\gamma c_{\gamma\beta\alpha} \right) X_\alpha \\
&= h \left(\frac{\partial}{\partial t} + \sum_{\alpha=1}^{r} b_\alpha X_\alpha \right).
\end{aligned}
\tag{9.5}
$$

Thus, $[Y, \bar{X}] = h\bar{X}$ is equivalent to

$$
\left(-\frac{\mathrm{d} f_0}{\mathrm{d} t} - h \right) \frac{\partial}{\partial t} + \sum_{\alpha=1}^{r} \left[\sum_{\beta,\gamma=1}^{r} b_\beta f_\gamma c_{\gamma\beta\alpha} + f_0 \frac{\mathrm{d} b_\alpha}{\mathrm{d} t} - \frac{\mathrm{d} f_\alpha}{\mathrm{d} t} - h b_\alpha \right]
$$
$$
X_\alpha = 0.
\tag{9.6}
$$

Since $\partial/\partial t, X_1, \ldots, X_r$ are linearly independent over $\mathbb{R}$, we obtain that $[Y, \bar{X}] = h\bar{X}$ if and only if (9.4) is fulfilled with h being an arbitrary t-dependent function such that $b_0(t) = -h(t)$ for every $t \in \mathbb{R}$. Hence, the elements of $\mathcal{S}_X^V$ are the particular solutions of (9.4) for arbitrary t-dependent functions b_0. $\qquad\square$

Definition 9.2. We call (9.4) the *symmetry system* of the Lie system (9.1) with respect to its VG Lie algebra V. We write Γ_X^V for the

t-dependent vector field associated with (9.4), that is,

$$\Gamma_X^V = b_0(t)\frac{\partial}{\partial f_0} + \sum_{\alpha=1}^r \left(f_0 \frac{\mathrm{d}b_\alpha}{\mathrm{d}t}(t) + b_0(t)b_\alpha(t) \right.$$

$$\left. + \sum_{\gamma,\beta=1}^r b_\beta(t) f_\gamma c_{\gamma\beta\alpha} \right) \frac{\partial}{\partial f_\alpha}. \tag{9.7}$$

Note that (9.1) may have different VG Lie algebras (see [108] for details). Let us prove that (9.4) is a Lie system.

Theorem 9.3. *The system Γ_X^V is a Lie system possessing a VG Lie algebra*

$$(A_1 \oplus_S A_2) \oplus_S V_L \simeq (\mathbb{R}^{r+1} \oplus_S \mathbb{R}^r) \oplus_S V/Z(V), \tag{9.8}$$

where

$$A_1 = \langle Z_0, \dots, Z_r \rangle \simeq \mathbb{R}^{r+1}, \quad A_2 = \langle W_1, \dots, W_r \rangle \simeq \mathbb{R}^r,$$
$$V_L = \langle Y_1, \dots, Y_r \rangle \simeq V/Z(V), \tag{9.9}$$

with

$$Y_\alpha = \sum_{\beta,\gamma=1}^r f_\beta c_{\beta\alpha\gamma}\frac{\partial}{\partial f_\gamma}, \quad W_\alpha = f_0\frac{\partial}{\partial f_\alpha}, \quad Z_0 = \frac{\partial}{\partial f_0}, \quad Z_\alpha = \frac{\partial}{\partial f_\alpha}, \tag{9.10}$$

with $\alpha = 1, \dots, r$, we write $A \oplus_S B$ for the semi-direct sum of the ideal A of $A + B$ with B, and $Z(V)$ is the center of the Lie algebra V.

Proof. By defining t-dependent functions

$$c_0 := b_0, \quad \bar{b}_0 := 0, \quad c_\alpha := b_0 b_\alpha, \quad \bar{b}_\alpha := \frac{\mathrm{d}b_\alpha}{\mathrm{d}t}, \quad \alpha = 1, \dots, r, \tag{9.11}$$

we can write

$$\Gamma_X^V(t,f) = \sum_{\alpha=0}^r c_\alpha(t)Z_\alpha(f) + \sum_{\alpha=1}^r [\bar{b}_\alpha(t)W_\alpha(f) + b_\alpha(t)Y_\alpha(f)],$$
$$f \in \mathbb{R}^{n+1}. \tag{9.12}$$

346   *A Guide to Lie Systems with Compatible Geometric Structures*

Hence, Γ_X^V is a t-dependent vector field taking values in the linear space $V^S := A_1 + A_2 + V_L$. Let us show that V^S is also a Lie algebra of vector fields. To do so, let us first prove that $V_L \simeq V/Z(V)$. Consider $Y_\alpha, Y_\beta \in V_L$. Recalling that $[X_\alpha, X_\beta] = \sum_{\gamma=1}^r c_{\alpha\beta\gamma} X_\gamma$, we obtain

$$
\begin{aligned}
[Y_\alpha, Y_\beta] &= \sum_{i,j,m,n=1}^r \left[c_{i\alpha j} f_i \frac{\partial}{\partial f_j}, c_{m\beta n} f_m \frac{\partial}{\partial f_n} \right] \\
&= \sum_{i,j,m,n=1}^r c_{i\alpha j} c_{m\beta n} \left(f_i \delta_j^m \frac{\partial}{\partial x_n} - f_m \delta_n^i \frac{\partial}{\partial f_j} \right) \\
&= \sum_{i,m,n=1}^r c_{i\alpha m} c_{m\beta n} f_i \frac{\partial}{\partial f_n} - \sum_{n,j,m=1}^r c_{n\alpha j} c_{m\beta n} f_m \frac{\partial}{\partial f_j} \\
&= \sum_{i,m,n=1}^r (c_{i\alpha m} c_{m\beta n} + c_{m\alpha n} c_{\beta im}) f_i \frac{\partial}{\partial f_n}. \tag{9.13}
\end{aligned}
$$

Using the Jacobi identity for the structure constants $c_{\alpha\beta\gamma}$, we see that

$$
\sum_{m,n=1}^r (c_{i\alpha m} c_{m\beta n} + c_{\alpha\beta m} c_{min} + c_{\beta im} c_{man}) = 0, \quad \forall i, \alpha, \beta = 1, \ldots, r. \tag{9.14}
$$

From this,

$$
\begin{aligned}
[Y_\alpha, Y_\beta] &= \sum_{i,m,n=1}^r (c_{i\alpha m} c_{m\beta n} + c_{m\alpha n} c_{\beta im}) f_i \frac{\partial}{\partial f_n} \\
&= - \sum_{i,m,n=1}^r c_{\alpha\beta m} c_{min} f_i \frac{\partial}{\partial f_n} = \sum_{m=1}^r c_{\alpha\beta m} Y_m. \tag{9.15}
\end{aligned}
$$

So, $Y_1, \ldots, Y_r$ span a Lie algebra. We can define a Lie algebra morphism $\phi : V \to V_L$ of the form $\phi(X_\alpha) = Y_\alpha$ for $\alpha = 1, \ldots, r$. The vector fields $Y_1, \ldots, Y_r$ do not need to be linearly independent. Let us show this. We can assume with no loss of generality that $X_1, \ldots, X_s$, with $s \leq r$, form a basis for $\ker \phi$. Since $\phi(X_\alpha) = 0$ for $\alpha = 1, \ldots, s$, we have $f_\beta c_{\beta\alpha\gamma} = 0$ for $\alpha = 1, \ldots, s$ and $\beta, \gamma = 1, \ldots, r$. Thus, we see that $[X_\alpha, X_\beta] = 0$ for $\alpha = 1, \ldots, s$ and $\beta = 1, \ldots, r$. This means that $X_\alpha \in Z(V)$. Conversely, we get by similar arguments that if $X \in Z(V)$, then $X \in \ker \phi$. Hence, $X \in \ker \phi$ if and only if $X \in Z(V)$. In consequence, $\ker \phi = Z(V)$ and $Y_1, \ldots, Y_r$ span a Lie algebra isomorphic to $V/Z(V)$.

It is obvious that A_1 is an ideal of $A_1 + A_2$. Moreover, as $[A_1, V_L] \subset A_1$ and $[A_2, V_L] \subset A_2$, then $A_1 \oplus_S A_2$ is an ideal of V^S. Consequently, V^S is a Lie algebra of the form (9.8). □

Definition 9.4. We say that the *Lie systems X_1 and X_2 are isomorphic* when they take values in two isomorphic VG Lie algebras V_1, V_2 and there exists a Lie algebra isomorphism $\phi : V_1 \to V_2$ such that $(X_2)_t = \phi((X_1)_t)$ for each $t \in \mathbb{R}$.

Proposition 9.5. *Given two isomorphic Lie systems X_1 and X_2 related to VG Lie algebras V_1, V_2, their symmetry systems relative to such Lie algebras are, up to a change of basis in V_1 and/or V_2, the same.*

9.2.2. Lie algebras of Lie symmetries for Lie systems

In this section, we study different Lie subalgebras of $\mathcal{S}_X^V$. Their interest resides in the fact that, when finite-dimensional, they can be integrated to form Lie group actions of symmetries for X. In turn, they can be employed to simplify the Lie system they are referred to.

Lemma 9.6. *The space of functions $C^\infty(\mathbb{R})$ can be endowed with a Lie bracket given by*

$$\{f, \bar{f}\}_\mathbb{R} = f\frac{\mathrm{d}\bar{f}}{\mathrm{d}t} - \bar{f}\frac{\mathrm{d}f}{\mathrm{d}t}, \quad \forall f, \bar{f} \in C^\infty(\mathbb{R}). \tag{9.16}$$

Proof. In order to prove that (9.16) is a Lie bracket, we must show that (9.16) is bilinear, antisymmetric and satisfies the Jacobi identity. From its definition (9.16) is clearly bilinear and antisymmetric. To see that (9.16) holds the Jacobi identity, we consider the map $\phi : f \in C^\infty(\mathbb{R}) \mapsto f\partial/\partial t \in \Gamma(\mathrm{T}\mathbb{R})$. Observe that ϕ is a linear isomorphism.

Moreover, it follows that

$$\phi(\{f,\bar f\}_{\mathbb{R}}) = \{f,\bar f\}_{\mathbb{R}}\frac{\partial}{\partial t} = \left(f\frac{\mathrm{d}\bar f}{\mathrm{d}t} - \bar f\frac{\mathrm{d}f}{\mathrm{d}t}\right)\frac{\partial}{\partial t} = \left[f\frac{\partial}{\partial t}, \bar f\frac{\partial}{\partial t}\right]$$
$$= [\phi(f),\phi(\bar f)], \tag{9.17}$$

for arbitrary $f,\bar f \in C^\infty(\mathbb{R})$. By using the Jacobi identity for vector fields on $\mathbb{R}$ with respect to the Lie bracket $[\cdot,\cdot]$, we obtain

$$\phi(\{\{f,\bar f\}_{\mathbb{R}},\bar{\bar f}\}_{\mathbb{R}} + \{\{\bar f,\bar{\bar f}\}_{\mathbb{R}},f\}_{\mathbb{R}} + \{\{\bar{\bar f},f\}_{\mathbb{R}},\bar f\}_{\mathbb{R}})$$
$$= [[\phi(f),\phi(\bar f)],\phi(\bar{\bar f})]$$
$$+ [[\phi(\bar f),\phi(\bar{\bar f})],\phi(f)] + [[\phi(\bar{\bar f}),\phi(f)],\phi(\bar f)]$$
$$= 0, \tag{9.18}$$

$\forall f,\bar f,\bar{\bar f} \in C^\infty(\mathbb{R})$. Since ϕ is a linear isomorphism, then

$$\{\{f,\bar f\}_{\mathbb{R}},\bar{\bar f}\}_{\mathbb{R}} + \{\{\bar f,\bar{\bar f}\}_{\mathbb{R}},f\}_{\mathbb{R}} + \{\{\bar{\bar f},f\}_{\mathbb{R}},\bar f\}_{\mathbb{R}} = 0, \tag{9.19}$$

and (9.16) satisfies the Jacobi identity giving rise to a Lie bracket on $C^\infty(\mathbb{R})$. Moreover, ϕ becomes a Lie algebra isomorphism. $\qquad\square$

Definition 9.7. Let X be a Lie system on N with a VG Lie algebra V and let $\mathfrak{W}$ be a non-empty set of t-dependent functions that form a Lie algebra with respect to the Lie bracket defined in (9.16). We call $\mathcal{S}^V_{X,\mathfrak{W}}$ the space

$$\mathcal{S}^V_{X,\mathfrak{W}} = \left\{ Y \in \mathcal{S}^V_X \mid Y = f_0\frac{\partial}{\partial t} + \sum_{\alpha=1}^r f_\alpha X_\alpha,\ f_0 \in \mathfrak{W} \right\}, \tag{9.20}$$

where $X_1,\dots,X_r$ is a basis for V.

Proposition 9.8. *The space of symmetries $\mathcal{S}^V_{X,\mathfrak{W}}$ is a Lie algebra of symmetries of X.*

Proof. Since $\mathfrak{W}$ and $\mathcal{S}^V_X$ are linear spaces, the linear combinations of elements of $\mathcal{S}^V_{X,\mathfrak{W}}$ belong to $\mathcal{S}^V_{X,\mathfrak{W}}$. So, this space becomes a vector space. Moreover, given two elements, $Y, Y^* \in \mathcal{S}_{X,\mathfrak{W}}^V$, their Lie bracket reads

$$
\begin{aligned}
[Y, Y^*] &= \left[f_0(t)\frac{\partial}{\partial t} + \sum_{\alpha=1}^{r} f_\alpha(t)X_\alpha, \; f_0^*(t)\frac{\partial}{\partial t} + \sum_{\beta=1}^{r} f_\beta^*(t)X_\beta \right] \\
&= \{f_0, f_0^*\}_{\mathbb{R}}\frac{\partial}{\partial t} + \sum_{\beta=1}^{r} \left[\left(f_0\frac{\mathrm{d}f_\beta^*}{\mathrm{d}t} - f_0^*\frac{df_\beta}{\mathrm{d}t} \right) X_\beta \right. \\
&\quad \left. + \sum_{\alpha,\gamma=1}^{r} f_\alpha f_\beta^* c_{\alpha\beta\gamma} X_\gamma \right] \\
&= \{f_0, f_0^*\}_{\mathbb{R}}\frac{\partial}{\partial t} + \sum_{\gamma=1}^{r} \left[\left(f_0\frac{\mathrm{d}f_\gamma^*}{\mathrm{d}t} - f_0^*\frac{\mathrm{d}f_\gamma}{\mathrm{d}t} \right) \right. \\
&\quad \left. + \sum_{\alpha,\beta=1}^{r} f_\alpha f_\beta^* c_{\alpha\beta\gamma} \right] X_\gamma.
\end{aligned} \tag{9.21}
$$

Since $\mathcal{S}_X^V$ is a Lie algebra and $Y, Y^* \in \mathcal{S}_X^V$, then $[Y, Y^*] \in \mathcal{S}_X^V$. As additionally $\{f_0, f_0^*\}_{\mathbb{R}} \in \mathfrak{W}$, then $[Y, Y^*] \in \mathcal{S}_{X,\mathfrak{W}}^V$. Hence, the Lie bracket of elements of $\mathcal{S}_{X,\mathfrak{W}}^V$ belongs to $\mathcal{S}_{X,\mathfrak{W}}^V$, which becomes a Lie algebra. $\qquad\square$

Corollary 9.9. *Given a Lie system X on N related to a VG Lie algebra V, the elements of $\mathcal{S}_{X,\mathfrak{W}}^V$ with*

1. $\mathfrak{W} = \{f_0 \in C^\infty(\mathbb{R}) \mid \mathrm{d}f_0/\mathrm{d}t = 0\}$,
2. $\mathfrak{W} = \{f_0 \in C^\infty(\mathbb{R}) \mid f_0 = 0\}$,

are finite-dimensional Lie algebras of vector fields. In the second case, $\mathcal{S}_{X,\mathfrak{W}}^V$ is isomorphic to V.

Proof. In both cases, $\mathfrak{W}$ is non-empty. In the first case, the functions with $\mathrm{d}f_0/\mathrm{d}t = 0$ are constant. These functions form an Abelian Lie algebra with respect to the Lie bracket $\{\cdot, \cdot\}_{\mathbb{R}}$. In view of Proposition 9.8, the space $\mathcal{S}_{X,\mathfrak{W}}^V$ is a Lie algebra.

In the second case, the function zero is also a zero-dimensional Lie algebra relative to $\{\cdot,\cdot\}_\mathbb{R}$. Since $f_0 = 0$ and using Proposition 9.8, we obtain that $\mathcal{S}_{X,\mathfrak{W}}^V$ can be understood as a Lie algebra of t-dependent vector fields taking values in V. To prove that $\mathcal{S}_{X,\mathfrak{W}}^V \simeq V$, let us consider the morphism which maps each t-dependent vector field with its value at $t = 0$, namely

$$\phi : \mathcal{S}_{X,\mathfrak{W}}^V \longrightarrow V$$
$$Z \mapsto Z_0.$$

Let $X_1, \ldots, X_r$ be a basis for V. From Lemma 9.1, we have that $(f_1(t), \ldots, f_r(t))$ is a particular solution of the system

$$\frac{\mathrm{d}f_\alpha}{\mathrm{d}t} = \sum_{\delta,\beta=1}^r b_\beta(t) f_\delta c_{\delta\beta\alpha}, \quad \alpha = 1, \ldots, r. \tag{9.22}$$

For each initial condition $f_\alpha(0) = c_\alpha \in \mathbb{R}$, with $\alpha = 1, \ldots, r$, i.e., by fixing Z_0, there exists a unique solution of the above system. Hence, there exists a unique t-dependent vector field Z of $\mathcal{S}_{X,\mathfrak{W}}^V$ with $Z_0 = \sum_{\alpha=0}^r c_\alpha X_\alpha$. Thus, ϕ is a bijection. Using that for two vector fields $Z_1, Z_2 \in \mathcal{S}_{X,\mathfrak{W}}^V$ we have $[Z_1, Z_2] \in \mathcal{S}_{X,\mathfrak{W}}^V$ and $[Z_1, Z_2]_t = [(Z_1)_t, (Z_2)_t]$, we see that ϕ is a Lie algebra morphism and $\mathcal{S}_{X,\mathfrak{W}}^V \simeq V$. $\square$

9.2.3. Applications to systems of ODEs and HODEs

Let us work out the symmetry systems and related Lie symmetries for some Lie systems of interest. In particular, we will illustrate that Proposition 9.5 enables us to determine simultaneously Lie symmetries for different Lie systems with isomorphic VG Lie algebras.

9.2.3.1. Lie symmetries for $\mathfrak{sl}(2, \mathbb{R})$-Lie systems

Let us obtain the symmetry systems and related Lie symmetries for $\mathfrak{sl}(2, \mathbb{R})$-Lie systems. This shall be used in the following subsections to obtain simultaneous Lie symmetries of isomorphic $\mathfrak{sl}(2, \mathbb{R})$-Lie systems appearing in the physics and/or mathematical literature.

Let us choose a basis of vector fields $\{X_1, X_2, X_3\}$ of $V \simeq \mathfrak{sl}(2, \mathbb{R})$ with commutation relations

$$[X_1, X_2] = X_1, \quad [X_1, X_3] = 2X_2, \quad [X_2, X_3] = X_3. \tag{9.23}$$

Every Lie system with VG Lie algebra V can be brought into the form

$$X = b_1(t)X_1 + b_2(t)X_2 + b_3(t)X_3, \tag{9.24}$$

for certain t-dependent functions b_1, b_2, and b_3. The Lie symmetries of $\mathcal{S}_X^V$ take the form

$$Y = f_0(t)\frac{\partial}{\partial t} + f_1(t)X_1 + f_2(t)X_2 + f_3(t)X_3, \tag{9.25}$$

where f_0, f_1, f_2, f_3 are some t-dependent functions to be determined. In view of (9.4) and the commutation relations (9.23), the symmetry system for X relative to V reads

$$\begin{cases} \dfrac{\mathrm{d}f_0}{\mathrm{d}t} = b_0(t), \\[2mm] \dfrac{\mathrm{d}f_1}{\mathrm{d}t} = f_0\dfrac{\mathrm{d}b_1}{\mathrm{d}t}(t) + f_1 b_2(t) - f_2 b_1(t) + b_0(t)b_1(t), \\[2mm] \dfrac{\mathrm{d}f_2}{\mathrm{d}t} = f_0\dfrac{\mathrm{d}b_2}{\mathrm{d}t}(t) + 2f_1 b_3(t) - 2f_3 b_1(t) + b_0(t)b_2(t), \\[2mm] \dfrac{\mathrm{d}f_3}{\mathrm{d}t} = f_0\dfrac{\mathrm{d}b_3}{\mathrm{d}t}(t) + f_2 b_3(t) - f_3 b_2(t) + b_0(t)b_3(t). \end{cases} \tag{9.26}$$

As stated in Theorem 9.3, this is a Lie system. Indeed, system (9.26) is related to the t-dependent vector field

$$\begin{aligned} \Gamma_X^{\mathfrak{sl}(2,\mathbb{R})} = {}& \frac{\mathrm{d}b_1(t)}{\mathrm{d}t}W_1 + \frac{\mathrm{d}b_2(t)}{\mathrm{d}t}W_2 + \frac{\mathrm{d}b_3(t)}{\mathrm{d}t}W_3 + b_0(t)Z_0 \\ & + b_0(t)b_1(t)Z_1 + b_0(t)b_2(t)Z_2 + b_0(t)b_3(t)Z_3 + b_1(t)Y_1 \\ & + b_2(t)Y_2 + b_3(t)Y_3, \end{aligned} \tag{9.27}$$

where

$$Z_\alpha = \frac{\partial}{\partial f_\alpha}, \quad \alpha = 0, 1, 2, 3, \quad W_\beta = f_0\frac{\partial}{\partial f_\beta}, \quad \beta = 1, 2, 3 \tag{9.28}$$

and

$$Y_1 = -f_2 \frac{\partial}{\partial f_1} - 2f_3 \frac{\partial}{\partial f_2}, \quad Y_2 = f_1 \frac{\partial}{\partial f_1} - f_3 \frac{\partial}{\partial f_3},$$

$$Y_3 = 2f_1 \frac{\partial}{\partial f_2} + f_2 \frac{\partial}{\partial f_3}. \tag{9.29}$$

These vector fields hold

$$[Y_1, Y_2] = Y_1, \quad [Y_1, Y_3] = 2Y_2, \quad [Y_2, Y_3] = Y_3. \tag{9.30}$$

Since $Z(V) = \{0\}$, then $V_L = \langle Y_1, Y_2, Y_3 \rangle$ is a Lie algebra isomorphic to $V/Z(V) \simeq \mathfrak{sl}(2, \mathbb{R})$ as stated in Theorem 9.3. The rest of commutation relations read

$$\begin{array}{llll}
[Y_1, Z_0] = 0, & [Y_1, Z_1] = 0, & [Y_1, Z_2] = Z_1, & [Y_1, Z_3] = 2Z_2, \\
[Y_2, Z_0] = 0, & [Y_2, Z_1] = -Z_1, & [Y_2, Z_2] = 0, & [Y_2, Z_3] = Z_3, \\
[Y_3, Z_0] = 0, & [Y_3, Z_1] = -2Z_2, & [Y_3, Z_2] = -Z_3, & [Y_3, Z_3] = 0.
\end{array} \tag{9.31}$$

Moreover,

$$\begin{array}{lll}
[Y_1, W_1] = 0, & [Y_1, W_2] = W_1, & [Y_1, W_3] = 2W_2 \\
[Y_2, W_1] = -W_1, & [Y_2, W_2] = 0, & [Y_2, W_3] = W_3, \\
[Y_3, W_1] = -2W_2, & [Y_3, W_2] = -W_3, & [Y_3, W_3] = 0
\end{array} \tag{9.32}$$

and

$$[Z_0, W_j] = Z_j, \quad [Z_i, W_j] = 0, \quad [W_i, W_j] = 0, \quad i, j = 1, 2, 3,$$

$$[Z_\alpha, Z_\beta] = 0, \quad \alpha, \beta = 0, \dots, 3. \tag{9.33}$$

Hence, $A_1 = \langle Z_0, Z_1, Z_2, Z_3 \rangle$ is an ideal of $A_1 + A_2$. And $A_1 + A_2$ is an ideal of $A_1 + A_2 + V_L$, with $A_2 = \langle W_1, W_2, W_3 \rangle$.

Example 9.1. We study a particular type of *first-order Riccati equation* [230, 325]

$$\frac{\mathrm{d}x}{\mathrm{d}t} = \eta(t) + x^2, \tag{9.34}$$

where η is an arbitrary t-dependent function.

It is well known that the Riccati equation is a Lie system with a VG Lie algebra isomorphic to $V^{\mathrm{Ric}} \simeq \mathfrak{sl}(2, \mathbb{R})$ [121]. Indeed, Equation

Lie Symmetry for Differential Equations 353

(9.34) has the associated t-dependent vector field $X^{\mathrm{Ric}} = X_3^{\mathrm{Ric}} + \eta(t)X_1^{\mathrm{Ric}}$, where

$$X_1^{\mathrm{Ric}} = \frac{\partial}{\partial x}, \quad X_2^{\mathrm{Ric}} = x\frac{\partial}{\partial x}, \quad X_3^{\mathrm{Ric}} = x^2\frac{\partial}{\partial x} \qquad (9.35)$$

satisfy the same commutation relations as X_1, X_2, X_3 in (9.23). In view of this, (9.34) is related to a t-dependent vector field taking values in a finite-dimensional Lie algebra of vector fields isomorphic to $\mathfrak{sl}(2, \mathbb{R})$. Then, (9.34) is an $\mathfrak{sl}(2, \mathbb{R})$-Lie system. Moreover, we can consider X^{Ric} as a particular case of system (9.24). By applying the results of the previous sections to generic $\mathfrak{sl}(2, \mathbb{R})$-Lie systems, we find that the symmetry system for X^{Ric} is of the form (9.26) with $b_1 = \eta, b_2 = 0$ and $b_3 = 1$, namely

$$\begin{cases} \dfrac{\mathrm{d}f_0}{\mathrm{d}t} = b_0(t), \\[2mm] \dfrac{\mathrm{d}f_1}{\mathrm{d}t} = f_0\dfrac{\mathrm{d}\eta}{\mathrm{d}t}(t) - \eta(t)f_2 + b_0(t)\eta(t), \\[2mm] \dfrac{\mathrm{d}f_2}{\mathrm{d}t} = 2f_1 - 2\eta(t)f_3, \\[2mm] \dfrac{\mathrm{d}f_3}{\mathrm{d}t} = f_2 + b_0(t). \end{cases} \qquad (9.36)$$

We can recover and generalize the results given in [182] by means of our approach. From their expressions in (9.36), we can differentiate $\mathrm{d}f_3/\mathrm{d}t$ twice and $\mathrm{d}f_2/\mathrm{d}t$ once. By substituting $\mathrm{d}^2 f_2/\mathrm{d}t^2$ in $\mathrm{d}^3 f_3/\mathrm{d}t^3$ and using the remaining equations in (9.36), we obtain that

$$\frac{\mathrm{d}^3 f_3}{\mathrm{d}t^3} = \frac{\mathrm{d}^3 f_0}{\mathrm{d}t^3} + 2\frac{\mathrm{d}f_1}{\mathrm{d}t} - 2\frac{\mathrm{d}\eta}{\mathrm{d}t}f_3 - 2\eta(t)\frac{\mathrm{d}f_3}{\mathrm{d}t}. \qquad (9.37)$$

By substituting the value of $\mathrm{d}f_1/\mathrm{d}t$ from (9.36) and using that $f_2 = \mathrm{d}f_3/\mathrm{d}t - b_0(t)$, we obtain the following equation for f_3 in terms of the coefficients of (9.34) and f_0

$$\frac{\mathrm{d}^3 f_3}{\mathrm{d}t^3} = \frac{\mathrm{d}^3 f_0}{\mathrm{d}t^3} + 4b_0\eta(t) + 2\frac{\mathrm{d}\eta}{\mathrm{d}t}f_0 - 2\frac{\mathrm{d}\eta}{\mathrm{d}t}f_3 - 4\eta(t)\frac{\mathrm{d}f_3}{\mathrm{d}t}. \qquad (9.38)$$

354   *A Guide to Lie Systems with Compatible Geometric Structures*

From this, we can retrieve the following corollary given in [182].

Corollary 9.10. *The Riccati equation* (9.34) *admits the Lie symmetry*

$$Y = f_0 \frac{\partial}{\partial t} - \frac{1}{2} \frac{\mathrm{d}^2 f_0}{\mathrm{d}t^2} \frac{\partial}{\partial x} - \frac{\mathrm{d}f_0}{\mathrm{d}t} x \frac{\partial}{\partial x}, \tag{9.39}$$

where

$$\frac{\mathrm{d}^3 f_0}{\mathrm{d}t^3} + 4b_0(t)\eta(t) + 2\frac{\mathrm{d}\eta}{\mathrm{d}t} f_0 = 0. \tag{9.40}$$

Proof. Since we are looking for Lie symmetries with $f_3 = 0$, Equation (9.38) reduces to (9.40). Moreover, by substituting $f_3 = 0$ in (9.36), we obtain that

$$\frac{\mathrm{d}f_3}{\mathrm{d}t} = b_0(t) + f_2 = 0, \qquad \frac{\mathrm{d}f_2}{\mathrm{d}t} = 2f_1, \tag{9.41}$$

which yields $f_2 = -\mathrm{d}f_0/\mathrm{d}t$ and $2f_1 = -\mathrm{d}^2 f_0/\mathrm{d}t^2$. Hence, the corollary follows. $\qquad\square$

Going back to general symmetries of (9.34), we can obtain some of its Lie symmetries by solving (9.36) for certain values of $\eta(t)$. These possibilities are summarized in Table A.5 in Appendix A.

Example 9.2. Reconsider the *Cayley–Klein Riccati equation* that appeared as (4.69) in Chapter 3. Its associated first-order system (4.70) was described in Chapter 3 by means of the t-dependent vector field

$$X^{\mathrm{CK}} = b_1(t)X_1^{\mathrm{CK}} + b_2(t)X_2^{\mathrm{CK}} + b_3(t)X_3^{\mathrm{CK}}, \tag{9.42}$$

where

$$X_1^{\mathrm{CK}} = \frac{\partial}{\partial x}, \quad X_2^{\mathrm{CK}} = x\frac{\partial}{\partial x} + y\frac{\partial}{\partial y}, \quad X_3^{\mathrm{CK}} = (x^2 + \iota^2 y^2)\frac{\partial}{\partial x} + 2xy\frac{\partial}{\partial y}, \tag{9.43}$$

satisfy the same commutation relations as X_1, X_2, X_3 in (9.23). Since X_1, X_2, X_3 span a Lie algebra $V \simeq \mathfrak{sl}(2, \mathbb{R})$, the vector fields

X_1^{CK}, X_2^{CK} and X_3^{CK} span a real Lie algebra, V^{CK}, which is isomorphic to the Lie algebra $\mathfrak{sl}(2, \mathbb{R})$ independently of the square of ι. Then, (9.42) is an $\mathfrak{sl}(2, \mathbb{R})$-Lie system.

Moreover, we can define the Lie algebra isomorphism $\phi : V^{CK} \to V$ mapping

$$\phi(X_\alpha^{CK}) = X_\alpha, \quad \alpha = 1, 2, 3, \tag{9.44}$$

where X_α, with $\alpha = 1, 2, 3$, are the generic vector fields in (9.23). Hence, we have that $\phi(X_t^{CK}) = X_t$ for every $t \in \mathbb{R}$ with X being the generic $\mathfrak{sl}(2, \mathbb{R})$-Lie system (9.24). Then, (9.42) is isomorphic to (9.24) and, in view of Proposition 9.5, the symmetry system of X^{CK} becomes (9.26).

Example 9.3. Let us consider the *quaternionic Riccati equation* with t-dependent real coefficients. Quaternions are the elements of the real vector space $\mathbb{H} \simeq \mathbb{R}^4$ of elements of the form $q = q_0 + q_1 i + q_2 j + q_3 k$, where $q_0, q_1, q_2, q_3 \in \mathbb{R}$ and with the standard sum of elements and multiplication by real numbers. We can define a multiplication of quaternions (see [149] for details) by assuming that real numbers commute with all quaternions and that the following operations are fulfilled:

$$i^2 = -1, \quad j^2 = -1, \quad k^2 = -1,$$
$$i \cdot j = -j \cdot i = k, \quad k \cdot i = -i \cdot k = j, \quad j \cdot k = -k \cdot j = i. \tag{9.45}$$

The quaternionic Riccati equation [374] takes the form

$$\frac{dq}{dt} = b_1(t) + a_1(t)q + qa_2(t) + qb_3(t)q, \tag{9.46}$$

where q and the t-dependent functions $a_1, a_2, b_1, b_3 : \mathbb{R} \to \mathbb{H}$ take values in $\mathbb{H}$ [149]. The existence of periodic solutions for particular cases of (9.46) has been studied in [77, 374] and, for real t-dependent coefficients, in [374]. We focus here on the latter case.

Writing q in coordinates, we obtain that (9.46) reads

$$
\begin{cases}
\dfrac{dq_0}{dt} = b_1(t) + b_2(t)q_0 + b_3(t)(q_0^2 - q_1^2 - q_2^2 - q_3^2), \\[2ex]
\dfrac{dq_1}{dt} = b_2(t)q_1 + 2b_3(t)q_0q_1, \\[2ex]
\dfrac{dq_2}{dt} = b_2(t)q_2 + 2b_3(t)q_0q_2, \\[2ex]
\dfrac{dq_3}{dt} = b_2(t)q_3 + 2b_3(t)q_0q_3,
\end{cases}
\tag{9.47}
$$

where $b_2(t) := a_1(t) + a_2(t)$. This system is related to the t-dependent vector field

$$
X^{\mathbb{H}} = b_1(t)X_1^{\mathbb{H}} + b_2(t)X_2^{\mathbb{H}} + b_3(t)X_3^{\mathbb{H}},
\tag{9.48}
$$

with

$$
X_1^{\mathbb{H}} = \frac{\partial}{\partial q_0}, \quad X_2^{\mathbb{H}} = q_0\frac{\partial}{\partial q_0} + q_1\frac{\partial}{\partial q_1} + q_2\frac{\partial}{\partial q_2} + q_3\frac{\partial}{\partial q_3},
$$

$$
X_3^{\mathbb{H}} = 2q_0\left(q_1\frac{\partial}{\partial q_1} + q_2\frac{\partial}{\partial q_2} + q_3\frac{\partial}{\partial q_3} \right) + (q_0^2 - q_1^2 - q_2^2 - q_3^2)\frac{\partial}{\partial q_0},
$$

$$
\tag{9.49}
$$

which satisfy the same commutation relations as X_1, X_2, and X_3 in (9.23). So, $X_1^{\mathbb{H}}, X_2^{\mathbb{H}}, X_3^{\mathbb{H}}$ span a real Lie algebra $V^{\mathbb{H}}$ isomorphic to $\mathfrak{sl}(2, \mathbb{R})$. Hence, the quaternionic Riccati equation (9.46) with t-dependent real coefficients is an $\mathfrak{sl}(2, \mathbb{R})$-Lie system isomorphic to (9.24) with respect to the Lie algebra isomorphism $\phi : V^{\mathbb{H}} \to V$ being given by

$$
\phi(X_\alpha^{\mathbb{H}}) = X_\alpha, \quad \alpha = 1, 2, 3.
\tag{9.50}
$$

Moreover, we have that $\phi(X_t^{\mathbb{H}}) = X_t$ for every $t \in \mathbb{R}$. Hence, the symmetry system for (9.46) is (9.26). If we assume for instance $f_0 = k \in \mathbb{R}$, and $b_1(t) = \eta(t)$, $b_2(t) = 0$ and $b_3(t) = 1$, we obtain that (9.46) is isomorphic to the Lie system (9.36). Hence, for certain values of $\eta(t)$ given in Table A.5 in Appendix A, we can derive several Lie symmetries for quaternionic Riccati equations.

Lie Symmetry for Differential Equations 357

Example 9.4. We now show that our theory in particular, and the whole theory of Lie systems in general, can be used to study autonomous systems of first-order ODEs. We consider the *generalized Darboux–Brioschi–Halphen system* (DBH system) [153]

$$\begin{cases} \dfrac{dw_1}{dt} = w_3 w_2 - w_1 w_3 - w_1 w_2 + \tau^2, \\[2mm] \dfrac{dw_2}{dt} = w_1 w_3 - w_2 w_1 - w_2 w_3 + \tau^2, \\[2mm] \dfrac{dw_3}{dt} = w_2 w_1 - w_3 w_2 - w_3 w_1 + \tau^2, \end{cases} \tag{9.51}$$

where

$$\tau^2 := \alpha_1^2 (\omega_1 - \omega_2)(\omega_3 - \omega_1) + \alpha_2^2 (\omega_2 - \omega_3)(\omega_1 - \omega_2)$$
$$+ \alpha_3^2 (\omega_3 - \omega_1)(\omega_2 - \omega_3) \tag{9.52}$$

and $\alpha_1, \alpha_2, \alpha_3$ are real constants.

When $\tau = 0$, system (9.51) retrieves the classical DBH system solved by Halphen [132, 153, 206] which appears in the study of triply orthogonal surfaces and the vacuum Einstein equations for hyper-Kähler Bianchi-IX metrics. For $\tau \neq 0$, the generalized DBH system can be considered as a reduction of the self-dual Yang–Mills equations corresponding to an infinite-dimensional gauge group of diffeomorphisms of a three-dimensional sphere [132].

It can be proven that (9.51) is an $\mathfrak{sl}(2, \mathbb{R})$-Lie system. Indeed, it is the associated system to the t-dependent vector field

$$X_t^{\mathrm{DBH}} = (w_3 w_2 - w_1 (w_3 + w_2) + \tau^2) \frac{\partial}{\partial w_1} + (w_1 w_3 - w_2 (w_1 + w_3)$$
$$+ \tau^2) \frac{\partial}{\partial w_2} + (w_2 w_1 - w_3 (w_2 + w_1) + \tau^2) \frac{\partial}{\partial w_3} = -X_3^{\mathrm{DBH}}. \tag{9.53}$$

This vector field spans a Lie algebra V^{DBH} of vector fields along with

$$X_1^{\mathrm{DBH}} = \frac{\partial}{\partial w_1} + \frac{\partial}{\partial w_2} + \frac{\partial}{\partial w_3},$$
$$X_2^{\mathrm{DBH}} = w_1 \frac{\partial}{\partial w_1} + w_2 \frac{\partial}{\partial w_2} + w_3 \frac{\partial}{\partial w_3} \tag{9.54}$$

satisfying the commutation relations (9.23). In consequence, $X_1^{\mathrm{DBH}}, X_2^{\mathrm{DBH}}$, and X_3^{DBH} span a three-dimensional Lie algebra of vector fields V^{DBH} isomorphic to $\mathfrak{sl}(2,\mathbb{R})$ and then X^{DBH} is an $\mathfrak{sl}(2,\mathbb{R})$-Lie system. Since X_1, X_2, X_3 admit the same structure constants as (9.23), the symmetry system for X^{DBH} becomes (9.26) with $b_1(t) = b_2(t) = 0$ and $b_3(t) = -1$, namely

$$\frac{\mathrm{d}f_0}{\mathrm{d}t} = b_0(t), \quad \frac{\mathrm{d}f_1}{\mathrm{d}t} = 0, \quad \frac{\mathrm{d}f_2}{\mathrm{d}t} = -2f_1, \quad \frac{\mathrm{d}f_3}{\mathrm{d}t} = -f_2 - b_0(t). \quad (9.55)$$

Hence, for $b_0(t) = 0$, we obtain

$$Y = t_0\frac{\partial}{\partial t} + \lambda_1 X_1 - (2\lambda_1 t - \lambda_2)X_2 + (\lambda_1 t^2 - \lambda_2 t + \lambda_3)X_3, \quad (9.56)$$

with $\lambda_1, \lambda_2, \lambda_3, t_0 \in \mathbb{R}$. Evidently, these vector fields span a Lie algebra of Lie symmetries isomorphic to $\mathfrak{sl}(2,\mathbb{R})$ for $t_0 = 0$. For $b_0(t) = c_0$, we obtain

$$Y = (c_0 t + t_0)\frac{\partial}{\partial t} + \lambda_1 X_1 - (2\lambda_1 t - \lambda_2)X_2$$
$$+ (\lambda_1 t^2 - (\lambda_2 + c_0)t + \lambda_3)X_3, \quad (9.57)$$

with $\lambda_1, \lambda_2, \lambda_3, t_0, c_0 \in \mathbb{R}$. Finally, for $b_0(t) = c_0 t$, we get

$$Y = \left(t_0 + \frac{c_0 t^2}{2}\right)\frac{\partial}{\partial t} + \lambda_1 X_1 - (2\lambda_1 t - \lambda_2)X_2$$
$$+ \left[\lambda_1 t^2 - \left(\lambda_2 + \frac{c_0 t}{2}\right)t + \lambda_3\right]X_3, \quad (9.58)$$

with $\lambda_1, \lambda_2, \lambda_3, t_0, c_0 \in \mathbb{R}$. Since $\mathfrak{W} = \langle 1, t, t^2\rangle$ is a Lie algebra with respect to the Lie bracket, $\{\cdot, \cdot\}_{\mathbb{R}}$, we obtain in view of Proposition 9.8 that

$$\mathcal{S}_{X,\mathfrak{W}}^V = \{Y \in \mathcal{S}_X^V | f_0 \in \langle 1, t, t^2\rangle\}, \quad (9.59)$$

is a Lie algebra of Lie symmetries. By choosing appropriately the constant coefficients of the above vector fields and setting $\tau = 0$, we recover the Lie algebra of symmetries isomorphic to $\mathfrak{sl}(2,\mathbb{R})$ for classical DBH systems [295].

Lie Symmetry for Differential Equations 359

The study of systems of HODEs (higher-order ODEs) through Lie systems implies the addition of extra variables in order to express a higher-order system as a first-order one. The introduction of these extra variables frequently results in the obtainment of non-Lie point symmetries, as we shall exemplify in forthcoming examples. Certain non-local Lie symmetries can be identified with the prolongations of certain vector fields.

Lemma 9.11. *Given a Lie algebra of Lie point symmetries V, its prolongations form a Lie algebra $\widehat{V}$ isomorphic to the former.*

Proof. Consider the map

$$\Phi : \Gamma(\mathrm{T}(\mathbb{R} \times N)) \to \Gamma(\mathrm{T}[\mathbb{R} \times \mathrm{T}^p N]) \\ X \mapsto \widehat{X}, \tag{9.60}$$

mapping sections of $\mathrm{T}(\mathbb{R} \times N)$, i.e., vector fields on $\mathbb{R} \times N$, onto sections of $\mathrm{T}(\mathbb{R} \times \mathrm{T}^p N)$, i.e., vector fields on the manifold $\mathbb{R} \times \mathrm{T}^p N$, where $\mathrm{T}^p N$ is the p-order tangent space. Recall that, roughly speaking, $\mathrm{T}^p N$ is the space of equivalence classes of curves in N with the same Taylor expansion up to order p. It admits a differentiable structure induced by the variables $x_1, \ldots, x_n$ of the coordinate systems of N and the induced variables $x_i^{k)}$ describing the derivatives in terms of t up to order p of the coordinates of a curve within N. Given the so-called *contact 1-forms* $\theta_i^k = \mathrm{d}x_i^{k)} - x_i^{k+1)}\mathrm{d}t$ with $i = 1, \ldots, n$ and $k = 0, \ldots, p-1$ on $\mathbb{R} \times \mathrm{T}^p N$, we say that $\widehat{X}$ is the prolongation of $X \in \Gamma(\mathrm{T}(\mathbb{R} \times N))$ to $\mathbb{R} \times \mathrm{T}^p N$ if and only if every $\mathcal{L}_{\widehat{X}}\theta_i^k$ belongs to the contact ideal spanned by all the contact forms and $\mathrm{J}^p_{\pi*}\widehat{X} = X$, with $\mathrm{J}^p_{\pi*}$ being the tangent map to the projection $\mathrm{J}^p_\pi : (t, x, x^{k)}) \in \mathbb{R} \times \mathrm{T}^p N \mapsto (t, x) \in \mathbb{R} \times N$ with $k = 1, \ldots, p$. This implies that Φ is $\mathbb{R}$-linear and injective. Additionally,

$$\mathcal{L}_{[\widehat{X_1}, \widehat{X_2}]}\theta_i^k = \left(\mathcal{L}_{\widehat{X_1}} \mathcal{L}_{\widehat{X_2}} - \mathcal{L}_{\widehat{X_2}} \mathcal{L}_{\widehat{X_1}} \right) \theta_i^k \tag{9.61}$$

belongs to the ideal of the contact forms because $\mathcal{L}_{\widehat{X_1}}$ and $\mathcal{L}_{\widehat{X_2}}$ do so. Moreover,

$$\mathrm{J}^p_{\pi*}[\widehat{X_1}, \widehat{X_2}] = [\mathrm{J}^p_{\pi*}\widehat{X_1}, \mathrm{J}^p_{\pi*}\widehat{X_2}] = [X_1, X_2]. \tag{9.62}$$

Hence, $[\widehat{X_1}, \widehat{X_2}]$ must be the prolongation of $[X_1, X_2]$, i.e., $[\widehat{X_1}, \widehat{X_2}] = \widehat{[X_1, X_2]}$. In this way, Φ is a Lie algebra morphism. Obviously, given a Lie algebra of Lie point symmetries V, its prolongations form a Lie algebra $\widehat{V} = \Phi(V)$ isomorphic to V. $\qquad\square$

Definition 9.12. We call $\widehat{V}$ the Lie algebra whose elements are the prolongations of the elements in V.

Example 9.5. As an example of a second-order Lie system possessing a Vessiot–Gulberg Lie algebra isomorphic to $\mathfrak{sl}(2, \mathbb{R})$, we recall the *second-order Kummer–Schwarz equation* given in (3.68) in Chapter 3. In form of a first-order system, it took the expression

$$
\begin{cases}
\dfrac{dx}{dt} = v, \\[2mm]
\dfrac{dv}{dt} = \dfrac{3}{2}\dfrac{v^2}{x} - 2c_0 x^3 + 2b_1(t)x,
\end{cases}
\tag{9.63}
$$

on $T\mathbb{R}_0$, with $\mathbb{R}_0 = \mathbb{R} - \{0\}$. Such a system is associated with the t-dependent vector field $M^{\mathrm{KS}} = M_3^{\mathrm{KS}} + \eta(t)M_1^{\mathrm{KS}}$, with vector fields

$$
M_1^{\mathrm{KS}} = 2x\frac{\partial}{\partial v}, \quad M_2^{\mathrm{KS}} = x\frac{\partial}{\partial x} + 2v\frac{\partial}{\partial v},
$$
$$
M_3^{\mathrm{KS}} = v\frac{\partial}{\partial x} + \left(\frac{3}{2}\frac{v^2}{x} - 2c_0 x^3\right)\frac{\partial}{\partial v}
\tag{9.64}
$$

satisfying the same commutation relations as the vector fields X_1, X_2, X_3 in (9.23) and they therefore span a VG Lie algebra isomorphic to $\mathfrak{sl}(2, \mathbb{R})$. Since the basis $M_1^{\mathrm{KS}}, M_2^{\mathrm{KS}}, M_3^{\mathrm{KS}}$ have the same structure constants as the $\mathfrak{sl}(2, \mathbb{R})$-Lie systems analyzed in Section 9.5, we can define, for instance, a Lie algebra morphism ϕ mapping $M_1^{\mathrm{KS}}, M_2^{\mathrm{KS}}, M_3^{\mathrm{KS}}$ to the basis $X_1^{\mathrm{Ric}}, X_2^{\mathrm{Ric}}, X_3^{\mathrm{Ric}}$ for Riccati equation (9.34). In such a case, (9.63) maps to X^{Ric} and, in view of Proposition 9.5, the symmetry system for (9.63) is of the form of the symmetry system (9.36). As a consequence, the particular solutions f_0, f_1, f_2, and f_3 for Riccati equations, detailed in Table A.5, are valid for (9.63) as well.

9.3. Lie Symmetries for Aff($\mathbb{R}$)-Lie Systems

In this section, we aim to obtain elements of $\mathcal{S}_X^V$ for Aff($\mathbb{R}$)-Lie systems. We choose a basis of vector fields $\{X_1, X_2\}$ of $V \simeq$ Aff($\mathbb{R}$) with $[X_1, X_2] = X_1$, and express the most general Lie system with VG Lie algebra V as $X = a(t)X_1 + b(t)X_2$. Let us now look for its Lie symmetries of the form

$$Y = f_0(t)\frac{\partial}{\partial t} + f_1(t)X_1 + f_2(t)X_2. \tag{9.65}$$

The symmetry condition gives rise to a symmetry system

$$\begin{cases} \dfrac{\mathrm{d}f_0}{\mathrm{d}t} = b_0(t), \\ \dfrac{\mathrm{d}f_1}{\mathrm{d}t} = f_0\dfrac{\mathrm{d}a}{\mathrm{d}t}(t) + a(t)b_0(t) + b(t)f_1 - a(t)f_2, \\ \dfrac{\mathrm{d}f_2}{\mathrm{d}t} = f_0\dfrac{\mathrm{d}b}{\mathrm{d}t}(t) + b(t)b_0(t), \end{cases} \tag{9.66}$$

associated with the t-dependent vector field

$$\begin{aligned} \Gamma_X^{\text{Aff}(\mathbb{R})} &= b_0(t)Z_0 + \frac{\mathrm{d}a}{\mathrm{d}t}W_1 + \frac{\mathrm{d}b}{\mathrm{d}t}W_2 + b_0(t)a(t)Z_1 \\ &\quad + b_0(t)b(t)Z_2 + a(t)Y_1 + b(t)Y_2, \end{aligned} \tag{9.67}$$

where

$$Y_1 = -f_2\frac{\partial}{\partial f_1}, \quad Y_2 = f_1\frac{\partial}{\partial f_1}, \quad W_1 = f_0\frac{\partial}{\partial f_1},$$
$$W_2 = f_0\frac{\partial}{\partial f_2}, \quad Z_\alpha = \frac{\partial}{\partial f_\alpha}, \tag{9.68}$$

with $\alpha = 0, 1, 2$. Since, $[Y_1, Y_2] = Y_1$, then $V_L = \langle Y_1, Y_2 \rangle$ gives rise to a Lie algebra isomorphic to Aff($\mathbb{R}$). Moreover,

$$\begin{aligned} &[Z_0, W_1] = Z_1, \quad [Z_0, W_2] = Z_2, \quad [Z_i, W_j] = 0, \quad [Z_\alpha, Z_\beta] = 0, \\ &[Y_1, W_1] = 0, \quad [Y_1, W_2] = W_1, \quad [Y_1, Z_0] = 0, \quad [Y_1, Z_1] = 0, \\ &[Y_1, Z_2] = Z_1, \quad [Y_2, W_1] = -W_1, \quad [Y_2, W_2] = 0, \quad [Y_2, Z_0] = 0, \\ &[Y_2, Z_1] = -Z_1, \quad [Y_2, Z_2] = 0, \end{aligned}$$
$$\tag{9.69}$$

with $i, j = 1, 2$ and $\alpha, \beta = 0, 1, 2$. In this way, $A_1 = \langle Z_0, Z_1, Z_2 \rangle$ is an ideal of $A_1 + A_2$, with $A_2 = \langle W_1, W_2 \rangle$ and $A_1 \oplus_S A_2$ an ideal of $A_1 + A_2 + V_L$. Hence, system (9.66) possesses a VG Lie algebra

$$(A_1 \oplus_S A_2) \oplus_S V_L \simeq (\mathbb{R}^3 \oplus_S \mathbb{R}^2) \oplus_S \text{Aff}(\mathbb{R}). \tag{9.70}$$

We can solve (9.66) when $b_0(t) = 0$. In this case, $f_0 = k \in \mathbb{R}$ and (9.66) reduces to a trivial equation for f_2 and a linear one for f_1. The general solution reads

$$f_1(t) = \left[\int_0^t \left[k \frac{da}{dt'}(t') - a(t')(kb(t') + c_1) \right] e^{-\int_0^{t'} b(t'')dt''} dt' + c_2 \right] e^{\int_0^t b(t')dt'},$$

$$f_2(t) = kb(t) + c_1, \tag{9.71}$$

where c_1 and c_2 are integration constants.

Example 9.6. In order to illustrate $\text{Aff}(\mathbb{R})$-Lie systems through a physical example, we reconsider the *Buchdahl equation* [65, 133, 141], which was analyzed in Chapter 3 as a first-order system

$$\begin{cases} \dfrac{dx}{dt} = y, \\ \dfrac{dy}{dt} = a(x)y^2 + b(t)y. \end{cases} \tag{9.72}$$

Recall that we have proven there that (9.72) is a Lie system [23]. Indeed, (9.72) describes the integral curves of the t-dependent vector field

$$X^{BD} = v \frac{\partial}{\partial x} + (f(x)v^2 + a_2(t)v) \frac{\partial}{\partial v} = X_1 - a_2(t)X_2, \tag{9.73}$$

where

$$X_1^{BD} = v \frac{\partial}{\partial x} + f(x)v^2 \frac{\partial}{\partial v}, \quad X_2^{BD} = -v \frac{\partial}{\partial v}, \tag{9.74}$$

satisfy $[X_1, X_2] = X_1$. That is, X is an $\text{Aff}(\mathbb{R})$-Lie system. By applying Theorem 9.3, we see that the Lie symmetries of (9.72) of the form

$$Y = f_0(t)\frac{\partial}{\partial t} + f_1(t)X_1 + f_2(t)X_2 \tag{9.75}$$

are determined by the first-order system (9.66) with $a(t) = 1$ and $b(t) = -a_2(t)$, i.e.,

$$\begin{cases} \dfrac{\mathrm{d}f_0}{\mathrm{d}t} = b_0(t), \\[2mm] \dfrac{\mathrm{d}f_1}{\mathrm{d}t} = b_0(t) - a_2(t)f_1 - f_2, \\[2mm] \dfrac{\mathrm{d}f_2}{\mathrm{d}t} = -f_0\dfrac{\mathrm{d}a_2}{\mathrm{d}t} - b_0(t)a_2(t), \end{cases} \tag{9.76}$$

which is associated with the t-dependent vector field

$$\Gamma_X^{\mathrm{BD}} = b_0(t)Z_0 - \frac{\mathrm{d}a_2}{\mathrm{d}t}W_2 + b_0(t)Z_1 - a_2(t)b_0(t)Z_2 + Y_1 - a_2(t)Y_2, \tag{9.77}$$

where the vector fields $Z_0, Z_1, Z_2, Y_1, Y_2, W_2$ are those detailed in (9.68) and have the commutation relations (9.69). Hence, these vector fields span a Lie algebra (9.70). Therefore, we can obtain the Lie symmetries for this system of the form (9.65) by substituting $a(t) = 1$ and $b(t) = -a_2(t)$ in (9.71).

9.4. Lie Symmetries for PDE Lie Systems

Let us consider a system of PDEs

$$\frac{\partial x_i}{\partial t_l} = X_{li}(t, x), \quad i = 1, \dots, n, \quad l = 1, \dots, s, \tag{9.78}$$

where the $X_{li} : \mathbb{R}^s \times \mathbb{R}^n \mapsto T(\mathbb{R}^s \times \mathbb{R}^n) \in X(t, x)$ are arbitrary functions. Its particular solutions are sections $x : t := (t_1, \dots, t_s) \in \mathbb{R}^s \mapsto x(t) \in \mathbb{R}^s \times \mathbb{R}^n$ of the bundle $(\mathbb{R}^s \times \mathbb{R}^n, \mathbb{R}^s, \pi_s : (t, x) \in \mathbb{R}^s \times \mathbb{R}^n \mapsto t \in \mathbb{R}^s)$. In particular, we recover the simple case of ODEs when $s = 1$.

Let us assume $t \in \mathbb{R}^s$. We call t-*dependent vector field* X on N a map $X : \mathbb{R}^s \times N \to TN$ such that $\tau \circ X = \pi_s$. We call $\bar{X}^l$, with

364   *A Guide to Lie Systems with Compatible Geometric Structures*

$l = 1, \ldots, s$, the *autonomization* of X with respect to t_l, i.e.,

$$\bar{X}^l(t, x) := \frac{\partial}{\partial t_l} + X(t, x). \tag{9.79}$$

Definition 9.13. A first-order system of a number $n \times s$ PDEs of the form (9.78)

$$\frac{\partial x_i}{\partial t_l} = \sum_{\alpha=1}^{r} b_{\alpha l}(t) X_\alpha(x), \quad l = 1, \ldots, s, \quad i = 1, \ldots, n, \tag{9.80}$$

is a *PDE Lie system* if

- There exist vector fields $X_1, \ldots, X_r$ on N spanning an r-dimensional real Lie algebra V satisfying that each t-dependent vector field $X_l(t, x) = \sum_{i=1}^{n} X_{li}(t, x) \cdot \partial/\partial x_i$, where $l = 1, \ldots, s$, can be written in the form

$$X_l(t, x) = \sum_{\alpha=1}^{r} b_{\alpha l}(t) X_\alpha(x), \quad l = 1, \ldots, s, \tag{9.81}$$

  for certain t-dependent functions $b_{\alpha l}$.
- Let $c_{\alpha\beta\gamma}$ be the structure constants of $X_1, \ldots, X_r$, i.e., $[X_\alpha, X_\beta] = \sum_{\gamma=1}^{r} c_{\alpha\beta\gamma} X_\gamma$. Then,

$$\frac{\partial b_{\gamma k}}{\partial t_l}(t) - \frac{\partial b_{\gamma l}}{\partial t_k}(t) + \sum_{\alpha,\beta=1}^{r} b_{\alpha l}(t) b_{\beta k}(t) c_{\alpha\beta\gamma} = 0, \tag{9.82}$$

  with $l \neq k = 1, \ldots, s, \quad \gamma = 1, \ldots, r$.

If these conditions are satisfied, we call V a *Vessiot–Guldberg Lie algebra for the PDE Lie system* (9.78).

Theorem 9.14. *Let V be a finite-dimensional real Lie algebra of vector fields on $\mathbb{R}^n$ with a basis $X_1, \ldots, X_r$. Given a PDE Lie system as* (9.80), *then*

$$Y(t, x) = \sum_{\beta=1}^{r} f_\beta(t) X_\beta(x) \tag{9.83}$$

is a Lie symmetry of (9.80) *if and only if $[\bar{X}^l, Y] = 0$ for every l, where $\bar{X}^l$ is the l-autonomization given by* (9.79) *of the t-dependent* *vector field*

$$X_l(t, x) := \sum_{\alpha=1}^{r} b_{\alpha l}(t) X_\alpha(x), \quad l = 1, \ldots, s. \tag{9.84}$$

Proof. The coordinate systems $\{t_1, \ldots, t_s\}$ on $\mathbb{R}^s$ and $\{x_1, \ldots, x_n\}$ on $\mathbb{R}^n$ induce a coordinate system on the fiber bundle $J^1\pi \simeq \mathbb{R}^s \times \mathbb{R}^n \times \mathbb{R}^{ns}$ with respect to the projection $\pi : (t_l, x_j) \in \mathbb{R}^s \times \mathbb{R}^n \mapsto t_l \in \mathbb{R}^s$ of the form $t_l, x_j, x_{j,l}$, where $x_{j,l} = \partial x_j / \partial x_l$ with $1 \leq j \leq n$ and $1 \leq l \leq s$. A vector field $Y = \sum_{k=1}^{n} \eta_k(t, x) \partial / \partial x_k$, $t \in \mathbb{R}^s$ and $x \in \mathbb{R}^n$ is a Lie symmetry of (9.80) if and only if

$$\widehat{Y} F_l^i = 0, \qquad i = 1, \ldots, n, \ \ l = 1, \ldots, s \tag{9.85}$$

on the submanifold $\mathcal{S} = \cap_{i=1}^{n} \cap_{l=1}^{s} (F_l^i)^{-1}(0)$, with $\widehat{Y} : J^1\pi \to T(J^1\pi)$ being the prolongation of Y, namely

$$\widehat{Y} = \sum_{k=1}^{n} \left[\eta_k \frac{\partial}{\partial x_k} + \sum_{q=1}^{s} \left(\frac{\partial \eta_k}{\partial t_q} + \sum_{j=1}^{n} \frac{\partial \eta_k}{\partial x_j} x_{j,q} \right) \frac{\partial}{\partial x_{k,q}} \right], \tag{9.86}$$

and

$$F_l^i = x_{i,l} - \sum_{\alpha=1}^{r} b_{\alpha l}(t) X_\alpha(x), \quad i = 1, \ldots, n, \quad l = 1, \ldots, s,$$

$$X_\alpha = \sum_{i=1}^{n} X_\alpha^i(x) \frac{\partial}{\partial x_i}, \quad \alpha = 1, \ldots, r. \tag{9.87}$$

By assumption, $\eta_k(t, x) = \sum_{\beta=1}^{r} f_\beta(t) X_\beta^k(x)$ for $k = 1, \ldots, n$. So,

$$\widehat{Y} = \sum_{k=1}^{n} \sum_{\beta=1}^{r} \left[f_\beta(t) X_\beta^k \frac{\partial}{\partial x_k} \right.$$

$$\left. + \sum_{q=1}^{s} \left(\frac{\partial f_\beta(t)}{\partial t_q} X_\beta^k + \sum_{j=1}^{n} f_\beta(t) \frac{\partial X_\beta^k}{\partial x_j} x_{j,q} \right) \frac{\partial}{\partial x_{k,q}} \right]. \tag{9.88}$$

Substituting this in $\widehat{Y}F_l^i = 0$, we obtain

$$
\sum_{k=1}^{n}\sum_{\beta=1}^{r}\left[f_\beta(t)X_\beta^k\left(-\sum_{\alpha=1}^{r}b_{\alpha l}(t)\frac{\partial X_\alpha^i}{\partial x_k}\right)\right.
$$

$$
\left.+\sum_{q=1}^{s}\left(\frac{\partial f_\beta(t)}{\partial t_q}X_\beta^k+\sum_{j=1}^{n}f_\beta(t)\frac{\partial X_\beta^k}{\partial x_j}x_{j,q}\right)\delta_q^l\delta_k^i\right]
$$

$$
=\sum_{\beta=1}^{r}\left[-\sum_{k=1}^{n}\sum_{\alpha=1}^{r}f_\beta(t)b_{\alpha l}(t)X_\beta^k\frac{\partial X_\alpha^i}{\partial x_k}+\frac{\partial f_\beta(t)}{\partial t_l}X_\beta^i\right.
$$

$$
\left.+\sum_{j=1}^{n}f_\beta(t)\frac{\partial X_\beta^i}{\partial x_j}x_{j,l}\right]=0. \tag{9.89}
$$

Restricting the above expression to the submanifold $\mathcal{S} = \cap_{i=1}^{n}\cap_{l=1}^{s}$ $(F_l^i)^{-1}(0)$ and renaming indexes appropriately, we obtain

$$
\sum_{\beta=1}^{r}\left[\frac{\partial f_\beta(t)}{\partial t_l}X_\beta^i+\sum_{k=1}^{n}\sum_{\alpha=1}^{r}\left(f_\beta(t)\frac{\partial X_\beta^i}{\partial x_k}b_{\alpha l}(t)X_\alpha^k\right.\right.
$$

$$
\left.\left.-f_\beta(t)b_{\alpha l}(t)X_\beta^k\frac{\partial X_\alpha^i}{\partial x_k}\right)\right]=0. \tag{9.90}
$$

Hence,

$$
\sum_{\beta=1}^{r}\left[\frac{\partial f_\beta(t)}{\partial t_l}X_\beta^i+\sum_{\gamma=1}^{r}\sum_{k=1}^{n}f_\beta(t)b_{\gamma l}(t)\left(\frac{\partial X_\beta^i}{\partial x_k}X_\gamma^k-X_\beta^k\frac{\partial X_\gamma^i}{\partial x_k}\right)\right]=0,
$$

$$
\tag{9.91}
$$

whose right-hand side becomes, for each fixed l, the coefficients in the basis $\partial/\partial x^i$, with $i = 1,\dots,n$, of $[\bar{X}^l, Y]$. So, the above amounts to

$$
\left[\bar{X}^l, Y\right] = \left[\frac{\partial}{\partial t_l} + \sum_{k=1}^{n}\sum_{\gamma=1}^{r} b_{\gamma l}(t) X_\gamma^k \frac{\partial}{\partial x_k}, \sum_{i=1}^{n}\sum_{\beta=1}^{r} f_\beta(t) X_\beta^i \frac{\partial}{\partial x_i}\right]
$$

$$
= \left[\frac{\partial}{\partial t_l} + \sum_{\gamma=1}^{r} b_{\gamma l}(t) X_\gamma, \sum_{\beta=1}^{r} f_\beta(t) X_\beta\right] = 0. \tag{9.92}
$$

Then, Y is a Lie symmetry of (9.80) if and only if the condition

$$
\left[\bar{X}^l, Y\right] = 0 \tag{9.93}
$$

is satisfied for $l = 1, \ldots, s$. $\square$

Theorem 9.15. *Given a Lie symmetry of the form (9.83) for the system (9.80), the coefficients $f_1(t), \ldots, f_r(t)$ satisfy a PDE Lie system admitting a VG Lie algebra $V^S \simeq V/Z(V)$, where we recall that V is a VG Lie algebra for (9.80).*

Proof. Let $X_1, \ldots, X_r$ be a basis for V with structure constants $c_{\alpha\beta\gamma}$. From Theorem 9.14, we have

$$
\left[\bar{X}^l, Y\right] = \left[\frac{\partial}{\partial t_l} + \sum_{\alpha=1}^{r} b_{\alpha l}(t) X_\alpha, \sum_{\delta=1}^{r} f_\delta(t) X_\delta\right]
$$

$$
= \sum_{\delta=1}^{r} \left(\frac{\partial f_\delta}{\partial t_l} X_\delta + \sum_{\alpha=1}^{r} b_{\alpha l}(t) f_\delta [X_\alpha, X_\delta]\right)
$$

$$
= \sum_{\pi=1}^{r} \left(\frac{\partial f_\pi}{\partial t_l} + \sum_{\alpha=1}^{r}\sum_{\delta=1}^{r} b_{\alpha l}(t) f_\delta c_{\alpha\delta\pi}\right) X_\pi = 0. \tag{9.94}
$$

Since $X_1, \ldots, X_r$ are linearly independent over $\mathbb{R}$ and the coefficients of the above expression are only t-dependent, we get that the above amounts to

$$\frac{\partial f_\pi}{\partial t_l} = \sum_{\alpha,\delta=1}^{r} b_{\alpha l}(t) f_\delta c_{\delta\alpha\pi}, \quad \pi = 1, \dots, r, \quad l = 1, \dots, s. \qquad (9.95)$$

To prove that this is a PDE Lie system, we define the vector fields

$$Y_\alpha = \sum_{\delta,\pi=1}^{r} c_{\delta\alpha\pi} f_\delta \frac{\partial}{\partial f_\pi}, \quad \alpha = 1, \dots, r. \qquad (9.96)$$

We have already proven that $[Y_\alpha, Y_\beta] = \sum_{\delta=1}^{r} c_{\alpha\beta\delta} Y_\delta$ in Theorem 9.3. So, these vector fields span a Lie algebra isomorphic to $V/Z(V)$ (for a proof of this fact, follow the same line of reasoning as in Theorem 9.3). In terms of these vector fields, we see that (9.95) is related to the t-dependent vector fields $X_l(t, x) = \sum_{\alpha=1}^{r} b_{\alpha l}(t) X_\alpha(x)$, with $l = 1, \dots, s$. Additionally, to be a PDE Lie system, the above system (9.95) must satisfy the condition

$$\sum_{\alpha,\mu=1}^{r} \left(\frac{\partial b_{\alpha\pi}}{\partial \kappa} - \frac{\partial b_{\alpha\kappa}}{\partial \pi} + \sum_{\delta,\epsilon=1}^{r} b_{\delta\kappa} b_{\epsilon\pi} c_{\delta\epsilon\alpha} \right) f_\mu c_{\mu\alpha\sigma} = 0, \qquad (9.97)$$

with $\kappa \neq \pi = 1, \dots, r$, $\sigma = 1, \dots, r$. The expression in brackets vanishes due to the integrability condition for system (9.80). Hence, (9.95) is a PDE Lie system. We call (9.95) the *symmetry system* for (9.80) relative to V. $\qquad \square$

Definition 9.16. Given a PDE Lie system X with a VG Lie algebra V, we call $\mathcal{S}_X^V$ the *space of Lie symmetries* of X that are also t-dependent vector fields taking values in V.

We can straightforwardly prove that the space $\mathcal{S}_X^V$ is a Lie algebra.

9.4.1. *Lie symmetries for $\mathfrak{sl}(2, \mathbb{R})$-PDE Lie systems*

An $\mathfrak{sl}(2, \mathbb{R})$-PDE Lie system is a PDE Lie system admitting a VG Lie algebra isomorphic to $\mathfrak{sl}(2, \mathbb{R})$. Let us obtain the elements of $\mathcal{S}_X^V$ for this case. Let us choose a basis of vector fields $\{X_1, X_2, X_3\}$ for V satisfying the same commutation relations as in (9.23). Let us write a general PDE Lie system whose autonomization for a fixed value l is

$$\bar{X}^l = \frac{\partial}{\partial t_l} + b_{1l}(t)X_1 + b_{2l}(t)X_2 + b_{3l}(t)X_3, \qquad (9.98)$$

with $t = (t_1, \ldots, t_s) \in \mathbb{R}^s$, $1 \leq l \leq s$ and a certain type of possible Lie symmetry $Y = f_1(t)X_1 + f_2(t)X_2 + f_3(t)X_3$, where $f_1(t), f_2(t), f_3(t)$ are t-dependent functions to be determined by the symmetry condition (9.93). This leads us to the system of s first-order PDEs

$$\begin{cases} \dfrac{\partial f_1}{\partial t_l} = b_{2l}(t)f_1 - b_{1l}(t)f_2, \\[2mm] \dfrac{\partial f_2}{\partial t_l} = 2(b_{3l}(t)f_1 - b_{1l}(t)f_3), \\[2mm] \dfrac{\partial f_3}{\partial t_l} = b_{3l}(t)f_2 - b_{2l}(t)f_3, \end{cases} \qquad (9.99)$$

with $l = 1, \ldots, s$. Expressed in terms of t-dependent vector fields, for a fixed value of l

$$\Gamma_l^{\mathfrak{sl}(2,\mathbb{R})} = b_{1l}(t)Y_1 + b_{2l}(t)Y_2 + b_{3l}(t)Y_3, \qquad l = 1, \ldots, s, \qquad (9.100)$$

where

$$Y_1 = -f_2\frac{\partial}{\partial f_1} - 2f_3\frac{\partial}{\partial f_2}, \quad Y_2 = f_1\frac{\partial}{\partial f_1} - f_3\frac{\partial}{\partial f_3},$$
$$Y_3 = 2f_1\frac{\partial}{\partial f_2} + f_2\frac{\partial}{\partial f_3} \qquad (9.101)$$

close the commutation relations in (9.23). So, they span a Lie algebra isomorphic to $\mathfrak{sl}(2, \mathbb{R})$.

Example 9.7. Let us consider the *partial Riccati equation*, i.e., the PDE system

$$\frac{\partial x}{\partial t_1} = b_{11}(t) + b_{21}(t)x + b_{31}(t)x^2,$$

$$\frac{\partial x}{\partial t_2} = b_{12}(t) + b_{22}(t)x + b_{32}(t)x^2, \qquad (9.102)$$

with the t-dependent coefficients satisfying the appropriate integrability condition (9.97). Such systems appear in the study of

WZNW equations and multidimensional Toda systems [167]. Observe that the partials $\partial x/\partial t_1$ and $\partial x/\partial t_2$ are related to the t-dependent vector fields $X_{t_1}^{\mathrm{pRic}} = b_{11}(t)X_1^{\mathrm{pRic}} + b_{21}(t)X_2^{\mathrm{pRic}} + b_{31}(t)X_3^{\mathrm{pRic}}$ and $X_{t_2}^{\mathrm{pRic}} = b_{12}(t)X_1^{\mathrm{pRic}} + b_{22}(t)X_2^{\mathrm{pRic}} + b_{32}(t)X_3^{\mathrm{pRic}}$, with

$$X_1^{\mathrm{pRic}} = \frac{\partial}{\partial x}, \quad X_2^{\mathrm{pRic}} = x\frac{\partial}{\partial x}, \quad X_3^{\mathrm{pRic}} = x^2\frac{\partial}{\partial x} \tag{9.103}$$

satisfying the commutation relations (9.23). That is, the vector fields $\langle X_1^{\mathrm{pRic}}, X_2^{\mathrm{pRic}}, X_3^{\mathrm{pRic}} \rangle$ span a VG Lie algebra $V^{\mathrm{pRic}} \simeq \mathfrak{sl}(2,\mathbb{R})$. Since we assume that the functions $b_{ij}(t)$ with $i = 1,2,3$ and $j = 1,2$ satisfy (9.97), we get that (9.102) is a PDE Lie system.

Let us look for Lie symmetries of the form $Y = f_1(t)X_1^{\mathrm{pRic}} + f_2(t)X_2^{\mathrm{pRic}} + f_3(t)X_3^{\mathrm{pRic}}$ for (9.102). In view of Theorem 9.15, such Lie symmetries are solutions of the system of PDEs

$$\begin{cases} \dfrac{\partial f_1}{\partial t_j} = b_{2j}(t)f_1 - b_{1j}(t)f_2, \\[2mm] \dfrac{\partial f_2}{\partial t_j} = 2(b_{3j}(t)f_1 - b_{1j}(t)f_3), \\[2mm] \dfrac{\partial f_3}{\partial t_j} = b_{3j}(t)f_2 - b_{2j}(t)f_3, \end{cases} \tag{9.104}$$

with $j = 1,2$. This resulting system can be interpreted in terms of the t-dependent vector fields $\Gamma_j^{\mathrm{pRic}} = b_{1j}(t)Y_1 + b_{2j}(t)Y_2 + b_{3j}(t)Y_3$, with $j = 1,2$ and (9.101). These vector fields have the same structure constants as the $X_1^{\mathrm{pRic}}, X_2^{\mathrm{pRic}}, X_3^{\mathrm{pRic}}$. Therefore, (9.102) is a PDE Lie system with a Vessiot–Guldberg Lie algebra isomorphic to $\mathfrak{sl}(2,\mathbb{R})$.

Appendix A

Table A.1. The GKO classification of the $8 + 20$ finite-dimensional real Lie algebras of vector fields on the plane and their most relevant characteristics.

#	Primitive	Basis of vector fields X_i	Dom
P_1	$A_\alpha \simeq \mathbb{R} \ltimes \mathbb{R}^2$	$\{\partial_x, \partial_y\}, \alpha(x\partial_x + y\partial_y) + y\partial_x - x\partial_y, \quad \alpha \geq 0$	$\mathbb{R}^2$
P_2	$\mathfrak{sl}(2, \mathbb{R})$	$\{\partial_x, x\partial_x + y\partial_y\}, (x^2 - y^2)\partial_x + 2xy\partial_y$	$\mathbb{R}^2_{y \neq 0}$
P_3	$\mathfrak{so}(3)$	$\{y\partial_x - x\partial_y, (1 + x^2 - y^2)\partial_x + 2xy\partial_y\},$ $2xy\partial_x + (1 + y^2 - x^2)\partial_y$	$\mathbb{R}^2$
P_4	$\mathbb{R}^2 \ltimes \mathbb{R}^2$	$\{\partial_x, \partial_y\}, x\partial_x + y\partial_y, y\partial_x - x\partial_y$	$\mathbb{R}^2$
P_5	$\mathfrak{sl}(2, \mathbb{R}) \ltimes \mathbb{R}^2$	$\{\partial_x, \partial_y\}, x\partial_x - y\partial_y, y\partial_x, x\partial_y$	$\mathbb{R}^2$
P_6	$\mathfrak{gl}(2, \mathbb{R}) \ltimes \mathbb{R}^2$	$\{\partial_x, \partial_y\}, x\partial_x, y\partial_x, x\partial_y, y\partial_y$	$\mathbb{R}^2$
P_7	$\mathfrak{so}(3, 1)$	$\{\partial_x, \partial_y\}, x\partial_x + y\partial_y, y\partial_x - x\partial_y, (x^2 - y^2)\partial_x$ $+ 2xy\partial_y, 2xy\partial_x + (y^2 - x^2)\partial_y$	$\mathbb{R}^2$
P_8	$\mathfrak{sl}(3, \mathbb{R})$	$\{\partial_x, \partial_y\}, x\partial_x, y\partial_x, x\partial_y, y\partial_y, x^2\partial_x$ $+ xy\partial_y, xy\partial_x + y^2\partial_y$	$\mathbb{R}^2$

#	Imprimitive	Basis of vector fields X_i	Dom
I_1	$\mathbb{R}$	$\{\partial_x\}$	$\mathbb{R}^2$
I_2	$\mathfrak{h}_2$	$\{\partial_x\}, x\partial_x$	$\mathbb{R}^2$
I_3	$\mathfrak{sl}(2, \mathbb{R})$ (type I)	$\{\partial_x\}, x\partial_x, x^2\partial_x$	$\mathbb{R}^2$
I_4	$\mathfrak{sl}(2, \mathbb{R})$ (type II)	$\{\partial_x + \partial_y, x\partial_x + y\partial_y\}, x^2\partial_x + y^2\partial_y$	$\mathbb{R}^2_{x \neq y}$
I_5	$\mathfrak{sl}(2, \mathbb{R})$ (type III)	$\{\partial_x, 2x\partial_x + y\partial_y\}, x^2\partial_x + xy\partial_y$	$\mathbb{R}^2_{y \neq 0}$
I_6	$\mathfrak{gl}(2, \mathbb{R})$ (type I)	$\{\partial_x, \partial_y\}, x\partial_x, x^2\partial_x$	$\mathbb{R}^2$
I_7	$\mathfrak{gl}(2, \mathbb{R})$ (type II)	$\{\partial_x, y\partial_y\}, x\partial_x, x^2\partial_x + xy\partial_y$	$\mathbb{R}^2_{y \neq 0}$
I_8	$B_\alpha \simeq \mathbb{R} \ltimes \mathbb{R}^2$	$\{\partial_x, \partial_y\}, x\partial_x + \alpha y\partial_y, \quad 0 < |\alpha| \leq 1$	$\mathbb{R}^2$

(Continued)

Table A.1. (*Continued*)

#	Imprimitive	Basis of vector fields X_i	Dom
I_9	$\mathfrak{h}_2 \oplus \mathfrak{h}_2$	$\{\partial_x, \partial_y\}, x\partial_x, y\partial_y$	$\mathbb{R}^2$
I_{10}	$\mathfrak{sl}(2,\mathbb{R}) \oplus \mathfrak{h}_2$	$\{\partial_x, \partial_y\}, x\partial_x, y\partial_y, x^2\partial_x$	$\mathbb{R}^2$
I_{11}	$\mathfrak{sl}(2,\mathbb{R}) \oplus \mathfrak{sl}(2,\mathbb{R})$	$\{\partial_x, \partial_y\}, x\partial_x, y\partial_y, x^2\partial_x, y^2\partial_y$	$\mathbb{R}^2$
I_{12}	$\mathbb{R}^{r+1}$	$\{\partial_y\}, \xi_1(x)\partial_y, \ldots, \xi_r(x)\partial_y, \quad r \geq 1$	$\mathbb{R}^2$
I_{13}	$\mathbb{R} \ltimes \mathbb{R}^{r+1}$	$\{\partial_y\}, y\partial_y, \xi_1(x)\partial_y, \ldots, \xi_r(x)\partial_y, \quad r \geq 1$	$\mathbb{R}^2$
I_{14}	$\mathbb{R} \ltimes \mathbb{R}^r$	$\{\partial_x, \eta_1(x)\partial_y\}, \eta_2(x)\partial_y, \ldots, \eta_r(x)\partial_y, \quad r \geq 1$	$\mathbb{R}^2$
I_{15}	$\mathbb{R}^2 \ltimes \mathbb{R}^r$	$\{\partial_x, y\partial_y\}, \eta_1(x)\partial_y, \ldots, \eta_r(x)\partial_y, \quad r \geq 1$	$\mathbb{R}^2$
I_{16}	$C_\alpha^r \simeq \mathfrak{h}_2 \ltimes \mathbb{R}^{r+1}$	$\{\partial_x, \partial_y\}, x\partial_x + \alpha y\partial_y, x\partial_y, \ldots, x^r\partial_y, r \geq 1,$ $\alpha \in \mathbb{R}$	$\mathbb{R}^2$
I_{17}	$\mathbb{R} \ltimes (\mathbb{R} \ltimes \mathbb{R}^r)$	$\{\partial_x, \partial_y\}, x\partial_x + (ry + x^r)\partial_y, x\partial_y, \ldots, x^{r-1}\partial_y, \quad r \geq 1$	$\mathbb{R}^2$
I_{18}	$(\mathfrak{h}_2 \oplus \mathbb{R}) \ltimes \mathbb{R}^{r+1}$	$\{\partial_x, \partial_y\}, x\partial_x, x\partial_y, y\partial_y, x^2\partial_y, \ldots, x^r\partial_y, \quad r \geq 1$	$\mathbb{R}^2$
I_{19}	$\mathfrak{sl}(2,\mathbb{R}) \ltimes \mathbb{R}^{r+1}$	$\{\partial_x, \partial_y\}, x\partial_y, 2x\partial_x + ry\partial_y, x^2\partial_x + rxy\partial_y, x^2\partial_y, \ldots, x^r\partial_y, \quad r \geq 1$	$\mathbb{R}^2$
I_{20}	$\mathfrak{gl}(2,\mathbb{R}) \ltimes \mathbb{R}^{r+1}$	$\{\partial_x, \partial_y\}, x\partial_x, x\partial_y, y\partial_y, x^2\partial_x + rxy\partial_y, x^2\partial_y, \ldots, x^r\partial_y, \quad r \geq 1$	$\mathbb{R}^2$

Note: The vector fields which are written between brackets form a modular generating system. The functions $\xi_1(x), \ldots, \xi_r(x)$ and 1 are linearly independent over $\mathbb{R}$ and the functions $\eta_1(x), \ldots, \eta_r(x)$ form a basis of solutions for an r-order differential equation with constant coefficients [5, p. 470–471]. Finally, $\mathfrak{g} = \mathfrak{g}_1 \ltimes \mathfrak{g}_2$ stands for the semi-direct sum (as Lie algebras) of $\mathfrak{g}_1$ by $\mathfrak{g}_2$. Dom stands for the domain of the Lie algebra of vector fields.

Table A.2. The classification of the $4+8$ finite-dimensional real Lie algebras of Hamiltonian vector fields on $\mathbb{R}^2$.

#	Primitive	Hamiltonian functions h_i	ω	Lie–Hamilton algebra
P_1	$A_0 \simeq \mathfrak{iso}(2)$	$y,\ -x,\ \frac{1}{2}(x^2+y^2),\ 1$	$\mathrm{d}x \wedge \mathrm{d}y$	$\overline{\mathfrak{iso}(2)}$
P_2	$\mathfrak{sl}(2,\mathbb{R})$	$-\dfrac{1}{y},\ -\dfrac{x}{y},\ -\dfrac{x^2+y^2}{y}$	$\dfrac{\mathrm{d}x \wedge \mathrm{d}y}{y^2}$	$\mathfrak{sl}(2,\mathbb{R})$ or $\mathfrak{sl}(2,\mathbb{R}) \oplus \mathbb{R}$
P_3	$\mathfrak{so}(3)$	$\dfrac{-1}{2(1+x^2+y^2)},\ \dfrac{y}{1+x^2+y^2},$ $-\dfrac{x}{1+x^2+y^2},\ 1$	$\dfrac{\mathrm{d}x \wedge \mathrm{d}y}{(1+x^2+y^2)^2}$	$\mathfrak{so}(3)$ or $\mathfrak{so}(3) \oplus \mathbb{R}$
P_5	$\mathfrak{sl}(2,\mathbb{R}) \ltimes \mathbb{R}^2$	$y,\ -x,\ xy,\ \frac{1}{2}y^2,\ -\frac{1}{2}x^2,\ 1$	$\mathrm{d}x \wedge \mathrm{d}y$	$\overline{\mathfrak{sl}(2,\mathbb{R}) \ltimes \mathbb{R}^2} \simeq \mathfrak{h}_6$

#	Imprimitive	Hamiltonian functions h_i	ω	Lie–Hamilton algebra
I_1	$\mathbb{R}$	$\int^y f(y')\mathrm{d}y'$	$f(y)\mathrm{d}x \wedge \mathrm{d}y$	$\mathbb{R}$ or $\mathbb{R}^2$
I_4	$\mathfrak{sl}(2,\mathbb{R})$ (type II)	$\dfrac{1}{x-y},\ \dfrac{x+y}{2(x-y)},\ \dfrac{xy}{x-y}$	$\dfrac{\mathrm{d}x \wedge \mathrm{d}y}{(x-y)^2}$	$\mathfrak{sl}(2,\mathbb{R})$ or $\mathfrak{sl}(2,\mathbb{R}) \oplus \mathbb{R}$
I_5	$\mathfrak{sl}(2,\mathbb{R})$ (type III)	$-\dfrac{1}{2y^2},\ -\dfrac{x}{y^2},\ -\dfrac{x^2}{2y^2}$	$\dfrac{\mathrm{d}x \wedge \mathrm{d}y}{y^3}$	$\mathfrak{sl}(2,\mathbb{R})$ or $\mathfrak{sl}(2,\mathbb{R}) \oplus \mathbb{R}$
I_8	$B_{-1} \simeq \mathfrak{iso}(1,1)$	$y,\ -x,\ xy,\ 1$	$\mathrm{d}x \wedge \mathrm{d}y$	$\overline{\mathfrak{iso}(1,1)} \simeq \mathfrak{h}_4$
I_{12}	$\mathbb{R}^{r+1}$	$-\int^x f(x')\mathrm{d}x',\ -\int^x f(x')\xi_j(x')\mathrm{d}x'$	$f(x)\mathrm{d}x \wedge \mathrm{d}y$	$\mathbb{R}^{r+1}$ or $\mathbb{R}^{r+2}$
I_{14A}	$\mathbb{R} \ltimes \mathbb{R}^r$ (type I)	$y,\ -\int^x \eta_j(x')\mathrm{d}x',\ 1 \notin \langle \eta_j \rangle$	$\mathrm{d}x \wedge \mathrm{d}y$	$\mathbb{R} \ltimes \mathbb{R}^r$ or $(\mathbb{R} \ltimes \mathbb{R}^r) \oplus \mathbb{R}$
I_{14B}	$\mathbb{R} \ltimes \mathbb{R}^r$ (type II)	$y,\ -x,\ -\int^x \eta_j(x')\mathrm{d}x',\ 1$	$\mathrm{d}x \wedge \mathrm{d}y$	$\overline{(\mathbb{R} \ltimes \mathbb{R}^r)}$
I_{16}	$C^r_{-1} \simeq \mathfrak{h}_2 \ltimes \mathbb{R}^{r+1}$	$y,\ -x,\ xy,\ -\dfrac{x^{j+1}}{j+1},\ 1$	$\mathrm{d}x \wedge \mathrm{d}y$	$\overline{\mathfrak{h}_2 \ltimes \mathbb{R}^{r+1}}$

Note: For I_{12}, I_{14A} and I_{16}, we have $j = 1, \ldots, r$ and $r \geq 1$; in I_{14B}, the index $j = 2, \ldots, r$.

374　　　　　　　　　　　　　　　　*Appendix A*

Table A.3.　Specific Lie–Hamilton systems according to the family given in the classification of Table A.1.

LH algebra	#	LH systems
$\mathfrak{sl}(2,\mathbb{R})$	P_2	Milne–Pinney and Kummer–Schwarz equations with $c > 0$
		Complex Riccati equation
$\mathfrak{sl}(2,\mathbb{R})$	I_4	Milne–Pinney and Kummer–Schwarz equations with $c < 0$
		Split-complex Riccati equation
		Coupled Riccati equations
		Planar diffusion Riccati system for $c_0 = 1$
$\mathfrak{sl}(2,\mathbb{R})$	I_5	Milne–Pinney and Kummer–Schwarz equations with $c = 0$
		Dual-Study Riccati equation
		Harmonic oscillator
		Planar diffusion Riccati system for $c_0 = 0$
$\mathfrak{h}_6 \simeq \overline{\mathfrak{sl}(2,\mathbb{R}) \ltimes \mathbb{R}^2}$	P_5	Dissipative harmonic oscillator
		Second-order Riccati equation in Hamiltonian form
$\mathfrak{h}_2 \simeq \mathbb{R} \ltimes \mathbb{R}$	$\mathrm{I}_{14A}^{r=1}$	Complex Bernoulli equation
		Generalized Buchdahl equations
		Lotka–Volterra systems

Note: All of these systems have t-dependent real coefficients.

Appendix A

Table A.4. Functions $1, \xi_1(x), \ldots, \xi_r(x)$ are linearly independent and $\eta_1(x), \ldots, \eta_r(x)$ form a basis of solutions for a system of r-order linear differential equations with constant coefficients.

#	Lie algebra	Basis of vector fields X_i	Jacobi
P$_1$	$A_\alpha \simeq \mathbb{R} \ltimes \mathbb{R}^2$	$\partial_x, \partial_y, \alpha(x\partial_x + y\partial_y) + y\partial_x - x\partial_y,$ $\alpha \geq 0$	$(\alpha = 0)$ Pois.
P$_2$	$\mathfrak{sl}(2,\mathbb{R})$	$\partial_x, x\partial_x + y\partial_y, (x^2 - y^2)\partial_x + 2xy\partial_y$	Poisson
P$_3$	$\mathfrak{so}(3)$	$y\partial_x - x\partial_y, (1 + x^2 - y^2)\partial_x + 2xy\partial_y,$ $2xy\partial_x + (1 + y^2 - x^2)\partial_y$	Poisson
P$_4$	$\mathbb{R}^2 \ltimes \mathbb{R}^2$	$\partial_x, \partial_y, x\partial_x + y\partial_y, y\partial_x - x\partial_y$	No
P$_5$	$\mathfrak{sl}(2,\mathbb{R}) \ltimes \mathbb{R}^2$	$\partial_x, \partial_y, x\partial_x - y\partial_y, y\partial_x, x\partial_y$	Poisson
P$_6$	$\mathfrak{gl}(2,\mathbb{R}) \ltimes \mathbb{R}^2$	$\partial_x, \partial_y, x\partial_x, y\partial_x, x\partial_y, y\partial_y$	No
P$_7$	$\mathfrak{so}(3,1)$	$\partial_x, \partial_y, x\partial_x + y\partial_y, y\partial_x - x\partial_y,$ $(x^2 - y^2)\partial_x + 2xy\partial_y,$ $2xy\partial_x + (y^2 - x^2)\partial_y$	No
P$_8$	$\mathfrak{sl}(3,\mathbb{R})$	$\partial_x, \partial_y, x\partial_x, y\partial_x, x\partial_y, y\partial_y, x^2\partial_x$ $+ xy\partial_y, xy\partial_x + y^2\partial_y$	No
I$_1$	$\mathbb{R}$	∂_x	$(0, \partial_x)$, Pois.
I$_2$	$\mathfrak{h}_2$	$\partial_x, x\partial_x$	$(0, \partial_x)$
I$_3$	$\mathfrak{sl}(2,\mathbb{R})$ (type I)	$\partial_x, x\partial_x, x^2\partial_x$	$(0, \partial_x)$
I$_4$	$\mathfrak{sl}(2,\mathbb{R})$ (type II)	$\partial_x + \partial_y, x\partial_x + y\partial_y, x^2\partial_x + y^2\partial_y$	Poisson
I$_5$	$\mathfrak{sl}(2,\mathbb{R})$ (type III)	$\partial_x, 2x\partial_x + y\partial_y, x^2\partial_x + xy\partial_y$	Poisson
I$_6$	$\mathfrak{gl}(2,\mathbb{R})$ (type I)	$\partial_x, \partial_y, x\partial_x, x^2\partial_x$	No
I$_7$	$\mathfrak{gl}(2,\mathbb{R})$ (type II)	$\partial_x, y\partial_y, x\partial_x, x^2\partial_x + xy\partial_y$	No
I$_8$	$B_\alpha \simeq \mathbb{R} \ltimes \mathbb{R}^2$	$\partial_x, \partial_y, x\partial_x + \alpha y\partial_y, \quad 0 < |\alpha| \leq 1$	$(\alpha = -1)$ Pois.
I$_9$	$\mathfrak{h}_2 \oplus \mathfrak{h}_2$	$\partial_x, \partial_y, x\partial_x, y\partial_y$	No
I$_{10}$	$\mathfrak{sl}(2,\mathbb{R}) \oplus \mathfrak{h}_2$	$\partial_x, \partial_y, x\partial_x, y\partial_y, x^2\partial_x$	No
I$_{11}$	$\mathfrak{sl}(2,\mathbb{R}) \oplus \mathfrak{sl}(2,\mathbb{R})$	$\partial_x, \partial_y, x\partial_x, y\partial_y, x^2\partial_x, y^2\partial_y$	No
I$_{12}$	$\mathbb{R}^{r+1}$	$\partial_y, \xi_1(x)\partial_y, \ldots, \xi_r(x)\partial_y$	$(0, \partial_y)$, Pois.
I$_{13}$	$\mathbb{R} \ltimes \mathbb{R}^{r+1}$	$\partial_y, y\partial_y, \xi_1(x)\partial_y, \ldots, \xi_r(x)\partial_y$	$(0, \partial_y)$
I$_{14}$	$\mathbb{R} \ltimes \mathbb{R}^r$	$\partial_x, \eta_1(x)\partial_y, \eta_2(x)\partial_y, \ldots, \eta_r(x)\partial_y$	Poisson
I$_{15}$	$\mathbb{R}^2 \ltimes \mathbb{R}^r$	$\partial_x, y\partial_y, \eta_1(x)\partial_y, \ldots, \eta_r(x)\partial_y$	No
I$_{16}$	$C_\alpha^r \simeq \mathfrak{h}_2 \ltimes \mathbb{R}^{r+1}$	$\partial_x, \partial_y, x\partial_x + \alpha y\partial_y, x\partial_y, \ldots,$ $x^r\partial_y, \quad \alpha \in \mathbb{R}$	$(\alpha = -1)$ Pois.
I$_{17}$	$\mathbb{R} \ltimes (\mathbb{R} \ltimes \mathbb{R}^r)$	$\partial_x, \partial_y, x\partial_x + (ry +$ $x^r)\partial_y, x\partial_y, \ldots, x^{r-1}\partial_y$	No
I$_{18}$	$(\mathfrak{h}_2 \oplus \mathbb{R}) \ltimes \mathbb{R}^{r+1}$	$\partial_x, \partial_y, x\partial_x, x\partial_y, y\partial_y, x^2\partial_y, \ldots, x^r\partial_y$	No
I$_{19}$	$\mathfrak{sl}(2,\mathbb{R}) \ltimes \mathbb{R}^{r+1}$	$\partial_x, \partial_y, x\partial_y, 2x\partial_x + ry\partial_y, x^2\partial_x$ $+ rxy\partial_y, x^2\partial_y, \ldots, x^r\partial_y$	No
I$_{20}$	$\mathfrak{gl}(2,\mathbb{R}) \ltimes \mathbb{R}^{r+1}$	$\partial_x, \partial_y, x\partial_x, x\partial_y, y\partial_y, x^2\partial_x$ $+ rxy\partial_y, x^2\partial_y, \ldots, x^r\partial_y$	No

Note: $\mathfrak{g}_1 \ltimes \mathfrak{g}_2$ stands for the semi-direct sum of $\mathfrak{g}_1$ by $\mathfrak{g}_2$, i.e., $\mathfrak{g}_2$ is an ideal of $\mathfrak{g}_1 \ltimes \mathfrak{g}_2$.

376 Appendix A

Table A.5. Lie symmetries for Riccati equations (9.34) for different $\eta(t)$.

$\eta(t)$	$f_3(t)$
$\dfrac{k}{at+b}$	$k + c_1(at+b)J^2\left(1, 2\sqrt{(at+b)/a^2}\right) + c_2(at+b)Y^2\left(1, 2\sqrt{at+b/a^2}\right)$ $+ c_3(at+b)J\left(1, 2\sqrt{t+b/a}\right)Y\left(1, 2\sqrt{t+b/a}\right)$
$\dfrac{k}{(at+b)^2}$	$-\dfrac{kat}{b} + c_1\left(\dfrac{at+b}{a}\right)^{\frac{a+\sqrt{a^2-4k}}{a}} + c_2\left(\dfrac{at+b}{a}\right)^{-\frac{-a+\sqrt{a^2-4k}}{a}} + c_3(at+b),$
$at+b$	$k + c_1\mathrm{Airy}_A\left(-\dfrac{at+b}{a^{2/3}}\right)^2 + c_2\mathrm{Airy}_B\left(-\dfrac{at+b}{a^{2/3}}\right)^2$ $+c_3\mathrm{Airy}_A\left(-\dfrac{at+b}{a^{2/3}}\right)\mathrm{Airy}_B\left(-\dfrac{at+b}{a^{2/3}}\right)$

Note: We assume $f_0 = k \in \mathbb{R}$. We have $f_2 = \mathrm{d}f_3/\mathrm{d}t$ and $f_1 = k\eta(t) - \int \eta(t)f_3\mathrm{d}t$. Airy_A and Airy_B denote the Airy and Bairy functions. J, Y are the Bessel functions of first and second kinds. Function f_3 follows from (9.40).

Bibliography

[1] L. Abellanas, L. Martínez Alonso, A general setting for Casimir invariants, *J. Math. Phys.* **16**, 1580–1584 (1975).

[2] M.J. Ablowitz, P.A. Clarkson, *Solitons, Nonlinear Evolution Equations and Inverse Scattering,* London Mathematical Society, Lecture Notes Series, Vol. 149, Cambridge University Press, Cambridge (1991).

[3] R. Abraham, J.E. Marsden, *Foundations of Mechanics,* Benjamin/ Cummings publishing, Reading (1978).

[4] R. Abraham, J.E. Marsden, *Foundations of Mechanics*, Addison–Wesley, Redwood City (1987).

[5] M. Ackerman, R. Hermann, *Sophus Lie's Transformation Group Paper,* Math. Sci. Press., Brookline (1975).

[6] I.D. Ado, The representation of Lie algebras by matrices, *Uspehi Matem. Nauk* (N.S.) **2**, 159–173 (1947) (Russian); *Amer. Math. Soc. Translation* **1949**, 1–21 (1949) (English transl.).

[7] J.L. Allen, F.M. Stein, Classroom Notes: On solutions of certain Riccati differential equations, *Amer. Math. Monthly* **71**, 1113–1115 (1964).

[8] W.F. Ames, *Nonlinear superposition for operator equations*, in: *Proceedings of an International Symposium on Nonlinear Equations in Abstract Spaces*, Academic Press, New York (1978), pp. 43–66.

[9] J.M. Ancochea, R. Campoamor-Stursberg, L. Garcia Vergnolle, Solvable Lie algebras with naturally graded nilradicals and their invariants, *J. Phys. A* **39**, 1339–1355 (2006).

[10] R.L. Anderson, A nonlinear superposition principle admitted by coupled Riccati equations of the projective type, *Lett. Math. Phys.* **4**, 1–7 (1980).

[11] I.M. Anderson, M.E. Fels, P.J. Vassiliou, Superposition formulas for exterior differential systems *Adv. Math.* **221**, 1910–1963 (2009).

[12] R.L. Anderson, J. Harnad, P. Winternitz, Group theoretical approach to superposition rules for systems of Riccati equations, *Lett. Math. Phys.* **5**, 143–148 (1981).

[13] R.L. Anderson, J. Harnad, P. Winternitz, Superposition principles for nonlinear differential equations, in: *Proceedings of NATO Advanced Study Institute on Nonlinear Phenomena in Physics & Biology*, Plenum Press, New York (1981), pp. 573–576.

378   *Bibliography*

[14] R.L. Anderson, J. Harnad, P. Winternitz, Systems of ordinary differential equations with nonlinear superposition principles, *Phys. D* **4**, 164–182 (1982).

[15] R.L. Anderson, P. Winternitz, *A nonlinear superposition principle for Riccati equations of the conformal type*, in: *Proceedings of 9th International Colloquium on Group Theoretical Methods in Physics*, Lecture Notes in Physics **115**, Springer, Berlin (1980), pp. 165–169.

[16] R.M. Angelo, E.I. Duzzioni, A.D. Ribeiro. Integrability in time-dependent systems with one degree of freedom, *J. Phys. A* **45**, 055101 (2012).

[17] R.M. Angelo, W.F. Wreszinski, Two-level quantum dynamics, integrability and unitary not gates, *Phys. Rev. A* **72**, 034105 (2005).

[18] Í. Aslan, The Exp-function approach to the Schwarzian Korteweg de Vries equation, *Comp. Math. Appl.* **59**, 2896–2900 (2010).

[19] F. Avram, J.F. Cariñena, J. de Lucas, A Lie systems approach for the first passage-time of piecewise deterministic processes, in: *Modern Trends of Controlled Stochastic Processes: Theory and Applications*, Luniver Press, Frome (2010), pp. 144–160.

[20] A. Awane, k-symplectic structures, *J. Math. Phys.* **33**, 4046–4052 (1992).

[21] J.A. de Azcárraga, F.J. Herranz, J.C. Pérez Bueno, M. Santander, Central extensions of the quasi-orthogonal Lie algebras, *J. Phys. A* **31**, 1373–1394 (1998).

[22] A. Ballesteros, A. Blasco, F.J. Herranz, N-dimensional integrability from two-photon coalgebra symmetry, *J. Phys. A: Math. Theor.* **42**, 265205 (2009).

[23] A. Ballesteros, A. Blasco, F.J. Herranz, J. de Lucas, C. Sardón, Lie–Hamilton systems on the plane: Properties, classification and applications, *J. Differential Equations* **258**, 2873–2907 (2015).

[24] A. Ballesteros, A. Blasco, F.J. Herranz, F. Musso, O. Ragnisco, (Super) integrability from coalgebra symmetry: formalism and applications, *J. Phys.: Conf. Ser.* **175**, 012004 (2009).

[25] A. Ballesteros, R. Campoamor-Stursberg, E. Fernandez-Saiz, F.J. Herranz, J. de Lucas, Poisson-Hopf algebra deformations of Lie-Hamilton systems, *J. Phys. A* **51**, 065202 (2018).

[26] A. Ballesteros, R. Campoamor-Stursberg, E. Fernandez-Saiz, F.J. Herranz, J. de Lucas, A unified approach to Poisson-Hopf deformations of Lie-Hamilton systems based on $\mathfrak{sl}(2)$, in: *Quantum Theory and Symmetries with Lie Theory and Its Applications in Physics Volume I*, Springer, Singapore, (2018) pp. 347–366.

[27] A. Ballesteros, J.F. Cariñena, F.J. Herranz J. de Lucas, C. Sardón, From constants of motion to superposition rules of Lie–Hamilton systems, *J. Phys. A* **46**, 285203 (2013).

[28] A. Ballesteros, M. Corsetti, O. Ragnisco, N-dimensional classical integrable systems from Hopf algebras, *Czech. J. Phys.* **46**, 1153–1165 (1996).

[29] A. Ballesteros, F.J. Herranz, M.A. del Olmo, M. Santander, Quantum structure of the motion groups of the two-dimensional Cayley–Klein geometries, *J. Phys. A* **26**, 5801–5823 (1993).

Bibliography

[30] A. Ballesteros, F.J. Herranz, P. Parashar, $(1+1)$ Schrödinger Lie bialgebras and their Poisson–Lie groups, *J. Phys. A* **33**, 3445–3465 (2000).

[31] A. Ballesteros, O. Ragnisco, A systematic construction of completely integrable Hamiltonian from coalgebras, *J. Phys. A* **31**, 3791–3813 (1998).

[32] J. Beckers, L. Gagnon, V. Hussin, P. Winternitz, Nonlinear differential equations with superposition formulas and Lie superalgebras, *Lett. Math. Phys.* **13**, 113–120 (1987).

[33] J. Beckers, L. Gagnon, V. Hussin, P. Winternitz, Superposition formulas for nonlinear superequations, *J. Math. Phys.* **31**, 2528–2534 (1990).

[34] J. Beckers, V. Hussin, P. Winternitz, Complex parabolic subgroups of G_2 and nonlinear differential equations, *Lett. Math. Phys.* **11**, 81–86 (1986).

[35] J. Beckers, V. Hussin, P. Winternitz, Nonlinear equations with superposition formulas and the exceptional group G_2. I. Complex and real forms of $\mathfrak{g}_2$ and their maximal subalgebras, *J. Math. Phys.* **27**, 2217–2227 (1986).

[36] J. Beckers, V. Hussin, P. Winternitz, Nonlinear equations with superposition formulas and the exceptional group G_2. II. Classification of the equation, *J. Math. Phys.* **28**, 520–529 (1987).

[37] G.M. Beffa, Moving frames, geometric Poisson brackets and the KdV-Schwarzian evolution of pure spinors, *Ann. Inst. Fourier (Grenoble)* **61**, 2405–2434 (2011).

[38] A.A. Bekov, Integrable cases and trajectories in the Gylden-Meshcherskivi problem, *Soviet Astronom.* **33**, 71–78 (1989).

[39] I.M. Benn, R.W. Tucker, *An Introduction to Spinors and Geometry with Applications in Physics*, Adam Hilger, Ltd., Bristol (1987).

[40] L.M. Berkovich, Gylden-Mevsvcerskivi problem, *Celestian Mech.* **24**, 407–429 (1981).

[41] L.M. Berkovich, Transformation of Sturm–Liouville differential equations, *Funct. Anal. Appl.* **16**, 190–192 (1982).

[42] L.M. Berkovich, Canonical forms of ordinary linear differential equations, *Arch. Math. (Brno)* **24**, 25–42 (1988).

[43] L.M. Berkovich, The generalized Emden-Fowler equation, Symmetry in nonlinear mathematical physics, Vol. 1, 2 *Natl. Acad. Sci. Ukraine, Inst. Math., Kiev* (1997), pp. 155–163.

[44] L.M. Berkovich, Discussion of Lie's nonlinear superposition theory, in: *Proceedings of the International Conference of Modern Group, Analysis for the New Millennium (MOGRAN '00)*, Ufa (2000).

[45] L.M. Berkovich, Method of factorization of ordinary differential operators and some of its applications, *Appl. Anal. Discrete Math.* **1**, 122–149 (2007).

[46] L.M. Berkovich, F.L. Berkovich, Transformation and factorization of second order linear ordinary differential equations and its implementation in REDUCE, *Univ. Beograd. Publ. Elektrotehn. Fak. Ser. Mat.* **6**, 11–24 (1995).

[47] L.M. Berkovich, N.H. Rozov, Transformations of linear differential equations of second order and adjoined nonlinear equations, *Arch. Math. (Brno)* **33**, 75–98 (1997).

[48] G. Birkhoff, *Hydrodynamics: A Study in Logic, Fact and Similitude*, Princeton University Press, Princeton (1950).

380   *Bibliography*

[49] G. Birkhoff, Sound waves in fluids, *Appl. Num. Math.* **3**, 3–24 (1987).

[50] A. Blasco, F.J. Herranz, J. de Lucas, C. Sardón, Lie–Hamilton systems on the plane: Applications and superposition rules, *J. Phys. A* **48**, 345202 (2015).

[51] D. Blázquez-Sanz, *Differential Galois Theory and Lie-Vessiot Systems*, VDM Verlag (2008).

[52] D. Blázquez-Sanz, J.J. Morales-Ruiz, Local and Global Aspects of Lie's Superposition Theorem, *J. Lie Theory* **20**, 483–517 (2010).

[53] D. Blázquez-Sanz, J.J. Morales-Ruiz, Lie's reduction method and differential Galois theory in the complex analytic context, *Discrete Contin. Dyn. Syst. — Series A* **32**, 353–379 (2012).

[54] G.W. Bluman, S. Kumei, *Symmetries and Differential Equations*, Springer, New York (1989).

[55] A.V. Bocharov, V.N. Chetverikov *et al.*, *Symmetries and Conservation Laws for Differential Equations of Mathematical Physics*, Translations of Mathematical Monographs 182, American Mathematical Society, Providence, R.I. (1999).

[56] O.I. Bogoyavlenskii, Breaking solitons in $2 + 1$ dimensional integrable equations, *Russian Math. Surv.* **45**, 1–86 (1990).

[57] F. Boniver, P.B. Lecomte, A remark about the Lie algebra of infinitesimal conformal transformations of the Euclidean space, *Bull. London Math. Soc.* **32**, 263–266 (2000).

[58] T.C. Bountis, V. Papageorgiou, P. Winternitz, On the integrability of systems of nonlinear ordinary differential equations with superposition principles, *J. Math. Phys.* **27**, 1215–1224 (1986).

[59] T.C. Bountis, V. Papageorgiou, P. Winternitz, On the integrability and perturbations of systems of ODEs with nonlinear superposition principles, *Phys. D* **18**, 211–212 (1986).

[60] M. Boutin, *On invariants of Lie group actions and their application to some equivalence problems*, Thesis (Ph.D.), University of Minnesota, ProQuest LLC, Ann Arbor (2001).

[61] R.H. Boyer, R.W. Lindquist, Maximal Analytic Extension of the Kerr Metric, *J. Math. Phys.* **8**, 265–281 (1967).

[62] R.W. Brockett, Systems theory on group manifolds and coset spaces, *SIAM J. Control Optim.* **10**, 265–284 (1972).

[63] R.W. Brockett, Lie theory and control systems defined on spheres. Lie algebras: applications and computational methods, *SIAM J. Appl. Math.* **25**, 213–225 (1973).

[64] R.W. Brockett, L. Dai, Non-holonomic kinematics and the role of elliptic functions in constructive controllability, in: *Nonholonomic Motion Planning*, Kluwer, Norwell (1993).

[65] H.A. Buchdahl, A relativistic fluid spheres resembling the Emden polytrope of index 5, *Astrophys. J.* **140**, 1512–1516 (1964).

[66] H. Bursztyn, A brief introduction to Dirac manifolds, in: *Proceedings of the 2009 Villa de Leyva Summer School, Geometric and Topological Methods for Quantum Field Theory*, Cambridge University Press, Cambridge (2013), pp. 4–38.

[67] H. Bursztyn, O. Radko, Gauge equivalence of Dirac structures and symplectic groupoids, *Ann. Inst. Fourier (Grenoble)* **53**, 309–337 (2003).

[68] S. Callot, D. Hulin, J. Lafontane, *Riemannian Geometry,* Springer, Berlin (1987).

[69] R. Campoamor-Stursberg, Invariants of solvable rigid Lie algebras up to dimension 8, *J. Phys. A* **35**, 6293–6306 (2002).

[70] R. Campoamor-Stursberg, The structure of the invariants of perfect Lie algebras, *J. Phys. A* **36**, 6709–6723 (2003).

[71] R. Campoamor-Stursberg, Solvable Lie algebras with an $\mathbb{N}$-graded nilradical of maximal nilpotency degree and their invariants, *J. Phys. A* **43**, 145202 (2010).

[72] R. Campoamor-Stursberg, Low dimensional Vessiot-Guldberg-Lie algebras of second-order ordinary differential equations, *Symmetry* **8**, 3 (2016).

[73] R.Campoamor-Stursberg, A functional realization of $\mathfrak{sl}(3,\mathbb{R})$ providing minimal VG–Lie algebras of nonlinear second-order ordinary differential equations as proper subalgebras, *J. Math. Phys.* **57**, 063508 (2016).

[74] R. Campoamor-Stursberg, Low dimensional Vessiot-Guldberg-Lie algebras of second-order ordinary differential equations, *Symmetry* **8**, 15 (2016).

[75] R. Campoamor-Stursberg, Reduction by invariants and projection of linear representations of Lie algebras applied to the construction of nonlinear realizations, *J. Math. Phys.* **59**, 033502 (2018).

[76] R. Campoamor-Stursberg, Invariant functions of vector field realizations of Lie algebras and some applications to representation theory and dynamical systems, *J. Phys.: Conf. Ser.* **1071**, 012005 (2019).

[77] J. Campos, J. Mawhin, Periodic solutions of quaternionic-valued ordinary differential equations, *Ann. Mat. Pura Appl.* **185**, suppl., S109–S127 (2006).

[78] A. Cannas da Silva, *Lectures on Symplectic Geometry*, Lecture Notes in Mathematics, 1764, Springer-Verlag, Berlin-Heidelberg (2001).

[79] J.F. Cariñena, Sections along maps in geometry and physics. Geometrical structures for physical theories, I, *Rend. Sem. Mat. Univ. Pol. Torino* **54**, 245–256 (1996).

[80] J.F. Cariñena, A new approach to Ermakov systems and applications in quantum physics, *Eur. Phys. J. Special Topics* **160**, 51–60 (2008).

[81] J.F. Cariñena, Recent advances on Lie systems and their applications, *Banach Center Publ.* **113**, 95–110 (2017).

[82] J.F. Cariñena, F. Avram, J. de Lucas, A Lie systems approach for the first passage-time of piecewise deterministic processes, in: *Modern Trends of Controlled Stochastic Processes: Theory and Applications*, Luniver Press, Frome, UK, (2010), pp. 144–160.

[83] J.F. Cariñena, J. Clemente-Gallardo, J.A. Jover-Galtier, J. de Lucas, Application of Lie systems to quantum mechanics: Superposition rules, in: *60 years Alberto Ibort Fest Classical and Quantum Physics*, Springer, Nature Switzerland, 2019.

[84] J.F. Cariñena, J. Clemente-Gallardo, A. Ramos, Motion on Lie groups and its applications in control theory, *Rep. Math. Phys.* **51**, 159–170 (2003).

[85] J.F. Cariñena, J. Grabowski, J. de Lucas, Quasi-Lie schemes: theory and applications, *J. Phys. A* **42**, 335206 (2009).

[86] J.F. Cariñena, J. Grabowski, J. de Lucas, Lie families: theory and applications, *J. Phys. A* **43**, 305201 (2010).

[87] J.F. Cariñena, J. Grabowski, J. de Lucas, Superposition rules for higher-order systems and their applications, *J. Phys. A* **45**, 185202 (2012).

[88] J.F. Cariñena, J. Grabowski, J. de Lucas. Quasi-Lie families, schemes, invariants and their applications to Abel equations, *J. Math. Anal, Appl.* **430**, 648–671 (2015).

[89] J.F. Cariñena, J. Grabowski, J. de Lucas, Quasi-Lie schemes for PDEs, *Int. J. Geom. Methods Mod. Phys.* **16**, 1950096 (2019).

[90] J.F. Cariñena, J. Grabowski, J. de Lucas, C. Sardón, Dirac-Lie systems and Schwarzian equations, *J. Differential Equations* **257**, 2303–2340 (2014).

[91] J.F. Cariñena, J. Grabowski, G. Marmo, *Lie-Scheffers Systems: A Geometric Approach*, Napoli Series on Physics and Astrophysics, Bibliopolis, Naples (2000).

[92] J.F. Cariñena, J. Grabowski, G. Marmo, Some physical applications of systems of differential equations admitting a superposition rule. *in: Proceedings of the XXXII Symposium on Mathematical Physics* (Toruń, 2000) *Rep. Math. Phys.* **48**, 47–58 (2001).

[93] J.F. Cariñena, J. Grabowski, G. Marmo, Superposition rules, Lie theorem and partial differential equations, *Rep. Math. Phys.* **60**, 237–258 (2007).

[94] J.F. Cariñena, J. Grabowski, A. Ramos, Reduction of time-dependent systems admitting a superposition principle, *Acta Appl. Math.* **66**, 67–87 (2001).

[95] J.F. Cariñena, P. Guha, J. de Lucas, A quasi-Lie schemes approach to the Gambier equation, *SIGMA* **9**, 026 (2013).

[96] J.F. Cariñena, P. Guha, M.F. Rañada, A geometric approach to higher-order Riccati chain: Darboux polynomials and constants of the motion, *J. Phys.: Conf. Ser.* **175**, 012009 (2009).

[97] J.F. Cariñena, P. Guha, M.F. Rañada, Geometrical and dynamical aspects of nonlinear higher-order Riccati systems, arXiv:1507.00512.

[98] J.F. Cariñena, A. Ibort, G. Marmo, G. Morandi, *Geometry from Dynamics, Classical and Quantum*, Springer, Dordrecht (2015).

[99] J.F. Cariñena, P.G.L. Leach, J. de Lucas, Quasi-Lie schemes and Emden-Fowler equations, *J. Math. Phys* **50**, 103515 (2009).

[100] J.F. Cariñena, C. López, Group theoretical perturbative treatment of nonlinear Hamiltonians on the dual of a Lie algebra, *Rep. Math. Phys.* **43**, 43–51 (1999).

[101] J.F. Cariñena, J. de Lucas, Lie systems and integrability conditions of differential equations and some of its applications, in: *Differential Geometry and its applications*, World Sci. Publ., Hackensack (2008), pp. 407–417.

[102] J.F. Cariñena, J. de Lucas, Recent applications of the theory of Lie systems in Ermakov systems, *SIGMA* **4**, 031 (2008).

[103] J.F. Cariñena, J. de Lucas, Applications of Lie systems in dissipative Milne-Pinney equations, *Int. J. Geom. Meth. Mod. Phys.* **6**, 683–699 (2009).

[104] J.F. Cariñena and J. de Lucas, A nonlinear superposition rule for solutions of the Milne-Pinney equation, *Phys. Lett. A* **372**, 5385–5389 (2008).

[105] J.F. Cariñena, J. de Lucas, Quantum Lie systems and integrability conditions, *Int. J. Geom. Meth. Mod. Phys.* **6**, 1235–1252 (2009).

[106] J.F. Cariñena, J. de Lucas, Applications of Lie systems in dissipative Milne–Pinney equations, *Int. J. Geom. Methods Mod. Phys.* **6**, 683 (2009).

[107] J.F. Cariñena, J. de Lucas, Superposition rules and second-order Riccati equations, *J. Geom. Mech.* **3**, 1–22 (2011).

[108] J.F. Cariñena, J. de Lucas, Lie systems: theory, generalisations, and applications, *Dissertationes Math.* **479**, 1–162 (2011).

[109] J.F. Cariñena, J. de Lucas, Superposition rules and second-order differential equations, in: *XIX International Fall Workshop on Geometry and Physics*, AIP Conf. Proc. **360**. Amer. Inst. Phys., New York (2011), pp. 127–132, arXiv: 1102.1299.

[110] J.F. Cariñena, J. de Lucas, Quasi-Lie schemes and second-order Riccati equations, *J. Geom. Mech.* **3**, 1–22 (2011).

[111] J. F. Cariñena, J. de Lucas, Superposition rules and second-order differential equations, in: *Proceedings of the XIX International Fall Workshop on Geometry and Physics*, AIP Conference Proceedings 1360, American Institute of Mathematics (2011), pp. 127–132.

[112] J.F. Cariñena, J. de Lucas, Integrability of Lie systems through Riccati equations, *J. Nonlinear Math. Phys.* **18**, 29–54 (2011).

[113] J.F. Cariñena, J. de Lucas, A. Ramos, A geometric approach to integrability conditions for Riccati Equations, *Electron. J. Diff. Equations.* **122**, 1 (2007).

[114] J.F. Cariñena, J. de Lucas, A. Ramos, A geometric approach to time operators of Lie quantum systems, *Int. J. Theor. Phys.* **48**, 1379–1404 (2009).

[115] J.F. Cariñena, J. de Lucas, M.F. Rañada, Nonlinear superpositions and Ermakov systems, *Differential Geometric Methods in Mechanics and Field Theory*, Academia Press, Genth (2007), pp. 15–33.

[116] J.F. Cariñena, J. de Lucas, M.F. Rañada, Recent applications of the theory of Lie systems in Ermakov systems, *SIGMA Symmetry Integrability Geom. Methods Appl.* **4**, 031 (2008).

[117] J.F. Cariñena, J. de Lucas, M.F. Rañada, Integrability of Lie systems and some of its applications in physics, *J. Phys. A* **41**, 304029 (2008).

[118] J.F. Cariñena, J. de Lucas, M.F. Rañada, Lie systems and integrability conditions for t-dependent frequency harmonic oscillators, *Int. J. Geom. Methods Mod. Phys.* **7**, 289–310 (2010).

[119] J.F. Cariñena, J. de Lucas, M.F. Rañada, A geometric approach to integrability of Abel differential equations, *Int. J. Theor. Phys.* **50**, 2114–2124 (2011).

[120] J.F. Cariñena, J. de Lucas, C. Sardón, A new Lie system's approach to second–order Riccati equations, *Int. J. Geom. Methods Mod. Phys.* **9**, 1260007 (2012).

384   *Bibliography*

[121] J.F. Cariñena, J. de Lucas, C. Sardón, Lie–Hamilton systems: theory and applications, *Int. J. Geom. Methods Mod. Phys.* **10**, 1350047 (2013).

[122] J.F. Cariñena, G. Marmo, J. Nasarre, The non-linear superposition principle and the Wei-Norman method, *Int. J. Mod. Phys. A* **13**, 3601–3627 (1998).

[123] J.F. Cariñena, G. Marmo, M. Rañada, Non-symplectic symmetries and bi-Hamiltonian structures of the rational harmonic oscillator, *J. Phys. A* **35**, L679–L686 (2002).

[124] J.F. Cariñena, J. Nasarre, Lie–Scheffers systems in optics, *J. Opt. B Quantum Semiclass. Opt.* **2**, 94–99 (2000).

[125] J.F. Cariñena, A. Ramos, Integrability of the Riccati equation from a group-theoretical viewpoint, *Int. J. Modern Phys. A* **14**, 1935–1951 (1999).

[126] J.F. Cariñena, A. Ramos, Applications of Lie systems in quantum mechanics and control theory, in: *Classical and Quantum integrability, Banach Center, Publ.* **59**, Warsaw (2003), pp. 143–162.

[127] J.F. Cariñena, A. Ramos, Riccati equation, factorization method and shape invariance, *Rev. Math. Phys.* **12**, 1279–1304 (2000).

[128] J.F. Cariñena, A. Ramos, A new geometric approach to Lie systems and physical applications, *Acta Appl. Math.* **70**, 43–69 (2002).

[129] J.F. Cariñena, A. Ramos, Lie systems in control theory, in: *Contemporary trends in Non-Linear Geometric Control Theory and Its Applications*, World Scientific, Singapore (2002).

[130] J.F. Cariñena, A. Ramos, Lie systems and connections in fibre bundles: applications in quantum mechanics, in: *Differential geometry and its applications*, Matfyzpress, Prague (2005), pp. 437–452.

[131] J.F. Cariñena, M.F. Rañada, M. Santander, Lagrangian formalism for non-linear second-order Riccati systems: one-dimensional integrability and two-dimensional superintegrability, *J. Math. Phys.* **46**, 062703 (2005).

[132] S. Chakravartya, R. Halburd, First integrals of a generalized Darboux–Halphen system, *J. Math. Phys.* **44**, 1751–1762 (2003).

[133] V.K. Chandrasekar, M. Senthilvelan, M. Lakshmanan, On the complete integrability and linearization of certain second-order nonlinear ordinary differential equations, *Proc. R. Soc. Lond. Ser. A Math. Phys. Eng. Sci.* **461**, 2451–2476 (2005).

[134] S. Chandrasekhar, K.S. Thorne, The Mathematical Theory of Black Holes, *Amer. J. Phys.* **53**, 1013 (1985).

[135] V. Chari, A. Pressley, *A Guide to Quantum Groups* Cambridge University Press, Cambridge (1994).

[136] S. Charzyński, M. Kuś, Wei-Norman equations for a unitary evolution, *J. Phys. A* **46**, 265208 (2013).

[137] E.S. Cheb-Terrab, A.D. Roche, An Abel ordinary differential equation class generalizing known integrable classes, *European J. Appl. Math.* **14**, 217–229 (2003).

[138] A. Chiellini, Alcune ricerche sulla forma dell'integrale generale dell'equazione differenziale del primo ordine $y' = c_0 y^3 + c_1 y^2 + c_2 y + c_3$, *Rend. Semin. Fac. Sci. Univ. Cagliari* **10**, 16 (1940).

[139] A. Chiellini, Sui sistemi di Riccati, *Rend. Sem. Fac. Sci. Univ. Cagliari* **18**, 44 (1948).

[140] J.S.R. Chisholm, A.K. Common, A class of second-order differential equations and related first-order systems, *J. Phys. A* **20**, 5459–5472 (1987).

[141] J.L. Cieśliński, T. Nikiciuk, A direct approach to the construction of standard and non-standard Lagrangians for dissipative-like dynamical systems with variable coefficients, *J. Phys. A* **43**, 175205 (2010).

[142] J.N. Clellanda, P.J. Vassiliou, A solvable string on a Lorentzian surface, *Differential Geom. Appl.* **33**, 177–198 (2014).

[143] J. Clemente-Gallardo, On the relations between control systems and Lie systems, in: *Groups, Geometry and Physics*, Monogr. Real Acad. Ci. Exact. Fís.-Quím. Nat. Zaragoza, 29, Acad. Cienc. Exact. Fís. Quím. Nat. Zaragoza, Zaragoza (2006), pp. 65–78.

[144] W.J. Coles, Linear and Riccati systems, *Duke Math. J.* **22**, 333–338 (1955).

[145] W.J. Coles, A Note on matrix Riccati systems, *Proc. Amer. Math. Soc.* **12**, 557–559 (1961).

[146] W.J. Coles, Matrix Riccati differential equations, *J. Soc. Indust. Appl. Math.* **13**, 627–634 (1965).

[147] A.K. Common, E. Hessameddini, M. Musette, The Pinney equation and its discretization, *J. Phys. A* **29**, 63–43 (1996).

[148] R. Conte, Singularities of differential equations and integrability, in: *An Introduction to Methods of Complex Analysis and Geometry for Classical Mechanics and Non-linear Waves*, Frontières, Gif–sur- Yvette (1994), pp. 49–143.

[149] J.H. Conway, D.A. Smith, *On Quaternions and Octonions: Their Geometry, Arithmetic and Symmetry*, A.K. Peters, Ltd., Natick (2003).

[150] T. Courant, *Dirac manifolds*, Ph.D. Thesis, University of California, Berkeley, (1987).

[151] T.J. Courant, Dirac manifolds, *Trans. Amer. Math. Soc.* **319**, 631–661 (1990).

[152] T. Courant, A. Weinstein, Beyond Poisson structures, in: *Action Hamiltoniennes de groupes. Troisième théorème de Lie (Lyon, 1986), Travaux en Cours Vol.* **27**. Hermann, Paris (1988), pp. 39–49.

[153] G. Darboux, Sur la théorie des coordinneés cuvilignes et les systémes, orthogonaux. *Ann. Ec. Norm. Supér.* **7**, 101–150 (1878).

[154] H.T. Davis, *Introduction to nonlinear differential and integral equations*, Dover Publications, New York (1962).

[155] A. Degasperis, D.D. Holm, A.N.W. Hone, A new integral equation with peakon solutions, *Theor. Math. Phys.* **133**, 1463–1474 (2002).

[156] M. Delgado, The Lagrange–Charpit Method, *SIAM Rev.* **39**, 298–304 (1997).

[157] P.A.M. Dirac, Generalized Hamiltonian dynamics, *Canadian J. Math.* **2**, 129–148 (1950).

[158] I. Dorfman, Dirac structures of integrable evolution equations, *Phys. Lett. A* **125**, 240–246 (1987).

386 Bibliography

[159] I. Dorfman, Krichever--Novikov equations and local symplectic structures, *Sov. Math. Dokl.* **38**, 340–343 (1989).

[160] C. Duval, P. Lecomte, V. Ovsienko, Conformally equivariant quantization: existence and uniqueness, *Ann. Inst. Fourier (Grenoble)* **49**, 1999–2029 (1999).

[161] A. Echeverría-Enríquez, M.C. Muñoz-Lecanda, *Remarks on multisymplectic reduction*, *Rep. Math. Phys.* **81**, 415–424 (2018).

[162] A. Echeverría-Enríquez, M.C. Muñoz-Lecanda, N. Román-Roy, Reduction of presymplectic manifolds with symmetry, *Rev. Math. Phys.* **11**, 1209–1247 (1999).

[163] L. Edelstein-Keshet, *Mathematical Models in Biology. SIAM Classics in Applied Mathematics*, Vol. 46, SIAM, Philadelphia (2005).

[164] M.A. Egorov, Some properties of the matrix Riccati equation, *Akad. Nauk SSSR Inst. Prikl. Mat. Preprint* **147**, 20 (1990).

[165] A.I. Egorov, *Riccati Equations*, Russian Academic Monographs 5, Pensoft Publ., Sofia-Moscow (2007).

[166] P.G. Estévez, F.J. Herranz, J. de Lucas, C. Sardón, Lie symmetries for Lie systems: Applications to systems of ODEs and PDEs, *Appl. Math. Comput.* **273**, 435–452 (2016).

[167] L.A. Ferreira, F.J. Gomes, A.V. Razumov, M.V. Saveliev, A.H. Zimerman, Riccati-type equations, generalized WZNW equations and multidimensional Toda systems, *Commun. Math. Phys.* **203**, 649–666 (1999).

[168] Z. Fiala, Evolution equation of Lie-type for finite deformations, time-discrete integration, and incremental methods, *Acta Mech.* **226**, 17–35 (2015).

[169] R. Flores-Espinoza, Monodromy factorization for periodic Lie systems and reconstruction phases, in: AIP Conf. Proc., *Geometric Methods in Physics*, Vol. 1079, Amer. Inst. Phys., New York (2008), pp. 189–195.

[170] R. Flores-Espinoza, Periodic first integrals for Hamiltonian systems of Lie type, *Int. J. Geom. Meth. Mod. Phys.* **8**, 1169–1177 (2011).

[171] R. Flores-Espinoza, J. de Lucas, Y.M. Vorobiev, Phase splitting for periodic Lie systems, *J. Phys. A* **43**, 205208 (2010).

[172] R. Florez-Espinoza, Y.M. Vorobiev, On dynamical and geometric phases of time-periodic linear Euler equations, *Russ. J. Math. Phys.* **12**, 326–349 (2005).

[173] J. Fris, V. Mandrosov, Y.A. Smorodinsky, M. Uhlir, P. Winternitz, On higher symmetries in quantum mechanics, *Phys. Lett.* **16**, 354–356 (1965).

[174] L. Gagnon, J. Beckers, V. Hussin, and P. Winternitz, Nonlinear differential equations with superposition laws for the $\mathfrak{osp}(1,2)$ superalgebra, in: *Proceedings of 15th Internat. Coll. on Group Theor. Methods in Physics*, World Scientic, Singapore, Philadelphia (1987), pp. 435–439.

[175] L. Gagnon, V. Hussin, P. Winternitz, Nonlinear equations with superposition formulas and the exceptional group III. The superposition formulas, *J. Math. Phys.* **29**, 2145–2155 (1988).

Bibliography 387

[176] I.A. García, J. Giné, J. Llibre, Liénard and Riccati differential equations related via Lie algebras, *Discrete Contin. Dyn. Syst. Ser. B* **10**, 485–494 (2008).

[177] C.F. Gauss, *Disquisitiones generales circa superficies curvas,* Typis Dieterichianie, Gottingae (1827).

[178] C.F. Gauss, *General Investigations of Curves Surfaces of 1827 and 1825,* The Princeton University Library, Princeton (1902).

[179] C.F. Gauss, *General Investigations of Curved Surfaces — Unabridged,* Watchmaker Publishing, Princeton (2007).

[180] A. Gheorghe, Quantum and classical Lie systems for extended symplectic groups, *Rom. Journ. Phys.* **58**, 1436–1445 (2013).

[181] J. Golenia, On the Bäcklund transformations of the Riccati equation: the differential-geometric approach revisited, *Rep. Math. Phys.* **55**, 341–349 (2005).

[182] C.A. Gómez, A. Salas, Special symmetries to standard Riccati equations and applications, *Appl. Math. Comp.* **216**, 3089–3096 (2010).

[183] A. González-López, N. Kamran, P.J. Olver, Lie algebras of vector fields in the real plane, *Proc. London Math. Soc. (3)* **64**, 339–368 (1992).

[184] M. Gopal, *Modern Control Systems Theory,* Halsted Press New York, New York (1984).

[185] P.R. Gordoa, Algebraic and differential nonlinear superposition formulas , *Theor. Math. Phys.* **137**, 1430–1438 (2003).

[186] P.R. Gordoa, J.M. Conde, A linear algebraic nonlinear superposition formula, *Phys. Lett. A* **295**, 287–298 (2002).

[187] K. Grabowska, J. Grabowski, Dirac algebroids in Lagrangian and Hamiltonian mechanics, *J. Geom. Phys.* **61**, 2233–2253 (2011).

[188] J. Grabowski, J. de Lucas, Mixed superposition rules and the Riccati hierarchy, *J. Differential Equations* **254**, 179–198 (2013).

[189] X. Grácia, J. de Lucas, M. Muñoz Lecanda, S. Vilariño, *Multisymplectic structures and invariant tensors for Lie systems, J. Phys. A* **52**, 215201 (2019).

[190] B. Grammaticos, A. Ramani, S. Lafortune, The Gambier mapping, revisited, *Phys. A* **253**, 260–270 (1998).

[191] B. Grammaticos, A. Ramani, P. Winternitz, Discretizing families of linearizable equations, *Phys. Lett. A* **245**, 382–388 (1998).

[192] V. Gritsev, P. Barmettler, E. Demler, Scaling approach to quantum nonequilibrium dynamics of many-body systems, *New J. Phys.* **12**, 113005 (2010).

[193] N.A. Gromov, *Contractions and Analytical Continuations of the Classical Groups. Unified Approach,* (Syktyvkar: Komi Scienfic Center) (in russian) (1992).

[194] C. Grosche, G.S. Pogosyan, A.N. Sissakian. Path integral discussion for Smorodinsky-Winternitz potentials. I. Two- and three-dimensional Euclidean space, *Fortschr. Phys.* **43**, 453–521 (1995).

[195] B. Gruber, L. O'Raifeartaigh, S theorem and construction of the invariants of the semisimple compact Lie algebras, *J. Math. Phys.* **5**, 1796–1804 (1964).

[196] A.M. Grundland, D. Levi, On higher-order Riccati equations as Bäcklund transformations, *J. Phys. A* **32**, 3931–3937 (1999).

[197] A.M. Grundland, J. de Lucas, A Lie systems approach to the Riccati hierarchy and partial differential equations, *J. Differential Equations* **263**, 299–337 (2017).

[198] A.M. Grundland, J. de Lucas, On the geometry of the Clairin theory of conditional symmetries for higher-order systems of PDEs with applications, *Diff. Geom. Appl.* **67**, 101557 (2019).

[199] A.M. Grundland, L. Martina, and G. Rideau, Partial differential equations with differential constraints, in: *Advances in Mathematical Sciences: CRM's 25 years*, CRM Proc. Lecture Notes, 11, Amer. Math. Soc., Providence (1997), pp. 135–154.

[200] A.M. Grundland, G. Rideau, *Conditional symmetries for 1st order systems of PDES in the context of the Clairin method*, in: *Modern group theor. methods phys.*, Kluwer Acad. Publ., Dordrecht (1995), pp. 167–178.

[201] P. Guha, A.G. Choudhury, B. Grammaticos, Dynamical studies of equations from the Gambier family, *SIGMA* **7**, 028 (2011).

[202] A. Guldberg, Sur les équations différentielles ordinaires qui possèdent un système fondamental d'intégrales, *C.R. Math. Acad. Sci. Paris* **116**, 964–965 (1893).

[203] C. Günther, The polysymplectic Hamiltonian formalism in field theory and calculus of variations. I. The local case, *J. Differential Geom.* **25**, 23–53 (1987).

[204] B. Hall, *Lie groups, Lie algebras, and representations. An Elementary Introduction*, Graduate Texts in Mathematics, Vol. 222, Springer, Cham (2015).

[205] F. Haas, J. Goedert, Lie point symmetries for reduced Ermakov systems, *Phys. Lett. A* **332**, 25–34 (2004).

[206] G. Halphen, Sur un système d'èquations différentielles, *C.R. Acad. Sci. Paris* **92**, 1101–1103 (1881).

[207] J. Harnad, R.L. Anderson, P. Winternitz, Superposition principles for matrix Riccati equations, *J. Math. Phys.* **24**, 1062–1072 (1983).

[208] M. Havlícec, S. Posta, P. Winternitz, Nonlinear superposition formulas based on imprimitive group action, *J. Math. Phys.* **40**, 3104–3122 (1999).

[209] M. Havlíček, S. Posta, and P. Winternitz, Superposition formulas based on nonprimitive group actions, in *Backlund and Darboux transformations. The Geometry of Solitons* CRM Proceedings and Lecture Notes, 29, Providence (2001), pp. 225–231.

[210] R.M. Hawkins, J.E. Lidsey, Ermakov-Pinney equation in scalar field cosmologies, *Phys. Rev. D* **66**, 023523 (2002).

[211] R. Hermann, *Cartanian Geometry, Nonlinear Waves, and Control Theory.* Part A, Math. Sci. Press, Brookline (1979).

[212] F.J. Herranz, A. Ballesteros, Superintegrability on three-dimensional Riemannian and relativistic spaces of constant curvature, *SIGMA* **2**, 010 (2006).

[213] F.J. Herranz, A. Ballesteros, M. Santander, T. Sanz-Gil, Maximally super-integrable Smorodinsky-Winternitz systems on the N-dimensional sphere and hyperbolic spaces, in: *Superintegrability in classical and quantum systems*, CRM Proc. Lecture Notes **37**, Amer. Math. Soc., Providence, p. 75–89 (2004).

[214] F.J. Herranz, J. de Lucas, C. Sardón. Jacobi-Lie systems: fundamentals and low-dimensional classification, *Discrete Contin. Dyn. Syst.*, (Dynamical systems, differential equations and applications. 10th AIMS Conference. Suppl.) (2015), pp. 605–614.

[215] F.J. Herranz, J. de Lucas, M. Tobolski, Lie–Hamilton systems on curved spaces: A geometrical approach, *J. Phys. A* **50**, 495201 (2017).

[216] F.J. Herranz, M. Santander, Casimir invariants for the complete family of quasisimple orthogonal algebras, *J. Phys. A: Math. Gen.* **30**, 5411–5426 (1997).

[217] A.N.W. Hone, Exact discretization of the Ermakov–Pinney equation *Phys. Lett. A* **263**, 347–354 (1999).

[218] V. Hussin, J. Beckers, L. Gagnon, P. Winternitz, Nonlinear equations with superposition principles, the Lie group $G(2)$ and its maximal subgroups, in: *Proc. 17th Internat. Coll. Group Theor. Methods in Physics* p. 492–495, Sainte-Adele, 1988. World Scientific, Singapore (1989).

[219] A. Ibort, Multisymplectic geometry: generic and exceptional, in: *Proceedings of the IX Fall Workshop on Geometry and Physics*, Publicationes de la RSME **3**, 79–88 (2000).

[220] A. Ibort, T. Rodriguez De La Peña, R. Salmoni, Dirac structures and reduction of optimal control problems with symmetries, arXiv:1004.1438.

[221] N.H. Ibragimov, Vessiot-Guldberg-Lie algebra and its application in solving nonlinear differential equations, in: *Proceedings of 11th National conference Lie group analysis of differential equations*, Samara, Russia, (1993).

[222] N.H. Ibragimov, Primer of group analysis, Znanie, No. 8, Moscow, 1989 (in Russian); *Introduction to Modern Group Analysis*, Tau, Ufa (2000).

[223] N.H. Ibragimov, Discussion of Lie's nonlinear superposition theory, in: *MO-GRAM 2000, Modern Group Analysis for the New Millenium,* USATU Publishers, Ufa (2000), pp. 116–119.

[224] N.H. Ibragimov, Memoir on integration of ordinary differential equations by quadrature, *Archives of ALGA* **5**, 27–62 (2008).

[225] N.H. Ibragimov, Utilization of canonical variables for integration of systems of first-order differential equations, *ALGA* **6**, 1–18 (2009).

[226] N.H. Ibragimov *et al.*, *CRC Handbook of Lie Group Analysis of Differential Equations. Vol. 2. Applications in Engineering and Physical Sciences*, CRC Press, Boca Raton (1995).

[227] N. Ibragimov, A. Gainetdinova, Three-dimensional dynamical systems admitting nonlinear superposition with three-dimensional Vessiot-Guldberg-Lie algebras, *Appl. Math. Lett.* **52**, 126–131 (2016).

[228] N. Ibragimov, A. Gainetdinova, Classification and integration of four-dimensional dynamical systems admitting non-linear superposition, *Int. J. Non-Linear Mechanics* **90**, 50–71 (2017).

[229] N.H. Ibragimov, M.C. Nucci, Integration of third-order ordinary differential equations by Lie's method: Equations admitting three-dimensional Lie algebras, *Lie Groups and Their Applications* **1**, 2 (1994).

[230] E.L. Ince, *Ordinary Differential Equations*, Dover Publications, New York (1956).

[231] A. Inselberg, On classification and superposition principles for nonlinear operators, Thesis (Ph.D.), University of Illinois at Urbana-Champaign, ProQuest LLC, Ann Arbor, (1965).

[232] A. Inselberg, On classification and superposition principles for nonlinear operators, *AF Grant 7-64, Tech. Report 4, Electr. Engr. Res. Lab.*, Univ. of Illinois, Urbana (1965).

[233] A. Inselberg, Superpositions for nonlinear operators. I. Strong superposition rules and linearizability, *J. Math. Anal. Appl.* **40**, 494–508 (1972).

[234] M.V. Ioffe, H.J. Korschb, Nonlinear supersymmetric (Darboux) covariance of the Ermakov–Milne–Pinney equation, *Phys. Lett. A* **311**, 200–205 (2003).

[235] K. Itô, *Encyclopedic Dictionary of Mathematics*, Vol. I-IV, Second edition, The MIT Press, Kingsport, MA (1987).

[236] Z. Jin, H. Maoan, L. Guihua, The persistence in a Lotka–Volterra competition systems with impulsive, *Chaos, Solitons and Fractals* **24**, 1105–1117 (2005).

[237] S.E. Jones, W.F. Ames, Nonlinear superpositions, *J. Math. Anal. Appl.* **17**, 484–487 (1967).

[238] M. Jotz, T. Ratiu, Induced Dirac structures on isotropy-type manifolds, *Transform. Groups* **16**, 175–191 (2011).

[239] R.E. Kalman, On the general theory of Control systems, in: *Proc. First Intern. Congr. Autom.*, Butterworth, London (1960), pp. 481–493.

[240] S. Khan, T. Hussain, A.H. Bokhari, G. Ali Khan, Conformal Killing vectors of plane symmetric four dimensional Lorentzian manifolds, *Eur. Phys. J.* **75**, 523 (2015).

[241] A. Kirillov, Local Lie algebras, *Uspekhi Mat. Nauk.* **31**, 57–76 (1976).

[242] A. Korn, Zwei Anwendungen der Methode der sukzessiven Annäherungen, *Schwrz Abhandlungen*, 215–219 (1916).

[243] B. Komrakov, A. Churyunov, B. Doubrov, Two-dimensional homogeneous spaces, *Pure Mathematics* **17**, 1–142 (1993).

[244] L. Königsberger, Über die einer beliebigen differentialgleichung erster Ordnung angehörigen selb-ständigen Transcendenten, *Acta Math.* **3**, 1–48 (1883) (in German).

[245] E.S. Kryachko, Few Sketches on Connections Between the Riccati and Ermakov–Milne–Pinney Equations, *Int. J. Quantum Chem.* **109**, 2897–2902 (2009).

[246] N. Kudryashov, A. Pickering, Rational solutions for Schwarzian integrable hierarchies, *J. Phys. A: Math. Gen.* **31**, 9505–9518 (1998).

Bibliography

[247] M. Kuna, J. Naudts, On the von Neumann equation with time-dependent Hamiltonian. Part I: Method, arXiv:0805.4487v1.

[248] M. Kuna, J. Naudts, On the von Neumann equation with time-dependent Hamiltonian. Part II: Applications, arXiv:0805.4488v1.

[249] S. Lafortune, P. Winternitz, Superposition formulas for pseudounitary matrix Riccati equations, *J. Math. Phys.* **37**, 1539–1550 (1996).

[250] N. Lanfear, R.M. López, S.K. Suslov, Exact wave functions for generalized harmonic oscillators, J. Russ. Laser Research **32**, 352–361 (2011).

[251] J. Lange, J. de Lucas, A symplectic approach to Schrödinger equations in an infinite-dimensional setting, preprint.

[252] J. Lange, J. de Lucas, Geometric models for Lie-Hamilton systems on $\mathbb{R}$? *Mathematics* **7**, 1053 (2019).

[253] J.A. Lázaro-Camí, J.P. Ortega, *Superposition rules and stochastic Lie-Scheffers systems*, Ann. Inst. H. Poincaré Probab. Stat. **45**, 910–931 (2009).

[254] P.G.L. Leach, K. Andriopoulos, The Ermakov equation: A commentary, *Appl. Anal. Discrete Math.* **2**, 146–157 (2008).

[255] P.G.L. Leach, K.G. Govinder, On the uniqueness of the Schwarzian and linearization by nonlocal contact transformation, *J. Math. Anal. Appl.* **235**, 84–107 (1999).

[256] P.B.A. Lecomte, V.Y. Ovsienko, Projectively equivariant symbol calculus, *Lett. Math. Phys.* **49**, 173–196 (1999).

[257] M. de León, I. Méndez, M. Salgado, Regular p-almost cotangent structures, *J. Korean Math. Soc.* **25**, 273–287 (1988).

[258] M. de León, I. Méndez, M. Salgado. p-almost cotangent structures, *Boll. Un. Mat. Ital.* A **7**, 97–107 (1993).

[259] M. de León, M. Salgado, S. Vilariño, *Methods of Differential Geometry in Classical Field Theories*, World Scientific Publishing Co. Pte. Ltd., Hackensack (2016).

[260] M. de León, C. Sardón, A geometric Hamilton–Jacobi theory for a Nambu–Poisson structure, *J. Math. Phys.* **58**, 033508 (2016).

[261] M. de León, S. Vilariño, Lagrangian submanifolds in k-symplectic settings, *Monatsh. Math.* **170**, 381–404 (2013).

[262] J.J. Levin, On the matrix Riccati equation, *Proc. Amer. Math. Soc.* **10**, 519–524 (1959).

[263] M.M. Lewandowski, J. de Lucas, Geometric features of Vessiot–Guldberg Lie algebras of conformal and Killing vector fields on $\mathbb{R}^2$, *Banach Center Publ.* **113**, 243–262 (2017).

[264] P. Libermann, Ch.M. Marle, Symplectic geometry and analytical mechanics, *Mathematics and its Applications*, Vol. 35, D. Reidel Publishing Co., Dordrecht, (1987).

[265] A. Lichnerowicz. Les variétés de Poisson et leurs algèbres de Lie associées, *J. Differential Geometry* **12**, 253–300 (1977) (in French).

[266] L. Lichtenstein, Zur Theorie der konfermen Abbildung: konforme Abbildung nichtanalytischer, singularitätenfreier Flächenstücke auf ebene Gebiete. *Bull. Internat. Acad. Sci. Cracovie, CI. Sci. Math. Nat. Ser. A*, 192–217 (1916).

392   *Bibliography*

[267] S. Lie, Theorie der Transformationsgruppen I, *Math. Ann.* **16**, 441–528 (1880).

[268] S. Lie, Über die Integration durch bestimmte Integrale von einer Klasse linearer partieller Differentialgleichungen, *Arch. for Math.* **6**, 328–368 (1881).

[269] S. Lie, Allgemeine Untersuchungen über Differentialgleichungen, die eine continuirliche endliche Gruppe gestatten, *Math. Ann.* **25**, 71–151 (1885) (in German).

[270] S. Lie, *Theorie der Transformationgruppen* Vol. 2, Teubner, Leipzig (1890).

[271] S. Lie, *Theorie der Transformationgruppen* Vol. 3, Teubner, Leipzig (1893).

[272] S. Lie. Theorie der Transformationsgruppen III, *Math. Ann.* **16**, 441–528 (1893) (in German).

[273] S. Lie, Sur une classe d'équations différentialles qui possèdent des systèmes fundamentaux d'intégrales, *C.R. Math. Acad. Sci. Paris* **116**, 1233–1236 (1893) (in French).

[274] S. Lie, *On Differential Equations Possessing Fundamental Integrals*, Leipziger Berichte. Reprinted in Ges. Abhandl., Bd. 4, paper VI, (1893), pp. 307–313 (in German).

[275] S. Lie, *Theorie der Transformationsgruppen Dritter Abschnitt, Abteilung I. Unter Mitwirkung von Dr. F. Engel*, Teubner, Leipzig (1893).

[276] S. Lie, G. Scheffers, *Vorlesungen über continuierliche Gruppen mit geometrischen und anderen Anwendungen*, Teubner, Leipzig (1893) (in German).

[277] J. Llibre, D. Peralta-Salas, A note on the first integrals of vector fields with integrating factors and normalizers, *SIGMA* **8**, 035 (2012).

[278] J. Llibre, C.Valls, Liouvillian first integrals of quadratic-linear polynomial differential systems, *J. Math. Anal. Appl.* **379**, 188–199 (2011).

[279] P.G. Luan, C.S. Tang, Lewis-Riesenfeld approach to the solutions of the Schrödinger equation in the presence of a time-dependent linear potential, *Phys. Rev. A* **71**, 014101 (2005).

[280] J. de Lucas, C. Sardón, On Lie systems and Kummer–Schwarz equations, *J. Math. Phys.* **54**, 033505 (2013).

[281] J. de Lucas, M. Tobolski, S. Vilariño, A new application of k-symplectic Lie systems, *Int. J. Geom. Methods Mod. Phys.* **12**, 1550071 (2015).

[282] J. de Lucas, M. Tobolski, S. Vilariño, Geometry of Riccati equations over normed division algebras, *J. Math. Anal. Appl.* **440**, 394–414 (2016).

[283] J. de Lucas, S. Vilariño, *k-symplectic Lie systems: theory and applications*, J. Differential Equations **258**, 2221–2255 (2015).

[284] M. Maamache. Ermakov systems, exact solution, and geometrical angles and phases, *Phys. Rev. A* **95**, 936 (1995).

[285] A. Marino, Topological methods, variational inequalities and elastic bounce trajectories, *Atti Accad. Naz. Lincei Cl. Sci. Fis. Mat. Natur. Rend. Lincei (9) Mat. Appl.* **22**, 269–290 (2011).

[286] M. Mariton, P. Bertrand, A homotophy algorithm for solving coupled Riccati equations, *Potim. Contr. Appl. Met.* **6**, 351–357 (1985).

[287] J.E. Marsden, S. Pekarsky, S. Shkoller, M. West, Variational methods, multisymplectic geometry and continuum mechanics, *J. Geom. Phys.* **38**, 253–284 (2001).

[288] L. Menini, A. Tornambé, Nonlinear superposition formulas: Some physically motivated examples, in: *2011 50th IEEE Conference on Decision and Control and European Control Conference* (2011), pp. 1092–1097.

[289] E. Merino, Geometría k–simpléctica y k–cosimpléctica, in: Aplicaciones a las teorías clásicas de campos, Publicaciones del Dtp. de Geometría y Topología **87**, Universidad de Santiago de Compostela, Santiago de Compostela (1997).

[290] L. Michel, P. Winternitz, Families of transitive primitive maximal simple Lie subalgebras of diff(n), in: *Advances in Mathematical Sciences: CRM's 25 years*, CRM Proc. Lecture Notes, 11, Amer. Math. Soc., Providence (1997), pp. 451–479.

[291] J. Napora, The Moser type reduction of integrable Riccati differential equations and its Lie algebraic structure, *Rep. Math. Phys.* **46**, 211–216 (2000).

[292] J.C. Ndogmo, F.M. Mahomed, On certain properties of linear iterative equations, *Center European J. Math.* **56**, 34–36 (2013).

[293] S. Nikitin, Control synthesis for Čaplygin polynomial systems, *Acta Appl. Math.* **60**, 199–212 (2000).

[294] M. Nowakowski, H.C. Rosu, Newton's laws of motion in the form of a Riccati equation, *Phys. Rev. E* **65**, 047602 (2002).

[295] M.C. Nucci, Jacobi last multiplier and Lie symmetries: A novel application of an old relationship, *J. Nonlin. Math. Phys.* **12**, 284–304 (2005).

[296] A. Odzijewicz, A.M. Grundland, The superposition principle for the Lie type first-order PDEs, *Rep. Math. Phys.* **45**, 293–306 (2000).

[297] B. Øksendal, *Stochastic Differential Equations: An Introduction with Applications*, Springer-Verlag, Berlin (1985).

[298] M.A. del Olmo, M.A. Rodríguez, P. Winternitz, Simple subgroups of simple Lie groups and nonlinear differential equations with superposition principles, *J. Math. Phys.* **27**, 14–23 (1986).

[299] M. A. del Olmo, M.A. Rodriguez, and P. Winternitz, Integrability, chaos and nonlinear superposition formulas for differential matrix Riccati equations, in: *Proccedings of Second International Conference on Quantum Chaos*, Lecture Notes in Physics, 263, Springer, Berlin 372–378 (1986).

[300] M.A. del Olmo, M.A. Rodríguez, P. Winternitz, Superposition formulas for rectangular matrix Riccati equations, *J. Math. Phys.* **28**, 530–535 (1987).

[301] M.A. del Olmo, M.A. Rodriguez, and P. Winternitz, Superposicion no lineal en ecuaciones diferenciales ordinarias, in: *Ecuaciones Diferenciales y Aplicaciones*, Secretariado de Publicaciones, Valladolid (1987), pp. 307–312.

[302] P.J. Olver, *Applications of Lie Groups to Differential equations. Graduate Texts in Mathematics*, 107, Springer–Verlag, New York (1993).

[303] A.V. Oppenheim, Superposition in a class of nonlinear systems, *IEEE Int. Convention Record* **1964**, 171–177 (1964).

[304] V. Ovsienko, S. Tabachnikov, *Projective differential geometry old and new: from the Schwarzian derivative to cohomology of diffeomorphism groups,* Cambridge University Press, Cambridge (2005).

[305] V. Ovsienko, S. Tabachnikov, What is the Schwarzian derivative, *Notices of the AMS* **56**, 34–36 (2009).

[306] R.S. Palais, Global formulation of the Lie theory of transformation groups, *Mem. Amer. Math. Soc.* **22**, 1–123 (1957).

[307] J. Patera, R.T. Sharp, P. Winternitz, Invariants of real low-dimensional Lie algebras, *J. Math. Phys.* **17**, 986–94 (1976).

[308] A.V. Penskoi, P. Winternitz, Discrete matrix Riccati equations with superposition formulas, *J. Math. Anal. Appl.* **294**, 533–547 (2004).

[309] A.M. Perelomov, V.S. Popov, Casimir operators for semisimple Lie groups, *Math. USSR–Izvestija* **2**, 1313–1335 (1968).

[310] N.P. Petrov, The dynamical Casimir effect in a periodically changing domain: a dynamical systems approach, *J. Opt. B: Quantum Semiclass. Opt.* **7**, S89 (2005).

[311] G. Pietrzkowski, Explicit solutions of the $\mathfrak{a}_1$-type Lie–Scheffers system and a general Riccati equation, *J. Dyn. and Control Sys.* **18**, 551–571 (2012).

[312] E. Pinney, The nonlinear differential equation $y'' + p(x)y' + cy^{-3} = 0$, *Proc. Amer. Math. Soc.* **1**, 681 (1950).

[313] J. Ramírez, M.S. Bruzón, C. Muriel, M.L. Gandarias, The Schwarzian Korteweg–de Vries equation in (2+1) dimensions, *J. Phys. A* **36**, 1467–1484 (2003).

[314] A. Ramos, *Sistemas de Lie y sus aplicaciones en Física y Teoría de Control*, PhD Thesis, University of Zaragoza, arXiv:1106.3775.

[315] A. Ramos, A connection approach to Lie systems, in: *Proceedings of the XI Fall Workshop on Geometry and Physics, Publ. R. Soc. Mat. Esp.* **6**, 235–239 (2004).

[316] A. Ramos, New links and reductions between the Brockett nonholonomic integrator and related systems *Rend. Semin. Mat. Univ. Politec. Torino* **64**, 39–54 (2006).

[317] D.W. Rand, P. Winternitz, Nonlinear superposition principles: a new numerical method for solving matrix Riccati equations, *Comput. Phys. Comm.* **33**, 305–328 (1984).

[318] P.R.P. Rao, Classroom notes: The Riccati differential equation, *Amer. Math. Monthly* **69**, 995–996 (1962).

[319] P.R.P. Rao, V.H. Ukidave, Some separable forms of the Riccati equation, *Amer. Math. Monthly* **75**, 38–39 (1968).

[320] J.R. Ray, J.L. Reid, Ermakov systems, Noether's theorem and the Sarlet-Bahar method, *Lett. Math. Phys.* **4**, 235–240 (1980).

[321] R. Redheffer, Steen's equation and its generalizations, *Aequationes Math.* **58**, 60–72 (1999).

[322] W.T. Reid, A matrix differential equation of Riccati type, *Amer. J. Math.* **68**, 237–246 (1946).

[323] J.L. Reid, G.L. Strobel, The nonlinear superposition theorem of Lie and Abel's differential equations, *Lettere al Nuovo Cimento della Societa Italiana di Fisica* **38**, 448–452 (1983).

[324] S. Rezzag, R. Dridi, A. Makhlouf, Sur le principe de superposition et l'équation de Riccati, *C.R. Math. Acad. Sci. Paris.* **340**, 799–802 (2005).

Bibliography   395

[325] J. Riccati, Animadversiones in aequationes differentiales secundi gradus, *Actorum Eruditorum, quae Lipsiae publicantur, Supplementa* **8**, 66–73 (1724).

[326] B. Riemann, *Über die Hypothesen welche der Geometrie zu Grunde liegen*, Abhandlungen der Königlichen Gesellschaft der Wissenschaften zu Göttingen **30**, (1868) (German).

[327] T. Rodrigues de la Peña, *Reducción de principios variacionales con simetría y problemas de control óptimo de Lie-Scheffers-Brockett*, PhD Thesis, Universidad Carlos III de Madrid, Madrid (2009).

[328] C. Rogers, Application of a reciprocal transformation to a two-phase Stefan Problem, *J. Phys. A* **18**, L105–L109 (1985).

[329] C. Rogers, W.K. Schief, P. Winternitz, Lie-theoretical generalizations and discretization of the Pinney equation, *J. Math. Anal. Appl.* **216**, 246–264 (1997).

[330] C. Rogers, W.F. Shadwick, *Bäcklund Transformations and Their Applications*, Academic Press, New York (1982).

[331] B.A. Rozenfeld, *A History of Non-Euclidean Geometry*, Springer, New York, (1988).

[332] W. Rudin, *Real and Complex Analysis*, McGraw Hill, New-York (1987).

[333] T. Rybicki, On automorphisms of a Jacobi manifold, *Univ. Iagel. Acta Math.* **38**, 89–98 (2000).

[334] S.Y. Sakovich, On Miura transformations of evolution equations, *J. Phys. A* **26**, L369–L373 (1993).

[335] C. Sardón, Lie systems, Lie symmetries and reciprocal transformations, PhD Thesis, University of Salamanca, arXiv:1508.00726.

[336] W. Sarlet, Further generalization of Ray–Reid systems, *Phys. Lett. A* **82**, 161–164 (1981).

[337] W. Sarlet, F. Cantrijn, A generalization of the nonlinear superposition idea for Ermakov systems, *Phys. Lett. A* **88**, 383–387 (1982).

[338] D. Schuch, Riccati and Ermakov Equations in time-dependent and time-independent quantum systems, *SIGMA.* **4**, 043 (2008).

[339] A. Sebbar, A. Sebbar, Eisenstein series and modular differential equations, *Canad. Math. Bull.* **55**, 400–409 (2012).

[340] S. Shnider, P. Winternitz, Classification of systems of nonlinear ordinary differential equations with superposition principles, *J. Math. Phys.* **25**, 3155–3165 (1984).

[341] S. Shnider, P. Winternitz, Nonlinear equations with superposition principles and the theory of transitive primitive Lie algebras, *Lett. Math. Phys.* **8**, 69–78 (1984).

[342] J. Śniatycki, Multisymplectic reduction for proper actions, *Canad. J. Math.* **56**, 638–654 (2004).

[343] M. Sorine, P. Winternitz, Superposition laws for solutions of differential matrix Riccati equations arising in control theory, *IEEE Trans. Automat. Control* **30**, 266–272 (1985).

[344] G.L. Strobel, J.L. Reid, Nonlinear superposition rule for Abel's equations, *Phys. Lett. A* **91**, 209–210 (1982).

396   *Bibliography*

[345] A. Stubhaug, *The Mathematician Sophus Lie: It was the Audacity of My Thinking*, Springer, Berlin (2002).

[346] E. Suazo, S.K. Suslov, J.M. Vega-Guzmán, The Riccati equation and a diffusion-type equation, *New York J. Math.* **17**, 225–244 (2011).

[347] E. Suazo, K.S. Suslov, J.M. Vega-Guzmán, The Riccati system and a diffusion-type equation, *Mathematics 2014* **2**, 96–118 (2014).

[348] N. Sultana, Explicit parametrization of Delaunay surfaces in space forms via loop group methods, *Kobe J. Math.* **22**, 71–107 (2005).

[349] H. Sussmann, Orbits of families of vector fields and integrability of systems with singularities, *Bull. Amer. Math. Soc.* **79**, 197–199 (1973).

[350] S. Tabachnikov, On zeros of the Schwarzian derivative, in: *Topics in singularity theory*, *Amer. Math. Soc. Transl. Ser. 2*, Vol. 180, Amer. Math. Soc., (1997), pp. 229–239.

[351] A.H. Taub, A characterization of conformally flat spaces, *Bull. Amer. Math. Soc.* **55**, 85–89 (1949).

[352] G. Temple, A superposition principle for ordinary nonlinear differential equations, in: *Lectures on Topics in Nonlinear Differential Equations*, David Taylor Model Basin, Carderock, Report 1415 (1960), pp. 1–15.

[353] A. Tongas, D. Tsoubelis, P. Xenitidis, Integrability aspects of a Schwarzian PDE, *Phys. Lett. A* **284**, 266–274 (2011).

[354] A. Trautman, *Teoria grup*, Skrypt FUW, (2011). (Polish), available at http://www.fuw.edu.pl/ amt/skr4.pdf.

[355] D.P. Tsvetkov, A periodic Lotka–Volterra system, *Serdica Math. J.* **22**, 109–116 (1996).

[356] A. Turbiner, P. Winternitz, Solutions of nonlinear ordinary differential and difference equations with superposition formulas, *Lett. Math. Phys.* **50**, 189–201 (1999).

[357] K. Ueno, Automorphic systems and Lie-Vessiot systems, *Publ. Res. Inst. Math. Sci.* **8**, 311–334 (1972).

[358] I. Vaisman. *Lectures on the geometry of Poisson manifolds*, Progress in Mathematics, 118, Birkhäuser Verlag, Basel (1994).

[359] V.S. Varadarajan, *Lie groups, Lie algebras, and their representations*, Graduate texts in Mathematics, 108, Springer, New York (1984).

[360] A.M. Vershik, D.P. Zhelobenko (editors), *Representation of Lie Groups and Related Topics*, Advanced Studies in Contemporary Mathematics, 7, Gordon and Breach Science Publishers, New York (1990).

[361] P.J. Vassilou, Cauchy problem for a Darboux integrable wave map system and equations of Lie type, *SIGMA* **9**, 024 (2013).

[362] M.E. Vessiot, Sur une classe d'équations différentielles, *Ann. Sci. École Norm. Sup.* **10**, 53–64 (1893) (in French).

[363] M.E. Vessiot, Sur une classe d'équations différentielles, *C.R. Math. Acad. Sci. Paris* **116**, 959–961 (1893) (in French).

[364] M.E. Vessiot, Sur les systèmes d'équations différentielles du premier ordre qui ont des systèmes fondamentaux d'intégrales, *Ann. Fac. Sci. Toulouse Sci. Math. Sci. Phys.* **8**, H1–H33 (1894) (in French).

Bibliography 397

[365] E. Vessiot, Sur quelques équations différentielles ordinaires du second ordre, *Ann. Fac. Sci. Toulouse Sci. Math. Sci. Phys.* **9**, F1–F26 (1895) (in French).

[366] M.E. Vessiot, Sur la recherche des équations finies d'un groupe continu fini de transformations, et sur les équations de Lie, *Ann. Fac. Sci. Toulouse Sci. Math. Sci. Phys.* **10**, C1–C26 (1896) (in French).

[367] M.E. Vessiot, Sur une double généralisation des équations de Lie, *C.R. Math. Acad. Sci. Paris* **125**, 1019–1021 (1897) (in French).

[368] E. Vessiot, *Méthodes d'intégration élémentaires*, in: *Encyclopédie des sciences mathématiques pures et appliquées, 2,*, Gauthier-Villars & Teubner, (1910), pp. 58–170 (in French).

[369] R.M. Wald, *General Relativity,*. University of Chicago Press, Chicago (1984).

[370] N. Weaver, Sub-Riemannian metrics for quantum Heisenberg manifolds, *J. Operator Theory* **43**, 223–242 (2000).

[371] J. Wei, E. Norman, Lie algebraic solution of linear differential equations, *J. Math. Phys.* **4**, 575–581 (1963).

[372] J. Wei, E. Norman, On global representations of the solutions of linear differential equations as a product of exponentials, *Proc. Amer. Math. Soc.* **15**, 327–334 (1964).

[373] S. Weinberg, *Gravitation and Cosmology: Principles and Applications of the General Theory of Relativity*, John Wiley and Sons, New York (1972).

[374] P. Wilczyński, Quaternionic-valued ordinary differential equations. The Riccati equation, *J. Differential Equations* **247**, 2163–2187 (2007).

[375] P. Winternitz, Nonlinear action of Lie groups and superposition principles for nonlinear differential equations, *J. Phys. A* **114**, 105–113 (1982).

[376] P. Winternitz, Lie groups and solutions of nonlinear differential equations, in: *Nonlinear phenomena (Oaxtepec, 1982)*, Lecture Notes in Phys., 189, Springer, Berlin, p. 263–331 (1983).

[377] P. Winternitz, Comments on superposition rules for nonlinear coupled first order differential equations, *J. Math. Phys.* **25**, 2149–2150 (1984).

[378] P. Winternitz, Nonlinear difference equations with superposition formulae, in: *International Conference on Symmetries and Integrability of Difference Equations*, LMS Lect. Notes Ser, 255, Cambridge University Press, Canterbury (1998), pp. 275–284.

[379] P. Winternitz, Bäcklund transformations as nonlinear ordinary differential, or difference equations with superposition formulas, in *Backlund and Darboux transformations: the Geometry of Solitons*, CRM Proceedings and Lecture Notes, 29, Providence (2001), pp. 429–436.

[380] P. Winternitz, Lie groups, singularities and solutions of nonlinear partial differential equations, in: *Direct and inverse methods in nonlinear evolution equations*, Lecture Notes in Phys., 632, Springer, Berlin (2003), pp. 223–273.

[381] P. Winternitz, Y.A. Smorodinski, M. Uhlír, I. Fris, Symmetry groups in classical and quantum mechanics, *Soviet J. Nuclear Phys.* **4**, 444–450 (1967).

[382] P. Winternitz, L. Šnobl, *Classification and Identification of Lie Algebras*, American Mathematical Society, Providence (2014).

[383] I.M. Yaglom, *A simple non-euclidean geometry and its physical basis, in: An Elementary Account of Galilean Geometry and the Galilean Principle of Relativity*, Springer-Verlag, New York (1979).

[384] L. You-Ning, H. Hua-Jun, Nonlinear dynamical symmetries of Smorodinsky–Winternitz and Fokas–Lagerstorm systems, *Chinese Phys. B* **20**, 010302 (2011).

[385] W.M. Zhang, D.H. Feng, R. Gilmore, Coherent states: theory and some applications, *Rev. Mod. Phys.* **62**, 867–927 (1990).

[386] V.V. Zharinov, *Lecture notes of geometrical aspects of partial differential equations,* Series on Soviet and East European Mathematics, 9, World Scientific, River Edge (1992).

[387] N.T. Zung, Action-angle variables on Dirac manifolds, arXiv.org:1204.3865 (2013).

Author Index

A

Abel, Niels Henrik, 5

B

Ballesteros, Angel, 16
Blázquez Sanz, David, 15

C

Cariñena, José Fernando, 3, 12–13, 15
Campoamor-Stursberg, Rutwig, 19

G

Gainetdinova, Aliya, 19
Grabowski, Janusz, 3, 13, 15
Grundland, Alfred Michel, 9, 11–13, 20, 122
Guldberg, Alf, 2, 5–6, 69, 72, 74

H

Harnad, John, 9
Herranz, Francisco José, 21
Hussin, Veronique, 9

I

Ibragimov, Nail, 19

J

Jacobi, 375

K

Königsberger, Leo, 5, 72, 74

L

Lie, Sophus, 2–3, 5–6, 8, 30, 69, 72, 74

M

Marmo, Giuseppe, 3, 12–13
Morales-Ruiz, Juan José, 15

R

Ramos, Arturo, 20

S

Scheffers, Georg, 6–7
Shnider, S., 9, 21, 82

V

Vessiot, Ernest, 2–3, 5–6, 8, 69, 72, 74, 79, 96

W

Winternitz, Pavel, 3, 9–12, 14, 16, 21–22, 82, 96

Subject Index

A

abbreviated Lie–Scheffers theorem, 78
Abel equation, 16
admissible function, 52, 55, 232, 290, 305
affine differential equation, 5
almost-Dirac manifold, 54
anchor, 55
annihilator, 40
associated codistribution, 40
associated distribution, 40
associated system, 38
automorphic Lie system, 8, 15, 23, 81–82
autonomization, 38, 141, 343, 364, 369

B

Bäcklund transformation, 2, 12, 21, 121, 298
basic superposition rule, 117
Bernoulli equation, 79, 223
bi-Dirac–Lie system, 26, 245–246, 248
bi-Lie–Hamilton system, 159
bi-symplectic structure, 249
Bose–Einstein condensate, 192
Brownian motion, 22
Buchdahl equation, 17, 202, 222, 342, 362
Burgers' equation, 21, 229

C

Caldirola–Kanai oscillator, 2
canonical form, 19
canonical coordinates of the second kind, 109
Ĉaplying polynomial system, 269
Cartan's criterion, 337
Cartesian product bundle, 83
Casimir, 127
Casimir codistribution, 52
Casimir distribution, 133
Casimir element, 4, 17, 22, 49, 149–150, 153, 156–157, 160, 166, 305, 308, 330–332, 340
Casimir function, 49, 55, 130, 133, 140, 145, 149, 235, 242
Casimir invariant, 150, 244
Casimir operators, 20
Casimir tensor field, 3, 17, 308–309, 330
Cauchy–Riemann conditions, 314
Cayley–Klein geometries, 210
Cayley–Klein Riccati equation, 25, 191, 209, 211, 214–215, 222, 342, 354
centrally extended Euclidean algebra, 179, 182
centrally extended Poincaré algebra, 188
centrally extended Schrödinger Lie algebra, 179, 185

402 Subject Index

characteristic distribution, 55, 57–58
characteristic system, 95
classical DBH system, 357–358
classical field theory, 262
classical Lie symmetry, 341
classical mechanics, 2
classical physics, 15
Classical XYZ Gaudin Magnet, 134
classification of Lie algebras, 10
Clifford unit, 210
coalgebra, 22
coalgebra approach, 170
coalgebra method, 150, 155
coalgebra symmetric system, 153
coassociative, 50
complex Bernoulli equation, 25,
    222–223, 225, 374
complex classical Lie algebra, 10
complex Riccati equation, 24,
    191–192, 198, 217, 224
complex number, 210
conditional symmetry, 1
conformal polynomial vector field, 308
conformal form, 27
conformal vector field, 307–309,
    312–314, 316–319, 332, 336
conformally flat, 312, 319
conformally equivalent, 312
conformally flat Euclidean metric, 317
conformally flat Riemannian metric,
    316
connection, 1, 88–89, 119
constants of motion, 4, 17, 20, 135,
    143–144, 150, 164, 241, 244, 246
contact of order p, 41, 43
contact 1-form, 359
continuous Heisenberg group, 59, 252,
    296
contraction, 150
control system, 4, 269, 271
control theory, 9, 15
coproduct, 50, 150, 152
Cosmology, 341
cotangent bundle, 34
cotangent manifold, 34
coupled Riccati equations, 23, 191,
    214, 245, 256, 261, 374

Courant–Dorfman bracket, 54, 57
CRM, 9
CRM school, 9–11, 22
curvature tensor, 329

D

deformed Lie–Hamilton systems, 20
derivation, 32–33
derived Poisson algebra, 262–263, 289
diagonal prolongation, 83–85, 91,
    152–153, 159, 162, 166, 237–240,
    242–243, 267, 301–306
difference equation, 18, 22
diffusion equation, 261
diffusion models, 17
diffusion-type equation, 229
Dirac manifold, 54, 57–58, 231–232,
    236, 240, 246, 251
Dirac structure, 3, 54, 227, 238–240,
    248, 279
Dirac–Lie Hamiltonian, 25, 231–235,
    240–242, 246–248
Dirac–Lie system, 23, 25–26, 227,
    231–237, 239, 241–243, 245,
    247–249, 251, 261–262, 279–280,
    306
discrete differential equation, 1
discretization, 11, 23
dissipative harmonic oscillator, 191,
    219, 222, 374
dissipative Milne–Pinney equation,
    15, 164
dissipative quantum oscillator, 231
distributional method, 306
domain, 171, 310, 372
double-Clifford, 211
double-Clifford Riccati equation, 191
Drinfel'd double, 3, 13
Dual-Study Riccati equation, 25, 212,
    374

E

Einstein equations, 307
Emden–Fowler equation, 14, 16
Ermakov system, 155, 162
Ermakov–Ince invariant, 3, 17

Subject Index

Ermakov-like system, 16
Euclidean algebra, 179, 182
Euclidean space, 29
Euler equation, 143
Euler–Lagrange equations, 262
extended tangent bundle, 54

F

financial mathematics, 15
flat connection, 14
flat metric, 323
flat pseudo-Riemannian metric, 308
flat Riemannian metric, 327
flow, 35
foliated Lie system, 13
four-symplectic Lie system, 275
four-symplectic structure, 270, 272
forward Dirac map, 237

G

Galois theory, 8
gauge distribution, 56
gauge equivalent, 61, 248
gauge vector field, 52–53, 55, 58
Gauss's theorem egregium, 29–30
germ, 33
generalized Buchdahl equation, 191,
    199, 224–225, 374
generalized codistribution, 39
generalized distribution, 39, 171
generalized
    Darboux–Brioschi–Halphen system,
    342, 357
generating modular system, 175
generic, 71
generic point, 168, 171, 310
GKO classification, 10, 19, 24, 26,
    168, 170, 174, 179–181, 195–198,
    223, 256, 307–308, 319–320, 324,
    371
global superposition rule, 69, 71–72,
    90, 98
good Hamiltonian function, 59,
    253
good Hamiltonian vector field, 59
Grassmann-valued differential
    equation, 21

H

Ω-Hamiltonian function, 263,
    283–285, 288, 290, 292, 294–295
Ω-Hamiltonian vector field, 279,
    284–285, 287–288, 292, 294–295
Hamilton equation, 132, 159, 162, 222
Hamilton–De Donder–Weyl field
    equations, 262
Hamiltonian function, 50, 52, 59
Hamiltonian vector field, 51, 53, 59,
    309, 338, 373
harmonic oscillator, 2, 214, 374
harmonic oscillator algebra, 188
Heisenberg group, 299–300
Heisenberg–Weyl Lie algebra, 185
Hermitian Hamiltonian operator, 338
higher-order Lie system, 342
higher-order ordinary differential
    equation, 3
higher-order prolongation, 46
higher-order tangent space, 41
HODE Lie system, 96, 249
homogeneous linear differential
    equation, 67
homogeneous space, 10
horizontal distribution, 119
horizontal lift, 119
hyper-Kähler Bianchi-IX metric, 357
hypergeometric function, 203

I

imprimitive, 169, 193, 197, 206, 208,
    311, 371–373
imprimitive Lie algebra, 185
incomplete Gamma function, 203
indefinite metric, 318
indefinite orthogonal Lie algebra, 315
infinitesimal transformation, 44
inhomogeneous differential equation,
    77
integrable distribution, 120
integrability, 136, 242
integrability by quadratures, 78
integrability condition, 122
integral curve, 31, 35, 38

404 Subject Index

integrating factor, 24, 172, 174–175, 181
invariant distribution, 310, 317–318, 332–336
involutive, 52

J

Jacobi bracket, 253, 255
Jacobi manifold, 26, 58, 60, 251, 254, 256–259
Jacobi structure, 3, 253
Jacobi–Lie Hamiltonian, 254–256
Jacobi–Lie system, 26, 251–253, 255–257
jet bundle, 42
jet, 43

K

k-cotangent structures, 61
k-Hamiltonian, 278
k-Hamiltonian function, 289, 304
k-Hamiltonian system, 283
k-Hamiltonian vector field, 283, 289, 298, 303
k-polysymplectic manifold, 61
k-polysymplectic structure, 60
k-symplectic geometry, 287
k-symplectic Hamiltonian function, 261
k-symplectic Lie–Hamiltonian structure, 26, 296–297, 299–300
k-symplectic Lie system, 20, 261–263, 276, 278–281, 283, 289, 296–302
k-symplectic manifold, 61–63, 299
k-symplectic l-th orthogonal, 62
k-symplectic structure, 3–4, 26, 60, 261–263, 268–269, 272–273, 278–280, 283, 294, 298, 302, 304, 306
k-symplectic vector field, 261
Kerr metric, 312
Killing form, 337
Killing Lie algebra, 309

Killing vector field, 21, 27, 307–309, 312, 315–316, 320, 323, 325–329, 331–332, 336, 338, 340
Kirillov–Kostant–Souriau Poisson algebra, 19
Korteweg-de Vries (KdV) equation, 21, 121, 249
Kummer's problem, 105
Kummer–Schwarz equation, 374

L

L-Hamiltonian function, 55
L-Hamiltonian vector field, 55, 58
l-symplectic submanifold, 62, 298
Lagrange–Charpit equations, 114
Lax pair, 341
Leibniz rule, 38, 48, 51, 288
Lewis–Riesenfeld invariant, 157
Lewis–Riesenfeld method, 143, 145
Lie algebra, 39
Lie algebroid, 54
Lie condition, 6, 75
Lie family, 16
Lie group action, 36
Lie integral, 142–144
Lie point symmetry, 359
Lie symmetry, 138, 140, 244, 246–247, 341–343, 347, 351, 358, 361, 363, 367–368, 370, 376
Lie system, 1, 3, 5, 7–9, 13, 23, 65, 74, 76, 78–79, 103, 253, 341, 344, 347, 350
Lie's condition, 76–77, 93
Lie–Hamilton algebra, 127–129, 170, 176–178, 180–183, 185–190, 207, 210, 222, 373
Lie–Hamilton no-go theorem, 228
Lie–Hamilton system, 18–19, 22–25, 123–124, 127–131, 133–136, 138–143, 144, 147, 149–150, 152–153, 155–159, 167, 170, 176–182, 191, 195, 199, 201–202, 207, 209, 214–217, 221–222, 225, 228–229, 231, 247, 251, 267, 279, 309, 374

Subject Index

Lie–Hamiltonian structure, 123, 127–128, 129–133, 135, 137–138, 140–144, 147, 149–150, 152, 157, 159, 176–178, 183–184, 231, 248

Lie–Scheffers theorem, 6–7, 9, 11, 13–14, 74–78, 83, 85–86, 90, 92–93, 267

Lie–Scheffers system, 6

linear differential equation, 5

linear superposition rule, 68

local rectifying coordinates, 257

locally Riemannian, 329

Lotka–Volterra system, 201, 222, 225, 274, 374

low-dimensional VG Lie algebra, 18

M

mth coproduct, 50

manifold, 29

manifold structure, 43

Mathews–Lakshmanan oscillator, 16

matrix Riccati equation, 1, 11–12, 79, 168, 170

method of characteristics, 93, 95, 114, 243

metric, 313–314, 319, 323, 326, 329

Milne–Pinney equation, 2, 11, 14, 16, 22, 24, 96, 105–106, 163, 165, 192, 194, 215, 309, 337, 374

minimal Lie algebra, 40, 171–172, 179–180, 182–183, 188, 190

mixed superposition rule, 14

modular generating system, 24, 173–174, 176, 180–181, 183, 185–186, 188–189, 372

momentum map, 1, 242

multi-imprimitive, 311, 317–318, 322

multidimensional Toda system, 370

multisymplectic form, 18

multisymplectic Lie system, 4, 22

multisymplectic reduction, 4

N

Navier-Stokes equations, 21

nilpotent Lie algebra, 150

no-go theorem, 25

no-go theorem for k-symplectic Lie systems, 280

non-classical Lie symmetry, 341

non-Lie point symmetry, 359

non-trigonometric Hamiltonian system, 17

nonlinear superposition rule, 67

O

odd-primitive, 2, 281

one-imprimitive, 311

orthogonal trajectories to spheres, 8

P

ϵ-parametric group of diffeomorphisms, 36

ϵ-parametric group of transformations, 44

p-order prolongation, 47

p-tangent vector, 41

p-jet, 43

partial differential equation, 1, 30–31, 42, 118–122, 266, 272

partial Riccati equation, 342, 369

partial superposition rule, 14

PDE Lie system, 11, 14, 16, 20, 27, 342, 363–364, 367–368, 370

perfect Lie algebra, 140

periodic solution, 355

planar differential equation, 79

planar diffusion Riccati equation, 191

planar diffusion Riccati system, 216, 374

planar Riccati equation, 124

planar vector field, 168

plasma dynamics, 192

Plasma Physics, 341

plate-ball system, 271

Pochhammer symbol, 203

Poincaré algebra, 179, 188

Poincaré lemma, 173

Poisson algebra, 4, 18, 26, 48–49, 51, 55, 133–134, 141–142, 147,

149–150, 152–153, 158, 240–241, 253, 262, 287, 289–290, 299
Poisson algebra morphism, 48, 50
Poisson bivector, 51, 141, 156, 159, 309
Poisson bracket, 48, 51, 142–143, 146, 158, 285, 294, 299–300
Poisson coalgebra, 24, 50, 150
Poisson manifold, 51, 57–59, 141, 228, 251, 375
Poisson structure, 3, 25, 51, 123–124, 133, 150, 170–171, 176, 180, 227, 262
Poisson–Hopf algebra, 2
polynomial Lie integral, 144–145, 147–148
polysymplectic form, 188, 279, 288–291, 303–304
polysymplectic Lie–Hamiltonian structure, 295–296
polysymplectic manifold, 287, 289, 292
polysymplectic structure, 61, 283, 286
Pontryagin bundle, 54
positive Borel subalgebra, 224
presymplectic annihilator, 62
presymplectic form, 25, 58, 229, 261, 273, 275, 278, 283, 285, 290, 293–294, 299, 304
presymplectic geometry, 287
presymplectic manifold, 52, 57, 232, 246, 289
primitive, 169, 281, 311, 317, 371, 373
primitive Lie algebra, 180
projective t-dependent Schrödinger equation, 309
projective Riccati equation, 11
projective Schrödinger equations, 309
prolongation, 359–360
pseudo-Riemannian metric, 27, 307–308, 311, 319
pseudo-Riemannian manifold, 311
pseudo-orthogonal Lie algebra, 315

Q

quadratic polynomial Lie–Hamilton system, 205
quadratic polynomial Lie system, 204–205
quadratic polynomial model, 24
quadratic polynomial system, 191, 204
quadratic-linear polynomial system, 204
quaternionic Riccati equation, 355
quantum deformation, 20
quantum group, 18
quantum mechanics, 2–3, 15
quantum Lie system, 253
quantum system, 15
quasi-Lie scheme, 16
quaternionic Riccati equation, 342, 356

R

rectifying coordinate system, 258
reduced Ermakov system, 231
reduction of Lie systems, 8
Reeb vector field, 58
regular, 39, 310
Riccati diffusion system, 272
Riccati equation, 1, 4, 9, 69, 72, 74, 78, 94, 103, 105, 126–127, 155, 157–158, 267, 273, 294, 341, 352, 360, 376
Riccati hierarchy, 12, 17, 79
Riccati systems, 229, 267
Riemann structures, 18
Riemannian metric, 317, 319, 329, 340

S

$\mathfrak{sl}_2$, 20
$\mathfrak{sl}_3$, 19
s-primitive, 281
scalar curvature, 315
Schouten–Nijenhuis bracket, 51
Schrödinger equation, 338
Schwarzian derivative, 23, 105, 114, 263

Subject Index

Schwarzian equation, 79, 243, 245, 247, 249, 261, 263, 265, 267, 294, 303, 306

Schwarzian Korteweg–de Vries (KdV) equation, 26, 248–249

second-order ordinary differential equations, 5

second-order Gambier equation, 104

second-order Kummer–Schwarz equation, 17, 23–24, 104, 106, 123, 132, 159, 191, 195, 197, 214–215, 265, 342, 360

second-order ordinary differential equations, 308

second-order Riccati equation, 2, 8, 15, 75, 100, 103, 128, 130, 191, 220

second-order Riccati equation in Hamiltonian form, 374

second-order tangent bundle, 111

self-dual Yang–Mills equations, 357

sign function, 108

sine-Gordon equation, 121

singular confinement, 11

smallest Lie algebra, 40, 75

Smorodinsky–Winternitz oscillator, 24, 123, 125, 136, 139–140, 162, 192, 214

SODE Lie system, 19, 96–99, 360

space of p-jets, 43

split-complex Riccati equation, 25, 191, 211, 214, 374

split-complex numbers, 210

stable, 281

Stefan-Sussmann distribution, 309

stochastic differential equation, 22

stochastic Lie systems, 18

stochastic models, 2

strong comomentum map, 137

Study number, 211

super-superposition rules, 11, 21

superdifferential equations, 1

superequations, 11

superintegrability, 136, 242

superposition formula, 10, 67

superposition principles, 67

superposition rule, 1, 3–5, 8–9, 13–14, 17, 21, 65, 67, 73, 75, 77–78, 85, 88–90, 92–93, 96, 99, 103, 118–120, 122, 126, 150, 155, 157–158, 162, 164, 240, 242, 245–246, 301, 306

symmetric algebra, 49, 145, 156, 160

symmetric tensor algebra, 330

symmetric tensor field, 330–331

symmetrizer map, 49, 156

symmetry, 30

symmetry distribution, 138

symmetry systems, 344, 347, 350, 360

symplectic form, 3, 13, 124, 169–170, 172–174, 176, 181, 187, 189, 200, 202, 208, 223, 286, 338, 340

symplectic invariant, 4, 17

symplectic manifold, 50

T

t-independent constants of motion, 40–41, 133, 135–136, 138, 141, 144, 149, 153, 240, 242, 299, 301, 305–306

t-dependent vector field, 37–38

t-dependent Lotka–Volterra systems, 225

t-dependent superposition rule, 14

t-dependent vector field, 65, 84, 89

tangent bundle, 34

tangent manifold, 33

tangent space, 31–32

tangent vector, 32

tensor field invariant, 22

third-order Kummer–Schwarz equation, 17, 23, 104, 111, 114, 228, 232, 243, 263

Toda lattices, 21

transitive primitive Lie algebras, 9–10

trigonometric system, 165

two-photon Lie algebra, 128, 179, 191, 221

two-photon Lie–Hamilton system, 218–219

two-symplectic structure, 274

U

undecoupling Lie systems, 10
universal enveloping algebra, 19, 48, 330

V

vacuum Einstein equations, 357
vector bundle, 39
vector fields, 1, 31, 34
Vessiot–Guldberg Lie algebra, 6, 10, 65–66, 75, 77–78, 82, 93, 105, 121, 124, 126, 128, 168–169, 179, 196, 198, 211, 213–214, 216–217, 222, 227, 229, 231, 239, 251, 253, 256–257, 261–262, 268, 270–271, 276, 294, 303, 309, 316, 319–320, 323–326, 328–330, 332–333, 336, 341, 343–345, 347–352, 360, 367, 370
viral infection, 191
viral infection model, 208
volume form, 172

W

Wei–Norman equation, 231
Wei–Norman method, 15, 80
Whitney sum, 54
Witt algebra, 314
WZNW equations, 370

Z

zero curvature connection, 89